Measuring and Reasoning

In *Measuring and Reasoning*, Fred L. Bookstein examines the way ordinary arithmetic and numerical patterns are translated into scientific understanding, showing how the process relies on two carefully managed forms of argument:

- Abduction: the generation of new hypotheses to accord with findings that were surprising on previous hypotheses, and
- Consilience: the confirmation of numerical pattern claims by analogous findings at other levels of measurement.

These profound principles include an understanding of the role of arithmetic and, more importantly, of how numerical patterns found in one study can relate to numbers found in others. They are illustrated through numerous classic and contemporary examples arising in disciplines ranging from atomic physics through geosciences to social psychology.

The author goes on to teach core techniques of pattern analysis, including regression and correlation, normal distributions, and inference, and shows how these accord with abduction and consilience, first in the simple setting of one dependent variable and then in studies of image data for complex or interdependent systems. More than 200 figures and diagrams illuminate the text.

The book can be read with profit by any student of the empirical natural or social sciences and by anyone concerned with how scientists persuade those of us who are not scientists why we should credit the most important claims about scientific facts or theories.

Fred L. Bookstein is Professor of Morphometrics, Faculty of Life Sciences, University of Vienna, Austria; Professor of Statistics at the University of Washington, Seattle; and an emeritus Distinguished Research Professor at the University of Michigan. Since 1977 he has produced some 300 books, chapters, articles, and videotapes on various aspects of these methods and their applications in studies of normal and abnormal craniofacial growth in humans and other mammals, studies in the neuroanatomy and behavior of schizophrenia and fetal alcohol spectrum disorders, and evolutionary studies of hominids and ammonoids. He is especially interested in how statistical diagrams can convey the valid numerical patterns that characterize complicated systems like continental drift or fetal alcohol brain damage to the broad modern public. The figures in this book include many of his current favorites along these lines.

Measuring and Reasoning

Numerical Inference in the Sciences

FRED L. BOOKSTEIN

Department of Anthropology, University of Vienna
Department of Statistics, University of Washington

CAMBRIDGE
UNIVERSITY PRESS

32 Avenue of the Americas, New York, NY 10013-2473, USA

Cambridge University Press is part of the University of Cambridge.

It furthers the University's mission by disseminating knowledge in the pursuit of education, learning, and research at the highest international levels of excellence.

www.cambridge.org
Information on this title: www.cambridge.org/9781107024151

© Fred L. Bookstein 2014

This publication is in copyright. Subject to statutory exception and to the provisions of relevant collective licensing agreements, no reproduction of any part may take place without the written permission of Cambridge University Press.

First published 2014

Printed in the United States of America

A catalog record for this publication is available from the British Library.

Library of Congress Cataloging in Publication data
Bookstein, Fred L., 1947– author.
Measuring and reasoning : numerical inference in the sciences / Fred L. Bookstein.
pages cm
Includes bibliographical references and index.
ISBN 978-1-107-02415-1 (hardback)
1. Statistical hypothesis testing. I. Title.
QA277.B66 2014
519.5′4–dc23   2013030370

ISBN 978-1-107-02415-1 Hardback

Cambridge University Press has no responsibility for the persistence or accuracy of URLs for external or third-party Internet Web sites referred to in this publication and does not guarantee that any content on such Web sites is, or will remain, accurate or appropriate.

For my daughters Victoria Bookstein and Amelia Bookstein Kyazze, who teach me how to work continually toward improving the world even while accommodating its constraints and ironies; and

for my grandsons Asher, Avi, and Keo and my granddaughter Yvette, although none of them can read yet, in hopes that they may be assigned a section or two of this when they get to college in 2027 or thereabouts; but mostly

for my dear wife Ede, who has encouraged me to meditate on these themes, and on the teaching of them, ever since the original seminars in our Ann Arbor living room 27 years ago. This book is the better for all your singing.

Contents

Analytical Table of Contents		*page* xi
Preface		xix
Epigraphs		xxvii

Part I The Basic Structure of a Numerical Inference — 1

1 Getting Started — 3
- 1.1 Our Central Problem: What Is a Good Numerical Inference? — 3
- 1.2 The Sinking of the *Scorpion* — 5
- 1.3 Prospectus — 11

2 Consilience as a Rhetorical Strategy — 17
- 2.1 Continental Drift — 18
- 2.2 E. O. Wilson's View of Consilience — 28
- 2.3 Some Earlier Critiques — 37
- 2.4 The Issue of Heterogeneity — 42
- 2.5 The Graphics of Consilience — 51
- 2.A Mathematics or Physics? Consilience and Celestial Mechanics — 58
- 2.B Historical Note: From Weiss through Kuhn to This Book — 70

3 Abduction and Strong Inference — 73
- 3.1 Example: Global Warming Is Anthropogenic — 74
- 3.2 That Hypothesis Wins That Is the Only One That Fits the Data — 85
- 3.3 Numerical Inferences Are the Natural Home of Abductive Reasoning — 94
- 3.4 Strong Inference — 100
- 3.5 Summary of Part I — 105
- 3.A Appendix: Update on the Rhetoric of Climate Change — 106

Part II A Sampler of Strategies — 113

4 The Undergraduate Course — 115
- E4.1 John Snow on the Origin of Cholera — 117

L4.1	The Logic and Algebra of Least Squares	129
E4.2	Millikan and the Photoelectric Effect	143
L4.2	Galton's Machine	150
E4.3	Jean Perrin and the Reality of Atoms	175
L4.3	Likelihood-Based Statistical Inference	188
E4.4	Ulcers Are Infectious	201
L4.4	Avoiding Pathologies and Pitfalls	206
E4.5	Consilience and the Double Helix	214
L4.5	From Correlation to Multiple Regression	220
E4.6	On Passive Smoking	253
L4.6	Plausible Rival Hypotheses	263
E4.7	Milgram's Obedience Experiment	266
E4.8	Death of the Dinosaurs	280
	Interim Concluding Remark	284

Part III Numerical Inference for General Systems   289

5 Abduction and Consilience in More Complicated Systems   291
- 5.1 Analysis of Patterns in the Presence of Complexity   291
- 5.2 Abduction and Consilience in General Systems   301
- 5.3 Information and Information-Based Statistics   331
- 5.4 A Concluding Comment   344

6 The Singular-Value Decomposition: A Family of Pattern Engines for Organized Systems   346
- 6.1 The Hyperbolic Paraboloid   346
- 6.2 The Singular-Value Decomposition   349
- 6.3 Principal Components Analysis   357
- 6.4 Partial Least Squares and Related Methods for Associations Among Multiple Measurement Domains   366
- 6.5 Another Tableau: Dissimilarities as Primary Data   381
- 6.6 Concluding Comment   399

7 Morphometrics, and Other Examples   402
- 7.1 Description by Landmark Configurations   406
- 7.2 Procrustes Shape Distance   410
- 7.3 Procrustes Shape Coordinates and Their Subspaces   412
- 7.4 Procrustes Form Distance   419
- 7.5 The Thin-Plate Spline in 2D and 3D   421
- 7.6 Semilandmarks: Representation and Information Content of Curving Form   426
- 7.7 Putting Everything Together: Examples   437
- 7.8 Other Examples   461

Part IV	**What Is to Be Done?**		479
8	**Retrospect and Prospect**		481
	8.1	Abduction, Consilience, and the Psychosocial Structures of Science	481
	8.2	Implications for Inference Across the Sciences	491
	8.3	What, Then, IS to Be Done?	493
References			501
Index			519

Color plates follow page 116

Analytical Table of Contents

Preface	xix
Epigraphs	xxvii

Part I. The Basic Structure of a Numerical Inference — 1

Demonstrations of the book's central rhetorical style. Three preliminary exemplars of particularly compelling logic. Relation of the author's recommended methodology to two classics of pragmatism, E. O. Wilson's version of "consilience" and the abductive syllogism formalized by C. S. Peirce.

1 Getting Started — 3

First encounter with the book's central question and with our suggested mode of answering it. An introductory example with a bittersweet ending. Prospectus for the rest of this essay.

    1.1 Our Central Problem: What Is a Good Numerical Inference? — 3

    Introduction of the central puzzle of this book: how quantitative arguments persuade when they persuade, and when this persuasion is justified.

    1.2 The Sinking of the *Scorpion* — 5

    Our first example, a true detective story of a particularly intriguing shipwreck, demonstrates our rhetorical method of inspecting inferences from numbers and instrument readings.

    1.3 Prospectus — 11

    Outline of the rest of the book: dozens more examples carefully deconstructed, interwoven with lectures.

2 Consilience as a Rhetorical Strategy — 17

The logic and rhetoric of quantitative inference together constitute a tractable special case of scientific inference, explicable by most of the pragmatic epistemologies currently applying to the natural and social sciences.

- 2.1 Continental Drift — 18
  Triumph of a graphical demonstration after 50 years of trying.

- 2.2 E. O. Wilson's View of Consilience — 28
  In our context, E. O. Wilson's pragmatism of consilience simplifies to the reasonable expectation that measured numbers should agree with other measured numbers under some circumstances.

- 2.3 Some Earlier Critiques — 37
  The numerical version of consilience relates in interesting ways to earlier critiques of empirical inference by Thomas Kuhn and Harry Collins, and to Robert Merton's views on the moral structure of science.

- 2.4 The Issue of Heterogeneity — 42
  In statistics, random walks violate one's intuitions about randomness; in biology, heterogeneity interferes with the praxis of consilience by making clear how very ambiguous the program of "reductionism" is.

- 2.5 The Graphics of Consilience — 51
  The Periodic Table; graphical models of the force of correlation.

- 2.A Mathematics or Physics? Consilience and Celestial Mechanics — 58
  How Newton's law of gravitation exemplifies the way that classical mechanics intentionally blurs the line between mathematics and physics, between the necessary and the contingent.

- 2.B Historical Note: From Weiss through Kuhn to This Book — 70
  History of the arguments of this book, from Paul Weiss and Ralph Gerard through Thomas Kuhn to the present.

3 Abduction and Strong Inference — 73
The power of numerical inference can often be cast into the model of abduction, a syllogistic form invented by Charles Sanders Peirce in the 1870s to describe the logic by which hypotheses are inferred from unexpected patterns in data in order to account for them.

- 3.1 Example: Global Warming Is Anthropogenic — 74
  A third introductory example: the culminating inference from the 2007 report of the Intergovernmental Panel on Climate Change, demonstrating the anthropogenic origins of global warming and hence the dependence of its future values on collective human actions.

- 3.2 That Hypothesis Wins That Is the Only One That Fits the Data — 85
  Powerful arguments like the IPCC's generally not only demonstrate patterns but also argue that they are consistent with no hypothesis except their own. This form of inference, abduction,

is closely related to other tenets of American pragmatism, the philosophy that Peirce founded.

- 3.3 Numerical Inferences Are the Natural Home of Abductive Reasoning 94
  The abduction syllogism fits a great many exemplars of quantitative inference from mathematics through biology and into the practical social sciences.
- 3.4 Strong Inference 100
  There is a close relation to the context of multiple hypotheses put forward as the best scientific method by T. C. Chamberlin in 1890 and by John Platt in 1964.
- 3.5 Summary of Part I 105
- 3.A Appendix: Update on the Rhetoric of Climate Change 106
  Both the science of climate change and the dissemination of that science are in a state of rapid flux. Some developments since the drafting of the original version of this chapter are briefly reviewed.

Part II. A Sampler of Strategies 113

A practical curriculum along these lines, beginning simply and ending at the state of the art for uncomplicated data.

4 The Undergraduate Course 115
  The most important themes of elementary statistical method can be taught using the preceding principles as scaffolding. Key to subsections: E, examples; L, lectures.

- E4.1 John Snow on the Origin of Cholera 117
  - E4.1.1. The basic abduction
  - E4.1.2. Two modern analogues

  - L4.1 The Logic and Algebra of Least Squares 129
    - L4.1.1. Sums of squares
    - L4.1.2. Precision of averages, and weighted averages
    - L4.1.3. Least-squares fits to linear laws
    - L4.1.4. Regression as weighted averaging

- E4.2 Millikan and the Photoelectric Effect 143
  - E4.2.1. The role of lines; the role of certainty
  - E4.2.2. The unreasonable effectiveness of mathematics in the natural sciences

  - L4.2 Galton's Machine 150
    - L4.2.1. The large-scale pattern of binomial coefficients
    - L4.2.2. Ratio of chances of heads to its maximum
    - L4.2.3. Notation: the Normal distribution

>   L4.2.4. Other origins of the Normal distribution
>   L4.2.5. The role of the Normal distribution in statistics
>   L4.2.6. Galton's quincunx; regression revisited
>   L4.2.7. Appendix to section L4.2: Stirling's formula

E4.3 Jean Perrin and the Reality of Atoms   175
   E4.3.1. Multiple routes to the same quantity
   E4.3.2. More detail about Brownian motion
   E4.3.3. The Maxwell distribution is exact

>   L4.3 Likelihood-Based Statistical Inference   188
>   L4.3.1. Bayes' Theorem
>   L4.3.2. The Gauss–Laplace synthesis
>   L4.3.3. Significance testing: decisions about theories
>   L4.3.4. The contemporary critique

E4.4 Ulcers Are Infectious   201
   E4.4.1. Chronicle of a preposterous proposition
   E4.4.2. Evolution of Koch's postulates

>   L4.4 Avoiding Pathologies and Pitfalls   206
>   L4.4.1. The Langmuir–Rousseau argument
>   L4.4.2. An unflawed example from Frank Livingstone

E4.5 Consilience and the Double Helix   214
   E4.5.1. The *Nature* version
   E4.5.2. The true version

>   L4.5 From Correlation to Multiple Regression   220
>   L4.5.1. From causation to correlation
>   L4.5.2. The algebra of multiple regression: three interpretations
>   L4.5.3. Three special cases of multiple regression
>   L4.5.4. The daunting epistemology of multiple regression

E4.6 On Passive Smoking   253
   E4.6.1. Passsive smoking and heart attacks
   E4.6.2. Passive smoking and lung cancer

>   L4.6 Plausible Rival Hypotheses   263

E4.7 Milgram's Obedience Experiment   266
   E4.7.1. Calibration as consilience
   E4.7.2. Critique of pseudoquantitative psychology

E4.8 Death of the Dinosaurs   280
   E4.8.1. Both kinds of abductions
   E4.8.2. The 1992 confirmation

Interim concluding remark   284

| | **Part III. Numerical Inference for General Systems** | 289 |

The key to extensions of these methods into more complicated organized systems is to require that sets of measurements be as carefully designed as samples or interventions are.

| 5 | **Abduction and Consilience in More Complicated Systems** | 291 |

The tools we used for numerical reasoning in the simpler systems need to be reformulated for applications that are more complex.

    5.1 Analysis of Patterns in the Presence of Complexity   291

Three diverse examples of the discovery of patterns in organized systems: stars, proteins, congenital brain damage. On choosing the largest signal, and fields where that matters.

    5.2 Abduction and Consilience in General Systems   301

The search for meaning in analyses of organized systems has to scale at least as powerfully as the range of models for meaninglessness. These models are not conveyed well by the notation of data matrices. Some reinterpretations are introduced for the bell curve, and alternatives to it are reviewed: random walks, networks, directions in hyperspace, fields over pictures, and deformations *of* pictures.

    5.3 Information and Information-Based Statistics   331

One central innovation of late 20th-century statistics comes to our rescue. Information is a formal mathematical quantity obeying exact theorems having considerable relevance to the statistical praxis of complex systems. An example from applied medical imaging.

    5.4 A Concluding Comment   344

| 6 | **The Singular-Value Decomposition: A Family of Pattern Engines for Organized Systems** | 346 |

One pattern formalism is sufficient for many of the routine pattern engines we apply to adequately quantified data representing organized systems.

    6.1 The Hyperbolic Paraboloid   346

The crucial geometric object for multivariate pattern analysis is neither the plane nor the ellipsoid, but this particular doubly ruled surface.

    6.2 The Singular-Value Decomposition   349

The key to one style of description of organized complexity: Any rectangular data set can be reconstituted as the sum of a stack of hyperbolic paraboloids. An example is presented involving viremia in experimental monkeys.

6.3 Principal Components Analysis 357
Deriving patterns from ellipsoids, and for what kinds of measurements this makes sense. Pattern spaces for variables. An example from human evolution.

6.4 Partial Least Squares and Related Methods for Associations Among Multiple Measurement Domains 366
This fundamental and flexible analytic approach to multiple complex measurement domains is a modification of some ideas of Sewall Wright from the 1920s, whereby path analysis becomes a model for other approaches to explanation than the multiple regression to which it was applied in Chapter 4.

6.5 Another Tableau: Dissimilarities as Primary Data 381
The technique of principal coordinates analysis reconstructs variable values from Euclidean distances among cases. Applications include studies of distances, discrimination, and the geometry of covariance matrices.

6.6 Concluding Comment 399

7 Morphometrics, and Other Examples 402
Morphometrics, the applied statistics of biological shape, is, so far, the most highly ramified and carefully articulated application domain for the ideas put forward here.

7.1 Description by Landmark Configurations 406
From curves to points. The First Rule: think for a century before measuring.

7.2 Procrustes Shape Distance 410
The fundamental geometric construction of morphometrics begins by changing the object of measurement from one form to two, thereby freeing the investigator from any dependence on particular Cartesian coordinate systems.

7.3 Procrustes Shape Coordinates and Their Subspaces 412
What "shape measurement" means in this context. The Second Rule: show how the new variables account for the old distances. Structure of the corresponding linear statistical space.

7.4 Procrustes Form Distance 419
To get back to biology, in most settings, requires that size be put back into the data even after the geometers have taken it out. The Third Rule: if size is part of the answer, it must be part of the question.

Analytical Table of Contents   xvii

7.5 The Thin-Plate Spline in 2D and 3D   421
Human cognitive evolution to the rescue: legible diagrams of form comparisons. The Fourth Rule: rely on the integrative power of your visual brain for the import of spatially differentiated explanations.

7.6 Semilandmarks: Representation and Information Content of Curving Form   426
Bringing in all the rest of the information about form, courtesy of the centrality of the classical anatomical knowledge bases as formalized in morphometric templates. The essence of the procedure can be understood just from the problem of sliding one single landmark.

7.7 Putting Everything Together: Examples   437
How the preceding rules combine to deal with patterns in human evolution, allometric growth, endophrenology. Large-scale and small-scale descriptions. Permutation testing. Two distinct ways of thinking about symmetry.

7.8 Other Examples   461
Unfortunate things can happen when these techniques are applied outside the realm of organized systems. Low-dose fetal alcohol behavioral teratology; Karl Pearson on the inheritance of "civic virtue." Guttman scaling and other approaches to structured data domains.

Part IV. What Is to Be Done?   479

8 Retrospect and Prospect   481
The tension between pragmatism and organized skepticism persists. The rhetoric of numerical consilience and abduction is suggested as one way of letting us discern, and then communicate, the difference between weakly and strongly supported findings, so as to help one's colleagues and the public comprehend the uncertainties of science as a human endeavor.

8.1 Abduction, Consilience, and the Psychosocial Structures of Science   481
The power of the pattern engines has gotten well ahead of the power of the conventional forms of "confirmation." The underlying processes are cognitive, and the strengthening of these methods involves a renewed attention to the scientist's modes of narration in the presence of complexity. Another example from fetal alcohol science.

8.2 Implications for Inference Across the Sciences   491
   The principal differences between analyses of simple systems and analyses of organized complexity, which pertain to the general topic of agreement between data and expectation, owe mainly to differences in the representation of uncertainty between the traditions.

8.3 What, Then, IS to Be Done?   493
   For numerical inference to take its appropriate role in 21st-century science, there will need to be major changes in the curriculum of statistics wherever it is taught.

References   501

Index   519

Preface

For many years I have been concerned with the process by which scientists turn arithmetic into understanding. The word "process" here is meant in at least four different senses at the same time – algebraic, logical, cognitive, social. These correspond to at least four different disciplines that have hitherto considered parts of the puzzle in isolation, and so my arguments will jump back and forth among these four ways of speaking. There seems to be no standard name for this nexus in the literature of any of the fields that touch on it, and in the 20 or so years that I've been watching closely, as far as I know no popular book or scholarly monograph has appeared that focuses on these topics at any level. That this book has materialized in your lap in some copyrighted form (codex, Kindle, netbook, whatever) is evidence that editors and reviewers believe the gap to have been worth filling.

Before I was trying to consider these issues all together, I was trying to think them through separately. Over 35 years as a professional statistician and biometrician I have been employed to transform arithmetic into understanding in five main areas of scientific application: first, craniofacial biology, then gross neuroanatomy (the National Institutes of Health's "Human Brain Project"), then image-based anatomy of the whole organism (NIH's "Visible Human Project"), and, most recently, physical anthropology and organismal theoretical biology. Along the way there have been diversions: the science of fetal alcohol damage, analysis of war wounds to the heart, studies of hominid fossils. But I suspect nowadays that this material may be crucial to a broader range of applications than just those I've threaded it through personally. Likewise, the methodology seems richer and potentially more fruitful than any single scholar could oversee. It is time to weave the arguments together into a focused and coherent narrative capable of seeding further developments leveraging the efforts of others, not just me.

The volume you hold is not a scholarly monograph in the strict sense, an argument about one thing (a **mono**-graph), but instead an essay of somewhat experimental structure offering a set of core arguments drawn from interpretations of an idiosyncratic selection of main readings. This structure is different from all my earlier books but is in line with a spate of recent essays, for instance, Bookstein (2009a,b,c; 2013a,b), aimed at the less narrowly disciplinary reader. So the spirit of the expositions here is considerably less formal than the material being exposited. The first public presentations of these ideas outside of a classroom came at the conference "Measuring Biology" hosted by the Konrad Lorenz Institute for Evolution and Cognition Research, Altenberg, Austria, in 2008. (There is a review of the historical context of this conference in Appendix 2.B of Chapter 2.) Reactions of the people in the room to my argument were reassuring. Perhaps, I thought, the time had come to transport the argument out of its previous niche of course syllabus or colloquium presentations for delivery in the larger arena of printed dissemination, where a broader spectrum of colleagues and browsers might provide thoughtful, searching commentaries.

In this repackaged version the argument seems not to face much competition. It puzzles me that hardly any scholars pursuing social studies of science are looking at this arena. The formulas accompanying this topic are technically difficult from time to time, that is true, but never as hard as, say, philosophy of physics. Rather, they are *constructively* demanding in ways that only an insufficiently wise or overly strategic graduate student in science studies would choose to circumvent. It may also be the case that those aspects of science studies are most interesting to one's colleagues in science studies just where Nature or Clio is mumbling incomprehensibly, not, as in many of the examples here, where her message is clear and unequivocal. Still, whatever the reason, issues at the foundations of quantitative reasoning inhabit a true lacuna everywhere (except in the physical sciences, where things are too simple, owing to the miracle of exact laws).

In short, I think a book like this one is obviously needed, enough that I'm surprised nobody has written one before this. Readers of course decide things like this (the obviousness of the need, I mean) for themselves. For that frame of mind to be suitable, among its core stances should be one sketched effectively by University of Michigan philosopher Jack Meiland 30 years ago under the suggestively simplistic rubric of "college thinking." College thinking entails a constant concern for why one adopts the inferential heuristics and shortcuts one adopts: why one believes what one believes, why one reasons as one does, how one justifies one's own logical conclusions to oneself prior (it is hoped) to justifying them to others. In the area of "applied probability" this has been the concern of generations of cognitive scientists, with a Nobel Prize for one of them, Daniel Kahneman; but over here, where we deal with inferences about scientific claims rather than singular human decisions, there seems to be no matching literature at all.

At the core of scientific methodology have always lain concerns like these about the validity of empirical inferences. The subtype that is inference from *quantitative* data, our topic in this book, is easier to manage than the general case, yet serves to illustrate most of the themes from the broader context. Its study can never be begun too early, nor can students ever be trained to be too skeptical. Evidently the essential iconoclasm of topics like mine – the continual interruption of one's colleagues' arguments by the repeated challenge "Just *why* do you believe that that claim follows from your data, and how are the opposing or competing claims impeached?" – is ill-suited to the Intro. – Materials & Methods –Results – Discussion format of today's modal scientific article, as it requires threading across multiple exemplars. You don't win points from the anthropological reviewer, for instance, when you note that the way you are managing an inference in hominid evolution is exactly the same as the way somebody from the other side of campus handled an inference in the ecology of urban agriculture a couple of years back. Your reviewer cares only about anthropology, more's the pity. Then the issue of how numbers persuade independent of discipline is not a reasonable candidate for a peer-reviewed article in any of the disciplines so benefited. There remains mainly the possibility of a book-length argument like this one that claims citizenship in no field except its own. Compare Nassim Taleb's 2008 bestseller *The Black Swan,* likewise a book of applied quantitative reasoning (and likewise concerned with heuristics and their biases). I don't write as wittily as Taleb, whom I would like to meet some day, but the breadths of our nonoverlapping magisteria (his in finance and the madness of crowds, mine in the comparative natural sciences) seem to be commensurate.

This is not a book of "statistical methods," then, but instead a complement, or perhaps an antidote, to the whole shelfful of those. Most of the books on that shelf don't tell you where the numbers that are the objects of algebraic manipulation by the formulas there originated. Consider, for instance, the otherwise superb textbook *Statistics* by the late David Freedman and colleagues, the book we in Seattle use for our undergraduate service course – it never breathes a single hint about any of the rhetorics here at all. There are nevertheless some domains that, out of necessity or perhaps just by virtue of the vicissitudes of intellectual history, manifest a real concern for the empirical origin of the quantities on which further reasoning is based. The list of such fields includes epidemiology (Rothman et al., 2008), "multivariate calibration" (Martens and Næs, 1989), and observational astronomy (Harwit, 1981). In contrast, our topic is related only tangentially to a burgeoning scholarly literature on causation; see, for example, Pearl (2000/2009), who, like Freedman, pays remarkably little attention to the actual empirical origin of variables and their empirical information content.

My precursors are rather the great unclassifiable monographs about scientific reasoning as an intellectual or cognitive style: books such as Karl Pearson's *Grammar of Science* (1892/1911), E. T. Jaynes's posthumous *Probability Theory: The Logic of Science* of 2003, or, halfway in-between these, E. B. Wilson's *Introduction to Scientific Research,* still in print, from 1952. In its philosophy of science, though not its concern for arithmetic, my approach aligns well with that of Ziman (1978). Our topic has some ties to the "theory of knowledge" that is a component of today's International Baccalaureate curriculum for advanced high schools and Gymnasiums, and hence overlaps with the content of textbooks such as van de Lagemaat's *Theory of Knowledge for the IB Diploma* (2006). For multivariate analysis, a fundamental component of advanced protocols in this domain, W. J. Krzanowski's *Principles of Multivariate Analysis* of 1988 (revised edition, 2000) is a useful handbook. General biometrical discussions that show some awareness of these issues include two books by Richard Reyment (R. A. Reyment, *Multidimensional Palaeobiology,* 1991, and R. A. Reyment and K. G. Jöreskog, *Applied Factor Analysis in the Natural Sciences,*

1993), *Model Selection and Multimodel Inference* by K. P. Burnham and D. R. Anderson (2002), Joseph Felsenstein's *Inferring Phylogenies* (2004), and my own *Morphometric Tools for Landmark Data* of 1991. But, in general, explicit discussions at adequate length of crucial issues in the rhetoric of quantitative inference are quite rare. Null-hypothesis statistical significance testing, by contrast, though appallingly widely known, is no kind of substitute argument at all, and is dismissed quite rudely throughout this volume, especially at Section L4.3.4, except for examples where the null is actually true.

The sociologist and methodologist Otis Dudley Duncan (1984), setting out a curriculum for applications of a similar rhetoric to a neighboring area, describes the strategy as one of "beads on a string... or perhaps nuts in a fruitcake." Regardless of the spatial/culinary metaphor, we can agree that our field is not "foundations of statistics," nor "foundations of metrology," nor philosophy of science per se. It is instead a collection of interrelated tropes, insights, and exemplars that I believe cumulate to an argument worth making independent of any intellectual or academic specialization. The text here does not require any subject-area knowledge by way of prerequisite. The puzzler page of *Saturday Review,* a once-popular American literary magazine, used to describe its most enthusiastic participants as having minds "well-furnished but not overstuffed," meaning widely read in the public arguments (mainly, but not entirely, political or aesthetic) of the past and present, but not pedantic on any single issue in particular; the same would apply to my intended reader. It would be helpful if you were sturdily independent of mind, at least outside of mathematics and physics, and if you had at least a mild curiosity for history of ideas and at least a modest acquaintance with some of the major ironies of philosophy of science as expressed in the 14 epigraphs to follow. A few of the examples in the book are practical enough that personal circumstances might be among the factors motivating a closer reading of these sections than others – sharing air with a smoker (Section 4.6), drinking alcohol during a pregnancy (Section 7.7), living in Seattle in an inadequately buttressed building (Section 3.3). Parts of this text will be within the reach of the good (meaning curious and open-minded, but also skeptical) high school student. Other fragments are appropriate for an undergraduate, and still others are appropriate as part of the catechism that is a doctoral dissertation defense. Few will find all of this easy going, but every reader should be able to follow the main line of argument at some level of engagement.

There are many diagrams here, some mine and some the original authors', some about facts and some about formulas. There are also many equations, perhaps more than some readers would like but none that are not mandated by the exposition in their vicinity. Part of my theme is that the formalisms of mathematics, even if they were your bête noire in the course of high school or college algebra or calculus, have by now become your friendly guide in exploring these domains of applied epistemology of science. A prior course in probability theory is not a disadvantage. A prior course in statistics probably *is* a disadvantage, as my views both contradict most of what you are likely to have been taught there and emphasize issues, like where measurements come from, that are not generally part of the syllabus of courses like those, but that end up framing those other syllabi in rather unfavorable terms. The better your grade in that introductory statistics course, the more you will need to unlearn.

The material here is flexible enough for more than one pedagogical context. I have used it as part of the core of a liberal arts education (College Honors program, University of Michigan, 1986–2004), as part of a graduate program in the life sciences (University of Vienna, from 2004 on), as part of an upperclass undergraduate course in linear modeling for applied mathematics and statistics majors (University of Washington, 2011), and as part of the freshman induction/indoctrination into the idea of college (University of Washington, 2007–2013). The graduate students work over the more advanced tools and the more complex examples of Part III along with discipline-customized versions of the forward-looking prophecies in Chapter 8. For American undergraduates, a semester's course on these principles can draw entirely from Chapters 1 through 4. Yet when Parts III and IV are included these same lectures and readings drive a sophisticated version of the same material at Vienna that challenges biology students right up through the predoctoral year.

Although I have never taught in a philosophy department, or in a science studies program for that matter, I suspect that the applied epistemology of numbers may likewise be a useful venue for introducing both the philosophy of science and the study of its social/cognitive aspects in general. Perhaps, in spite of what Platt avers (Section 3.4) about "mathematical" versus "logical" boxes, this is the tamest part of empirical scientific reasoning, the part that looks like physics in a particularly crucial cognitive aspect: it uses not the number line but real paper or real space as the sandbox for exploring forms of empirical understanding and modeling.

Thanks are due several institutions: the Honors College of the University of Michigan; the Faculty of Life Sciences of the University of Vienna; the Early Start Program of the College of Arts and Sciences at the University of Washington; Kanti Mardia's continuing Leeds Applied Statistical Workshop (which has listened with remarkable politeness to early versions of many of these ideas over the past decade); and the Konrad Lorenz Institute, Altenberg, Austria, along with its director, Gerd Müller. Still in Austria, I would like to thank Horst Seidler, currently Dean of the Faculty of Life Sciences, who in his previous role as Chair of the Department of Anthropology supported my early efforts to develop this material, both the epistemology and the morphometrics, into a curriculum with support from Austrian Federal Ministry of Science and Research project GZ200.093_I-VI_2004, "New Perspectives in Anthropological Studies." The examples from fetal alcohol science in Chapters 7 and 8 were supported by diverse grants from the erstwhile National Institute on Alcohol Abuse and Alcoholism (a division of the U.S. National Institutes of Health) to the Fetal Alcohol and Drug Unit (Ann Streissguth, founding director and principal investigator for most of the grants; Therese Grant, current director) in the Department of Psychiatry and Behavioral Sciences of the University of Washington. Major funding for presentations to one particularly diverse audience coalesced recently with the generous support of the European Union's Marie Curie program (Sixth European Framework Programme MRTMN-CT-2005-019564, "EVAN" [the European Virtual Anthropology

Network], G. W. Weber, coordinator; F. Bookstein, task administrator). The work on random walks is supported at present by U.S. National Science Foundation grant DEB–1019583 to Joe Felsenstein and me.

I should acknowledge the 400 or so students who have taken my courses under titles like "Numbers and Reasons" over the quarter-century that these readings and their deconstructions have accrued. Beyond their ranks have been my audiences at the Konrad Lorenz Institute in 2006 and 2008, at the international meetings on Partial Least Squares PLS'07 (Ås, Norway) and PLS'09 (Beijing), at most of the annual conventions of the American Association of Physical Anthropology between 2000 and 2013, and at my Rohlf Medal Lecture at Stony Brook University on October 24, 2011. In encouraging me to pursue the work on Karl Pearson reported at the end of Chapter 7, English professor James Winn taught me a lot about close reading of texts the year I was Faculty Fellow at the Institute for the Humanities he directed at the University of Michigan before moving to Boston.

Warmest personal thanks are owed (and hereby offered) to Werner Callebaut, Katrin Schaefer, Verena Winiwarter, Philipp Mitteroecker, and Martin Fieder (all of Vienna); to Paul O'Higgins (York, UK); and to Michael Perlman, Joe Felsenstein, and Paul Sampson (Seattle) for the repeated discussions during which they negotiated the pedagogy and phrasing of much of this conceptual material with me. For innumerable conversations about numerical methods and methodologies, especially geometric morphometrics (the extended exemplar in Chapter 7), I am grateful to F. James Rohlf (Stony Brook University), Dennis Slice (Florida State University), Charles Oxnard (Western Australia and York), and Richard Reyment (Uppsala University, emeritus). For more than 20 years of endlessly patient software and hardware support I am deeply grateful to my long-term collaborator Bill Green of Bellingham, Washington, developer of `Edgewarp`.

I am grateful to all the authors and institutions who granted me permission to reproduce their graphical materials as part of this new text. To the best of our ability, specific acknowledgments according with the copyright owners' requirements are conveyed caption by caption.

The entire manuscript of this book was closely read in an earlier draft by my colleagues Clive Bowman (London, England) and Hermann Prossinger (Vienna, Austria), and it is far better by virtue of their efforts. I also thank three anonymous reviewers for the publisher and the gentle but principled editing of Lauren Cowles. Whatever solecisms, eccentricities, blunders, and gaps of logic or philosophy remain are, of course, entirely my own fault.

Many of the readings that underlie this syllabus were called to my attention by others. I'd welcome suggested additional or alternate readings and reinterpretations from any reader of this book. Please send your suggestions along with any other comments to `flbookst@uw.edu` or `fred.bookstein@univie.ac.at`.

Preparation of a "big book," even though a matter of a myriad of details over the year or so prior to publication, is otherwise a background task running over decades during which one's job is always to be doing something else. I am grateful for readers like you who are willing to foreground these thoughts, even for a little while. I hope that my text explains some things that you may have wondered about before, and points out other issues that you now wonder why you *didn't* wonder about before. In short, I hope these arguments of mine are worth the time you will spend letting them change the way you appreciate these remarkably widespread forms of scientific thinking.

Fred L. Bookstein
Vienna and Seattle, August 5, 2013

Epigraphs

Here are collected a few wry aphorisms evoked from time to time in my main text. They appear to be fundamentally in agreement about how good science works, but they also highlight the risks involved.

1. It has been called the interocular traumatic test; you know what the data mean when the conclusion hits you between the eyes.
   – W. Edwards, H. Lindman, and L. J. Savage, 1963, p. 217

2. Trust in consilience is the foundation of the natural sciences.
   – E. O. Wilson, 1998, p. 11

3. I have sought knowledge. I have wished to understand the hearts of men. I have wished to know why the stars shine. And I have tried to apprehend the Pythagorean power by which numbers hold sway above the flux. A little of this, but not much, I have achieved.
   – Bertrand Russell, "What I Have Lived For,"
   prologue to his *Autobiography* (1967)

4. Science, no less than theology or philosophy, is the field for personal influence, for the creation of enthusiasm, and for the establishment of ideals of self-discipline and self-development. No man becomes great in science from the mere force of intellect, unguided and unaccompanied by what really amounts to moral force. Behind the intellectual capacity there is the devotion to truth, the deep sympathy with nature, and the determination to sacrifice all minor matters to one great end.
   – Karl Pearson, 1906, pp. 1–2

5. The enormous usefulness of mathematics in the natural sciences is something bordering on the mysterious, and there is no rational explanation for it.... It is not at all natural that 'laws of nature' exist, much less that man is able to discover them.... The miracle of the appropriateness of mathematics for the formulation of the laws of physics is a wonderful gift which we neither understand nor deserve.
   – Eugene Wigner, 1960, pp. 2, 5, 14

6. When you cannot express it in numbers, your knowledge is of a meagre and unsatisfactory kind.
   – Sir William Thompson (Lord Kelvin), 1889,
   as cited by Kuhn (1961, p. 161)

7. The route from theory or law to measurement can almost never be travelled backwards. Numbers gathered without some knowledge of the regularity to be expected almost never speak for themselves. Almost certainly they remain just numbers.
   – Thomas Kuhn, 1961, pp. 174–175

8. Identical twins are much more similar than any microscopic sections from corresponding sites you can lay through either of them.
   – Paul Weiss, 1956,
   as quoted by Gerard (1958, p. 140)

9. The first starting of a hypothesis and the entertaining of it, whether as a simple interrogation or with any degree of confidence, is an inferential step which I propose to call *abduction* [or *retroduction*]. I call all such inference by the peculiar name, *abduction*, because its legitimacy depends upon altogether different principles from those of other kinds of inference. The form of inference is this:
   - The surprising fact, C, is observed;
   - But if A were true, C would be a matter of course,
   - Hence, there is reason to suspect that A is true.

   – C. S. Peirce, 1903,
   as transcribed in Buchler, ed., 1940, p. 151

10. The time is past, if ever there was such a time, when you can just discover knowledge and turn it loose in the world and assume that you have done good.

    – Wendell Berry, 2000, p. 145

11. It is wrong, always, everywhere, and for anyone, to believe anything upon insufficient evidence.

    – William Kingdon Clifford, "The Ethics of Belief," 1877

12. The force of reason is a social force.... The problem of cognitive order *is* the problem of social order.

    David Bloor, 2011, pp. 3–4

13. But things got better. One day, a few months later, with a machine rattling off the results of a test I was trying to devise, I first felt the joy of being a scientist. Although I had not yet discovered anything, I realized that I had developed a reliable way to measure what I wanted to measure. Science consists largely of measurement. When the metaphorical ruler – what we call an assay – is in hand, results and happiness generally follow.

    Harold Varmus, 2009, p. 37

14. Die [wissenschaftliche] Tatsächlichkeit... liegt in einer solchen Lösung des Problems, die unter gegebenen Verhältnissen bei kleinster Denkwillkürlichkeit den stärksten Denkzwang bietet. Auf diese Weise stellt diese Tatsache ein stilgemäßes Aviso des Denkwiderstandes vor. Da das Trägertum des Denkstiles dem Denkkollektiv zukommt, können wir sie kurz als "denkkollektives Widerstandsaviso" bezeichnen.

    – Ludwik Fleck, 1935, p. 129

    In the standard translation:

    [Scientific] factuality... consists in just this kind of solution to the problem of minimizing thought caprice under given conditions while maximizing thought constraint. The fact thus represents a stylized signal of resistance in thinking. Because the thought style is carried by the thought collective, this "fact" can be designated in brief as the signal of resistance by the thought collective.

    – Ludwik Fleck, 1979, p. 98

    In other words, namely, mine,

14a. A scientific fact is a socially imposed constraint on speculative thought.

    – free précis/translation by the author

Part I

The Basic Structure of a Numerical Inference

1

Getting Started

1.1 Our Central Problem: What Is a Good Numerical Inference?

You know this move. You have seen it dozens of times, or thousands if you have been a scientist or professor for long enough. A text or a PowerPoint slide presents some arithmetic computation based on measured data – group averages, or a regression slope, or the range of some measurement, or a bar chart or time series – and then the author continues, "So I have shown...." or, more conspiratorially, "We have thus shown...." or, more didactically, "It follows that..." where the "..." in all cases is some proposition more general or otherwise more assertive than the scope of the data actually reported. The qualitative assertion may deal with a cause, or a consequence, or a generalization about past, hypothetical present, or future under some eutopian or dystopian policy. Or the sentence may be a bit more sophisticated, with the same import: "Hence our null hypothesis is rejected" or, the form with the most internal evidence of thoughtfulness, "Hence Professor Smith's theory is false."

Here are a few examples unsystematically extracted and rephrased as single sentences from later in this book. "It followed that the submarine was traveling east." "The two continental plates are moving apart at a rate of about four centimeters per year." "Cholera is caused by some morbid material in your drinking water." "There was an anthrax epidemic in Sverdlovsk, Russia, in 1979, that originated in a secret biowarfare factory." "Einstein's law of the photoelectric effect is correct." "Atoms exist." "Stomach ulcers are caused by a bacillus, *Helicobacter pylori*." "The form of DNA is a double helix pairing adenine with thymine and guanine with cytosine." "Environmental tobacco smoke raises your chances of a heart attack or of lung cancer by just about one-fourth even after considering differences in diet." "Civic worth is inherited just as much as eye color." "This is the skull of Oliver Cromwell." "This brain was damaged before birth by its exposure to high levels of alcohol."

There are many more general tropes having this form. It is fun to estimate the propensity of stereotyped research summaries in this family by using a Google search in its "sentence completion" mode. If you type "Boys have more" (including the quotation marks) into Google, a search returns more than 6 million pages with this phrase. On one Tuesday morning in 2013, the first page of retrievals includes "Boys have more health problems in childhood than girls" (a study from *Acta Paediatrica* of Finnish children born in 1987); "Women expecting boys have more testosterone" (and therefore weaker immune systems, hence shorter life spans), ostensibly a health story from the *Daily Mail*, a British newspaper; "Boys have more cortical area devoted to spatial-mechanical functioning," from a University of Arizona keynote address about gender and mathematics achievement; and "Why do boys have more freedom than girls?," a finding that likely depends on details of the way the term "freedom" was operationalized. The completion template "Apes have less" retrieves almost 240,000 hits; "People with a high IQ," 410,000. It may be that this sort of binary reclassification semantic (more or less, higher or lower, no matter how much) is how computations get turned into propositions for the majority of contexts in the social and medical sciences, in popular wisdom, and in real life in the modern Western setting.

But how does this rhetorical and semantic process actually operate? How do we readers or listeners move from one sentence to the next across this divide, from displays of arithmetic to the cognitive focus on some inference? I am not questioning (at least, not in this book) the arithmetic, nor the fact that the numbers entering into the arithmetic arose from some sort of instrument reading that the experimenter could distort more or less at her own volition. Rather, I am questioning this central trope, usually unspoken, that guides us over the leap from arithmetic to inference – from routinized observations out of some empirical data set, with all of its accidents and contingencies, into an intentionally more abstract language of a truth claim. After this maneuver, the conversation has ceased to deal with "formulas" or "statistical methods" or other merely arithmetical concerns, and has now become a matter of qualitative assertions and their implications for human beliefs or human actions.

This book is concerned with the specific process by which a numerical statement about a particular data set is transformed into a more general proposition. That is the process I am calling "numerical inference." It is not syllogistic (though we see an attempt to make it so in the discussion of C. S. Peirce's ideas in Chapter 3). Cognitively, it would seem to be instantaneous, rather like the "Eureka!" moment reported by James Watson in his narrative of the discovery of the double helix (see at "Suddenly I became aware...," Section E4.5).[1] The inference itself seems intuitive. Epistemological discussions such as those of this book at best supply justifications after the fact or the mental discipline that renders that justification easier on a routine basis.

The basic inference itself – the claim that an essentially verbal proposition follows from an arithmetical manipulation of some measurements – is not intrinsically fallacious. Sometimes support for a proposition *does* follow from data – for instance, most of the examples in Chapter 4 are valid. But in almost all cases the inference went unjustified *in the course of the text in which it appears.* (For superb exceptions, see Perrin's argument for believing that atoms exist, Section E4.3, and also my last

---

[1] Throughout this book, crossreferences to extents less than an entire chapter are indicated by the word "Section" followed by a subsection or sub-subsection specification, such as "Section 5.3.1." In Chapter 4, after the brief introduction, subsections are further sorted as "lectures" with a heading beginning with "L4" or instead "examples" with headings beginning "E4." So crossreferences to this chapter include that letter, for instance, "Section L4.3.1" or "Section E4.3."

Chapter 4 exemplar, the argument by Alvarez et al. about one big reason for the extinction of the dinosaurs.)

There are ways of designing one's studies so that the numerical inference is more likely or less likely to be valid, and discussions of these ways make up much of the content of this book. But first it is necessary to foreground these conventionally tacit links inside conventional scientific narratives – to bring them out from the dark into the light so that their typical features can be noted and critiqued. Neither the natural sciences nor the social sciences could operate effectively without them. Our task is not to eliminate them, but to highlight them: first, to detect them when they appear, or to point out their omission at places where they should appear but do not; second, to suggest a prophylaxis by which their average cogency and reliability might be improved.

In other words: the central concern of this book is not for the *content* of a scientific argument but for its *flow* in an important special case, to wit, the rules and the logic of inferences from numbers or numerical measurements. The question is not "How do you state a conclusion once you have analyzed your numerical data?" For that there are a great many books at a great many levels of sophistication, of which the oldest in my collection is Huff's 1954 *How to Lie With Statistics*. Rather, the issue is under what circumstances such an inference can be trustworthy. *When* do inferences from numbers make scientific sense, and how?

Central domains for my examples here include the physical and biological/biomedical natural sciences and also the historical/biographical natural and social sciences. What these fields share is a respect for consilience together with the possibility of abduction. **Consilience** (here I anticipate the long introduction in Chapter 2) is the matching of evidence from data at different levels of observation or using essentially different kinds of measurement; **abduction**, the main topic of Chapter 3, is the sequence of finding something surprising in your data and then positing a hypothesis whose specific task is to remove that surprise. **Strong inference**, a sort of multiple abduction in parallel, is the wielding of a single data set to support one hypothesis at the same time it rejects several others that had previously been considered just as plausible.

But this has all been rather abstract. Before going any further, let us dive right into a particularly rich example from an unexpected quarter, namely, the ocean floor about 400 miles northwest of the Azores.

1.2 The Sinking of the *Scorpion*

The *Scorpion*, a submarine of the United States Navy, sank on May 21, 1968, with loss of all hands. I have no independent knowledge of this event, but proceed by deconstructing two easily accessed readings on the topic: the chapter "Death of a Submarine" in Sontag and Drew (1998), and the Wikipedia page http://en.wikipedia.org/wiki/USS_Scorpion_(SSN-589). For an alternative overview of this same episode, see McGrayne (2011, pp. 196–204).

The basic structure of the investigation here is easy to summarize. A submarine vanished, and the U.S. Navy wished to learn why so that similar accidents to her sister ships might be avoided. Although the data pertain to one historical event only, the inference is a generalization – to levels of risk across a class of vessels. What makes the case particularly interesting is that to begin with there were no data – those had to be constructed (not reconstructed) before any inferences could be grounded at all. A cycling was needed between potential inferences and the data resources that might render them plausible, thereby motivating further potential inferences suggesting the accrual of more new evidence in turn.

The story begins with a bald fact: an American submarine, the *Scorpion*, didn't show up at its base in Norfolk, Virginia, on May 27, 1968. The fact arrives unconnected to any narrative that could give it meaning. It has no associated numerical characteristics, but is merely a statement about the absence of something.

One has to begin somewhere, and the argument in this instance begins with the declared plan communicated by the submarine to its base prior to this journey. Its intended travel was along a fixed path, and we can assume (as did the Navy) that its own instruments were adequate to keep it on this path: a great circle (the shortest possible route) to be traversed underwater at a carefully monitored speed of 18 knots (about 20 miles per hour) all the way from the Azores to Norfolk. This intention counts as a preliminary "theory" in the discussion that follows: an explanatory structure, evidence-driven deviations from which remain to be discovered.

The *Scorpion* could have been anywhere on this 3000-mile track when something untoward happened. The task is to figure out where it was, and then *either* to account for what might have happened to it *or* to go take a look at the wreckage to see what causes of catastrophe might be inferrable from the image. Historically, this inference was in the hands of one man, Navy analyst John Craven, and it is in Craven's voice that Sontag and Drew tell the story. (For Craven's own retelling, see Craven, 2002.)

At this point a glimmer appears of the first of a series of inferences whose logical machinery concerns us in more detail in Chapters 2 and 3. The "track" is actually a path in space and time, a particular pairing of location with time that could perhaps be matched to instrument readings to produce evidence of particularly interesting moments that might engender a possible explanation or alter an explanation that had been proffered earlier. There is a tacit model here about the machine that was the submarine itself – that it was functioning within specifications (i.e., staying on that space-time track) until the moment of some disaster – and there are shortly to emerge some assumptions about other machines, namely, the listening devices that will highlight specific moments along this planned trajectory.

For our purposes, the main logical characteristic of machines is the predictability of their performance as long as they are "working." This presumption of predictability applies both to active machines like submarines and to the passive devices that make observations. That predictability is one of the deepest cruces in all of natural science, closely tied to the principle of consilience on which this book is ultimately based. As Wigner (1960, p. 5) put it, "the construction of machines, the functioning of which he can foresee, constitutes the most spectacular accomplishment of the physicist. In these machines, the physicist creates a situation in which all the relevant coordinates

1.2 The Sinking of the *Scorpion*

are known so that the behavior of the machine can be predicted. Radars and nuclear reactors are examples of such machines." And, I would add, attack submarines.

Along with this theory of submarine behavior – that in the absence of any other factor they tend to travel along great-circle routes – are additional facts that seem, but only seem, to be numerical in the same way. The *Scorpion*, for instance, was restricted to depths of less than 300 feet, owing to delays in some safety refittings; but this value of 300 will not enter into any inferences, it turns out. The sub had experienced a bad bout of wild accelerations and vibration some months earlier – a "corkscrewing problem" – but the specific physical parameters of that episode likewise will play no role in the inferences to come. Their irrelevance could not be asserted, though, until a valid theory had finally been asserted and verified. We are illustrating in passing a point from a discussion in Section 2.5 that much of the art of physics inheres in knowing what to measure and what to ignore.

A first step along the road to discovery was the simple scan of historical records to rule out collisions or battles. In the language of Section L4.6, these are "plausible rival hypotheses," easily ruled out by searches through databases whose details do not concern us here.

The numerical inference begins to take shape with the search for instrument records that might be informative. (There is a close analogy here to the reliance of evolutionary demography on massive databases, such as the Integrated Public Use Microdata Series [IPUMS], that have been accumulated by bureaucracies for other purposes entirely. See Huber et al., 2010.) One set, the Navy's own SOSUS (Sound Surveillance System), was systematically useless for Craven's purpose, by design: it was set to filter out all point events (like blasts) in order to highlight the periodic noises of Soviet machines. But an Office of Naval Research laboratory in the Canary Islands happened to have preserved a set of unfiltered records for the week in question, and within them were eight separate bursts of signals at amplitude well above background. Working backwards from time to place, this gave the Navy eight different locations to check for surface wreckage. Surveillance planes found nothing, or else there would be no story to tell here.

Now this is not a consilience yet (to anticipate the language of Chapter 2). Any event in the Canary Islands record for that week must match *some* location on the *Scorpion*'s track, and so before any grounded inference could begin there needed to be a coincidence (in the literal sense, co-incidence, meaning co-occurrence) between two channels of data, not just a map line and a microphone reading separately. Craven needed *another* source of information, as independent as possible. (We go more deeply into this notion of "independence" of information resources in Section 2.3, in connection with Collins's discussion of replications.) There were two more hydrophones in the Atlantic, intended to track Soviet nuclear tests: one off the peninsula of Argentia in Newfoundland, the other 200 miles away. This pair of positions was not entirely suitable – there were undersea mountains between them and the Azores – but when Craven laid the recordings from the two hydrophones over the Canary recordings, some of the blips lined up. The geodetic situation is approximately as rendered in Figure 1.1.

Figure 1.1. North Atlantic locations relevant to the *Scorpion* story. (A GoogleEarth figure from a viewpoint about 8000 miles above the water.) *Scorpion*'s planned track from the Azores to Norfolk, Virginia is the white line shown. The unnamed open circle along this line is the approximate location of "Point Oscar," the point on the track corresponding to the match of interesting loud sounds among the three microphones, one in the Canary Islands and two others in the vicinity of Argentia, Newfoundland.

At this point, there is an actual formal inference, of the style we review at length in Chapter 3. Here it is in Sontag and Drew's words:

> If the Argentia blips were worthless noise, then the plots would probably fall hundreds of miles or thousands of miles from the relatively tiny line of ocean that made up *Scorpion*'s track. But if the new data pinpointed *[sic]* any one of the eight events picked up in the Canary Islands on that tiny line, the acoustic matches would almost certainly have to be valid. (Sontag and Drew, 1998, p. 96)

And they matched, at one first moment and then 91 seconds later, 4 seconds after that, 5 seconds after that, and so on for 3 minutes and 10 seconds. Then silence.

The *inference*, then, is that these sounds recorded the destruction of the *Scorpion* from implosion. It is a qualitative inference from quantitative data – the very topic on which this book's argument is centered. This is the first inference: that these sounds are the sounds of *Scorpion*'s destruction at a particular place and a particular time.

1.2 The Sinking of the *Scorpion*

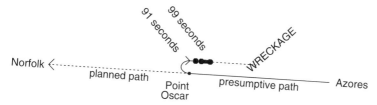

Figure 1.2. Schematic of Craven's understanding after he discovered that *Scorpion* was traveling eastward, not westward, over the series of detected blasts. On the diagram, 91 seconds is the interval between the first detected event and the second, much louder event, and 99 seconds is the interval between the second and the last of the series.

Thus the inference, a particularly powerful one in view of the *repeated* co-occurrences, but it must be validated. To verify the Alvarez argument we review in Section E4.8, somebody had to find a meteor crater of the appropriate size and age; to verify Craven's inference, somebody has to find the wreckage of the submarine. What to do? The submarine could be anywhere in a circle of radius roughly 10 miles around this point of the last implosion signal.

At this juncture three distinct numerical *enhancements* appear, each individually quite brilliant: a sharpening of the information from the instrumental record by filtering, and two separate sharpenings of the theory by human interview. The sharpening of the record is not described in my sources – I am assuming it represented a computation by differentials in the arrival times of the blast signals at the sites of what are by now three microphones (Canary Islands, Argentia, and the third one). From this Craven learned that *Scorpion* was moving *east*, not west, between explosions. Remember that its planned trajectory was *west*, directly west, from the Azores to Norfolk. *Trusting the instruments*, then, he asked what could make a submarine carry out a 180° turn in the middle of the ocean, the scenario diagrammed in Figure 1.2.

The universal answer, according to Craven's sources (mainly retired submarine commanders), was that this maneuver would be the result of a "hot torpedo," the accidental launch of a torpedo or the jettisoning of a defective and dangerous torpedo when there is no target in the vicinity except the submarine itself. It was a standard ploy (a trope, in the language of the torpedo's own guidance system) that instructs the torpedo not to pursue its own mother ship. This then constitutes a change in the theory (here, in the expected trajectory) based on new data.

But this is the wrong theory – it doesn't involve the sinking of a submarine. What else needs to have happened? Craven's sources suspected the "cook-off" of a torpedo – a fire in the torpedo room starting from a bad battery – followed by an explosion there that blew the hatches open, destroying the submarine's buoyancy. Simulations with humans showed that about 90 seconds after the simulated explosion, no matter what a commander did, the submarine would arrive at its implosion depth of about 2000 feet and the series of closely spaced blasts would begin. Recall that the actual interval between the first and the second blasts, from the hydrophone records, was 91 seconds.

This is the second inference: an actual historical sequence of events corresponding to the sequence of instrumented records. Craven now had to shift to a wholly quantitative *third* inference: given this scenario, where, exactly, would *Scorpion* come to rest on the seabed? This takes the form of a classic problem in prediction under uncertainty, along with an associated search strategy that was itself the subject of a book (Stone, 1975). The method for this was not remotely objective. A set of parameters was listed (the wreck's downward glide speed, the direction of glide, the glide slope, and so on). A bottle of Scotch whisky would be the reward for the expert whose predictions would best match the actual location of the sub when it was found. There resulted a prediction of the best (optimal) location to look and a corresponding search strategy. (We discuss the notion of "optimal" estimates of statistical parameters in Section 5.3.)

That was the third inference, leading to new data conveying its information most convincingly indeed. The wreckage of *Scorpion* was found on October 29 within 200 meters of where Craven and the experts had estimated its likeliest location. She lay at a depth of 11,000 feet (3.35 kilometers, just over 2 miles). We are not told who won the bottle of Chivas Regal.

This not only closes the inference about where *Scorpion* was – that is now an actual observation, no longer an inference – but also very nearly closes the inference about why it ended up there. If *Scorpion* is where a simulation based on the cook-off of a torpedo battery puts it, that becomes (see Chapter 3) quite strong evidentiary support for this particular causal theory. The Sontag–Drew chapter concludes with an investigation into all the other evidence for the bad-battery theory, including confidential Navy submarine corps memos, interviews with surviving submarine commanders, and finally (in data declassified only in 1998) pictures of the wreckage showing that the torpedo room compartment had not imploded (meaning that it was already balanced against hydrostatic pressure, i.e., flooded to begin with). The theory was actively rejected by the majority of Navy bureaucrats, because it implied that they themselves were responsible for the loss of the submarine, but this topic takes us too far from the theme of this book.

Let us review. We began with a one-bit fact – *Scorpion* was missing. Agreement (to be called "consilience" in Chapter 2) among the three channels of hydrophone data, as referred to the map of the planned trajectory, led to a location and then a new one-bit fact (reversal of direction) that, via multiple simulations, led to a presumed cause of the catastrophe accompanied by estimates of crucial continuous parameters of the wreckage's glide path. The simulations were confirmed by actual images of the wreckage, showing not only that *Scorpion* lay where the final theory would place her but that her physical form was consistent with one particular version of the detailed explanation of the wreck (explosion in the torpedo room). This is surely as far as you can expect one method (numerical inference) to get you in the reconstruction of historically unique events.

If this ebb and flow of argument seems familiar, it may be because the same cycling and recycling is embedded deep in popular culture as the method of detective stories. Technically obsessed police procedurals by authors such as Ed McBain or

Henning Mankell follow the same cycle: initial bafflement is followed by one clue, which constrains reasoning enough to suggest a more focused search for other clues, which eventually accumulate in a unique "theory" of the crime that is then confirmed via a confession by the perpetrator. But also scientific papers can follow this form: the origins of the Iceman Ötzi's clothes, or his breakfast, or his arrows, or his knife (Spindler, 1994; Fowler, 2001); the building of a room-sized three-dimensional model of a double helix with the physicochemical properties of DNA (Section E4.5); or, for that matter, any of the other examples reviewed in Chapter 4. Indeed the language of the detective story is explicit in such diverse texts as Jean Perrin's argument for the reality of atoms (Section E4.3) or the Alvarez group's demonstration of the extraterrestrial origin of the great extinction of 65 million years ago (Section E4.8). When arithmetic and algebra are done, what is left to do in order to close a numerical inference must be very much like the standard form of reasoning instantiated by the *Scorpion* narrative as it exemplifies the search for causes of historically singular events.

Other processes to which this form of inference bears more than a passing resemblance are medical diagnosis, where symptoms that are ordinarily qualitative or verbal need to be converted into numerical data for any formal diagnostic protocol to go forward, and historical epidemiology (as in John Snow's construction of the index case of the 1854 cholera epidemic in London, see at Section E4.1).

1.3 Prospectus

The previous section demonstrates the methods of this book as they apply throughout all the chapters that follow. It is based on close reading of texts and deconstruction of the author's voice (my own as well as that of others), but also a good deal of "constraint by the facts" and the theorems. In the case of *Scorpion*, just as Sontag and Drew could assume that their readers understood great circles and the precision of submarine navigation, I assumed you understood the rules of consilience in the graphical form that the original narrative took for granted (superimposition of a planned itinerary on a map, Figure 1.1, or of time-coded sound traces, Figure 1.2, on the same map). In some of the best examples, consilience arises likewise from the graphical superposition of patterns. The rules of superposition here are so familiar as to seem transparent, but when looked at in detail, they embody quite a bit of geometry (in *Scorpion*'s case, the understanding of what a shortest path on Earth's surface should look like).

This requirement that some aspect of the diagramming be already familiar – the tacit use of geometry to convey the essence of a pattern match – is essential to the genre. A persuasive claim of a numerical match cannot persuasively be floated atop a novel instrumental trace paired with another that is equally novel. This is one fairly common example of a general psychological restriction on the use of numerical inferences to drive discoveries: inference from a numerical consilience requires comfort as well as information. What Chapter 3 describes as the resolution of the surprise must be as satisfying and persuasive as the surprise itself was previously disturbing, and it must remain persuasive even after the relief of the tension associated with the surprise has been forgotten.

In the usual jargon, the topic of this book is ostensibly descriptive – this is how people have justified their inferences in public – but it is meant to be normative: those justifications have in fact *convinced* several generations of readers to accept the inferences as valid, and move on. The claim is not merely that this is how science *is* done, some of the time, but that this is how it *ought to be done* insofar as inferences from numbers are involved. It is remarkable how rarely this particular point is noted in the annals of scientific epistemology. Merton's four rules (Chapter 2), for instance, make no reference to the ways in which an inference from data is likely to become "valid," as this is a *social* process borne within the network, small or large, of readers of the arguments in question. Discussions by wise men like Wigner or Kuhn intentionally avoid the topic of extension to the less "lawful" sciences, such as paleontology, naval warfare, or social studies, partly because the generalizability of inferences like these, for example, the plausibility of a future practical application, tends to be weaker in those domains.

We nevertheless pursue this point as far as it can take us. The journey is in the form of dozens more worked examples, recounted in detail, logical step by logical step or diagram by diagram, in the course of their deconstruction. Two of these summarize themes from my own applied work. The remainder are the work of others and are treated in general with greater respect. (For several of them, immortality is assured anyway by the associated Nobel Prize.) These narratives are interwoven with a parallel series of formal mathematical or philosophical discussions, sketching how explanations get attached to numbers in specific domains, that arrive no later than the page at which they are required. The sequence of the examples is, roughly, the increasing difficulty of the rhetorical work that needs to be done to persuade the intended reader of something important. These arguments are simple and overpowering for the first few readings (Snow, submarines), but quite recondite by the end (Milgram, Law, Bookstein). Nevertheless the methodology being foregrounded here is unvarying – to present the information content of data in the way that best argues the dominance of your explanation over everybody else's. This is the underlying principle of *abduction* (Peirce, 1903), the central concern of Chapter 3 and the central logical machine of this entire domain as I read its design. Measurements matter only when they are evidence for something, or against something; otherwise, as Thomas Kuhn says in one of this book's epigraphs, "they remain just numbers."

Outline of what follows. The remaining text of this volume spirals down on most of the preceding themes in steadily increasing detail. We are in the first chapter of Part I, "The basic structure of a numerical inference," which continues with focused discussions of its two foundations, consilience and abduction. Chapter 2, "Consilience as a Rhetorical Strategy," reviews what is special about numerical reasoning in these scientific contexts: the reasonable expectation that information derived from measured numbers should agree with "the same information" (true states of nature) when derived from other measured numbers. The entomologist E. O. Wilson revivified this form of inference, and resuscitated the very 19th-century word "consilience," in a book of 1998 that informs the core of the chapter's concerns. Wilson's ideas are more controversial than they ought to be, perhaps because he pushed them into contexts (such as evolutionary psychology and the humanities) for which there is very little usable numerical evidence of any form, only verbal analogies between metaphors, which are a "meagre and unsatisfactory" form of knowledge indeed. I show how Wilson's themes relate to standing critiques of numerical inference by Thomas Kuhn and Harry Collins, but I stoutly defend their application to the domains discussed elsewhere in this book.

Consilience involves visual tools as much as numerical displays. The example that opens Chapter 2, an example much better known than the story of *Scorpion*, was based on convincing public presentation of diagrams from the very beginning: the demonstration that sea-floor spreading explains continental drift. Chapter 2 continues by reviewing a critique mounted against all of this by several people loosely describable as "theoretical biologists," such as Walter Elsasser, Ralph Gerard, and me: namely, that the task with which the *Scorpion* investigation began, the separation of signal from noise in the instrument records, is so difficult in biology as to be in principle impossible unless theory is present along with instrumentation from the very beginning. Unless, in other words, theory and data negotiate with one another – which is exactly the situation that Thomas Kuhn declared to be the case for physics. The proposition may apply to all of the consilient sciences, not just one, and may in fact supply a methodological foundation for reductionism when that consilience faces "downward."

If Chapter 2 represents the claim that Nature looks with a friendly face on attempts to argue from numerical agreement, then Chapter 3, "Abduction and Strong Inference," goes in greater depth into the logic by which we can intuit Nature's hidden views on the matter. This is the syllogism of *abduction* first dissected out of good scientific practice by the American C. S. Peirce in the 1870s and quickly turned into practical epistemology by the chemist T. C. Chamberlin, writing before 1900, in his "method of multiple working hypotheses." This paper was resurrected by John Platt in 1964 in his famous *Science* article "Strong inference" on the secret of making scientific progress rapidly: to have every scientific assertion be at the same time the refutation of somebody *else's* hypothesis, or of one of your own plausible possibilities. It is no accident that Peirce was the founder of the epistemology known as pragmatism. The methodology of consilience, in this special application to numerical inference, is completely consistent with that general approach to knowledge and in particular with the approach to human psychology that William James wove out of this foundation. But the exemplar of abductive reasoning introduced in this chapter is quite contemporary indeed: the demonstration in 2007 by the Intergovernmental Panel on Climate Change, mainly via one single summary graphic, that the available data overwhelmingly support the claim of an anthropogenic cause for global warming.

This ends the introductory material. Part II of the book, entitled "A Sampler of Strategies for Numerical Inference," is a loosely organized survey of diverse applications that require steadily more sophisticated algebraic approaches to arrive at the same kinds of inferences. Its single chapter, "The Undergraduate Course," is divided into 14 separate pieces that collectively amount to what in the United States would be considered a one-semester class for baccalaureates nearing the end of their college years. The examples here include John Snow's discovery of the mode of transmission of cholera, Robert Millikan's demonstration of the validity of Einstein's theory of the photoelectric effect (preceded by a lecture on the meaning of straight line fits to data), Jean Perrin's successful demonstration that atoms exist (preceded by a lecture on the origins of the Normal [Gaussian, bell-curve] model for noise), the discovery of the bacterial origin of stomach ulcers (preceded by a lecture about likelihood methods and null-hypothesis testing, and followed by a historical aperçu about pathological reasoning along these lines), a deconstruction by James Watson himself of the Watson–Crick paper on the double helix, a comparison of two papers on the effects of passive smoking (preceded by a lecture on correlation and regression), and then finally guides through two studies that need no additional lectures to gloss them: Stanley Milgram's experiment on human obedience (a discussion followed by a consideration of how his methodology differs from that of most of his colleagues in social psychology and allied fields), and the production by Luis and Walter Alvarez of the hypothesis (now a fact) that a meteorite caused the mass extinction at the Cretaceous–Tertiary boundary about 65 million years ago. (The word "fact" in this context is used in the sense of Fleck, 1935/1979: a socially imposed constraint on scientific speculation. See the discussion at Section 2.3.)

Besides their susceptibility to a close reading based on consilience and abduction, all of the examples so far share a simplicity in their approach to measurement – most of the arithmetic and most of the arguments focus on one empirical characteristic (one "variable") at a time. Where this composes a 173-page undergraduate course, Part III, "Numerical Analysis for General Systems," is more hierarchically organized as a graduate course in just three segments. Chapter 5 opens with a survey of ways in which the complexity of organized systems (a term of Warren Weaver's; see Weaver 1948) alters the complexity of the corresponding numerical displays. Rather than focusing on single numbers, a scientist studying a more complex system will typically be decrypting patterns in more than one dimension – not Gaussian bells but curves, clusters, or voids in geometrical ordinations. Intellectually, the focus has switched from models and patterns that are meaning*ful* to models or patterns of meaning*less*ness, or, if you will, the physics and psychology of randomness. Abduction and consilience both need to be modified to accord with the greater sophistication of models for randomness in these more complex domains, and the claims about surprise that drive abductions need to go forward in the more general context of information theory (see, e.g., Gleick 2011).

Chapter 6 introduces the singular-value decomposition (SVD), a startlingly powerful and flexible pattern engine appropriate for many systems like these. It is the SVD, for example, that allows scientists to track civilians who can keep both the "ideal form" of an animal, like a house cat, and the range of forms of actual house cats in the mind at the same time: large, small, fat, lean, tailed or tailless, male or female. The SVD builds explanatory models not out of ellipsoids or lines and planes but out of hyperbolic paraboloids, a somewhat less familiar geometrical object that represents the geometrization of multiplication. The central nineteenth-century theorem driving this is diagrammed in its three important special cases (principal components analysis, Partial Least Squares, principal coordinates analysis), each of which is conveyed by a variety of examples. The SVD family of pattern engines corresponds to a downgrading of the contents of data matrices from their usual centrality in statistics courses vis-a-vis a closer scrutiny of their row and column labels. For studies of organized systems, what *was* measured is often less important than what *could have been* measured. Measurements, in other words, need to come in organized systems of their own, an entire manifold of possible measurements in terms of which interpretation can go forward much more delicately.

Chapter 7 expands on this point of view at length, mostly in the context of one single emerging interdisciplinary methodology, modern morphometrics. A contemporary morphometric toolkit offers multiple formalisms for that approach to orderings of measurement spaces – not just linear combinations but nonlinear reindexings, and graphics that intentionally exploit human binocular vision to set the intuition working on the interiors of objects even though only their boundaries have been measured. Protocols for this exemplary methodology are distilled down into four Rules of Morphometrics that emphasize the origin of information content in the prior knowledge bases, such as classical comparative anatomy, from which the structural systems for these measurements derive. After several examples of successful investigations along this line, Part III closes with two examples of how similar attempts of putative abduction and consilience failed when the underlying measurement domains were not based on such powerful prior understandings. One of these is an attempt at reducing the sociological causes of fetal alcohol damage to psychology. The other, less well known to statisticians than it ought to be, is the great blunder of Karl Pearson, the central figure in the establishment of statistics as a modern intellectual discipline. Reviewing Pearson's efforts under these two headings (morphometrics, and innate biological characteristics), I conclude that he did roughly as well as I did. He almost solved the basic methodological problems of morphometrics way back in 1935, whereas in his attempts to get at an understanding of human behavior by correlation-like methods, he arrived at essentially no valid inferences at all, in spite of having invented all the methods involved in these metrological chains.

Part IV, which closes the book, is somewhat slyly entitled "What Is to Be Done?" Partly retrospect, partly prospect, the argument here returns to the creative tension between the pragmatism of abduction and the organized skepticism of consilience. Modern methods of pattern analysis (beginning with those in Chapter 6) have the power to launch abductions that the corresponding methods of consilient confirmation, which are always the more classical, simply cannot cope with. For instance, can we justify any serious penetration of these ideas into the multivariate social sciences? "Yes," Wilson (1998) answers, but in my view he is too optimistic, at least for the "pure" social sciences, which, in contrast to the interdisciplines like neuroeconomics, seem to be stuck on qualitative descriptions or other raw material on which a consilience machine can't get a grip. Kuhn said this more rudely in 1961 when he commented that either social sciences are not capable of having crises or they are nothing but crises – and he didn't really care which. In both domains, the natural and the social, there is a serious moral hazard of thinking that you have discovered something important and then haranguing your audience or your peer reviewers until they pretend to agree.

A closing envoi sends the reader off with every good wish.

As my Preface has already noted, the argument of this book is only partly about the primary cognitive processes of scientists – it is also about one particular juncture in the flow of their publications, the insufficiently formalized reasoning processes that lie unspoken between the "Results" and the "Discussion" sections of scientific papers and textbooks. Scientists are morally obliged to enlighten the public about protocols and uncertainties at *this* stage, just as much as at earlier stages such as literature reviews and "methods and materials" narratives. To enlighten the public, the scientists themselves must first be enlightened: hence this book. If it is to have a moral, perhaps that is the proposition that there *is* a systematic possibility of valid persuasive rhetoric for the domains in which theory and data can be aligned numerically, and that this possibility is worth a book-length essay as variegated as I hope this one is.

2

Consilience as a Rhetorical Strategy

The *Scorpion* story in Chapter 1 was technoscience journalism – a poignant historical anecdote that conveys one essential aspect of the way quantitative reasoning is used all across the sciences that use it. In this chapter we get down to the serious business of scientific methodology by retelling a similar story that arose in quite a different context: geophysics. Establishing the theory of *continental drift* as fact was inseparable from the discovery of *seafloor spreading*, the mechanism that provides the explanation for drift, and the connection of the mechanism to the pattern it accounts for is strictly quantitative. Many books are available regarding this now firmly established theory, including several on its intellectual history. This chapter begins by revisiting the core of the scientific inference in terms of *consilience*, the topic of this chapter, along with *abduction*, the topic of the next one.

William Whewell seems to have brought the word "consilience" into philosophy of science in 1840, noting, "The Consilience of Inductions takes place when an Induction, obtained from one class of facts, coincides with an Induction obtained from another different class. Thus Consilience is a test of the truth of the Theory in which it occurs." In its full generality like this the concept may seem forbiddingly subjective, slippery, even mystical. A lecturer like me would do well to begin instead with an example as accessible as the *Scorpion* story but dealing with a scientific topic rather than a discrete episode of military misfortune. In Section 2.1, I recount the central thread of the continental drift narrative, focusing on the way it anticipates our consilience theme to come. Following that story, which is mainly visual, I turn to one particularly powerful recent discussion of consilience, E. O. Wilson's widely read 1998 monograph of that title, which asserts that this notion is central to *all* of science. Section 2.2 reviews Wilson's claim in the general context of biological reductionism in which he couched it, but goes on to note how a restriction of the topic to *quantitative* inference, as this book proposes to do, obviates any need for reductionism and turns consilience from a philosophy for reductionism into a philosophy for numerical inference in a much broader context.

Section 2.3 reviews an assortment of standard arguments about rhetoric from the literature of quantitative science to show how they are all more or less consistent with this reading of Wilson. For applications in biology we need to attend to the important concept of *heterogeneity*; this is the subject of Section 2.4. We discuss some of the challenges engendered for quantitative reasoning by the complexity of some systems, particularly biological systems, and the corresponding complexity of their measurements. A closing commentary, Section 2.5, emphasizes the role of graphics and some other commonalities in the cognitive psychology of this domain. Following this section, the chapter offers two appendices on themes originating here. Appendix 2.A enlarges on the argument of Section 2.2 about "Newton's apple" to make a deeper point about the interchangeability of mathematics with physics in this highly specialized domain of quantitative science. It is this interchangeability that renders physics so odd a science and hence so poor a model for pragmatic investigations into scientific practices. Appendix 2.B, in a quite different style, enlarges on the intellectual setting of a brilliant quote from Paul Weiss embedded in the argument of Section 2.4. The Weiss quote came from a meeting reported in the printed record of a later meeting that was the direct predecessor of the 2008 conference at which the themes of this book had their first public airing; this appendix details the passage from the 1955 setting to that of 2008.

2.1 Continental Drift

The solution to the puzzle of the *Scorpion*'s fate, Section 1.2, hinged on a combination of numerical evidence from multiple sources. There was the line on the globe, from the planned track the submarine was going to take home. There was that series of loud underwater bangs, fortunately recorded by several microphones, that could be mapped back onto the planned track. Finally there was the dramatic reconstruction (i.e., the theory) of the underwater reality (the catastrophe aboard the submarine) that would account for precisely this combination of numerical measurements.

I noted in the course of that discussion that the reasoning we were following felt oddly similar to the logic of a good classic detective story. And so it does; but it is also a good prototype for the core logic of quantitative scientific inference, the central topic of this book. The inference could be foregrounded because the numerical evidence was in a familiar notation of map coordinates. Sontag and Drew's typical readers didn't need to be told what a chart was. They would imagine for themselves the pair of microphones off the coast of Newfoundland, the great circle in the Atlantic Ocean, and the series of black dots aligned eastward over one part of that track that I've clarified for you in my Figures 1.1–1.2.

But what if the data to be explained had not been restricted to one historically contingent event, instead conveying a globally stable circumstance that was not yet comprehensible? Individual sciences differ in the extent to which they will accept naked assertions of pattern without the possibility of corresponding explanations. In astronomy, for instance, instrument readings can be accepted years or even decades before their explanations become available (see Harwit [1981] or look up "Hanny's Voorwerp" in Wikipedia). It is possible that this tolerance for ambiguity is founded on the stringent limitations of the astronomer's empirical toolkit. As Harwit notes, astronomers have access to only a *very* restricted range of measurements, mostly intensity of electromagnetic radiation as a function of time, frequency, and position in the sky, along with density of fluxes of particles as observed above the atmosphere and, even more rarely, physical characteristics of meteorites (see Section E4.8), neutrinos, and, speculatively, gravitational waves.

Geophysics appears not to be like that. Here, acceptance of a pattern in data awaits a plausible theoretical account of the pattern as concocted by somebody. LeGrand (1988) tells this intertwined history of physical patterns and corresponding explanations in relatively full detail. From his narrative I extract, in an anachronistic ordering, the core of the argument in a way that maps onto the central concept, consilience, that drives the rest of this book.

This version of the story begins with Alfred Wegener, a German meteorologist who, in 1922, published the first effectively argued edition of a small book, *The Origin of Continents and Oceans*. Wegener wondered at how well maps of the continental shelf of Africa fit maps of the shelf of South America when they were moved toward each other by cutting out the intervening ocean floor on a spherical map of the planet. (A continental shelf is the region of relatively shallow ocean floor just off the coast of a continent.) Convinced by some very bad measurements purporting to show that Greenland was moving away from Scandinavia at a rate of about 11 meters per year, he assembled a map rather like Bullard's of half a century later, and then combed other literatures for supporting evidence – continuities of features such as mountain ranges, geological layers, fossils, and paleoclimatology across what are now the wide gaps of the North and South Atlantic. Briefly stated, Wegener's claim was that these continents *had* once been all together as a single landmass that broke into the pieces we now call the continents (plus Greenland), which then moved apart relatively recently (geologically speaking) owing to forces whose origin and dynamics remained uncertain, hence, "continental drift." Wegener called this single landmass "das Urkontinent"; since 1928 it has been known as *Pangaea* or Pangæa (and the surrounding ocean as Panthalassa). His basic theme can be conveyed via a later graphic, the "Bullard fit" of 1965 (Figure 2.1), which is a computed reassembly of most of Pangæa out of charts of today's continents along with their continental shelves. In this modern graphical setting, Wegener's claim would have been that given such an arrangement, according to which the continents fit into one another quite well *everywhere*, with surprisingly little blue (missing shelf) or red (overlapping shelf), then that must be where they had actually been located, once upon a time when there was no Atlantic Ocean.

Being generally pragmatic, geologists did not uniformly reject Wegener's claims out of hand. In contrast to the intellectual antagonists who continued the debate about "uniformitarianism versus contractionism" at a high level of generality, practical field geologists found that such a presumption of continuity across a once-absent Atlantic basin, even in the absence of any plausible explanation, often led to satisfactory accounts of local phenomena such as the specialized fauna of Australia or the further elaboration of pattern-matching across the South Atlantic. Data from a completely new instrumental channel, *paleomagnetism*, that became available in the 1950s not

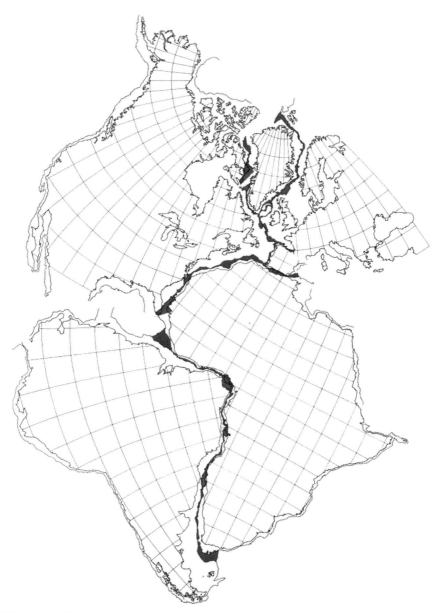

Figure 2.1. The "Bullard fit" of the continents around the Atlantic. Red: overlaps of the continental shelves at a depth of 3000 feet. Blue: gaps between the shelves. (From Bullard, Everett, and Smith, 1965, Figure 8. Both this original publication and the Raff and Mason article from which Figure 2.3 is extracted include an additional version of the crucial figure in enormously more detail than the downsampled versions I have reproduced here.) The language of color here refers to the original version of the figure reproduced in Plate 2.1. Republished with permission of The Royal Society (U.K.), from *Philosophical transactions of the Royal Society of London*, Bullard, Everitt, and Smith, 1965; permission conveyed through Copyright Clearance Center, Inc.

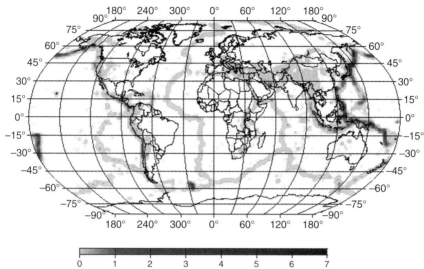

Figure 2.2. Earthquake densities per unit area, focus shallower than 70 km. (The data are a federation of multiple records; see http://earthquake.usgs.gov/earthquakes/eqarchives/epic/database.php. The figure is http://earthquake.usgs.gov/earthquakes/world/world_density.php, courtesy of the United States Geological Survey.)

only appeared to be in accord with hypotheses of drift but also seemed to demand such a hypothesis as the only possible explanation. (One of the physicists active in this development, Walter Elsasser, later became a biologist from whom we will hear in Section 2.4 on a completely different matter.) Certain rocks, at the time they are laid down, freeze a sample of Earth's magnetic field in vector form, "pointing" to the location of the north magnetic pole. On the assumption that there is only one north magnetic pole at a time, the continents can be reassembled in a way that maximizes the consistency of these different vector fields. When you do that, you can infer that Europe must have rotated about 24° around the pole with respect to North America over the last 200 million years, even if you have no idea how this could possibly have happened.

Yet another channel of evidence not available to Wegener was the global map of earthquake centers that was beginning to accrue by the early 1950s (a map based on more precise information from a new generation of seismographs). According to these new maps, earthquake centers were not randomly distributed over the globe, but instead arose mainly along curving lines typically aligned with the edges of continents or the bottoms of ocean trenches and, for shallow-focus earthquakes, along the mid-ocean ridges that were just beginning to be explored. (See Figure 2.2.) In the context of this book's themes, there is certainly an abduction to be launched from either of these two figures. Figure 2.1 immediately raises the question of why the continental shelves around the Atlantic fit together so well; Figure 2.2, why so much of the pattern of earthquake centers seems to comprise thin lines along the ridges up the middle of the large ocean basins.

Although Wegener himself had frozen to death in Greenland in 1930, in 1960 the challenge he had raised was still open. The continuing accretion of new data (from paleomagnetism, from earthquake focus maps) confirmed the plausibility of the original insight – yes, it looked like the continents had all been together in one big block at some point a few hundred million years ago – but continued to offer no route to a plausible *explanation*.

Turn now to the seemingly less dramatic picture in Figure 2.3. This is a map assembled from transects, images from a seagoing magnetometer of the sort that the ships tending submarines used to use to help the submariners figure out where they are and which way they are going without surfacing to look around. (Hence many of the records like these were kept secret for decades.) Today we understand it to be showing magnetic *reversals*, which are flips of Earth's north and south magnetic poles every few hundred thousand years (in response to a dynamic of its own that is still not well understood) that characterize the seabed as it is laid down and thereby modify Earth's current magnetic field positively or negatively to create the spatial pattern represented here by the contrast between black and white. When the magnetometers are towed behind vessels traversing a grid of the ocean, these traces reveal this pattern of *stripes* (hence its pet name, the "Zebra Diagram") over the underlying ocean floor.

But back in 1961 nothing of this mechanism was realized yet, and so the pattern, while striking, was quite incomprehensible. As Raff and Mason (1961, p. 1269) note with massive understatement, "These north-south linear features have aroused considerable interest and speculation.... There is as yet no satisfactory explanation of what sort of material bodies or physical configurations exist to give the very long magnetic anomalies."

It was an enhancement of this figure that stimulated a resolution of the challenge Wegener had posed of finding a mechanism capable of bearing the continents away from one another. The enhanced figure simplified the Zebra Diagram while incorporating crucial new information: the path of a new geophysical feature, the Juan de Fuca Ridge (see my Figure 2.4). This figure is Figure 3 of Wilson (1965), the first half of a two-part publication appearing over consecutive pages in the same issue of *Science*. My Figure 2.5, from the second of the pair of papers, shows the trace of the actual magnetic signal along transect "b" of Figure 2.4, *and it is symmetric*. By using the map to select a particular view of the raw instrument record, *together with information about the dates of these reversals* – which are data derived from other studies – a particularly powerful pattern claim was generated; this single claim managed to suggest a mechanism; and the suggestion, accepted rapidly, brought about what is now generally considered the revolution in geology from old-style geophysics to plate tectonics.

The underlying mechanism, which American schoolchildren are now taught in junior high school, is the mechanism of *seafloor spreading*. Magma (molten rock) upwells along the midlines of those ridges and then spreads laterally in both

2.1 Continental Drift

Figure 2.3. The "Zebra Diagram." The original caption reads, "Index anomaly map of the total magnetic field. The positive area of the anomalies is shown in black." The community's task was to account for the striking pattern of these "magnetic stripes." (From Raff and Mason, 1961, Figure 1.) Republished with permission of Geological Society of America, from *Geological Society of America bulletin*, Raff and Mason, 1961; permission conveyed through Copyright Clearance Center, Inc.

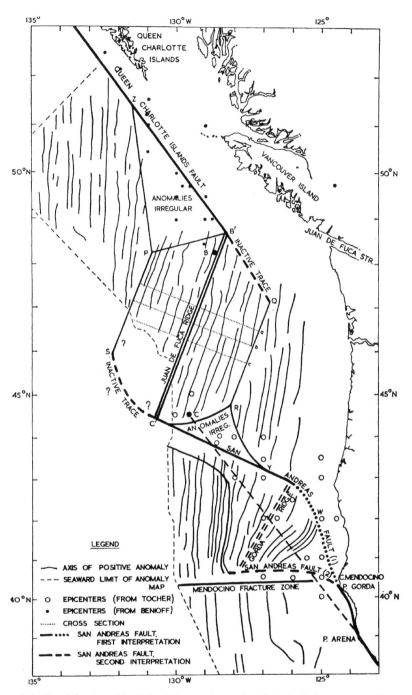

Figure 2.4. *Simplification* of the Zebra Diagram (preceding figure) admits a structural interpretation. (From Wilson, 1965, Figure 3.) Republished with permission of American Association for the Advancement of Science, from *Science 22 October 1965, Transform Faults, Oceanic Ridges, and Magnetic Anomalies Southwest of Vancouver Island*, J.T. Wilson; permission conveyed through Copyright Clearance Center, Inc.

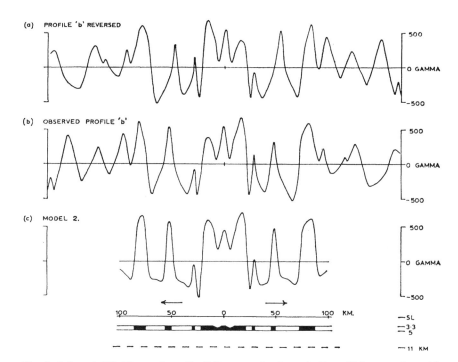

Fig. 3. (*a*) and (*b*) Observed profile "b" across the Juan de Fuca Ridge together with its mirror image about its midpoint, to demonstrate its symmetry. (*c*) Model and calculated anomaly for Juan de Fuca Ridge assuming a strongly magnetized basalt layer only. Black, normally magnetized material; unshaded material of this layer, reversely magnetized. Normal or reverse magnetization is with respect to an axial dipole vector; axial dipole dip taken as $+65°$. Effective susceptibility taken as ± 0.01, except for the central block, $+0.02$.

Figure 2.5. *Calibration* of one implication of the preceding figure: distance from the Juan de Fuca Ridge versus apparent age as indexed by those magnetic reversals. Note that this figure actually prints the same curve twice: once directly (east to west), and once mirrored (west to east). (From Vine and Wilson, 1965, Figure 3 with original caption.) Republished with permission of American Association for the Advancement of Science, from *Science 22 October 1965, Transform Faults, Oceanic Ridges, and Magnetic Anomalies Southwest of Vancouver Island*, J.T. Wilson; permission conveyed through Copyright Clearance Center, Inc.

directions. I summarize the quantitative inference here in LeGrand's words (1988, p. 211):

> [Fred] Vine and [Drummond] Matthews had not initially seen the implications for their model of an assumption of a uniform rate of [seafloor] spreading. Further, they had not drawn the inference that if the medial rift valley marked the center of an upwelling convection current which divided under the ridge, then the magnetic anomalies should ideally be distributed symmetrically on either side of the ridge. The first of these opened the possibility of applying dating techniques to the stripes themselves; the second, a direct comparison of the measured magnetic profiles of the two sides of a ridge: they should be mirror images of one another.... Vine and [J. Tuzo] Wilson late in 1965 applied their new ideas to the structure and magnetic patterns of the newly-described Juan de Fuca Ridge off Vancouver Island.... Graphical representations of some Juan de Fuca patterns

[my Fig. 2.5] appeared to show a high degree of bilateral symmetry about the ridge.... If spreading occurs at a uniform rate, then there should be a correlation between the widths of the normally- and reversely-magnetized strips and their ages. If the floor forms at 4 cm per year (2 cm on each side of the ridge), then a strip of floor 200 meters wide represents a record of 10,000 years of the earth's magnetic history. These ages should match time-scales based on continent rocks for reversals of the geomagnetic field.

The key rhetorical signal for consilience is contained in the auxiliary verb in that last sentence: these ages *should match* quantitative descriptors of the same entity, the same patch of ocean floor, derived from different instruments entirely. The magnetic reversals *seen on land* in rocks of the same age should match the corresponding patches of ocean floor under the assumption of some uniform rate of seafloor spreading. Consilience in the natural sciences is the assumption that measures of the same quantity by otherwise different machines or experiments should result in the same numerical patterns. As it happens, these paired records, one from land and some from the sea floor, match astonishingly well, and line up, too, with the texture of the map (Figure 2.3) in the same vicinity.

The combination of these two research traditions, continental drift/seafloor spreading and young-rock dating/magnetic reversals, resulted in a consilient display that was extraordinarily persuasive. It was the *symmetry* of this figure – emphasized by the brilliant idea of printing the same instrumental record twice in the same figure! – that ultimately persuaded skeptics to concede, beginning only a few months after the first public presentation of this figure calling attention to its implications. The corresponding revolution was "proclaimed" in November 1966, and eventually everybody whose opinion counted admitted the power of the new explanation, except for Harold Jeffreys, a statistician as well as a geophysicist, who in spite of living to the age of 97 never changed his views on continental fixity. (We encounter him in a more flexible context in Section L4.3.3.) There emerged the *plate tectonics* that turned geology into the "Earth science" by which it is now labeled in pedagogic contexts from elementary school up through graduate school. Figure 2.6 is a typical modern figure, fully standardized and updated through 2008.

Consilience, the topic of this chapter, is explicit in the original NOAA figure in the coloring of the areas of ocean floor – from red, youngest, to blue and ultimately purple, oldest – and particularly in the visually unmistakable symmetry of those colors with respect to the ridges along which the red patches run. Of course it is not enough that these data exist and run along the ridges (which were preexisting physical features of the ocean bottom that were known to be there prior to any measurements of magnetism). It is also the case that the colors are symmetric with respect to the ridges, and that the time scales perpendicular to the ridges match the time scales of magnetic reversals seen on the continents, *and no other theory appears able to account for this symmetry or this proportionality.*

In other words, where Wegener had only approximate matches of shapes or geographic patches on maps, today we see instead an astonishing, beautiful, and compelling *numerical* pattern, the proportionality of magnetically determined ocean floor age to distance from the ridge with the same proportionality constant on both sides of the ridge. Such a pattern would be expected as a matter of course if the seafloor

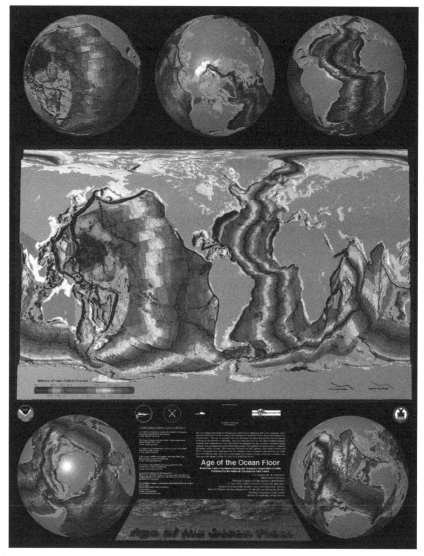

Figure 2.6. Extension of the combination of Figures 2.4 and 2.5 over the entire ocean floor. (Courtesy of the National Oceanic and Atmospheric Administration, U. S. Department of Commerce, from http://www.ngdc.noaa.gov/mgg/ocean_age/ocean_age_2008.html.) The areas corresponding to the continents or their shelves are to be ignored.

were indeed created anew continuously by upwelling of magma from the ridges at a fairly uniform rate of about two centimeters per year; hence there is reason to believe that seafloor spreading and its consequence, which is continental drift, are real phenomena. Both the equality and the proportionality highlighted in this little syllogism are consiliences, assertions that the same number or the same numerical pattern arises from more than one manipulation. We return to this form of inference, which is Peirce's syllogism of *abduction*, in Chapter 3.

2.2 E. O. Wilson's View of Consilience

The following argument, although of course wholly apocryphal as history, nevertheless is accurate to a few percent:

- Things like apples fall to Earth with an acceleration of about 32 feet per second per second. (This exposition uses English units, as the metric system had not yet been invented in Newton's lifetime.)
- If the moon goes around Earth once every 28 days, at every instant it appears to be "falling" (i.e., its path is curving downward with respect to the local gravitational vertical) with an acceleration of about 0.0085 feet, about one-tenth of an inch, per second per second. [In a day, it falls about $480000 \times (1 - \cos(360°/56))$ miles, about 6000 miles (see Figure 2.7). Because $d = \frac{1}{2}at^2$ for $t = 1$ day, the moon's acceleration a is evidently 12000 miles per day per day, which, since there are 86,400 seconds in a 24-hour day, is $5280 \times 12000/86400^2$ feet per second per second, 0.0085 feet or about 0.1 inch per second per second.]
- This is about 1/3600 of the acceleration experienced by the apple.
- The moon is about 60 times as far from the center of Earth as the apple was, and is falling with an acceleration about 3600 times smaller. So it looks as if gravitational acceleration might be proportional to the inverse square of distance, which is just what Newton stated.

This is a perfectly satisfactory and pleasingly petite example of consilience, the agreement of inferences from quite different systematic observations. It is, furthermore, a *numerical* or *quantitative* consilience, one conveyed by the agreement of numerical summaries of the two essentially different experiments. It is one prototype for the concerns explored in this book: we have measured "the same thing," acceleration radially downward, in two quite different experimental systems, the apple and the moon.

This is not mere sleight of hand (or pen). Important empirical regularities are intrinsic to the production of the consilience here: the implied commensurability of distances at all scales, from 0.1 inch to 240,000 miles and of course beyond these limits in both directions; the ease with which we understand a figure drawn as if we are looking straight down on the Earth–moon system from several million miles above Earth in a direction perpendicular to the orbital plane of the moon and on which the moon's orbit leaves a track like a figure skater's on freshly groomed ice; the flatness of the Euclidean model for space and the Galilean model for spacetime that, in contrast to the Einsteinian model, let us treat a picture of an arbitrarily large slice of the universe as if it were printed on paper, to a degree of approximation adequate for most of our needs. The whole proposition is roughly equivalent to the assertion that we can measure angles and compute trigonometric functions like sines and cosines at any geometric scale, from molecules out to galaxies, and always get the same answer. You have already let me do that when Figure 2.7 professed to inscribe a right angle inside the orbit of the moon. We muse further on "uniformitarian" concerns of this flavor when we discuss the ideas of Eugene Wigner in Section 2.5.

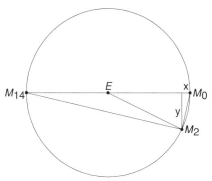

Figure 2.7. Guide figure for an argument from consilience for Newton's law of gravitation. E, Earth; M_0, M_2, M_{14}: three positions of the moon. (Subscripts count days after new moon.) The radius of the circle on which they lie is 240,000 miles. Recall from high school geometry that triangles inscribed on the diameter of a circle are right triangles, so that if M_0 and M_{14} are at opposite ends of a diameter, the triangle $M_0 M_2 M_{14}$ has a right angle at M_2. Recall also that central angles (like the angle at E here) are double the angles at the end of a diameter (like the angle at M_{14} here). In the interest of clarity of labeling, the angle drawn at E that, according to the text, represents one day's "fall" of the moon is not to scale: it actually shows two days' change, not one. We have $y \approx 6.28 \times 240000/28 \approx 54000$ miles, $x \approx 54000 \times \tan(360°/56) \approx 6000$ miles.

Here is another example, still from physics: an ordinary lever. (The suggestion is from Nash, 1963, pp. 50–54.) Imagine a set of weights that we can place on a lever arm at various distances from the fulcrum. The law of the lever, known in this form since Archimedes, is that "magnitudes are in equilibrium at distances reciprocally proportional to their weights": the lever balances when weights in a ratio w_1/w_2 are placed at distances at the *inverse* ratio, where $w_1 d_1 = w_2 d_2$ or $w_1/w_2 = d_2/d_1$. (As the principle that permits scales to be reproducible tools for assessing weight of a commodity, this law may be considered to be the fundamental fact of economics as well: Duncan, 1984.) In this computation, we have measured two different kinds of things, length and weight, using different instruments (rulers and counts), but we have measured them in the *same* system (albeit on opposite sides of the fulcrum). This, too, is a prototype for consilience, and of a similar mathematical structure to Newton's. For the apple–moon comparison, we had $a_1/a_2 = (d_2/d_1)^2$; for the lever, it is the same except that the exponent (the square) is gone. In passing I am making Wigner's point (from my fifth epigraph) regarding the "unreasonable" effectiveness of mathematics: here, the perseveration of the same notation, namely, a ratio of two distances, two times, two accelerations, . . . , in such diverse practical contexts.

Let me now radically change the level of abstraction of the argument here. According to E. O. Wilson, who wrote the 1998 book *Consilience* that restored the term to broad currency, the idea of consilience is an expression of the general Western urge toward "the Ionian enchantment," the unification of empirical knowledge (Wilson, 1998, p. 4). This is not an approach peculiar to the Harvard biology faculty where he resided at the time of writing, but is one recurrent theme of 20th-century philosophy of science, as expressed, for instance, in the Vienna-derived *Foundations of the Unity of Science*, originally titled *International Encyclopedia of Unified Science* (Neurath, 1970). Belief in the unity of the sciences is, at this level, the belief that the world is orderly in a way amenable to rational inspection. It is a metaphysics, not a physics, and, as Wilson startlingly notes a few pages later, it is another way of satisfying religious hunger: "It aims to save the spirit not by surrender but by liberation of the human mind."

Where Whewell and I emphasize the linking of fact-based inductions (in my case, quantitative abductions) across studies, Wilson (p. 10) emphasizes the "linking of fact-based theory across disciplines." And the only way to do this, he insists, is by the methods developed in the natural sciences, the methods that have "worked so well in exploring the material universe." He goes on (p. 11):

> To ask if . . . sound judgment will flow easily from one discipline to another is equivalent to asking whether, in the gathering of disciplines, specialists can ever reach agreement on a common body of abstract principles and evidentiary proof. I think they can. Trust in consilience is the foundation of the natural sciences.

When I originally read Wilson's book and assigned it to my Rhetoric of Evidence students at the University of Michigan at the turn of the century I puzzled over the apparent arrogance of statements like these. That arrogance proved a high barrier against embrace of Wilson's views over the broad realms of the social sciences and humanities, to the practitioners of which he was mainly addressing his invitation to join with him in a unified science of mind and world. Ceccarelli (2001) writes insightfully of the mistake Wilson made in reaching out to that audience when he couched his argument with such unambiguous condescension that his readers could not reinterpret it in ways that either salved or salvaged their own intellectual pride. By contrast, Ceccarelli notes, the great unifiers who came to biology earlier in the 20th century, intellectuals like Dobzhansky or Schrödinger, were careful to ensure by choice of words and arguments that methodologically diverse readers could recognize parts of their own prior training in the invitation to affiliate. Wilson failed to do this, Ceccarelli concludes, and thereby failed to establish any sort of movement beyond the boundaries of his own central interest at the time, sociobiology.

I think there is a better way to explore the impasse here. What characterizes the success of the natural sciences, from finding the *Scorpion* through Newton's laws and onward through most of the examples we will be encountering in later chapters, is not the mere "agreement on a common body of abstract principles and evidentiary proof" but the specifics of the rhetoric of that agreement. This is the domain of *quantitative* consilience, the subject of this book, and I will be arguing that wherever it is the ground for a scientific unification it seems to have a fair chance of working.

I come by this point of view by descent, having been taught it explicitly by the great economist Kenneth Boulding in his seminar entitled *General Systems* offered at the University of Michigan in the mid-1960s. In his uniquely interdisciplinary teaching Boulding emphasized the role of quantification in making sense of the competing claims of persons approaching the same subject from different disciplinary centers. He highlighted several of the fields that were just coming into mathematicization – game theory, psychology, econometrics, geography, the patterns that would shortly be formalized as chaos theory, the Arrow theorem, and dynamics of voting – and went on to note everything in their several rhetorics that could ease the interdisciplinary dialogue by focusing the disputants' attention on shared rules of *calculation*. Numerical reasoning was by far the best arena in which to come to interdisciplinary consensus about the meaning of empirical data.

If, to skip ahead a couple of chapters to Wilson's page 58, science is

> *the organized systematic enterprise that gathers knowledge about the world and condenses the knowledge into testable laws and principles* [italics Wilson's],

then the interdisciplinary sciences (the "unified sciences," such as plate tectonics or global warming) are those that can provide the necessary assurances of truth or practicality in the language of more than one discipline at the same time; and these will tend to have, among other desirable characteristics, a shared respect for numbers as evidence. Numbers are both our most effective ways of condensing knowledge (whether woven into the fabric of reality, as Wigner claims for physics, or instead just underpinning the language of statistical summaries like averages and regressions) and our best ways of initially *acquiring* such knowledge if we wish to persuade others that we are relatively free of bias, which is to say, intersubjective. It is no accident that the social sciences so often refer to their measurement protocols as "instruments," a metaphor evoking the reality that is the case throughout the natural sciences, where every machine has its own essence that is to a surprising extent independent of the will of the person who designed it or who is operating it. Consilience can work whenever machines are at hand to supply the numerical data on which more than one field can get to work. This requirement characterizes a good many disciplines and even more interdisciplines, totalling perhaps half of all the professional concerns of the classic "arts and sciences," along with many of the classic professions (for instance, medicine or engineering).

Hence part of the authority of numbers arises from the reliability of the machines that generate them, and the foundation of any numerical science rests no more on the "testable laws and principles" than on the machines by which we produce the numbers in terms of which those laws and principles are phrased. Over the centuries some of these machines have become so familiar that we forget that they *are* machines: telescopes and microscopes, for instance. The representation of images by inscriptions that are scalable, portable, and superimposable (Latour, 1990) was one of the major milestones in the development of science. Any radiologist or surgeon looking at a medical image is relying on a consilience like that, one that is too self-evident to come to our attention any longer: the agreement of geometry between a cadaver dissected in physical space and the computational output of a machine designed specifically to simulate essential aspects of that geometry on a computer screen. A computed tomography (CT) scan or magnetic resonance (MR) image is just the latest in a long line of graphical representations of the arrangement of objects and their parts that began with the development of perspective drawing and the printing technology of woodcuts in the Renaissance.

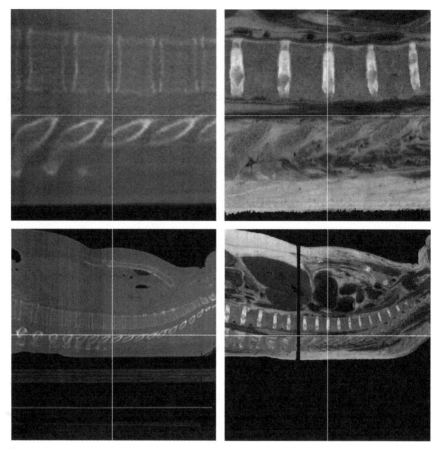

Figure 2.8. Example of a hidden consilience: CT geometry, left, matches physical tomographic geometry, right, of a solid organism (here, the Visible Female, Eve). (Data from the Visible Female project, National Library of Medicine. The images in the right-hand column are sections computed in Bill Green's software package Edgewarp. The scale of the lower image pair is one-fourth of that for the upper pair.) The panels in the right column are originally in true anatomical colors – see Plate 2.8.)

Figure 2.8, for instance, shows a photograph of one postmortem slice of the National Library of Medicine's Visible Female, "Eve," and, to its left, a reconstruction of the X-ray densities of the same slice, from the CT volume of the same specimen. (For the description of Eve, see http://www.nlm.nih.gov/research/visible/visible_human.html and links there. A freely accessible copy of this seven-gigabyte image along with tools for its navigation is part of the standard installation of William Green's program Edgewarp, see http://brainmap.stat.washington.edu/edgewarp/.) The image above is a fourfold enlargement of the central part of the image below; both are centered in the middle of the spinal canal at a convenient location. The CT volume (the gray-level image) is standard medical scanner output. The "physical" slice (the colored slice in the color insert) is itself a composite assembled from the middle line of each of the stack of 5700 individual digital photographs by copying them into the same screen window, each right up against its two neighbors. There is more Euclidean geometry here: the operation of the digital camera's image-capture circuitry, its collaboration with the physical optics of the camera's lens, the agreement between my computed zoom and your instinctive tactic of just moving your eye closer to the image. None of these steps change an object's shape, which is what "Euclidean geometry" is actually about. The closer you look, the more numerous the consiliences required to utter even the simplest empirical assertion involving quantitative data.

Cameras aside, there is consilience in these figures themselves, borne in the obvious agreement in geometry of the CT-visible structures (here, just the bones) between the two representations by two totally different machines, one a machine for physical slicing, not that different from what gives you your pastrami at the delicatessen, and the other a considerably more expensive clinical radiological device. Medical images would be useless if we could not count on consilience of geometry and also of image contents between different views of the same anatomy. Because the CT image is verified for cadavers (and also for the carefully designed artifacts known as *phantoms*) we trust in its geometry as applied to living creatures as well. We precondition ourselves for this trust by our familiarity with cameras, for which we were prepared by our familiarity with paintings, for which we were prepared by our familiarity with conventions of linear perspective that matured during the Renaissance, regarding which, finally, we can rely on Descartes's invention of *Cartesian coordinates*, the reduction of position itself to pairs or triples of numbers that (to argue anachronistically for a moment) 19th-century mathematicians and physicists found to obey the rules of vector arithmetic. This is the same miracle that characterizes maps (Latour, 1990), woodcuts, and other central aspects of the pictorialization of nature. There is a huge literature on this topic, spanning science and the graphic arts under the heading of *scientific visualization.* The ability to design machines that work routinely to reproduce Cartesian reality on paper this way is part of the predictability of machines in general, which is in turn, as Wigner (1960) notes, the greatest triumph of the physicist.

At least within the domain of "exact measurements," the Baconian physical sciences, an understanding of consilience adequate for many domains of practical exploitation has been around for considerably more than a century. Consider, for instance, the presumption of every physicist before Einstein, and every mathematician before Bernhard Riemann, that the physical universe is truly a three-dimensional Euclidean (flat) space, the sort of place where angles of triangles add to 180° and the squared hypotenuse of a right triangle is the sum of the squares of the lengths of the other two sides. This is what makes *geodesy* (surveying) possible: the certain knowledge that juxtaposition of any number of triangles leads to consistent results for distance measures and angles from start to end of the chain. When, for instance, British surveyors computed the height of the peak of a particularly tall mountain on the border between Tibet and Nepal by surveying its elevation from a circuit of triangles all around it, as part of the Great Trigonometric Survey of India, they expected it, like every other mountain peak, to *have* a location, and a height, too. On this assumption the peak of Mt. Everest could be quite precisely located in three dimensions a century before it was feasible to visit it in person. See Keay (2000).

Optics soon extended to photography, and so this presumption of exactitude soon extended from single instrument readings to extended imagery. Francis Galton, for instance, exploited the presumption when in 1877 he invented the method of *averaged photographs* (Pearson, 1914–1930, 2, pp. 287–325) and used it to search for an objective approach to "physiognomy," the correlation of psychic features with facial features. For instance, he averaged the faces of members of the Royal Corps of Engineers, and also the faces of "persons convicted of crimes of larceny without violence," and examined them to see whether, "on the average," the face connoted the extent of civic virtue across the poles of this comparison. These averaged pictures, along with many others, are explicitly reproduced in a wonderfully eccentric assortment of plates interleaved with Pearson's hagiographic words. The faces did not connote the extent of civic virtue, of course, but it was not *so* ridiculous an idea as all that. The physiognomic theme still comes up today in connection with birth defects such as fetal alcohol syndrome that affect both face and brain (see Figures 5.4 and 7.25). It is not Galton's fault that his hypotheses were fatuous, but it is his great virtue that the physics of *his* "medical imaging device" (an ordinary photographic enlarger modified so that a tilted mirror could be inserted along the optical path) was designed precisely so that if there were a finding it could be consiliently presented to the multiple disciplines, including criminology, that were concerned with the question he was asking (about a photographic approach to criminality). Only today is a different machine, the polygraph ("lie detector"), coming to be seen as a device nonconsilient for its intended purpose.

Wilson notes, on page 54, that "complexity is what interests scientists in the end, not simplicity. Reductionism is the way to understand it." I think this point of view is wrong, or, rather, incomplete because unsymmetrically phrased. What makes complex systems understandable is not reductionism per se but the consilience that comes from measurements at more than one level of the same system: measurements that agree between the holistic and the reduced levels. There is an example in Frank Livingstone's article on the geographic distribution of the sickle cell trait in West Africa that I discuss in Section L4.4. Livingston's tables are of the prevalence of a particular genotype, but his explanations are at the levels of whole organisms, indeed, whole cultures. The measurements span the levels of explanation, linking the genotypes to the historical migrations. Another example from Chapter 4, the discussion of the Cretaceous–Tertiary extinction by the Alvarez team, measures at the level of the individual iridium atom, but calibrates that measurement to a specific physical level in layered clays that are 65 million years old – the calibrations come from paleontology, sedimentary geology, and other holistic domains – and then the abduction is closed and the argument confirmed by an estimate of the size of a really, really big rock (a 10-kilometer asteroid) followed, 12 years later, by the discovery of a crater that could have been left by exactly that large an asteroid at exactly the date (65 million years ago) required for the explanation to work: a date, in turn, established by counting radioactive decays, which is to say, frequencies of an instrumental encounter with subatomic particles of a different species entirely. It is clear that these arguments, which are among the most powerful to be found anywhere in science, are not formally "reductionist," but they are certainly consilient, based on inferences combining measurements at many different levels. To adopt a phrase from a different discipline, they are instrumented *metrological chains.* Our plate tectonics story, Section 2.1, is ultimately the same sort of narrative. The measurements of the orientation of little magnetization vectors of microcrystals in tiny rock samples proved consilient with observations about the shapes of continents and the geology of mountain ranges and the patterns of paleoecology and the readings of underwater magnetometers at a scale of hundreds or thousands of kilometers.

In this connection, Wilson speaks (p. 68) of "consilience by reduction," meaning going down the levels, and then "consilience by synthesis," going up the levels. I suggest that in any context involving quantification these would be the selfsame inference as viewed sequentially first by the experts in one level of analysis, the members of one discipline, and then by the experts at the other level. What they share is, precisely, the equality of the measured numbers or the equivalence of the detected patterns, and both of these are *symmetric* relationships, not directional ones. Consider the anthrax example, Section E4.1. While the microbiologist could in principle confirm that most of the Sverdlovsk cemetery arrivals during April of 1979 arrived there owing to the presence of anthrax bacilli, it is the pattern in Figure 4.6, the alignment of the dead humans and the dead animals with the wind direction on April 2, 1979, that supplies the actual explanation the researchers sought. It is the agreement between the patterns, not the single measurement, that drives the reasoning in this case and most of the other cases I have assembled for your appreciation in this book. *To get equivalent answers from different versions of the same question:* this is the signal that the quantification was worth the instrumentation.

Wilson the biologist understands this even where Wilson the philosopher might not. He comments on page 81, for instance: "Consilience among the biological sciences is based on a thorough understanding of scale in time and space.... This conception of scale is the means by which the biological sciences have become consilient during the past fifty years." And, "the greatest challenge today ... is the accurate and complete description of complex systems." To be "accurate," a description has to be amenable to a concept such as accuracy; that means in practice that it must be quantitative. And any assertion of "completeness" needs to be challenged on both higher and lower levels, at both larger and smaller scales. As Wilson notes some pages later, "Complexity theory can be defined as the search for algorithms used in nature that display common features across many levels of organization." Again, the way we decide whether "common features" are being "displayed" is to quantify them in the context of the theories to which they might be congenial.

As Thomas Kuhn said, – I've already quoted this in the Epigraphs – "Numbers gathered without some knowledge of the regularity to be expected almost never speak for themselves. Almost certainly they remain just numbers." But the theory – the "regularity to be expected" – need not be at the same level as the measurements: at least, not all of them. Notice, too, how this dogma of consilience contradicts one lay understanding of the scientific method, the one encapsulated in the offhand comment of Harold Varmus that is this book's 13th epigraph. Varmus's remark here is ironic, of course. "Results and happiness" will follow only if the data are consilient with quantities measured earlier and with the scientist's preferred explanation. If not, perhaps the machine is malfunctioning, or the hypothesis may be foolish, or chance may have turned malicious on you.

In part because of this very iconoclasm, Wilson's perspectives on the sciences in which he is not an expert – the historical sciences, the social sciences, the humanities – are very refreshing. Wilson not only challenges the social sciences to become more consilient, after the fashion of the contemporary medical sciences, but supplies them with the prods that might be helpful should they wish to "pursue webs of causation across adjacent levels of organization" (p. 189). These include pointers to four thriving interdisciplinary endeavors (p. 192): sociobiology (Wilson's own invention), the environmental sciences, cognitive neuroscience, and human behavioral genetics. There will be examples of the first two of these later on in this book. Attention to consilience in these "webs of causation" is one crux of the scholarly literature about interdisciplines even in the hands of authors who do not explicitly cite Wilson as the exemplar. See, for instance, Bolhuis et al. (2011), who, after noting several failures of consilience (i.e., disagreements) between some current findings of evolutionary biology and the assumptions of contemporary evolutionary psychology, request that the latter field "change its daily practice" and "reconsider its scientific tenets." (But in speaking thus they replicated the error of Wilson's that Ceccarelli noted in the original. They presumed, however politely, that such a failure of consilience requires the nonbiological field to make all the adjustments.)

The praxis of consilience is not merely verbal; it makes strong demands on the ways that measures must behave. The argument along these lines is a standard topic of the rather specialized field usually called "foundations of measurement" (cf. Torgerson, 1958; Ellis, 1966; Krantz et al., 1970; Hand, 2004). These authors and their colleagues share a concern for the properties that numbers must have if they are to be useful for the purposes to which we wish to put them. If the theme is consilience, then there are likely to be at least two numbers under discussion, not just one, and the corresponding development can be a great deal less abstract. At the same time, the complexity of the consilience per se is typically less than that of the mensuration per se. We see in Chapter 4 that to fit a line to a set of data points involves examining all the data but yields only two constants (a slope, if the line is taken as going through the sample average, and an error variance). A third constant is generated if the sample average itself is not known a priori or is not scientifically meaningful, in which case one usually takes a stereotyped value of the height of the line such as its height where the abscissa is zero. (See the discussion of passive smoking in Section E4.6.) So the consilience between two fitted lines involves at most three quantities no matter how rich and ramified the data bases by means of which the two lines were separately fitted. In the statistics of shape (Chapter 7), no matter how many original anatomical points or curves were located, the statistical analysis is driven mainly by a single geometrical summary function, the Procrustes distance between pairs of such labeled point sets. (The resulting diagrams are in full detail, but *their* attributes are typically treated only qualitatively.) In short, *quantifying consilience between measures is simpler than the original quantifications were.* This is an instance of one general justification of statistics as a cognitive style: by supplying a basis in information theory for a wide range of conventional geometrical summaries, it facilitates abductions.

Measurement of the fundamental quantities of physics is normally justified by considerations of this kind. Take, for instance, the notion of temperature. As a characterization of the relations between extended bodies on its own, it needs only to fulfill some logical requirements – two bodies each at the same temperature as a third body must be at the same temperature themselves, and so forth – and for that we could measure using nearly any ordinal scale such as the volume of a column of mercury (the mercury-in-glass thermometer principle) or the displacement of a coiled metal strip under expansion (the principle of the thermostat). But if we need to speak of multiple levels, the way Wilson is recommending, our choices are restricted a great deal more (see Mach, 1896, in Ellis, 1966). Temperature enters into the so-called *Maxwell–Boltzmann distribution* of the states of molecules in a gas, for instance, in the form of the denominator of an exponent: the number of molecules of energy ε in a system at statistical equilibrium must be proportional to $e^{-\varepsilon/kT}$ where T is temperature and k is Boltzmann's constant. For this law to be applicable, T must be measured on a scale proportional to the average molecular kinetic energy; there is no longer any choice left. This way of measuring temperature (as the average energy of the molecules, more or less) is also the only way of quantifying it if the temperature of mixtures is to show the appropriate dependence on the temperature of the components, for instance, if mixing three gallons of water at 40° and one gallon of water at 80° is to result in four gallons of water at 50°.

Other contexts of consilience, such as those from the "softer sciences" not governed by exact laws, can place equally strenuous demands on our measurement formulas. For instance, in the course of the studies of fetal alcohol damage reviewed in Chapter 7, we intended that ordinary methods of linear regression (see Section L4.5.3) might be applied in such a way as to account for data on behavioral deficits in children by data on intake of alcohol reported by the mothers of those children while they were still pregnant. If those behavioral deficits were to be measured according to the rules of one discipline (here, psychometrics) as IQ scores or school achievement scores, the consilient alcohol dose could not be measured the way bartenders or liquor stores do, in units of, say, ounces of alcohol per day. The effect was not linear in those units, and so the ultimate explanatory purpose, that of explaining effects by causes in correlational terms and the adjustment of those explanations in path-analytic terms (Section L4.5.3.2), would not be able to go forward. To be a useful component of a consilient research program, the alcohol dose had to be scaled by its effects, as in Figure 6.14.

2.3 Some Earlier Critiques

Why should numbers (and their graphical patterns) be able to persuade us of scientific propositions at all?

There is an interesting hint in a book by Jean Perrin that we will be reviewing for its scientific import (the reality of atoms) in Section E4.3. In the course of demonstrating that little yellow balls of paint pigment suspended in water can be thought of as a model of the atmosphere, Perrin measured the vertical position of some 13,000 of them in a microscope mounted horizontally. His interest was in their density as a function of height over the bottom of the microscope – a height that balances random motion against their own weight, just like the atmosphere does – and he found the expected exponential decline in density:

> The readings gave concentrations proportional to the numbers 100, 47, 22.6, 12, which are approximately equal to the numbers 100, 48, 23, 11.1, which are in geometric progression. Another series [using larger grains but more closely spaced imaging planes] show respectively 1880, 940, 530, 305 images of grains; these differ but little from 1880, 995, 528, 280, which decrease in geometric progression. (Perrin, 1923, p. 103)

Notice this sudden change of rhetoric, from the empirical observation (the series 100, 47, 22.6, 12) to the fitted model that matches the theory (100, 48, 23, 11.1). Perrin is telling us here that *that fit is close enough* – that this particular rate of decline, in an experimental context that could a priori have shown any trend from uniformity to all piling up at the bottom – constituted sufficiently strong support for the qualitative inferences to follow.

This language occurs again and again in Perrin's book. After an explicit observation of Brownian motion of one of these yellow balls under his microscope (now mounted in the normal position) – Figure 4.19, which was among the very first images of Brownian motion ever published – he tabulates the distance moved by the particle in successive one-second intervals, compares it to the expected chi distribution under a Gaussian assumption, and declares the agreement to be "exact," whereas of course it is not: it is simply close enough. Similarly, Perrin's repeated experiments that find a range of values of Avogadro's number between 55×10^{22} and 80×10^{22} are likewise declared to be in "remarkable agreement... prov[ing] the rigorous accuracy of Einstein's formula and in a striking manner confirm[ing] the molecular theory."

We can turn again for guidance to that unusual 20th-century intellectual, Thomas Kuhn.[1] For a conference in 1958 on quantification across *all* the sciences, natural and social (see Appendix 2.B), Kuhn wrote an exegesis of the function of measurement in the physical sciences, eventually published as Kuhn (1961), that began with a close rhetorical study of the Perrin sort of phraseology. If the question is of the role of a published tabulation juxtaposing a modeled string of numbers (like "100, 48, 23, 11.1") to a measured set (like "100, 47, 22.6, 12"), the answer, Kuhn says, can only be that the juxtaposition is intended to teach the reader the meaning of the concept of reasonable agreement as it is used on the research frontier of the problem under discussion. "That," he says on page 166, "is why the tables are there [in the textbook]: they define 'reasonable agreement'."[2]

---

[1] Thomas S. Kuhn (1922–1996) began his career as a physicist and then as a teaching assistant at Harvard under James Conant as the Harvard program in general education was built in the 1950s (Fuller, 2001). His *Structure of Scientific Revolutions*, the all-time bestseller in history and philosophy of science, appeared in the Vienna-inspired "encyclopedia of unified science" in 1962 and as a separately published book at about the same time.

[2] Kuhn's text is illustrated by a joke to which no commentator has previously called attention as far as I know. In the sketch on page 163 of his article, the "machine output" that is being produced from theory

But there is a Manichaean quality to this role of quantification: it does not always result in satisfactory "agreement," and at its most powerful the *dis*agreement is a greater force for scientific progress than the agreement. To make this point, Kuhn began by coining the felicitous phrase "normal science" for science that examines the quantitative match between theory and evidence from the point of view of confirmation. Success, here, lies in the explicit demonstration of an agreement that was previously only implicit. No novelty is revealed, and, as Kuhn wisecracks (p. 171), we cannot say that such a finding "confirms" a theory, as different findings would not "infirm" the theory but would only derogate the experimental skills of the investigator, presumably a student, carrying out the experiment.

For instance, at some point during your formal education, surely you were graded in a laboratory science course by the extent to which your measurement came reasonably close to the textbook value (without exactly agreeing with it, which would indicate fraud). You were not testing the textbook's theory that day, only your own motor skills. A pedagogical criterion like this can actually be wielded in reverse to critique certain fields as prequantitative that would prefer not to be construed as such. A few years ago, I participated in a grant review of a project in cognitive neuroimaging, where I asked, in all innocence, what experiments could be run using a particular piece of equipment such that if a student's result failed to be in "reasonable agreement" with a textbook it would indicate the incompetence of the student rather than the variability of the phenomenon under study. To my surprise, the reply was that no such experiment could be envisioned using the equipment in question. It follows that there was no concept of consilience in that field yet – the combination of its theories and its data did not yet support quantitative inference, because according to my informants the images from this machine, however expensively obtained, could not actually be used to falsify anybody's theory.

In domains of the physical sciences where "normal measurement" usually does go right, *a persistent anomaly* – a continuing failure of consilience – is the strongest possible signal that an area is worth pursuing. The key to the craft of the most productive natural scientists is the intuition regarding which quantitative anomalies are to be passed over with a shrug, which not. "Isolated discrepancies," Kuhn notes, are much too common to pause for. Instruments may be out of adjustment, or some assumption may have drifted false, or a repeated measurement may simply not show the same numbers. But, Kuhn goes on, if the discrepancy falls well outside the usual range of "reasonable agreement," and if it cannot be resolved by adjustments of instrumentation or auxiliary assumptions, the community (for this is a *community* problem, not the problem of any single scientist) can resolve the problem in one of only a few ways: a new instrument, perhaps, or more likely, a new *qualitative* phenomenon that was indicated by a persistent and unresolvable quantitative anomaly blocking attempts at a consilient description. "Numbers," Kuhn notes on page 180, "register the departure from theory with an authority and finesse that no qualitative technique

---

and compared to measurement actually consists of the square roots of the sequence of prime numbers. That sequence certainly cannot be produced by any machine that is driven by a crank, as in the figure, and in any case it would not be treated as susceptible to any sort of empirical verification.

can duplicate.... I know of no fundamental theoretical innovation in natural science whose enunciation has not been preceded by clear recognition, often common to most of the profession, that something was [quantitatively] the matter with the theory then in vogue." He goes on: "No crisis is so hard to suppress as one that derives from a quantitative anomaly that has resisted all the usual efforts at reconciliation. A quantitative discrepancy proves persistently obtrusive to a degree that few qualitative anomalies can match,... but at its best it provides a razor-sharp instrument for judging the adequacy of proposed solutions." The example in Kuhn's next paragraph is Kepler's replacement of a scheme of circular planetary orbits by elliptical ones.

Kuhn concludes, very much in keeping with the spirit of this book:

> Measurement can be an immensely powerful weapon in the battle between two theories... It is for this function – aid in the choice between theories – and it alone, that we must reserve the word "confirmation" if "confirmation" is intended to denote a procedure anything like what scientists ever do. *In scientific practice the real confirmation questions always involve the comparison of two theories with each other and with the world*, not the comparison of a single theory with the world. In these three-way comparisons measurement has a particular advantage. (p. 184; italics mine)

Translating this into the language of consilience, the role of those textbook tables is to calibrate the scientist's own sense of how consilient a pair of values (one theoretical and one empirical, or both empirical from two different experiments) ought to be. It is when this comparison fails that science is at its most exhilarating, and only when a theory has been provided that restores the consilience does calm return. Improvements in instrumentation tighten these limits of consilience and thus often generate a Kuhnian crisis. As Kuhn summarizes his own argument, on page 185, "I know of no case in the development of science which exhibits a loss of quantitative accuracy as a consequence from the transition from an earlier to a later theory." In other words, the demands of consilience are continually increasing, at least in the natural sciences. It is in this sense that those sciences show directionality, that is to say, objective evidence of progress – not of knowledge, necessarily, but of precision of instrument readings.

Armed with this insight we can return to the discussion of Perrin's rhetoric with which we began this discussion. Numbers such as those that could "confirm" the molecular theory could do so only because Perrin knew that his measurements were accurate only to a few tens of percent. (The currently accepted value of Avogadro's number is 60.22798×10^{22}, a quantity that barely falls within the range of Perrin's own measured results.) What makes his values of 55×10^{22} to 80×10^{22} evidence for "remarkable agreement" rather than equally noteworthy disagreement is a function of the state of science and engineering at the time. If atoms did not exist, the values he computed for the number of atoms in a mole would vary as wildly as the experimental designs he exploited and the formulas for N in terms of the readings of remarkably diverse instruments. But to get numbers in a mere 50% range from experiments each of which had a measurement error of some tens of percent on its own is conceivable only if there were some real underlying value that they were all estimating with error, a value whose approximate magnitude had already been set 30 years earlier by measurements of argon vapor. In the course of his study Perrin actually focused not on this rhetoric of experimental agreement but on the associated hypotheses of Brownian motion that drove the data distributions. We return to these underlying concerns of his in Chapter 4. Meanwhile, the specific topic of multiple theories is the principal concern of my next chapter. Multiple theories are needed both for Peirce's method of abduction and for its extension, Chamberlin's praxis of strong inference.

Kuhn concludes his meditation with a very interesting "Appendix" responding to some comments from the conference audience to whom his paper was originally presented. One of these comments dealt with an issue to which we return late in Chapter 4, the issue of how quantification might work in the social sciences. Kuhn notes, on page 191, that "the concept of a crisis" (a persistent failure of consilience) "implies a prior unanimity of the group that experiences one. Anomalies, by definition, exist only with respect to firmly established expectations." In the social sciences, he notes, it is not at all obvious that any consensus exists anywhere strong enough to motivate the anxious search for resolution that characterizes crises in the physical sciences. A crisis, in other words, must be seen on a background of altogether routine previous agreement, and if that background is absent, there is no force to any particular quantitative disagreement (such as the discrepancies in the heights of the bars in a figure of David Buss's that I will be discussing in Section E4.7).

Many other aperçus into the cognitive processes of the scientist and his relationship to machines can be accommodated in this single fundamental insight of Kuhn's. His point is consistent with Robert Merton's famous normative system for scientific communities (Merton, 1942) in the form of the four principles spelling out "CUDOS" (Greek $\kappa\nu\delta o\varsigma$, "fame" or "honor"): C for communitarianism (originally, "communism"), U for universality, D for disinterest, and OS for organized skepticism. An ethical controversy has been ongoing around the D term now that scientists have lately been urged to take a financial interest in their own intellectual property, but the other parts persist unchanged, particularly the OS term, for which there is a corresponding structure, the combination of peer review with the expectation that the typical published article will include a methods discussion sufficiently detailed to permit replication if a reader is so inclined. The way you organize "organized skepticism," in other words, is to encourage explicit second-guessing, and the way you decide whether a replication is confirming or disconfirming typically involves a quantification somewhere. An entire subdiscipline of statistics has arisen to handle the issues that are generated by problems of this sort, the domain called "meta-analysis." We sketch this domain in Section E4.6; it includes formal statistical tests for deciding when one particular quantitative finding is discrepant with a consensus over a collection of several earlier findings.

The sociologist of science Harry Collins has some interesting comments on the extent to which a replication is worthy of being credited. Simply reading the same machine output, say, the strip charts, a second time would probably not count as a replication, but neither, he wisecracks, would be a study that attempted to duplicate the findings by haruspification (inspection of the entrails of sacrificed domestic animals) or a study by the original author's mother. Replications need to be somewhere in the interior of this simplex, but it turns out (Collins, 1985) to be difficult if not impossible to assert what is an acceptable replication without begging the question of the community within which that acceptability is being judged. It is easier, in practice, to attempt the annihilation of a new theory on first principles, such as its incompatibility with established laws of nature. The simple refutation of a claimed quantification on grounds of a truly absurd inference from the claimed magnitude of some effect is typically found for experiments that seem inconsilient with most other results. We have a brief discussion of this category, "pathological science," in Section L4.4 (see under Benveniste); another example pertinent to the same section is the comment quoted by Rousseau (1992) that if a certain measurement from a cold fusion experiment was to be believed, "everybody in the room would be dead." As with Wegener's premature discovery of continental drift, quantitative patterns, like the fit of continental shelves between Africa and South America, are typically not credited as fact unless and until there is some theory in place that would produce the empirical pattern "as a matter of course."

In connection with his later work on paradigm shifts, Kuhn noted that much of his thought had been anticipated to a startling extent by Ludwik Fleck's work of the 1930s on the history of the scientific study of syphilis. The transformation of syphilis from "scourge" to disease was inseparable from the discovery of the serological techniques for assessing the corresponding antibodies quantitatively. This made it possible, for the first time, for theories to confront one another in this domain. One of these theories (August Wassermann's) triumphed almost immediately thereafter – syphilis is an infection detectable by antibodies to the spirochete bacterium *Treponema pallidum*, not a punishment for immoral behavior. The insight converted the study of syphilis into microbiology and thence into medicine. Epigraph 14 sets out in detail Fleck's inference that the scientific fact is always this sort of group-based constraint on speculation. The constraint often takes the form of a mandatory consilience, possibly quantitative but always at least qualitative, with a preexisting understanding on the part of the community. What makes it mandatory is an aspect of the social structure of that community, namely, the capability of imposing a mandate. The message here is the same as David Bloor's in epigraph 12, which arises from a discussion of an entirely different kind of science (the technical evolution of the airplane wing): "The force of reason is a social force; . . . the problem of cognitive order *is* the problem of social order."

2.4 The Issue of Heterogeneity

The physicist Walter Elsasser turned in middle age to the concerns of theoretical biology, where he constructed a series of uniquely delicate arguments about the physical bases of a scientific biology. I am drawing mainly from his 1975 book *The Chief Abstractions of Biology*. Elsasser argues there that for such a domain as theoretical biology to exist, preliminary work is needed that overcomes existing "habits or conditioning of thought" resisting the necessary changes. As Fleck pointed out, usually this change of habit involves not only cognitive adjustments but actual social pressure. One of these core antinomies is the role of the concept of probability (for which this book will substitute descriptive statistics) in the construal of organisms. As Elsasser explains it,

> We are now able to bring into clear focus the difference between the traditional abstractions of physical science and the kind of abstract scheme that is suitable for biology. In physical science one deals invariably with definite abstract constructs, that is, sets of symbols, most commonly mathematical, in terms of which observed regularities can be represented.... Certain complications arise in modern atomic and molecular science whenever concepts of probability enter. This indicates that one no longer deals with individual objects but only with classes of these. Once the transition from objects to class is made, the abstract scheme of the description that uses probabilities will, in terms of classes, become as precise as the direct quantitative description of single objects in classical physics.
>
> But there is one exception, the description of objects that are of radically inhomogeneous structure and dynamics, which we identify with living objects. Here a property of the description that can otherwise be ignored will be of the essence: it is that for a somewhat complicated object the number of possible microscopic patterns compatible with a given set of known data (class parameters) becomes immense.... The description of a living object, therefore, implies an *abstract selection*. It is the selection of an immensely small subset from an immense number of possible descriptions.... One goes from theoretical physics into theoretical biology not by the method of superadding new, formal constructs (an approach usually described as vitalism) but by omitting an overwhelming fraction of an immense number of available descriptions. (p. 203; emphasis in original)

In physics, one claims that the description adequately characterizes the object – that is why laws work, after all[3]; but in biology, the description is "immensely indeterminate." (This is clearest at the microscopic level of detail: see, inter alia, Mayer, 1963.) The organism selects among these descriptions in ways we do not measure. There is a determinate part of the characterization of organisms, and an indeterminate part, and the statistical method works on the *indeterminate part only*, the part that is fully individualized rather than constrained by design. The statistician, in other words, deals with the part of the organism that is history-less, the sample of a finite number of equivalent alternative patterns out of the immense number that might have been encountered. In the context of Friedrich's *Der Wanderer über dem Nebelmeer*, the painting reproduced on the cover of this book, statistics serves for the analysis of the fog, at the same time that historical theories serve for the crags visible above the clouds in the middle distance and reductionist theories, so to speak, deal with the slopes of the mountains, their vegetation, etc.

Then the biometrician must take some care in the choice of what she measures. Pure historical accidents, for instance, actual genomic sequences, themselves submit mainly to historical explanations even when they are able to predict forward in

---

[3] There is an interesting extreme case, the sighting of objects at the extreme limits of visibility, such as Percival Lowell's reports regarding canals on Mars (Sheehan, 1988).

time to function or drug response. To account for organismal function in potentially consilient quantitative terms we need to concentrate not on data types suiting the -omic sciences but instead on measures of extent or capacity (lengths, volumes, states, flows) – see Riggs (1963) or Cook et al. (2011). I would include all the quantifications of morphometrics under that heading: positions of landmark points (Section 7.1) or shapes of curves as they express all of the tissue extents in-between (Section 7.6). Net measures like these ignore the uncountable richness of what the statistical physicist would call *microstates*, equivalent ways of arriving at the same measurements. In biology, statistics deals with the macrostates, the analogues of measureable thermodynamic quantities; the structural biologist, with the constraints of structure or texture that operate to stabilize the object across which these capacities or extents can be quantified.

One striking example of this dichotomy between the accidental or contingent and the essential or informative can be found in the first-year curriculum of any modern medical school. Barry Anson, one of the last of the old-style anatomists, published an atlas of human anatomy in 1950 (revised edition, 1963) that offered an unusual number of plates of "normal variants,"[4] many of which tabulate the frequencies of the observed types over a large sample of consecutively dissected cadavers from Anson's medical school anatomy classroom. It is clear from the provenance (these are normal variants, not matters for surgical correction) that we are working on the statistical side, Elsasser's indeterminate side, Friedrich's fog. If these structures made a difference for life or health, a reductionist approach might be a sensible one: a search for genetic bases, developmental–functional correlations, or clinical interventions. For instance, Hanahan and Weinberg (2011, p. 661) note that the insight dominating oncological research here in the second decade of the 21st century is the intrinsic heterogeneity of tumors, "organs whose complexity approaches and may even exceed that of normal healthy tissues." But from the biometric point of view these "normal variants" are like microstates of the molecules in a gas, variations that seem wholly irrelevant to quantitative summaries over organisms (while remaining relevant, of course, to the surgeon who needs not to nick them on the approach to a target, or the pharmacogenomicist who needs to synthesize a patient-specific therapeutic agent). Only today, half a century later than when Anson was drawing his plates, are embryologists beginning to explore the formalisms according to which we can separate these variants into the necessary versus the contingent, the genetic versus nongenetic or epigenetic versus nonepigenetic aspects of their variability. We cannot expect consilience with respect to just *any* aspect of biological structure, only with the lawful aspects.

The distinction I am making here is the quantitative version of one that is becoming familiar to the modern biologist under an alternate name, *plasticity*. A dictionary informs us that the word "plasticity" means the property of responding to load with

---

[4] The idea of using Anson's figures as evidence for a biomathematical argument is originally Williams's (1956).

a permanent change of form. In this it differs from *elasticity*, which is a temporary response. These constructs differ radically in the way they can be articulated to a quantitative biology. The equations of elasticity have to do with displacements from an equilibrium form, displacements that are likely to be linear in various matrices expressing the explicit forces of an experimental situation. The return to equilibrium, in other words, is in analogy to the role of the grand mean in the multivariate statistical formalisms that underlie modern biometrics. In contrast, a plastic system deforms *permanently*, and so loses the memory of its earlier state or its earlier developmental targets in the sheer factuality of its present condition. Neither biometrics in particular nor the natural sciences in general have any way of encoding this path dependence when the dimensionality of a system is higher than a handful of fingers. Plastic development, properly construed, is not a suitable class of investigations for statistical study. Biological statistics embraces not the plastic but the elastic, in which the memory of the original state is permanent: the mean as "ideal." See, in general, Schaefer and Bookstein (2009).

In plasticity phenomena, a form forgets its origins but remembers its history. The statistical methods suited to such studies thus are unlikely to be the same as the statistical methods arising in fields for entities that have no history, entities like random samples from an unchanging multivariate Gaussian distribution. But these Gaussians are the basic constructs of biometric models as they have been ever since Pearson and Weldon adopted Francis Galton's intuitions of regression around 1900 (see Section L4.3). Plasticity arises as the crags in Friedrich's painting, elasticity as the fog.

It is clear that in their branching the images of normal variants from Anson are classifying frozen plasticity. All variants on a page function in essentially the identical way, but the organism does not much care about these or a myriad of other options in deciding how to build itself or how to manage its physiology and its energy budget once it is built. Perhaps the great embryologist Paul Weiss said this best, in a conference transcript from 1956: "Identical twins are much more similar than are any microscopic sections from corresponding sites you can lay through either of them" (Gerard, 1958, p. 140). For an extended comment on the intellectual setting of Weiss's comment and its role in the inception of this book, please see Appendix 2.B to this chapter.

Random Walks

If there is so much information in heterogeneous structures that is refractory to quantification as a scientific strategy, how ought we to think about it? A particularly useful set of analogies and metaphors arises if we change our point of view about numbers from single values to structured *series* of values such as might arise from historical processes where only the final states (the crags of Friedrich's painting) have been preserved. One accessible variant of these, the univariate *isotropic random walk*, is the topic of a detailed treatment in Part III of this book, Section 5.2.3.2; but in its simplest features it is worth a preliminary introduction here.

One of the recurrent delights of life as a mathematical biologist is the freshness of the *mathematics*. I first encountered random walks (the subject of the following comments) in 1963, in an undergraduate probability course that was using William Feller's not-yet-classic textbook *Introduction to Probability Theory and Its Applications*. Chapter III, intentionally much simpler than the rest of his book, is called "Fluctuations in coin tossing and random walks." In the second edition (1957, p. 65) it opens as follows:

> This chapter serves two purposes. First, it will show that exceedingly simple methods may lead to far-reaching and important results. Second, in it we shall for the first time encounter theoretical conclusions which not only are unexpected but actually come as a shock to intuition and common sense. They will reveal that commonly accepted notions regarding chance fluctuations are without foundation and that the implications of the law of large numbers are widely misconstrued.

The set-up for Feller's demonstrations is close enough to that of the context of the historical/biological sciences that I am startled it isn't more widely incorporated in the pedagogy of interdisciplinary method. This is the notion of a *random walk*, the running sum of a series of independent identically distributed measurements. As Brownian motion it was discovered in the 1820s, but as a formal self-similar measurement protocol it is younger than statistics. It seems to have appeared first about 1900, in Louis Bachelier's doctoral thesis on financial statistics. His work was basically ignored until its rediscovery by Benoit Mandelbrot (1924–2010), who, beginning in the 1960s, extended it into the splendid applied and popularized mathematical graphics of *fractals* (Mandelbrot, 1982).

Feller introduces a collection of "preposterous" mathematical facts that are all versions of one underlying theorem, the "first arc sine law," that had recently been discovered. The topic here is the running excess of heads over tails in an arbitrarily long series of flips of a fair coin. We usually model it as a jagged curve of segments that move either 1 unit up or 1 unit down for every jump of one unit to the right, as in Figure 2.9. Feller's point is that as such a process is repeated a very large number of times, it produces patterns that need to be attributed to chance in a manner of which we humans prove curiously incapable. (Francis Galton himself, the inventor of the correlation coefficient, actually produced a physical model of this process, the *quincunx*, that we describe in Section L4.2. However, Galton and his student Pearson used it only to explore properties of the correlation coefficient itself, not properties of the sample paths as an ensemble.)

The distribution of a good many descriptions of the outcome is the mathematical function $f(\alpha) = \frac{2}{\pi} \sin^{-1} \sqrt{\alpha}$, as charted in Figure 2.10. This arcsine curve is about as different from the usual "bell curve" $\frac{1}{\sqrt{2\pi}} e^{-x^2/2}$ as any two plausibly realistic descriptors of processes could be. For instance, for the arcsine curve the lowest probability density is in the middle, and the highest at the very extremes of the range. When the cumulative normal distribution (the appropriate version of a bell curve) is plotted in the figure, it twists exactly the other way, steepest in the center of the graph and with asymptotes toward the horizontal at either end. Feller notes, for instance, that

2.4 The Issue of Heterogeneity

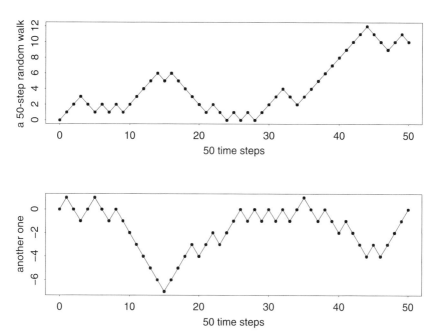

Figure 2.9. Microstructure of random walk: two examples, 50 steps each.

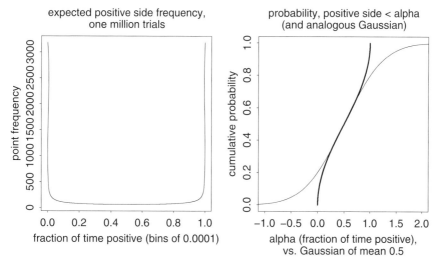

Figure 2.10. The arcsine law, one of the great elementary discoveries of 20th-century probability theory. (Left) The distribution of the fraction of time above the origin of a random walk is strikingly counterintuitive: the density rises without limit as one approaches the boundaries 0%, 100% of the range of this score. (Right) The cumulative distribution of this value is the "arcsine law" shown here in a heavy line (see text for formula). Its behavior is quite contrary to that of the Gaussian distribution that osculates it at the same mean and central density (light line).

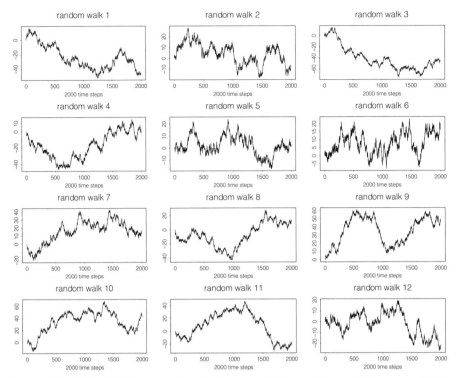

Figure 2.11. Twelve 2000-step versions of the process in Figure 2.9: cumulative sums of 2000 flips of a fair coin. These are far more variable than our intuition is expecting.

the distribution of the fraction of time on one side or the other of the horizontal axis follows the arcsine distribution. In accordance with the theorem, in a coin-tossing game that runs on and on for an entire year,

> ...*in more than one out of 20 cases the more fortunate player will be in the lead for more than 364 days and 10 hours.* Few people will believe that a perfect coin will produce preposterous sequences in which no change of lead occurs for millions of trials in succession, and yet this is what a good coin will do rather regularly. (Feller, 1957, p. 81; emphasis in original)

Also, the distribution of the location of either directional extremum of the walk (technically, the "first maximum," defined so as to break ties) has the same distribution, and so is far likelier to be near one end or the other than anywhere else.

Figure 2.11 shows 12 examples of a 2000-step random walk illustrating these points of Feller's. Any of these would be quite plausible as the outcome of the measurement of an evolutionary time series (or an ice core or other long-term record from a natural science). They have quite different vertical scales (recall that the distribution of the extreme values is not a bell curve). All of them "show" obvious trends, either from end to end or within some subrange of the horizontal index. Some seem to change level in the middle, never returning to an interval in which they began.

2.4 The Issue of Heterogeneity

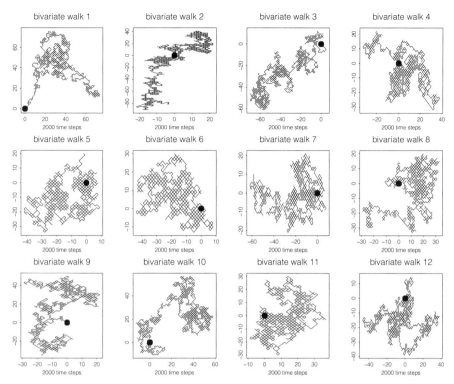

Figure 2.12. Two-dimensional equivalent of Figure 2.11: random walks on a plane. The underlying statistical process is homogeneous, but the viewing eye finds that almost impossible to believe in light of the obvious heterogeneities of outcome. Compare Figure 4.19. In every panel the black disk marks the starting location (0, 0) of the walk.

At such locations, under the influence of a persuasive but malicious master historian, we could be deluded into believing that the series has "changed its center" as a consequence of some imaginary event. *Trends and breaks like these are necessarily meaningless, no matter how strongly our intuition urges us to presume some meaning* (a scientific signal, if they were measured under experimental control).

Feller ended his lecture of 1957 with a one-dimensional walk like those in Figure 2.11, but natural scientists, especially evolutionary biologists and paleontologists, go on to consider the case of many measured variables. Figure 4.19 is among the earliest of these: a random walk (Brownian motion) as recorded by Perrin in the course of his study of "atmospheres of colossal molecules." Figure 2.12 simulates a dozen of *these*. We see at once that the intuition of meaning is even more compelling than when contemplating the preceding figure. Almost every walk breaks into "segments" that would surely be considered as typifying different categories of an "underlying selective regime," and we would be irresistibly tempted to account for the change of location (in some frames) or the change of direction of trend (in others) in terms of something happening in the time frame of the sequencing. The principle that extrema

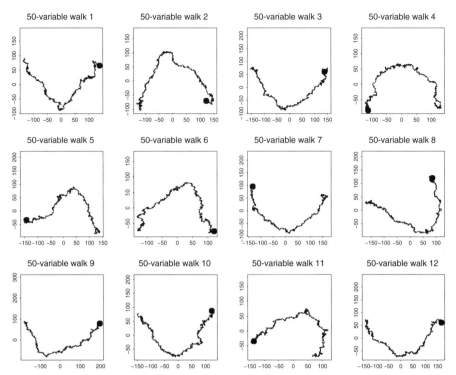

Figure 2.13. Fifty-dimensional equivalent of Figure 2.12, reduced to a best-fitting plane by the method of principal components analysis (see Section 6.3). Again the black disk marks the location assigned to the starting location of the walk. The true situation is suppressed by this inappropriate statistical approach.

of value are near extrema of time applies visually to *either* coordinate, so we see nearly every walk as "going somewhere" either horizontally or vertically.

Modern biometric statistics makes these problems even worse by the turn to one particular simplification, principal components (see Section 6.3). Principal components analysis is a tool for "dimension reduction" of a data set characterized by multiple measures of the same specimens. *Assuming the specimens to arise as independent samples of some population*, principal components analysis produces composite measurements (linear combinations of the measures actually made) that have the greatest variance for given sum of squared coefficients. Figure 2.13 shows the result of this particular tactic for a simulated data set of 50 independent dimensions each carrying out a random walk over the same time frame (and hence all failing to satisfy the requisites for a principal components analysis for exactly the same reason – successive observations are not independent). Think of this, if you like, as 50 genes all undergoing random drift. Unfortunately, the imputed structure always looks just about the same: a strong trend from end to end orthogonal to a roughly triangular trend reversing midway. I return to this topic in Section 7.8, at the very end of Part III, "The Graduate Course," or see the more technical discussion in Bookstein (2013b).

The connection between this theme and our main topic of heterogeneity is strikingly ironic. A random walk is *not* heterogeneous, but *homogeneous*. It looks heterogeneous only when analyzed at the wrong level of measurement, the cumulative state instead of the increments. But instruments measure the state. It requires a positive act of creative imagination to convert to the increment before proceeding with statistical analysis or pattern-matching. Imagine how often this might be the case for real data series – along any sort of time axis, from nanoseconds to billions of years, trends no matter how "statistically significant" might mean nothing at all in the sense that another run of the same process would yield outcomes as different from one another as the panels of Figure 2.12. One of the late Stephen Jay Gould's most profound arguments was the symbolic equivalent of this, when, in *Wonderful Life* (Gould, 1989), he noted that any rerun of the evolution of the chordates starting over again at the Burgess Shale would surely not result in anything remotely resembling *Homo sapiens* in general or Stephen Jay Gould in particular. (For a gently opposing view, see Conway Morris, 2004.)

2.5 The Graphics of Consilience

The objection raised in the last section to the basic strategy of quantitative consilience in biology – that it is based on a-priori knowledge of what is necessary and what is contingent, what is plastic and memoryless versus what is elastic and memory-laden – is no different in principle, only in preponderance, from what the physical scientist likewise has to cope with. Wigner (1960, pp. 4–5), in summarizing the role of mathematics in physics, notes that the physicist measures a vanishingly small fraction of the information available from, say, a kinetic system, but measures that fraction with obsessive attention to detail and fantastic accuracy.

> The regularity we are discussing [the law of falling bodies] is independent of so many conditions that could have an effect on it. It is valid no matter whether it rains or not, whether the experiment is carried out in a room or from the Leaning Tower, and no matter who drops the rocks.... The exploration of the conditions which do, and which do not, influence a phenomenon is part of the early experimental exploration of a field. It is the skill and ingenuity of the experimenter which show him phenomena which depend on a relatively narrow set of relatively easily realizable and reproducible conditions.... If there were no phenomena which are independent of all but a manageably small set of conditions, physics would be impossible.
>
> The laws of nature are all conditional statements and they relate to only a very small part of our knowledge of the world. Classical mechanics... gives the second derivatives of the positional coordinates of all bodies on the basis of the knowledge of the positions, etc., of these bodies. It gives no knowledge on the existence, the present positions, or velocities of these bodies.

Thus the job of the pursuer of quantitative consilience is not merely to imitate the formulas of physics, the search for principles of least action, constants, linear laws, and the like; it is also, and in practice more importantly, to find the parts of the other natural sciences or of the social sciences that look like they might be lawful in the way physics is, characterized by constants of some sort within samples or, even better, between samples. There is no reason to exclude the possibility of quantitative consilience from the biological or social sciences – for instance, we examine an informative agreement between regression coefficients in Section E4.6.2 – but it is likewise not to be presumed that anything you can measure is ipso facto a candidate for a quantitative consilience. Physics has had five centuries to winnow the range of its measurements. Biology may need a century of its own even after it decides (as it has, alas, not yet decided) that the contents of those great bioinformatic data repositories are not in fact the source of statistically useful information. These appear to be a domain solely of frozen plastic phenomena, Friedrich's crags; but consilience has to do with the remnant variability, which is to say, the fog in that same picture.

This issue – the limitation of the construal of consilience to only a part of the available empirical data – is a commonplace in most of the sciences we will consider. In the example of continental drift, for instance, we ignored everything about the emerging seafloor except its magnetism. In physics, as Wigner already noted, we do not model initial positions and velocities, only accelerations. Decisions about what to discard and what to take as data are just as fraught in the physical sciences as in the biological sciences. As a homely example, consider the familiar periodic table of the elements. Most readers will have seen one or another "standard version" of this chart, like that in Figure 2.14, with the symbols for the elements set out in a scheme of rows and columns corresponding to the highlights of their chemical properties.[5]

One general task of scientific visualization is the placing of the names of entities on a page in such a way as to effectively reproduce the relationships *among* them. Indeed, visual representations and visual metaphors are central to the cognitive economy of consilient quantitative reasoning in general. Though we can compare actual numbers by looking at their decimal representations, for any more extended set of measurements, like a table or a functional form, the easiest way to see whether two different quantitative structures align with one another or whether a set of measurements aligns with what a theory predicted is by looking at good diagrams. For continuously variable data readers are referred to the method of principal coordinates to be introduced in Section 6.4. But the periodic table is a tableau of constant data (counts of protons, counts of electrons shell by shell) measured without error, without variability sensu stricto. Still, the discovery of the periodic table was itself a manifestation of consilience, when Mendeleev noticed that elements sharing certain chemical properties could be ordered into a pattern of repeating columns consistent with increasing atomic number. Over the course of the 20th century these chemical properties have proved to derive from the properties of electron shells, which are distributed over the atoms in an even more orderly fashion. In two dimensions there are roughly two of these adjacencies to be cued, three in three dimensions. Note the extraordinary efforts that various designers have invested in indicating these commonalities.

---

[5] Readers who wish to learn more about this should turn first to Edward Mazurs's unique book of 1974, *Graphic Representations of the Periodic System During One Hundred Years.* Dozens more contemporary tables can be found via the links collected in http://www.meta-synthesis.com/webbook/35_pt/pt_database.php.

2.5 The Graphics of Consilience

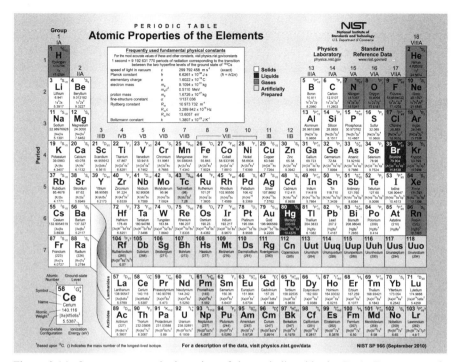

Figure 2.14. One current standard version of the periodic table, by R. A. Dragoset et al., National Institute of Standards and Technology, United States Department of Commerce.

In Figure 2.15, total atomic number is along the spiral, relevant outer electron shell is radius down the page, and alignments of valence run around the helix; in Figure 2.16, each new orbital series is introduced as a bud off the previous series. Both of these are quite differently informative than the version (Figure 2.14) that was likely on the wall of your high school chemistry lab.

Even as our understanding of the subatomic basis of chemistry continues to deepen, these graphs remain stable in their information content. There is general agreement among the cognizant community (here, the physical chemists) about how their numerical explanations come to interact with measurable reality. That specific topic has been part of discussions of the foundations of natural science for quite a long time, during which opinions have generally assorted themselves along a continuum between correlation and exactness in the corresponding cognitive stances. The extremes are well-represented by two early spokesmen, Karl Pearson and Henry Margenau, whose two contrary points of view I briefly review here.

You may know Karl Pearson as the inventor of the "Pearson correlation coefficient," the familiar r we will be discussing in Section L4.3, or as the grieving friend who, in the course of a published eulogy for the zoologist Walter F. R. Weldon, wrote the encomium to enthusiasm that appears among my Epigraphs. A pragmatic amateur philosopher of science, this same Pearson wrote a Victorian best-seller, *The Grammar of Science*, that included, in its third (1911) edition, an entire chapter on the subject of

54 2 Consilience as a Rhetorical Strategy

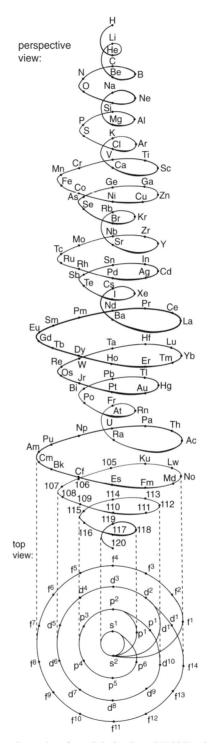

Figure 2.15. One modern alternative, from Schaltenbrand (1920), via Mazurs (1974).

2.5 The Graphics of Consilience 55

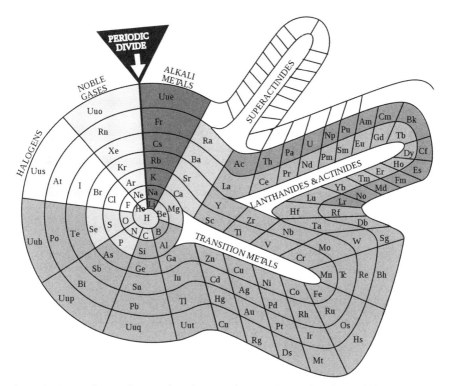

Figure 2.16. Another modern version, from Benfey (1960), emphasizing a different aspect of the situation.

"contingency and correlation." Here Pearson tries to extend another methodology of his invention, the method of contingency tables (including the χ^2 statistic qualifying their deviation from independence), from the social sciences to the natural sciences. To that end, indulging his penchant for rhetorical exaggeration, he denies the "absolute identity" even of atoms and physical constants in favor of "a certain average or statistical sameness." His picture of a physical law, then, is not the exact representation as a line that we will see in, for example, the discussion of Millikan's work in Section E4.2, but instead the figure I copy here as Figure 2.17, in which the law is "merely conceptual." On page 177 Pearson notes that in his view the quantitative scientist "sees variation as the fundamental fact in phenomena," a metaphysics that does indeed apply to some of the sciences we review, but not to all. "We cannot classify by sameness, but only by likeness" (p. 170) – in the clearest possible contradiction of Wigner, this is a declaration of the *kinds* of sciences for which the notion of exactness, of natural law, is likely to prove mainly a frustrating metaphor. Here he overstates, surely: consilience and meta-analysis are possible – two or more empirical studies can converge on a common value for the relation of two measures at different levels (in Wilson's cosmos) and thereby lead to very useful insights and even strong inferences about the underlying processes that must be at work.

Some additional clarity was arrived at 40 years on by the particularly thoughtful physicist Henry Margenau. Writing in 1950 of the metaphysics of physical science

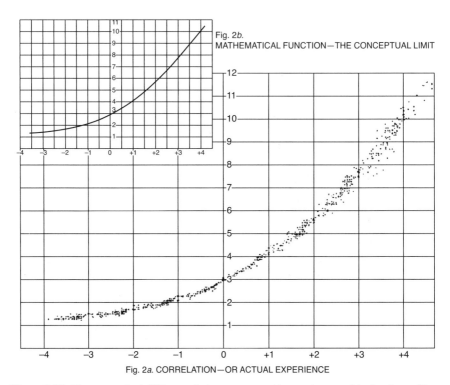

Figure 2.17. The conceptual difference between an exact law and an empirical pattern. (From Pearson, 1911.)

itself, he emphasizes the commonality among the sciences that use the correlation coefficient to describe the pairing of quantitative properties, but goes on to distinguish the particularly interesting case in which these correlations approach the value 1.0. For these situations, "speculation might arise as to whether a new kind of connection, not adequately expressible by means of a correlation coefficient, has not made its appearance. We suspect a 'law of nature.'" The view that science is limited to contemplation of correlations, attributed to "Mach, Mill, Pearson, and the Vienna circle," is "unassailable" but "does violence to certain sciences."

> Investigators bent on basic explanation are never satisfied with a statement of correlation coefficients. Their reaction to the discovery that the coefficient differs from 1 by less than 0.00001 is one of curious consternation; they feel the urge to probe more deeply, to *derive* this strange uniformity of experience from principles not immediately given.[6] A complete understanding of what in fact the workers in all exact sciences do is the central problem of today's philosophy of science. (Margenau 1950, p. 28)

Whereas for Pearson the variable is the fundamental, and the exact only a presumably unreliable idealization, to Margenau the exact is the fundamental, and the variable only the expression of fallible instrumentation.

---

[6] That is, to *abduce* something from it that would serve as its explanation. See Chapter 3.

2.5 The Graphics of Consilience

I submit that *this* is the domain of consilience: the probing of whatever "strange uniformity of experience" turns out to link the multiple levels of explanation that Wilson cherishes. Correlation is interesting, but the agreement of regression slopes between instrumentalities is even more interesting. Margenau goes on, indeed, to propose a very interesting figure (my Figure 2.18) for the relation of the scientist's conceptualizations to Nature; this is the topic of an extended discussion in his Chapter 5, "Metaphysical Requirements on Constructs." The requirement for a fruitful metaphysics centrally involves the presence of *multiple* connections between a conceptual scheme and reality – the multiplicity of those double lines leading out of Nature in the diagram. In their multiplicity this requirement explicitly relies on numerical consilience for its verification. He mentions five other dogmata as well (pp. 81, 89, 90, 92, 96). These are (1) "logical fertility" (by which he means the possibility of understanding universal propositions in a context of particular instances); (2) requirements of permanence and stability ("the theoretical components of experience are taken to be sharply defined and uniquely determinable, whereas the immediate ones are subject to defects of clarity"); (3) extensibility of constructs (regarding which I have already reviewed Margenau's chosen example, Newton's law of gravitation – also, he mentions with favor the possibility of biological reductionism, as of behavior to "the theoretical aspects of neurophysiology and chemistry"); (4) a requirement of causality ("constructs shall be so chosen as to generate causal laws," which are construed as relations of invariable succession between states of physical systems); and (5) "simplicity and elegance," which is to say, the possibility of mathematicization.

Even greater is the centrality of a concept quite like consilience that is the topic of the next chapter of Margenau's, the chapter entitled "Empirical Confirmation." What he means by this is not the confirmation of any proposition in particular but the methodology of confirming any of a broad class of propositions. "The meaning of agreement between theory and observation," Margenau writes, is inseparable from "the equivocality of observations," and so must reside specifically in "the distinction between the rational and the sensory" – measures of precision, standard errors and the Normal error curve, and "the all-important office of statistics in ordering and regularizing experience." You cannot "regularize" a measurement without a prior theoretical expectation of what its value ought to be. That expectation is invariably based on instrument measurements of its own, and in this way consilience worms its way into the foundations of physics just as thoroughly as Wilson asserts its role to be in the biological sciences and interdisciplinary bio-whatever sciences. It is here that the argument between Pearson and Margenau comes to rest. To the two different metaphysics correspond different sorts of sciences, and to the extent statistics offers concepts such as curve-fitting, precision, and error laws that accord with the exact sciences, it proves less amenable to the concepts of a purely correlation-driven world such as psychometrics.

Concluding comment. The measurements to which Margenau is referring are typically borne in geometrical properties of our familiar spacetime, and this property – the geometrical, indeed graphical nature of physical demonstrations – is, here in the 21st century, a commonplace. But this geometrization extends quite a bit further. For instance, astute readers have already noticed how the argument about continental

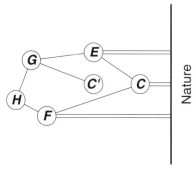

Figure 2.18. Graphical metaphor for the relation of physical law to reality: a first type, here C, E, F, of constructs, interact with empirical nature, while those of a second type (G, H) interact mainly with those of the first type. A construct like C', in contrast, is not likely to be helpful at all. (After Margenau, 1950.)

drift, Section 2.1, is actually carried by the figures, not the text, and the most important figure of all was not a raw instrumental record but one that was manipulated graphically (an ocean transect printed right atop its own reflection). This is not mere cognitive psychology. Our understanding of the printed page as metaphor for the real world is quantitative, not qualitative. A powerful engine for the success of scientific quantification in general, it is a carefully calibrated exploitation of ordinary plane and solid geometry. We are binocular primates, and navigating a graph uses mostly the same cognitive tools that we use in navigating a physical landscape.[7] The strength of morphometrics (Chapter 7) is in practice inseparable from the immediacy of its spatial diagrams and metaphors; and among the interdisciplinary psychosocial sciences those seem stronger that can rely on the power of diagrams in real space in this way – the studies that, for example, incorporate brain imaging along with variations of stimulus, response, or experience. Apparently consilience is very often a sturdy guide to truth in this physical world of ours, and we evolved to grasp the equivalent "propositions" in visuospatial form (buttressing a dwelling place) or kinetic form (throwing a spear, navigating a boat across the Mediterranean) long before anything like numbers had been invented. Much of the craft of quantification in science consists in the imitation of the craft of visual art: the display of numerical consiliences in graphical form.

Appendix 2.A Mathematics or Physics? Consilience and Celestial Mechanics

Section 2.3 very briefly mentioned Kepler's three laws of planetary motion. The core of the cognitive style that governs classical physics was set by Newton's demonstration

---

[7] When this fails it is often because our ordinary binocular intuition fails as well: for instance, in applications to the very high-dimensional "Euclidean" spaces of multivariate statistics, or the low-dimensional but curving spaces of Riemannian geometry. One of the greatest books of 20th-century mathematical popularization, David Hilbert's *Geometry and the Imagination* (original title, *Anschauliche Geometrie*; "anschaulich" means, in effect, concrete, demonstrative, descriptive, vivid, perspicuous), contains many tools for calibrating the imagination as it struggles with abstractions like these.

more than 300 years ago that one expression for gravitational force explains all three of Kepler's empirical generalizations.[8] The word "explain" here is used not in the statistical sense, explained variance or goodness of fit of some model (see Section L4.5.1), but in the far deeper sense according to which a phenomenon thought contingent, accidental, or arbitrary is revealed to be necessary, axiomatic, part of the fabric of science (or the universe, or the biosphere). If (and only if) gravity is an inverse-square law, the planet of a one-planet solar system will cycle around a periodic elliptical orbit with the sun at one focus, sweeping out equal areas in equal times. Without the inverse-square law, there would be no orbits, no life on earth, and no readers of this book: the matter is thus quite an important one.

The way Newton's equation accounts for Kepler's empirical summaries is the holotype of what we mean by an "exact science" even though we now know (since Einstein's work on general relativity) that Newton's law itself is not quite accurate – the planets would not follow *precisely* elliptical orbits even if they were so small that they had no discernible gravitational influences on one another. The discrepancy (which is small because the mass of the sun is small with respect to the distance of the planets from it) does not matter for the spirit of consilience that Wilson and I are foregrounding. Unexpectedly close agreement between a mathematical derivation and previously inexplicable empirical data is the strongest rhetoric in the whole world of science. By "close agreement" is meant an actual numerical match, within the measurement error known to characterize the instrumented data, of the "shape" of an empirical summary (here, the ellipse of a planetary orbit; in Section E4.3, the bell curve for speeds of gas molecules in any single direction) to the functional form generated by a formal mathematical manipulation beginning from abstract axioms. We will see this rhetoric – reasoning based on precision of a pattern match – repeated over and over in Parts II and III.

The foundations of quantification in the other natural sciences often follow a similar approach to the meaning of "exact," privileging a respect for the large-scale features of patterns over any particular parametric predictions. For instance, the new synthesis in evolutionary theory that coalesced around 1940, in the hands of Julian Huxley, Sewall Wright, Theodosius Dobzhansky, Ernst Mayr, and George Gaylord Simpson, among others, was founded on axioms setting out certain stochastic models of chromosome behavior in mitosis. Those models have proved unrealistic in a great many important aspects, and yet the foundations of the field of population genetics are not in question. Wright's masterwork *Evolution and the Genetics of Populations* emphasizes, in Chapter 6 of volume IV (Wright, 1978), the centrality of these fits between models and data, in this case by a detailed study of the shifting balance of two colors within a population of *Linanthus*, a diminutive flowering shrub, that ran along an Arizona highway for 200 miles. The data set had fascinated him for more than a quarter-century before that (see Wright, 1943a, 1943b, and the scientific biography by Provine, 1986). This preference is intriguingly complemented in the exact sciences by the "outrageously improbable correct prediction," exemplified by exact forecasting

---

[8] To Peirce, Kepler's reasoning is "the greatest piece of retroductive reasoning ever performed."

of eclipses or the presence of the planet Neptune exactly where Leverrier and Adams independently predicted it to appear in 1845 based on anomalies in the orbit of Uranus. But breakthroughs like these are uncommon; I can think of no similar example in any other natural science, unless perhaps the final example in Chapter 4, the existence of the Chicxulub crater, is considered such a prediction in respect of crater size (meteorite mass).

But let us turn our attention back to the 17th century. Newton worked out the consilience of his equation with Kepler's laws by himself in the late 1600s with the aid of the differential calculus he had just invented for exactly this purpose. The last component of the pedagogy here, the Coriolis force (one term in the formula for acceleration in a polar coordinate system), was formalized by Gaspard-Gustave Coriolis in 1835. The solution of Newton's differential equation by change of variables to $1/r$ is part of the standard textbook demonstrations of celestial mechanics, and must have appeared in hundreds of textbooks by now, some dating to well before the turn of the 20th century. The underlying theory (gravitation as an inverse-square law, lately modified by relativistic effects) has been confirmed better than anything else in the cognitive universe of today's scientist, most recently by its obvious successes at predicting the gravitational behavior of the artificial satellites that we have been launching off Earth since 1957, such as the satellites that drive your car's GPS device. My emphasis is on the consilience of the underlying mathematical model and its meaning as a model for quantitative scientific rhetoric in general. How can the nonspecialist understand why a planet travels an orbit that is so very nearly an ellipse? And what is the meaning of the word "why" in the preceding question?

I divide the rest of the exegesis here in seven sections. All involve some mathematics, but the sophistication required varies considerably from one to the next. Among the "mathematical facts" that it is useful to have encountered before are conic sections, polar coordinates, the curvature of circles, and one particular differential equation (the one satisfied by the cosine). Though all this mathematics is quite conventional, my pedagogy per se is an experiment in conveying the force of these mathematical arguments without requiring a reader's mastery of the corresponding symbol manipulations in any detail.

2.A.1 What Ellipses Are

The topic of this section is truly classical – the first four bullets here were already known to the ancient Greeks. There exists a family of curves, the *ellipses*, that are jointly characterized by *any* of the following properties:

- They are the curves cut by a plane intersecting a circular cylinder.
- They are also among the curves cut by a plane intersecting the ordinary sort of circular cone made of rolled-up paper coming to a point: hence, they are among the *conic sections*. (To wit, they are the curves cut by planes that intersect the cone only in a finite region.)

- They are the curves traced by a point whose distances to two fixed points, the *foci*, sum to a constant. (This is the basis of the familiar *string construction*, as the best way to get two distances to sum to a constant is to make them the two parts of one inextensible string.)
- They are the curves traced by a point whose distance to one point, a *focus*, is proportional (by a multiplier less than 1) to its distance from a fixed line, the *directrix*.
- They are the curves satisfying an equation of the general form $ax^2 + bxy + cy^2 + dx + ey + f = 0$ for which $a > 0$, $c > 0$, $b^2 - 4ac < 0$ and f is not too large. Here x and y are the ordinary Cartesian coordinates of points in a plane.
- They are the curves satisfying an equation $r = k/(1 + m\cos\theta)$, for some constant m, $0 \le m < 1$, in ordinary *polar coordinates*, the coordinates for a curve in terms of distance from a fixed point along with angle (from some defined meridian) out of that point. The constant m is often called the *eccentricity* of the ellipse.

The mathematicians have had hundreds of years to make sure that the six characterizations above are exactly equivalent, and they have not failed to do so. We will see that the easiest way to show how Kepler's ellipses follow from Newton's law of gravitation is the last of these, the assertion that the planet's distance from the sun satisfies the equation $1/r = 1 + m\cos\theta$ up to some scaling. But neither the cosine function here nor the reciprocal function has anything much to do with the physics of the situation. These and the other mathematical symbols to come, with two exceptions, are just a convenience furthering the mathematical demonstration that three centuries' worth of physics students have accepted. The exceptions are the eccentricity m in the equation, which is related to the total energy of the planetary system, and the exponent of -2 in the statement of the law of gravitation, which will turn out to be the essential tie between the mathematics and the physics.

2.A.2 What a Cosine Is

Like the situation for the ellipse, there is a huge installed base of equivalent characterizations for the mathematical function that is the cosine of an angle.

- You may remember from high-school trigonometry that the cosine of an angle θ less than $90°$ is the ratio of the shorter to the longer side of a right triangle at the angle θ (not the right angle).
- If you paint a dot on the top of the wheel of a car, and then drive the car in a straight line, the height of the dot above the axle on which the wheel is mounted, divided by the radius of the wheel, is the cosine of the angle that the wheel has rotated. (But the locus of the dot itself is a different curve, a *cycloid*.)
- The cosine function is one solution of the *differential equation*

$$\frac{d^2 f}{dt^2} + f = 0.$$

That is, it is equal to -1 times its own second derivative. (The same is true of the sine function, which can be thought of as the cosine of an angle 90° greater, but we do not need this.)

- For any real x, the cosine of x is the sum of the infinite series $1 - x^2/2! + x^4/4! - x^6/6! + \cdots$ where if θ was measured in degrees then x is the number $\theta \times \pi/180$, the angle in *radians*, and for any integer n the notation "$n!$" stands for $1 \times 2 \times \cdots \times n$, the product of the integers from 1 up through n, to be explored in greater depth in Chapter 4.

Notice, as for the conic sections, how wonderfully diverse these characterizations are. Mathematicians delight in proving the equivalence of such wildly different characterizations, just as Jean Perrin delighted in proving the empirical equivalence of 14 entirely different characterizations of Avogadro's number (Section E4.3).

2.A.3 Changing Coordinate Systems

Your high school or Gymnasium probably taught you how to transform points back and forth between Cartesian (x, y) and polar (r, θ) coordinate systems. For explaining Kepler's ellipses we will need a little more than that high school material. Specifically, we need two additional formulas, one for the effect of rotation on radial acceleration and one for the effect of rotation on tangential acceleration when the radius is changing.

The first of these you already know because you inhabit a human body: it is the *centrifugal force* exerted by any object that is rotating in a circle, the force pulling it away from the center of the circle. You have to pull on a rope (the *centripetal force*) to compensate for the centrifugal force and keep the object from flying off. Figure 2.19 sketches a plausible heuristic that should lead you to accept the assertion that the centripetal acceleration characterizing an object moving at any angular velocity $\dot{\theta}$ with constant r, in a polar coordinate system, is

$$-r\dot{\theta}^2$$

along the direction of the radius vector through the rotating point. (In this notation, the dot $(\cdot)$ over the name of any variable stands for its rate of change with respect to time.)

The second of these, the *Coriolis force*, you may already know as well, at least, if you follow those meteorologists on television. The Coriolis force is the force experienced by any object that is trying to move evenly, without radial acceleration, along a radius in a rotating polar coordinate system: for example, an air mass moving from north to south along a meridian upon this rotating earth of ours. The formula for this acceleration is the quantity $2\dot{r}\dot{\theta}$, for the reasons sketched in Figure 2.20 (which is modified from one of Richard Feynman's in Volume I, Chapter 19 of *The Feynman Lectures on Physics*). *Something* has to deflect that point from just continuing on in a straight line, and this will be a force perpendicular to the radius vectors, hence along the direction of changes in angle, and with a magnitude the calculation of which is

2.A Mathematics or Physics? Consilience and Celestial Mechanics

acceleration of circular motion

Figure 2.19. The acceleration required to move in a circle. (Compare Figure 2.7.) The osculating (best-fitting) parabola to the circle of radius 1 through $(0, 0)$ is simply the parabola $y = \frac{1}{2}x^2$, of second derivative $2 \times \frac{1}{2} = 1$. Thus the acceleration of the particle moving on that circle toward the center of that circle (the extent to which its path is bent away from the straight line that, according to Galileo, it "wants to follow") is 1 wherever such a parabola can be fitted to it, which is to say, everywhere. If the particle speeds up by a factor $\dot{\theta}$ the equation of the parabola picks up an additional factor $(\dot{\theta})^2$, and so does the acceleration. The equation in the text follows from scaling the radius of the circle. Heavy line: circle. Light line: osculating parabola at $(0, 0)$.

a matter of some delicacy. It arises as the sum of two pieces, each of magnitude $\dot{r}\dot{\theta}$. One piece is the (fictitious) inconstancy of an inertially constant velocity with time because rotation makes the direction of the velocity seem to change, and the other is the the change in the apparent velocity caused by displacement closer to or farther from the axis of rotation.

These are the accelerations attributed to a point simply by virtue of moving evenly in a polar coordinate system instead of the Cartesian one. In addition, the point might *actually* be accelerating: this adds a component $\ddot{r}$ to the acceleration in the direction of the radius variable, and $r\ddot{\theta}$ to the acceleration in the direction of the angle variable.

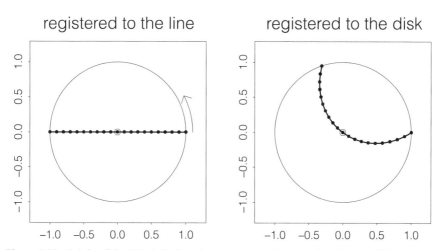

Figure 2.20. Origin of the "Coriolis force": apparent acceleration of uniform radial motion as viewed in a rotating polar system.

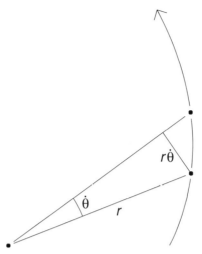

Figure 2.21. "Swept area": rationale for the formula $\frac{1}{2}r^2\dot{\theta}$.

(The notation ¨, two dots printed right atop a letter, stands for the second derivative of a variable, the rate of change of its rate of change.) So the net acceleration in the azimuthal direction is $r\ddot{\theta} + 2\dot{r}\dot{\theta}$, and in the radial direction, $\ddot{r} - r\dot{\theta}^2$. We use each of these equations below.

This is as good a time as any to write down the formula for one additional mathematical fact that will prove quite handy. Imagine (see Figure 2.21) that as some point moves in a polar coordinate system, we draw out the area covered by the vector that always connects it to the origin. This is called the *swept area*. A simple approximation of consecutive positions of this radius vector by two sides of a long, skinny triangle may convince you that the rate of change of this area is $\frac{1}{2}r^2\dot{\theta}$. (To convince yourself, notice that consecutive positions of the radius vector give you what is almost an isosceles triangle of which the height is r and the base is almost the distance moved perpendicular to the radius vector, namely, the rate $r\dot{\theta}$ times whatever small increment of time we are considering per little bit of area swept. Then just use the high school formula for area of a triangle, one-half times the base times the height.)

2.A.4 Newton's Law of Gravitation

These preliminaries completed, let us examine the single greatest example of this sort of reasoning, Newton's accounting for Kepler's three laws of planetary motion by one single differential equation.

It will be useful to launch the rest of this exposition by restating Kepler's laws in their modern form:

- In a dynamical system comprising a point (or sphere) of vanishingly small mass orbiting around a point of substantial mass, the small mass moves in an ellipse with the large mass at one focus.

- Within the plane of this ellipse, the moving planet sweeps out area around the focus at a constant rate.
- For multiple small masses around the same central mass, the orbital period is proportional to the 1.5 power of the mean distance from that central mass.

Newton's law of gravitation exploits a wholly different rhetoric from any of these. It is a differential equation, a statement about instantaneous forces (accelerations, really), not observed data. Every particle in the universe attracts every other (generates an acceleration along the line between them) according to a force along the direction of the line between them with a magnitude $F = -GM_1M_2/r^2$ proportional to a constant G times the product of their masses divided by the square of the distance between them. It is not too difficult to show that the law applies even to fairly large "particles," as long as they are spheres of uniform density. In a system of one really big particle (call it the "sun") and a whole lot of smaller ones ("planets"), then as long as the smaller particles don't get too close to one another it is adequate to some degree of approximation if the dynamics of the small particles are modeled, one by one, as if there were only two particles in the universe, the sun and the smaller planet. (What happens when they *do* get close accounts for most of the developments in celestial mechanics through the end of the 19th century, including the discovery of the outermost two planets, Uranus and Neptune.[9])

In the notation we have just been establishing, if we put a polar coordinate system in place with its origin at the center of the sun and with angle 0° placed anywhere convenient, this force leads to an acceleration that is

$$-GM/r^2$$

along the radius out of the sun. Here G is the universal gravitational constant, and M is the mass of the sun, but we don't need to know that for the manipulations to follow – their product is just a constant of its own, GM.

This *central force* is directed along the radius vector. In general, forces in a plane have two components. When the coordinate system is polar coordinates, one of these is radial (away from the origin) and the other, at right angles, is tangential (around the origin). And so it follows that the tangential component of the force must be zero – there is no "torque" to a gravitational attraction between point masses.

Now comes our first mathematical deduction:

- The tangential component of acceleration, under the central force model of gravitation, is zero.
- But, we noted earlier, this acceleration is equal to the "inertial" acceleration $r\ddot{\theta}$, the formula for a Cartesian system, plus the Coriolis acceleration $2\dot{r}\dot{\theta}$.
- Hence for a planet orbiting under a central force

$$r\ddot{\theta} + 2\dot{r}\dot{\theta} = 0.$$

[9] In the year 2006 some scientific body decided that Pluto is no longer to be deemed a planet.

- Multiply by r – this is harmless:
$$r^2\ddot{\theta} + 2r\dot{r}\dot{\theta} = 0.$$

But the usual rule $d(xy) = x\,dy + y\,dx$ (which in this notation would be $(xy)\dot{} = x\dot{y} + y\dot{x}$) identifies this with the time derivative of the quantity $r^2\dot{\theta}$.

Hence:

- **The time derivative of $r^2\dot{\theta}$ is 0 – this quantity is another constant, L.** When it is multiplied by the mass of the planet, it is the *angular momentum* of the planet around the sun.

But that is Kepler's Second Law! It follows from one part of Newton's law of gravitation (the force-direction part) by way of purely mathematical manipulations. Newton's law is a statement about how the world might be construed from moment to moment in terms of invisible forces. Kepler's second law follows; but that law was (1) already known to be true, and (2) surprising (not derivable from anything else in the cosmology or the astronomy of the time); hence (3) there is some reason to suspect Newton's law might be valid. This is, as always, the syllogism of *abduction* introduced by C. S. Peirce in the 1870's; it is the topic of Chapter 3 of this book.

So was this a mathematical manipulation, or a physical discovery? Both, evidently; the implied distinction does not actually make much sense. I have already quoted Eugene Wigner, and will continue to repeat, that fundamentally we do not understand why mathematics succeeds so well at describing the physical world.

Epistemological paradoxes aside, this constitutes real progress. For instance, it permits us to *eliminate a variable* in the equations to follow shortly. If we can get to an expression in r and $\dot{\theta}$, then the identity above lets us write $\dot{\theta}$ as the function L/r^2, where L is some constant (proportional to the planet's angular momentum), thereby removing all mention of $\dot{\theta}$ from the equation.

2.A.5 Another Change of Variable

We already changed variables from Cartesian coordinates to polar coordinates. Now we change variables again, leaving θ alone but replacing the radial distance variable r by its reciprocal $1/r$ multiplied by $p = L^2/GM$, a constant chosen purely to make the scale come out right in the final equations below. This means that wherever the equations had r in them, we now write, instead p/u, where u is "the new variable." The acceleration in Newton's equation is then $-GM/r^2 = -L^2 u^2/p^3$, and the function of r that substituted for $\dot{\theta}$, which was L/r^2, is now Lu^2/p^2. These both look unpromising, but be assured that all of the L's and p's will cancel out in the end. That is why the mathematical physicist would select this arbitrary-looking constant: so that everything cancels in the end.

Does this make any sense? Not physically: we do not experience the world in terms of reciprocals of distances to gravitating objects. Like many other symbol manipulations we will see in this book, this is a mathematical *stratagem*, a purely symbolic device to convince the student that the rearrangement or re-expression of one equation as another is actually valid. The substitution of p/u for r, then, is a device for *communication among mathematicians* and among the parts of physicists' brains that are mathematical: a way of getting to an inference (in this case, it will be Kepler's First Law) by a chain of reasoning that is shorter or less complex than otherwise. (One could do all of this in Cartesian coordinates, for instance – see Hyman, 1993 – and the answer comes out the same. But it is *much* more difficult to understand intuitively. You could say that it was *discovered* in one coordinate system but merely *confirmed* in the other, clumsier one.)

After an apology to the firmly nonmathematical reader – the pleasures of this sort of argument are somewhat rarefied – let us begin. By the usual (freshman calculus) formulas for differentials of functions, for any function X of time,

$$\dot{X} = \frac{dX}{dt} = \frac{dX}{d\theta}\frac{d\theta}{dt} = \frac{dX}{d\theta}\dot{\theta} = \frac{dX}{d\theta} \times Lu^2/p^2.$$

This is ultimately the reason for embracing the transform: the factors of u^2 that appear and disappear will greatly ease the interpretation of the radial acceleration equation. Continuing, put $X = r = p/u$ in the previous identity:

$$\dot{r} = \frac{d(p/u)}{dt} = \frac{d(p/u)}{d\theta} \times Lu^2/p^2 = -\frac{p}{u^2}\frac{du}{d\theta} \times Lu^2/p^2 = -L\frac{du}{d\theta}/p,$$

and so

$$\ddot{r} = \frac{d\dot{r}}{dt} = \frac{d\dot{r}}{d\theta} \times Lu^2/p^2 = \frac{d}{d\theta}\left(-L\frac{du}{d\theta}/p\right) \times Lu^2/p^2 = -L^2u^2\frac{d^2u}{d\theta^2}/p^3.$$

2.A.6 Getting to Kepler's First Law

Everything is ready now to work with the radial acceleration equation from Section 2.A.3 above. We had

$$\ddot{r} - r\dot{\theta}^2 = -GM/r^2 = -L^2u^2/p^3.$$

But

$$\ddot{r} = -L^2u^2\frac{d^2u}{d\theta^2}/p^3,$$

and

$$-r\dot{\theta}^2 = -(p/u)(Lu^2/p^2)^2 = -L^2u^2/p^3.$$

So the radial acceleration equation becomes

$$(-L^2u^2/p^3)\frac{d^2u}{d\theta^2} - (p/u)(Lu^2/p^2)^2 = -L^2u^2/p^3,$$

or, dividing through by $-L^2u^2/p^3$,

$$\frac{d^2u}{d\theta^2} + u = 1.$$

There are **no** arbitrary constants left in this equation! That was the whole point of the strange constant in the transform that defined u: to cancel out *both* the GM and the L. With no constants, this equation has an easy solution that has only two arbitrary constants of its own: the function $u = 1 + m \cos(\theta - \theta_0)$, where m is any positive decimal number and θ_0 any starting angle (the *perigee*, position of minimum r, which we just set to $0°$ and never mention again). The remaining constant m here has a physical meaning, the *eccentricity* of the orbit, which is a function of its energy as a function of the energy needed to escape the sun's gravitational field entirely.

So the result of all this manipulation is to produce an equation for u as a function of θ that is exquisitely simple, with only two constants of its own:

$$u = \frac{p}{r} = 1 + m \cos\theta.$$

And this is the equation of an ellipse, from the last bullet in Section 2.A.1, with the sun at one focus.

Hence Kepler's First Law follows from Newton's equation, too. (And so does the third, although its demonstration is not as instructive as the two preceding examples have been.)

2.A.7 Mathematics or Physics?

We have deduced Kepler's first two laws from Newton's one equation (and the third law can be deduced this way also). Have we "explained" anything?

Newton himself insisted not. "Hypotheses non fingo," he noted – "I do not pretend [feign] to hold hypotheses." The mechanism of gravitation was a mystery to him and remained a mystery until Einstein's theory of general relativity, in 1916, referred it to an even deeper property of the physical universe, the curvature of spacetime. But both for our purposes in this chapter on consilience and in this book as a whole it is not necessary that the reduction to an equation be to a mechanism: only that it be to some mathematizable system permitting free play with symbols along with the confirmation of inferences from that play back in the real world. Galton's demonstration (Section L4.2) that his quincunx is a machine for generating Gaussian distributions, Maxwell's demonstration of the Gaussian distribution of the velocities of gas molecules (which took a century to instrument convincingly), and even Eddington's evidence that light bends around the sun during a solar eclipse just as much as Einstein's equations of general relativity specify – all these are of the same logical species as this immortal demonstration of Newton's: mathematical arguments that can be verified by real machines. It is the *aesthetics* of demonstrations like these that drives quantitative inference where it is driven by the rules of quantity at all: the possibility that underneath fallible data there lurks an exact equation about *something* – the positions of the planets, the path of a ray of light, the stochastics of the chromosomes, the power law of the Internet and other Pareto distributions.

Of course it does not mean that there is a "gravitational physics of ellipses." In terms of the six alternate definitions with which this section began, the cylinder

2.A Mathematics or Physics? Consilience and Celestial Mechanics

and the cone have nothing to do with the issue, nor the equations in the Cartesian coordinate system. The polar coordinates are centered around one focus, but the other has no reality. The ellipse has two axes of symmetry, and while one of them extends from the point of closest approach of the planet to the sun ("perihelion") to the point of greatest distance ("aphelion"), the other, at 90° to the first, makes no difference for the equations or the planet.

No, the mathematical object that is an ellipse, per se, is not the issue here. What is real is the dynamics – the law of gravitation. Newton's Law then turns out to be, in a sense, a seventh "characterization of the ellipse" to go with the first six:

- Ellipses are the bounded curves traced by points moving around a fixed point under a centrally directed acceleration proportional to the inverse square of the distance from that fixed point.

I promised that I would come back eventually to the question Kepler might have called to Newton's attention (had their lives overlapped): Why are the orbits of the planets ellipses? That was the wrong question. Newton explained *what determines* the orbits of the planets – the inverse-square form of the law of gravitation – and the only cogent answer to the question of why the inverse-square is the notorious "anthropic answer": if it weren't inverse-square, there wouldn't be a planet capable of sustaining life on which we were resident to wonder about it. Likewise, one does not ask why "$r = 1/(1 + m \cos \theta)$" is the equation of an ellipse; it just *is*, as mathematicians have agreed for centuries.[10]

You should be asking, instead, "How can you most easily *convince me* that this is the equation of an ellipse?" That is a much more reasonable question. A classroom approach might start with the third characterization (the string construction, constancy of the sum of distances to two foci), write it out in Cartesian coordinates, and then switch to polar coordinates around either focus. The string construction, in turn, can be proved to be equivalent to the plane-cutting-cone and plane-cutting-cylinder characterizations by an elegant demonstration, the *Dandelin spheres*, discovered around 1830. Facility with multiple characterizations is one of the key skills of the competent mathematician (as recognizing the equivalence of patterns across such domains is one of the key skills of the applied statistician). For instance, once we actually *have* an ellipse, or its n-dimensional generalization, there are multiple characterizations of its axes or hyperplanes of symmetry; we shall explore the scientific uses of some of these in Chapter 6.

For the planets to orbit in ellipses, there must be an inverse-square law of gravitation. Mathematics or physics? We cannot draw the distinction here: that was always Wigner's point. The rhetoric of quantitative methods, as I am exploring it in this book, pursues the same maneuver as far as possible into other quantitative sciences as well. How far can we go into genetics, or neurological birth defects, or information theory of the DNA molecule, or the psychology of obedience, or the history of impacts

---

[10] Likewise, in Chapter 4, one does not ask *why* regressions on Galton's quincunx are linear. They just *are* – that's the way the mathematics works out.

with Earth, while still continuing to blur this boundary with mathematics – while continuing to confirm our axioms, which are mental constructions, via instrument readings and other real information-bearing objects? To be a quantitative science, it is not enough just to have a handful of numbers (e.g., human heights or IQ scores), even if it is quite a large handful (e.g., a brain image). It is required further that the mathematical properties of the quantities are consilient, to greater or lesser extent, with the empirically observed properties of those objects. To require the possibility of consilience is a very demanding requirement indeed; but otherwise our quantitative sciences are, as Kuhn said, "just numbers."

In this matter the argument from consilience differs from conventional arguments on "foundations of measurement" (Krantz and Suppes, Ellis, Hand). It is not enough that we manipulate measurements as if they were mathematical objects; it must be the case that the patterns observed empirically in those measurements match the mathematician's favorite properties of the mathematical objects, the way that Kepler's ellipses matched the solutions of Newton's differential equations. Otherwise the mathematization was merely a shorthand toward publishing, gaining us no actual understanding.

How do we use our own minds to assess the extent to which those claimed confirmations are valid? The core question of consilience is at the same time the core question of numerical inference: the extent to which Newton's intentional blurring of the line between mathematics and physics might be replicable in other sciences as well, and the nature of the language, replacing the geometry of ellipses, in which the equivalence might be argued and found to be persuasive.

Appendix 2.B Historical Note: From Weiss through Kuhn to This Book

Section 2.4 included, in passing, a quote from the embryologist Paul Weiss ("identical twins are much more similar than are any microscopic sections from corresponding sites you can lay through either of them") that aligned neatly with a point of Walter Elsasser's driving the argument about the metaphor in this book's cover painting, the difference between deterministic and stochastic in assessing the proper role of quantifications in science. The conference from which Weiss's comment is extracted has a relationship to the existence of this book that is worth sketching in some detail.

The proximal stimulus for the coalescence of this book as a whole was a workshop, "Measuring Biology," that took place at the Konrad Lorenz Institute for Evolution and Cognition Research, Altenberg, Austria, on September 11–14, 2008. Contributed papers were assembled as volume 4, number 1 and a bit of number 2, of the MIT Press journal *Biological Theory*. The stimulus for *that* workshop was the impending 50th anniversary of a very significant meeting held in November 1959 at the Social Sciences Research Council: the Conference on History of Quantification in the Sciences. The proceedings of this conference were published as Woolf (1961), and also as a special issue of *ISIS*, the journal for history of science. Of the contributions to that earlier conference volume of Woolf's, one is hugely important for sociology of science: Thomas Kuhn's essay on the function of measurement in modern physical science (Kuhn, 1961) that I discussed at length in Section 2.4.

But there was also a biological contribution in that volume: a chapter by Ralph W. Gerard entitled "Quantification in Biology" (Gerard, 1961). At the time, Gerard was at the Mental Health Research Institute at the University of Michigan, along with the psychologist Anatol Rapoport, my teacher Kenneth Boulding, and several other stalwarts of the general systems theory tradition. It would seem that Gerard reversed his assignment from Woolf. As he notes on page 204, he writes not on the role of quantification in biology, but on "the biology of quantification," quantification *as* biology – the sensory basis of quantities as they reduce to words, categories, or theories – and as a result he had virtually nothing to say on the actual theme of the meeting.

I suspect that was not what Woolf wanted from Gerard. The discussion that would suit the Woolf volume had already been published by Gerard three years earlier. It was his masterful "condensed transcript of the conference" from a meeting of the Biology Council of the (United States) National Academy of Sciences, a conference chaired by Paul Weiss in Lee, Massachusetts, in October 1955. Among the other 12 participants were Sewall Wright, Ernst Mayr, and George Gaylord Simpson. The proceedings, entitled "Concepts of Biology," appeared as Publication 560 of the Biology Council (Gerard, 1958); it is hard to find nowadays.

The declared purpose of the 1955 conference was to provide a renewed foundation for the methodology of the biosciences as a whole, in terms of a twofold scheme of *levels of organization* mapped over *methods*. This is set out quite literally in Gerard's text as a very large verbal table, pages 99–100. The rows are ordered by level, from molecule up to "total biota." The columns, "Methods," likewise constitute an ordered scale: "Identify, Modify, Isolate, Measure." The column stub under this last heading is telling:

> *Measure* – over finer units of mass, distance, time, and ever longer into past time. With precision use less material and disturb it less. Automatic recording, permanent record (photograph, kymograph), estimate of significance (statistics).

There is no reference here to the actual substance of biology; instead the phrases are about either instrumentation or inference. The entry in this column corresponding to the level of "Organ" is itself quite interesting: "Electricity measures – oscilloscopes; chemical measures – tracers; mechanical measures – strain gauges; functional tests – secretion, response to hormone growth." Electricity, tracers, strain – these are not biological concepts; the sole biological term here is "secretion." To Gerard, half a century ago, measurement in biology was a matter of its interfaces with the *other* natural sciences; nothing in biometrics was inherent in the biological sciences themselves. The interested reader is referred also to Gerard, 1957, his separately published summary of his own conference. No wonder that the world of "concepts of biology" seemed, in 1958, so quantitatively depauperate: the last specifically biometric innovations had been those of the 1920s, when multivariate analysis was brought into anthropometry. Not for ten years would monographs begin to appear on the first genuinely biometrical (as distinct from biostatistical) innovations for decades, the supersession of simple metric object interactions by imaging science (see, for instance, Ledley, 1965), and of course the modern period could not begin until the maturation of CT and MRI scanners toward the end of the century, and the machines for low-cost genomic sequencing even more recently.

Also "in the air" at the time of Woolf's meeting was Roger Williams's challenging polemic *Biochemical Individuality* (Williams, 1956), which emphasized the importance of variability over averages in any consideration of human physiology for medicine, and also called readers' attention (probably for the first time) to Anson's magnificent plates about normal anatomical variation. Furthermore, Olson and Miller's (1958) *Morphological Integration* had just appeared, intentionally reopening many issues in the classic fusion of multivariate statistics (particularly factor analysis) with comparative biology. Likewise, it would be reasonable to assume that Sewall Wright had already begun the magisterial revision of his life's work on biometrics that was to appear as Wright (1968), especially the retrospectives regarding biological frequency distributions, their statistical description, heterogeneity ("compound distributions"), and examples of "actual frequency distributions" univariate and multivariate.

The situation as of 1961 was thus left in the form of an unanswered question: *could* the thoroughgoing quantification at the root of the new systematics, the sheer brilliance of biometrical population genetics in the hands of people like Wright and Mayr, be extended to the rest of biology? Agreement had emerged on the epistemological roots of quantification as a source of rhetoric in physics, where statistics was fully integrated via its role in formalizing Boltzmannian thermodynamics; and the role of similarly exact quantities in structural biology had just become visible with the discovery that DNA had the structure of a double helix (1953). Yet at the same time it was becoming generally acknowledged that the role of quantification in biology was *not* to be analogous to that of physics, the provenance of steadily more and more "exact" values, but instead went better as a steady accumulation of the understanding of *variation*, its models and its significance. The main purpose of the workshop at Altenberg in 2008 was to revisit this antinomy of variance versus variability, heterogeneity versus lawfulness, at a respectful distance in time and in computer power. For all that has changed over the ensuing 50 years in instrumentation and computation, the main style of quantitative inquiry for biologists remains this persistent contention between concepts of variance (variation around a meaningful mean) and concepts of heterogeneity. There is thus a direct line connecting Kuhn's ideas to those published in this book, and the connection passes directly through the conference at which Paul Weiss stated the paradox most succinctly in the bon-mot I have already quoted, the one about the identical twins.

3

Abduction and Strong Inference

Chapter 2 reviewed diverse aspects of consilience, the general principle by which agreement of numerical values or matching of numerical patterns is turned into evidence for scientific assertions. But except for one example in each of the preceding two chapters we have not yet dealt with the step even earlier at which potential agreements were made the focus of our attention: the step at which some "surprising fact C is observed" from which a novel hypothesis will drain the surprise. The present chapter closes the logical circle by a detailed examination of this process, the launch of an *abduction*. It will prove to rely as much on human psychology as on any formal properties of numbers and the symbolic reasoning they convey.

As in earlier chapters, the first section reviews one specific example, here the demonstration by the Intergovernmental Panel on Climate Change (IPCC, 2007) that the rise of average global temperature beginning around 1960 is very likely the result of human action. I pay some attention to intellectual history, to publication details, and to several associated ironies. Section 3.2 introduces Charles Sanders Peirce, 19th-century American philosopher, and then, intentionally climbing higher on the scale of abstraction, presents in some detail the general syllogism he discovered, of which the episode in the introductory section serves as a particularly shining, if anachronistic, example. You will see how closely the IPCC's rhetoric matches that of Peirce's logical discovery and how this same logic can be traced forward in the work of 20th-century philosopher-epistemologists of science such as Ludwik Fleck and Thomas Kuhn. Section 3.3 shows the same penetrance by a review of multiple disciplines instead of multiple disciples. Finally, in Section 3.4 I review how another aspect of human psychology, this one a typical failure of rationality, led the 19th-century geologist Thomas Chamberlin to craft a strategy of numerical reasoning that better manages the impact of new data in real professional contexts. This is the method of *strong inference*, which demands there be more than one hypothesis under consideration at any time, lest you misapprehend consistency with your own preferred explanation as disconfirmation of everybody else's. The chapter closes with a brief review of Part I as a whole. Following this review there is a brief Appendix touching on some

74　　　　　　　　　　　　　　3 Abduction and Strong Inference

developments in the rhetoric of climate change since the publication of the IPCC report that supplies the chapter's principal example.

3.1 Example: Global Warming Is Anthropogenic

Here's an interesting paragraph from page 10 of *Climate Change 2007: the Physical Science Basis*, volume 1 of the three-volume *Fourth Assessment Report* (FAR) from the Intergovernmental Panel on Climate Change (IPCC).[1] The series of reports is specifically intended as "translational science," the statement of a consensus among professionals in terms of its implications for people and societies, and page 10 is from an introductory 18-page packet, in unusually large type, entitled *Summary for Policymakers*. (Such a summary would seem mandatory, as the full volume is one thousand double-column pages, and it is just the first volume out of three.) In the text extracts here, "TAR" stands for "Third Assessment Report," the summary published in 2001 that the 2007 volume was superseding. My selection was originally in an unusually large font set in red against an off-white background:

> Most of the observed increase in global average temperature since the mid-20th century is *very likely* due to the observed increase in anthropogenic greenhouse gas concentrations.[12] This is an advance since the TAR's conclusion that "most of the observed warming over the last 50 years is *likely* to have been due to the increase in greenhouse gas concentrations." Discernible human influences now extend to other aspects of climate, including ocean warming, continental-average temperatures, temperature extremes and wind patterns.

Footnote 12 (the IPCC's own, not mine) reads, "Consideration of remaining uncertainty is based on current methodologies." The italics are the original authors', not mine. *"Likely"* and *"very likely"* had been defined a few pages earlier in a footnote itself worth transcribing here:

> In this Summary for Policymakers, the following terms have been used to indicate the assessed likelihood, using expert judgement, of an outcome or a result: *Virtually certain* > 99% probability of occurrence, *Extremely likely* > 95%, *Very likely* > 90%, *Likely* > 66%, *More likely than not* > 50%, *Unlikely* < 33%, *Very unlikely* < 10%, *Extremely unlikely* < 5% (see Box TS.1 for more details).

"Anthropogenic" means "due to humans," so the claim here is that the increase in worldwide average temperatures is due to human activities as they affect atmospheric carbon dioxide, methane, and nitrous oxide. Or, more subtly, that the extent to which

---

[1] The IPCC report of 2007 is its fourth, there having been earlier summaries in 1990, 1996, and 2001. A fifth IPCC report, likewise in three fat volumes, is expected to be published in 2014, just after my book. Drafts of this major revision have been circulating on the Internet since early 2013, but neither the text nor the figures are scheduled to be finalized until late 2013. As far as I can tell from a perusal of the new draft, none of the arguments in this chapter need to be altered in any nontrivial way, although naturally all the time series have been updated to include seven more years of data.

the experts trust this claim has been increasing lately. The news in red is, precisely, a change in the italicized term, from "likely" to "very likely," meaning from below to above the 90% level of degree of belief. (In other words, the "likelihood" of those italicized words is not the likelihood of an event, but of a scientific hypothesis: quite a different matter.) The page from which I quoted begins with a heading (in big white Helvetica type on a blue background) that reads "Understanding and Attributing Climate Change," and correspondingly the likelihoods to which the italicized words are referring deal with the "attribution" of climate change, meaning the identification of "increase in anthropogenic greenhouse gas concentrations" as the cause of "most of the observed increase" in global temperatures.

If we turn to the IPCC's "Box TS.1," pages 22–23, which is a page and a half of boxed text inside the 73-page "Technical Summary," we are referred in turn to Section 6 of Chapter 1 of that document, "The IPCC assessments of climate change and uncertainties." The box is part of a commentary on methodology, without any footnotes to findings or to outside sources (such as precursors of this book), that differentiates between "confidence in scientific understanding" and "likelihoods of specific results." The latter conceptualization "allows authors to express high confidence that an event is extremely unlikely (e.g., rolling a die twice and getting a six both times), as well as high confidence that an event is about as likely as not (e.g., a tossed coin coming up heads)." That is not our topic here – the IPCC's issue is the probability of a hypothesis, not the cause of an event.

The most nuanced discussion of this deeper topic, "confidence in scientific understanding," is in IPCC's Chapter 2, pages 201–202, where the authors tabulate 16 potential contributors to global temperature change, from long-lived greenhouse gases and solar energy output to aircraft contrails, and for each the strength of evidence for the numerical claim of an effect on global temperature, the level of scientific consensus about the effect, the certainties and uncertainties of the received understanding of the effect, and, finally, a net "level of scientific understanding," which is rated "high" only for the greenhouse gases. For these gases, the "certainties" are concentrations from the recent past and the present, the "uncertainties" the preindustrial concentrations and the vertical distribution (from Earth's surface to the top of the atmosphere), and the basis for uncertainties comprises only the differences between models and between data sets. By comparison, the factors characterized by a "low level" of scientific understanding, which are the majority of the sixteen, mostly are characterized by a difficulty in separating anthropogenic from natural effects and may additionally manifest such problems as absence of data before 1990 (stratospheric methane), uncertainties of proxy data (solar irradiance), or confusion about atmospheric feedback mechanisms (volcanic aerosols).

Nevertheless, the conclusion conveyed by the IPCC's quoted text is defensible in spite of *all* these scientific uncertainties – that is the whole point of this part of the IPCC discussion. It is clear that the topic here is not the numerical content of the data, but their qualitative (verbal) meaning and, ultimately, their human (specifically, social-political) implications. The authors insist on converting quantitative statements into qualitative ones that are easier to export from the seminar room into the parliament – but these are not "sound bites," rather, simplified (yet not simplistic) versions of sound reasoning. The IPCC here offers us a superb example of the core rhetorical stratagem in this process, the conversion of measurements into differential support for one hypothesis over others. The present chapter covers the style and rhetoric of conversions like this one along with the ways that they relate to core concerns in the philosophy of science.

The Nested Claims: "Observed Increases" in Global Average Temperature and in Greenhouse Gases

Let us start simply, with the propositions that the IPCC grammatically nested within its main claim: "the observed increase in global average temperatures since the mid-20th century" and "the observed increase in anthropogenic greenhouse gas concentrations." These are evidently subordinated to the topic at issue, which is the connection between them, and so we should expect that, considered numerically, the contentiousness of these two clauses should be less than that of the relevance of the one for the other. Rhetorically, this subordination seems pretty clear in the relative complexity of discussions of the corresponding propositions, which, for our present purposes, can be reduced to issues for which the Interocular Trauma Test (ITT) is adequate. This is an intuitively very accessible statistical test whose modern version owes to a comment of Edwards et al. (1963) – for one canonical text, see the first epigraph of this book. Regarding increases in greenhouse gas concentrations, for instance, it is probably enough just to examine the celebrated *Keeling curve* of atmospheric carbon dioxide over Hawaii for 50 years, at left in my Figure 3.1, in a context of the same concentration as reconstructed over 10,000 years, Figure 3.1 (right). The word "unprecedented" certainly seems appropriate for what has been happening to atmospheric carbon dioxide lately. (Note how an important consilience has been explicitly embedded in this figure: the circannual pattern of carbon dioxide concentration, obvious in the sawteeth of the large-scale time series, which, when expanded graphically as in the inset at left, clearly aligns with the concentration of northern hemisphere photosynthesis in the spring and summer.)

The other subordinated fact, increases in global temperature over the same time period, is just as conducive to the ITT. It has certainly been warm lately (Figure 3.2), by comparison not just to conditions during our youth but to general conditions since, say, the Dark Ages (Figure 3.3). "Average Northern Hemisphere temperatures during the second half of the 20th century," the IPCC notes on page 9, "were very likely higher than during any other 50-year period in the last 500 years and likely the highest in at least the past 1,300 years."

> Associated with claims of this genre is a rhetorical irony all its own. The IPCC text goes on to note: "Some recent studies indicate greater variability in Northern Hemisphere temperatures than suggested in the TAR, particularly finding that cooler periods existed in the 12th to 14th, 17th and 19th centuries. Warmer periods prior to the 20th century are within the uncertainty range given in the TAR." That TAR commentary was based in part on a primary publication by Mann et al. (1998) stating, unwisely in retrospect, that "the 1990s are likely the warmest decade, and 1998 the warmest year, in at least a millennium." Uncertainties of historic reconstruction are too great to support envelope

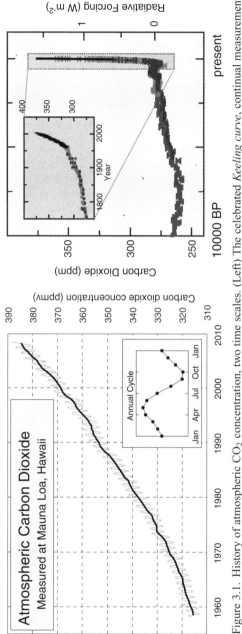

Figure 3.1. History of atmospheric CO_2 concentration, two time scales. (Left) The celebrated *Keeling curve*, continual measurements using the same method at the same locale on Mauna Loa, Hawaii, from 1958 right through today. From http://en.wikipedia.org/wiki/File:Mauna_Loa_Carbon_Dioxide-en.svg. (Right) The same, now over 10,000 years, mostly as reconstructed from a variety of physicochemical traces (IPCC, 2007, p. 3, with permission). The Keeling curve at left is the component drawn at far right here in the darkest line weight. In the 1PCC original publication, this curve is red. The boxed zoom-in indicates the part of this development corresponding to the interval from the Industrial Revolution to the present. The authors' point is that the last 50 years show a rate of increase of CO_2 density extreme even by historical comparisons of a century or so earlier.

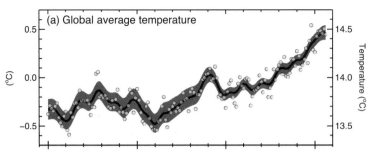

Figure 3.2. Global average temperatures, 1850–2000, and their smoothing over 10-year windows (IPCC, 2007, p. 6, with permission).

statements like that. The irony is that none of the actual implications of the IPCC's work depend on the truth (or not) of assertions of that type, as their great subject is trends, not individual values – hypotheses, not events. The principal job of rhetoric about extremes (I mean the pattern, not the numerical value per se) is not to convince the serious scientist but rather to confute the naïve skeptic. Scores for single years are no more central than scores for single regions, which, as we shall see presently, are strongly in keeping with generalizations of this type without necessarily agreeing about extremeness in any single instance (year or location).

In other words, the IPCC did itself a disservice in the TAR when it reverted to a rhetoric of extremes when all that was really needed was the consilience of trends I have been emphasizing, and it now says, in the FAR, that one should place "little confidence" in the statement of Mann's I quoted above.[2] For the FAR's own (quite technical) discussion of this controversy, see Section 6.6.1, pages 466–474 of IPCC, 2007. The subtlety of the rhetorical problem is captured well by their Figure 6.10 reproduced as my Figure 3.3. Uncertainties of those historical temperatures prior to 1900 are far greater than anything shown in Figure 3.2, the annual data for the last 150 years. The unusually fine new graphic in Figure 3.3 shows the central tendency of the historical reconstructions (the overlap of all the indirect estimates) by darkness of the shading in the lower panel. For the controversy about that peak around the year 1000, see Brumfiel (2006), or look under "hockey stick controversy" in Wikipedia (so named because this graph looks a bit like a hockey stick).

What is the Evidence?

Extremes are "events" in the sense of the IPCC's Box TS.1 discussed earlier. We need to turn now to an issue of "scientific understanding," namely, the nexus between the gases and the warming. This connection is principally borne in one single figure, of quite clever design, that appears in the IPCC Technical Summary (as Figure TS.23) and in situ in Chapter 9 of the document as Figure 9.5, where it bears the somewhat unwieldy title "Comparison between global mean surface temperature anomalies from observations and AOGCM simulations forced with (a) both anthropogenic

---

[2] It is remarkable how intemperate critiques can be when they do not understand the rhetorical basis of verbal summaries like these. Montford (2010) is not atypical of the counternarrative genre.

3.1 Example: Global Warming Is Anthropogenic

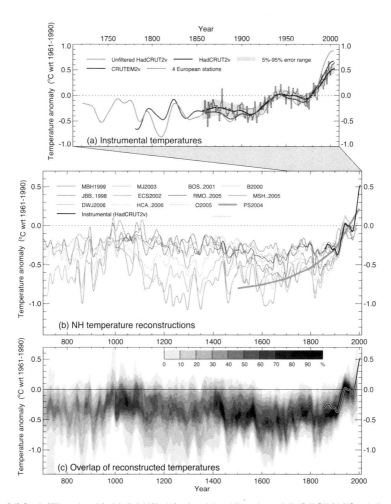

Figure 6.10. *Records of NH temperature variation during the last 1.3 kyr. (a) Annual mean instrumental temperature records, identified in Table 6.1. (b) Reconstructions using multiple climate proxy records, identified in Table 6.1, including three records (JBB..1998, MBH..1999 and BOS..2001) shown in the TAR, and the HadCRUT2v instrumental temperature record in black. (c) Overlap of the published multi-decadal time scale uncertainty ranges of all temperature reconstructions identified in Table 6.1 (except for RMO..2005 and PS2004), with temperatures within ±1 standard error (SE) of a reconstruction 'scoring' 10%, and regions within the 5 to 95% range 'scoring' 5% (the maximum 100% is obtained only for temperatures that fall within ±1 SE of all 10 reconstructions). The HadCRUT2v instrumental temperature record is shown in black. All series have been smoothed with a Gaussian-weighted filter to remove fluctuations on time scales less than 30 years; smoothed values are obtained up to both ends of each record by extending the records with the mean of the adjacent existing values. All temperatures represent anomalies (°C) from the 1961 to 1990 mean.*

Figure 3.3. Global average temperature again, this time over a 1200-year window. Note the difference between the "instrumental measurements" (upper frame) and the "reconstructions" (middle frame). The bottom frame renders the data from the middle as a superposition of ranges of the separate estimates – a brand-new kind of graphic computed especially for their Summary for Policymakers. I have included the IPCC's own caption to this figure as a hint of the complexity of the underlying information processing that went into its rendering. (From IPCC, 2007, p. 467, with permission.)

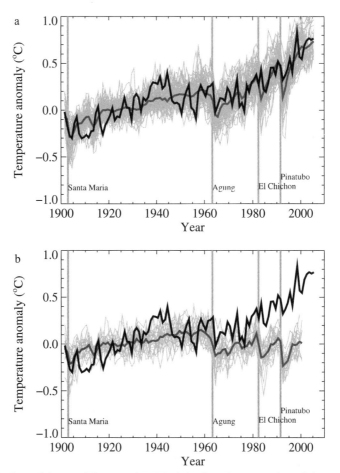

Figure 3.4. Central figure of the central IPCC abduction: demonstration of the perfect separation between two ensembles of partial differential equation computations corresponding to two classes of large-scale models, those including (upper panel) or not including (lower panel) the anthropogenic forcing functions of the greenhouse gases. (From IPCC, 2007, p. 684, with permission.) This figure is even more powerful when rendered in contrasting colors as in Plate 3.4.

and natural forcings and (b) natural forcings only." (AOGCM, atmospheric-oceanic general circulation models.) I reproduce this display as my Figure 3.4.

The black line is the same in both panels: the observed global mean surface temperature since 1900. A sudden upward trend appears to have begun about 1960. The shaded region of the bottom figure spans *all* the simulations assembled by the authors of this part of the study that include only what are called the "natural forcings" (mainly solar irradiation and volcanic eruptions). The ensemble shows a strong effect of the volcanic eruptions (the named vertical lines in the graph) but otherwise obviously fails to explain the anomaly – by the turn of the present century the heavy gray line in the bottom panel, average of all these simulations, misses the solid rising line completely. In the upper panel, by contrast, results from the ensemble of the models that include the anthropogenic components like greenhouse gases, ozone and soot neatly bracket the observed trend throughout the century, both before and after the acceleration of 1960, and their mean (the heavy gray line of this upper panel) tracks the real data (the heavy black line, which we have already seen in Figure 3.3) quite closely.

It further strengthens one's confidence in the inference here that the abduction appears to apply to every region of the planet separately. That is the subject of the multipanel Figure 3.5. (The IPCC offers two versions of this subaggregation. One is by continent; this one, by smaller regions.) The consilience is strengthened to the extent that the same pattern of features is seen in all or nearly all of the little frames separately. Here there are at least three such features: the match of the black line to the visual center of the lighter band, the clear separation of lighter and darker bands, and the emergence of that separation from a divergence of trend lines beginning around the year 1960.

> This exemplifies the principle of "small multiples" highlighted by Tufte (1983). Included is a subordinate abduction all its own: however likely or unlikely the observed net pattern, its surprise value can only increase if we see it over and over in otherwise arbitrary subsets like these. Yet embedded in this phrasing is a logical fallacy that has often entangled us in remarkable difficulties (see Perrow, 1984). The tacit assumption that *must* be checked is that the patterns observed in the subsets are "independent" in a probabilistic sense or at least in a logical sense: they have no common cause other than the claimed scientific process that is the topic of discussion. (Otherwise there would surely be much less surprise.) In the context of Figure 3.5, it would be as if there was something systematically affecting a measurement bias in all of the world's thermometers, or all of these regional coverage samples, at the same time. Such competing hypotheses must be taken very seriously if the apparent replication of the pattern finding is to count as actual additional support for the novel hypothesis that purports to make it "a matter of course."

How is this supposed to persuade us of something? We learn a good deal about the passage from numerical measurement to numerical inference by examining the IPCC authors' argument in Section 9.4, "Understanding of air temperature change during the industrial era," pages 683–705 of IPCC, 2007. This is the section from which my Figure 3.4 is actually taken – the data argument driving the "summary for policymakers" about the anthropogenicity of global temperature change – and the text in its vicinity (pages 683–684) explains the import of the figure in the greatest detail.

The authors begin with a reassurance: it is that "temperatures are continuing to warm," not that any historically unique process has begun during the six years since the TAR. "The annual global mean temperature for every year since the TAR has been among the 10 warmest years since the beginning of the instrumental record," and "the rate of warming over the last 50 years is almost double that over the last 100 years." (Although nothing actually follows from that for the theme of anthropogenesis, it

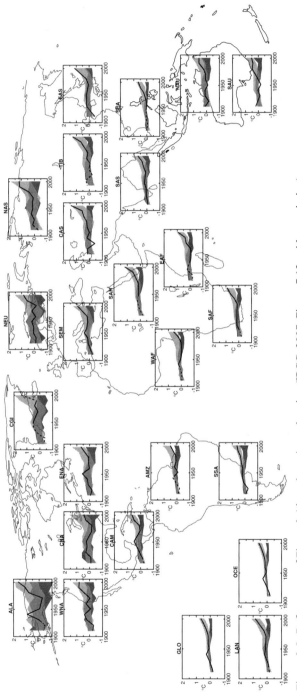

Figure 3.5. Refinements of Figure 3.4 by subcontinental region (IPCC, 2007, Figure 7.12, with permission).

3.1 Example: Global Warming Is Anthropogenic   83

is pertinent to the nested claim, the secular increase in global mean temperature, of Figures 3.4 and 3.5.)

> Simulations that incorporate anthropogenic forcings, including increasing greenhouse gas concentrations and the effects of aerosols, and that also incorporate natural external forcings provide a consistent explanation of the observed temperature record, whereas simulations that include only natural forcings do not simulate the warming observed over the last three decades... There is a clear separation [in Figure 3.4] between the simulations with anthropogenic forcing and those without.... The fact that climate models are only able to reproduce observed global mean temperature changes over the 20th century when they include anthropogenic forcings, and that they fail to do when they exclude anthropogenic forcings, is evidence for the influence of humans on global climate.

It is *this* reasoning that evidently justifies the authors' investment in all those simulations (and the summaries, and the graphics). Over the course of the several examples in Chapters 4 through 8 I will be arguing that it justifies most of the rest of numerical inference in the statistical natural and social sciences as well, to the extent that those inferences can be justified at all.

In keeping with the spacious stage on which they are reporting, the IPCC authors continue to recount additional observations consistent with this main numerical claim. For instance, "global mean temperature has not increased smoothly since 1900 as would be expected if it were influenced only by forcing from increased greenhouse gas concentrations" – there are also natural forcing functions, like volcanoes (see also Appendix 3.A in this chapter). The observed trend over the 20th century shows warming nearly everywhere, a trend also seen in the joint natural-anthropogenic simulations of Figure 3.5. Only the models with anthropogenic terms reproduce the trends by latitude. In sum, "the late 20th century warming is much more likely to be anthropogenic than natural in origin" – yet another version of the text from their page 10. This logic is the logic of *abduction*, C. S. Peirce's original contribution to the logic of science, to be reviewed at length in the next section.

The foregoing logic is enormously more powerful than any merely statistical inference from a conventional significance test for the effect size of the greenhouse gas concentrations.[3] The IPCC's point is not only how unlikely the data are on the hypothesis of *no* effect (though that implausibility – the "surprise" of the data – has to be a part of the reasoning), but also, and principally, how very reasonable are the findings given the assumption of *some* effect. Notice, for instance, that the models do not include a change-point around the year 1960, but the ensemble of orange simulations of Figure 3.4 finds it anyway. The change-point was not in the model, but in the data driving the simulations, and it passes through from forcing factor data to simulated outcome data because the equations are valid and have the right coefficients. This may be an example of what Lipton (2004) calls the "loveliness" of an explanation, which some of us would weigh even more highly than its "likeliness," its explanatory power. (There is, in other words, an aesthetic aspect to abduction,

---

[3] There is an extended discussion of this point at Section L4.3.4.

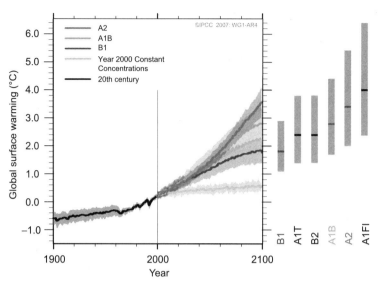

Figure SPM.5. *Solid lines are multi-model global averages of surface warming (relative to 1980–1999) for the scenarios A2, A1B and B1, shown as continuations of the 20th century simulations. Shading denotes the ±1 standard deviation range of individual model annual averages. The orange line is for the experiment where concentrations were held constant at year 2000 values. The grey bars at right indicate the best estimate (solid line within each bar) and the **likely** range assessed for the six SRES marker scenarios. The assessment of the best estimate and **likely** ranges in the grey bars includes the AOGCMs in the left part of the figure, as well as results from a hierarchy of independent models and observational constraints. {Figures 10.4 and 10.29}*

Figure 3.6. Multimodel averages and assessed ranges for mean global surface warming under various policy scenarios. (From IPCC, 2007, p. 14, with permission.)

as there is to most of the dominant approaches in the natural sciences. As Wigner [1960], p. 7 states, quoting Einstein, "the only physical theories we are willing to accept are the beautiful ones.")

But the impetus for this rhetoric, centered as it is about the simple statement that global temperature is sensitive to greenhouse gases (rather than any more technical, more numerical summary of these simulations or the underlying data driving them), has already been made plain in the Summary for Policymakers section nearly 700 pages earlier. We need to accept the main finding of the IPCC's Chapter 9 – the dominance of the anthropogenic effect on the observed global temperature trend – because this is the finding that drives the volume's principal advisory to "policymakers," namely, the dependence of predicted climate change on the historical record of atmospheric consequences of our collective human behavior (Figure 3.6). This is not presented in the form of a "warning" – stop producing greenhouse gases or horrible things will happen to you (the rhetoric of, for instance, Al Gore's *An Inconvenient Truth* documentary film and best-selling companion volume). The authors of this first volume of the Fourth Assessment Report are, rather, producing forecasts that, far from being extrapolations, state a dependence of the expected farther future on political decisions in the present and the near future. As the figure caption indicates, these are "scenarios," not forecasts, that fall under the heading of "unpredictability" rather than either "value uncertainty," the domain of standard errors, or "structural uncertainty," the domain of ignorance about actual system dynamics. I return to this issue of the practical context of numerical inferences in the closing section of this chapter and in Appendix 3.A.

Each scenario in this diagram, the IPCC explains on page 18, is an extended script describing a plausible future sociopolitical world. A1 represents rapid economic growth in a population peaking in 2050, either fossil-intensive (A1FI), nonfossil in energy sources (A1T), or balanced (A1B). A2 stands for a heterogeneous world with regional economic development and fragmented technological change. B1 modifies A1 in the direction of a service and information economy with clean, resource-efficient technologies. B2 emphasizes social and environmental sustainability, intermediate economic development, and slower technological change. There is also a curve at the bottom for "no increase in rate of greenhouse gas production," which is not actually a feasible future.

> In my opinion, this IPCC volume 1 of 2007 is the finest sustained system of statistical graphics ever published. In its scope and creativity its only match might be the Isotype system invented by Otto Neurath around 1930 for mass communication of social statistics about Vienna, Europe, and the world in graphical form. See Vossoughian, 2011. For a general aesthetic of statistical graphics, see the work of Edward Tufte, particularly *The Visual Display of Quantitative Information*, 1983, the first and still the most delightful and insightful volume in this privately published series.

3.2 That Hypothesis Wins That Is the Only One That Fits the Data

The abductive method the IPCC has elegantly exemplified for us is quite general – as ubiquitous as its complement, consilience. Credible numerical inferences not only need to match observed data patterns but also need to match them *uniquely*. That consilience underlies the IPCC's reasoning is abundantly clear – the simulations match one another, one set matches the data both globally and regionally, and also (though I have no space to review this matter here) the equations driving the simulations are hugely consilient with everything we know about the physics and the computer science of large-scale energy flows in the oceans and the atmosphere (those AOGCMs) with coefficients deriving from laboratory instruments, radiosondes, and every level of instrumentation in between. Yes, this IPCC report meets the standards of my Chapter 2. But apparently the interpretation of Figure 3.4 involves more, and that is my topic here.

For the next few pages most of the words will not be mine. The subject is an innovation in otherwise classical logic that has occupied better thinkers than I for well more than a century: the way we get from perceptions to hypotheses (or "hunches," or "guesses"), a passage first systematically explored by C. S. Peirce, and the way such cognitive progressions were justified then and are justified now.

A brief note on Charles Sanders Peirce. This greatest of all American philosophers was born in 1839, briefly ran the U.S. Geodetic Survey where he did fundamental work on the measurement of Earth's gravity, offended nearly everyone who tried to befriend him, and died in poverty in 1914 leaving tens of thousands of pages of unpublished manuscripts that are still being pored over a hundred years later. He had no gift for clear explication (for example, one of his sustained efforts is devoted to the introduction of three fundamental concepts named "Firstness," "Secondness," and "Thirdness," of which the method of abduction that concerns us is an instance) yet he managed to intuit the fundamental philosophical approach, *pragmatism*, that remains the unique American contribution to epistemology, and even to convince the equally great William James to share in the task of bringing it to the attention of a more general intelligentsia. For a biography, see Brent (1998). A small fraction of his papers and manuscripts have been issued in thick volumes as Peirce, *Collected Papers*, 1931–1958, and a smaller selection have been republished as Houser and Kloesel (1992, 1998). The word "Peirce" looks misspelled, and indeed the letter *i* is silent; the name is pronounced as if it were spelled "Perse" or "Purse" instead.

Peirce's notion of abduction. Early in his career, Peirce suspected the validity of a form of inference, *abduction*, which is distinctly different, logically speaking, from either of the standard forms with which you are probably acquainted already: *induction* (generalizing from multiple instances), and *deduction* (deriving the logical consequences of what is assumed known already). Scientific reasoning embraces all three – abduction, deduction, and induction – but only abduction creates new knowledge, new understanding.

> The first starting of a hypothesis and the entertaining of it, whether as a simple interrogation or with any degree of confidence, is an inferential step which I propose to call *abduction* [or *retroduction*, see later]. I call all such inference by the peculiar name, *abduction*, because its legitimacy depends upon altogether different principles from those of other kinds of inference. The form of inference is this:
>
> - The surprising fact, C, is observed;
> - But if A were true, C would be a matter of course,
> - Hence, there is reason to suspect that A is true.
>
> – Peirce, 1903 but probably mid-1870s

The verb form is "abduce," past tense, "abduced," as distinct from another verb, "abduct," that unfortunately has the same nominalization. Compare deduct-deduction, deduce-deduction. Adjective, "abductive."

Precursors and later formulations. There is a precursor of Peirce's syllogism in the classic Greek rhetorical method of *apagoge*, "an indirect argument which proves a thing by showing the impossibility or absurdity of the contrary." (But abduction is not proof, which requires additional evidence.) A more recent precursor is John Stuart Mill's suggestion about how to proceed in the presence of an initially chaotic and confusing situation, such as the maternal mortality that used to affect maternity wards through the mid-19th century. In the so-called "contrastive method," given an outcome of interest, arising apparently haphazardly, the investigator's task is to find a factor that is common to (most of) its occurrences and absent from (most of) its nonoccurrences. From this pattern an abduction follows that the factor thus uncovered is partially responsible for the difference in typical outcomes. (The abduction is the stronger the more similar are the resulting subsets on all the *other* candidates for an explanation.) Then systematically control that factor, and examine whether you have controlled the outcome of interest as well. In the case of those maternity wards,

3.2 That Hypothesis Wins That Is the Only One That Fits the Data

Dr. Ignaz Semmelweis noticed that the mothers who died of puerperal fever typically had been examined by physicians who had been touching corpses, and suggested that all physicians should wash their hands with chlorinated lime before touching any living patient. In this he was entirely correct, even though his contemporaries did not do him the honor of conceding that fact.

Or (Josephson, 1996, p. 5):

D is a collection of data (facts, observations, givens).
H explains D (would, if true, explain D).
No other hypothesis can explain D as well as H does.
Therefore, H is probably true.

Or this, from Peter Lipton's book of 2004 on "inference to the best explanation," which apparently does not ever use the word "abduction" but defines it anyway (well, two-thirds of it; the first clause, about surprise, is overlooked), thus:

> According to the model of Inference to the Best Explanation, our explanatory considerations guide our inferences. Beginning with the evidence available to us, we infer what would, if true, provide the best explanation of that evidence. These sorts of explanatory inferences are extremely common. The sleuth infers that the butler did it, since this is the best explanation of the evidence before him. The doctor infers that his patient has measles, since this is the best explanation of the symptoms. The astronomer infers the existence and motion of Neptune, since that is the best explanation of the observed perturbations of Uranus.... Given our data and our background beliefs, we infer what would, if true, provide the best of the competing explanations we can generate of those data (so long as the best is good enough for us to make any inference at all). (Lipton, 2004, pp. 1, 56)

George Pólya (Pólya György, 1887–1985) was a Hungarian-born mathematician and mathematics educator. His book *How to Solve It* (1945) lay at the core of my training for mathematics problem competitions in Michigan in the early 1960s. His later, more formal treatise *Mathematics and Plausible Reasoning* of 1954 is a neglected masterpiece dealing with the central role of heuristics – of *guessing*, really – in the successful approaches to problem-solving all over the natural sciences that exploit mathematics as well as in intermediate and advanced mathematics itself. The following alternate versions of abduction are all taken from Pólya, 1954:

A implies B
B very improbable in itself
B true
Therefore, A very much more credible

or

A implies B
B without A hardly credible
B true
Therefore, A very much more credible

[compare: "A implies B; B almost certain anyway; B true; **Therefore**, A very little more credible" – this is, alas, a prototype of the usual null-hypothesis statistical significance test, see Section L4.3.4]

or

B with A readily credible
B without A less readily credible
B true
Therefore, A more credible

or even

A implies B_{n+1}
B_{n+1} very different from the formerly verified consequences B_1, B_2, ... B_n of A
B_{n+1} true
Therefore, A much more credible.

What these all have in common, Pólya notes in italics, page 2:9, is that *"the verification of a consequence counts more or less according as the consequence is less or more probable in itself."* In other words, "the verification of the most surprising consequence is the most convincing."[4]

Abduction in the IPCC Report. Here is how this abstracted, formalized logic maps onto the reasoning actually put forward by the IPCC as reviewed in Section 3.1:

- By the right end of the horizontal axis in Figure 3.4, the years around the turn of the present century, we see a perfect separation between the outputs of two types of simulations. Both types incorporate most of what we know about the natural background factors affecting global temperature (volcanoes, solar output, etc.), but one type (in the upper panel) also includes trends in greenhouse gases such as that for CO_2 already displayed in Figure 3.1. Moreover, the average (heavy gray line) of the simulations summarized in the upper panel tracks the actual global mean temperature data remarkably closely, whereas the average for the simulations those omitting greenhouse and other anthropogenic factors, those summarized in the lower panel, clearly does not. This pattern is genuinely startling – simulations do not ordinarily fit data remotely this well, and single factors rarely account for a sharp change of trend like this one, at just the right time and with the right slope.
- But if anthropogenic factors were actually responsible for (the bulk of) the observed temperature trend, the agreement shown here would be expected as a

---

[4] This aperçu of Pólya's is particularly interesting insofar as his actual domain of application is mathematical proof, for which the notion of "data" is not really a propos. As would Imre Lakatos (1976) a few years later, Pólya is thinking of mathematics as if it were an experimental science, with generalizations serving as the "theories" that summarize the "facts" that represent individual cases. For an abduced hypothesis the mathematician will usually reserve the word "conjecture," and of course the mathematician does not infer theorems by either deduction or induction; she *proves* them. As Pólya says, "Mathematical facts are first guessed and then proved."

3.2 That Hypothesis Wins That Is the Only One That Fits the Data

matter of course. The simulations omitting that central causal factor should then naturally fail to fit the data wherever that factor is particularly salient, whereas simulations including that factor should come close just to the extent that that factor was being estimated and modeled correctly at all times, both when that factor is salient and when it is not.
- Hence, this diagram, combining information from hundreds of simulations based on partial differential equations that deal with data over more than a hundred years, provides a startlingly powerful *quantitative* argument for the *qualitative* claim that human activities, as reflected in trends of greenhouse gases, are the principal cause of global warming.

This claim by the IPCC authors was promptly submitted to a variety of formal tests (Section 9.4.1.4) that actually detected a *moderation* of the greenhouse gas forcings by other effects both anthropogenic and natural, such that the fifth (low-end) percentile of greenhouse-attributable warming is greater than the warming actually observed over the period 1950–2000. This is not a criticism of the explanation arrived at, but a subdivision into greenhouse effects (positive factors) against other anthropogenic factors (e.g., total aerosols, land use effects on surface albedo) that tend to lower temperatures rather than raise them. The detection of the anthropogenic term is highly robust over a range of formal inference methods that replace the ITT and the abductive wording by specific algebraic formulas. For instance, 10-year extrapolations converge on observed 10-year changes across alternate modes of extrapolation only if based on anthropogenic factors, not otherwise.[5] The effect of greenhouse gases is dominant not only over aerosols, another anthropogenic factor, but also over fluctuations in solar irradiance, a natural forcing function. The effects that are specific to anthropogenic factors alone, such as contrasts between land and ocean or the latitude gradient in the Northern Hemisphere, closely match "almost all" of the warming observed from 1946 to 1995. As I already mentioned, the region-by-region models (Figure 3.5) generally show the same features, such as separation of upper and lower ensembles and the match of the upper ensemble average curve to the data, that were already reviewed at the global level in Figure 3.4. This ability to "simulate many features of the observed temperature changes and variability at continental and sub-continental scales...provides stronger evidence of human influence on climate than was available to the TAR" (p. 694). As always, that there is far better agreement between simulations and observations when the models include anthropogenic forcings is interpreted as detection of an "anthropogenic warming signal," which might thus be reported regionally instead of or in addition to the global effect that is necessarily somewhat more abstract.

Two text boxes indicate the limits of this rhetoric. FAQ 9.1 asks, "Can individual extreme events be explained by greenhouse warming?" and gently steers the reader over to a language of "risks," such as very hot summers, rather than explanations of

---

[5] This kind of examination of a causal claim, in terms of changes across time, is sometimes called an investigation of "Granger causality."

unique events. Switzerland in 2003, for example, had a mean temperature more than 5 standard deviations from its historical average, an event exceeded fewer than once in 30 million trials on a Gaussian hypothesis: clearly something is changing in the system controlling Swiss summer temperatures. But a subsequent text box, FAQ 9.2, asks, "Can the warming of the 20th century be explained by natural variability?" and answers, quite patiently, that it cannot – this is the plain meaning of Figures 3.4 and 3.5, the global and the regional abductions we are pursuing. The difference between these two senses of the word "explain," the historically singular and the systematic, is made as clear as any non-didactic text could make it. Regarding the latter trope, it is not only the 20th-century data that are relevant, but those of earlier eras as well – the models omitting the anthropogenic factors cannot reproduce the late 20th-century trend, but they *can* fit the trends prior to our time:

> Confidence in these estimates is increased because prior to the industrial era, much of the variation they show in Northern Hemisphere average temperatures can be explained by episodic cooling caused by large volcanic eruptions and by changes in the sun's output, [while] the remaining variation is generally consistent with the variability simulated by climate models in the absence of [both] natural and human-induced external factors.... [But] the estimated climate variability caused by natural factors is small compared to the strong 20th-century warming. (1PCC, 2007, p. 703)

The upshot of all this detail is to buttress the abduction of Figure 3.4 against challenges from other hypotheses or methodological criticisms. The criticisms would work not on clause 2 of the abduction, the assertion that the hypothesis under scrutiny (anthropogenic factors as the dominant cause of global warming) would lead to the observed data "as a matter of course," but on clause 1, the statement that the data are surprising on any other model. The supplementary information we have been reviewing makes the surprise, so to speak, even greater: the models fit the data in every region, the residual variance from the model is constant over a millennium, and so forth. Consistency of the data with the anthropogenic-and-natural-forcing models is no longer the issue; it is the inconsistency with everybody else's models that benefits from this repeated buttressing.

In connection with this purpose, page 14 of Josephson (1996) supplies some helpful notes. The judgment of likelihood associated with an abductive conclusion should depend on the following considerations: how decisively A surpasses the alternatives; how good A is by itself, independently of considering the alternatives; judgments of the reliability of the data; how thorough the search was for alternative explanations; pragmatic considerations, such as the costs of being wrong or the benefits of being right; how strong the need is to come to a conclusion at all, especially considering the possibility of seeking further evidence.[6] Taken together, these sound like advisories to some algorithm or notes on a software package, and in fact the topic of abduction is most commonly encountered these days in the literature of artificial

---

[6] Earlier, Peirce, 1908 (p. 321): "Retroduction does not afford security. The hypothesis must be tested." Or, Peirce 1901 (p. 357): "A hypothesis ought, at first, to be entertained interrogatively. Thereupon, it ought to be tested by experiment as far as practicable."

intelligence – computer-aided diagnosis, troubleshooting, and the like. (This is ironic, as Peirce insisted that all inference had to be a conscious process.) As Chapter 1 already noted, it is also found in the popular detective literature. See, for example, Sebeok and Umiker-Sebeok, 1979.

How does abduction come about? Abduction begins with a surprise, and reduces it to an obvious inference post hoc. This specific component, the role of surprise, is omitted from all the earlier approaches to scientific discovery from which Peirce was concerned to differentiate this one. The key, in my view, is its origin in human psychology. That is Peirce's view as well. In fragmentary notes dated from 1891 to 1903 (Buchler, 1940, pp. 302–305), he considers its origin in the "perceptual judgments" that we are constantly making, both consciously and unconsciously, whenever we are awake:

> All belief might involve expectation as its essence.... To say that a body is hard, or red, or heavy, or of a given weight, or has any other property, is to say that it is subject to law and therefore is a statement referring to the future.[7] ... Abductive inference shades into perceptual judgment without any sharp line of demarcation between them; or, in other words, our first premisses, the perceptual judgments, are to be regarded as an extreme case of abductive inferences, from which they differ in being absolutely beyond criticism. The abductive suggestion comes to us like a flash.[8] It is an act of *insight*, although of extremely fallible insight. It is true that the different elements of the hypothesis were in our minds before; but it is the idea of putting together what we had never before dreamed of putting together which flashes the new suggestion before our contemplation.
>
> On its side, the perceptive judgment is the result of a process, although of a process not sufficiently conscious to be controlled, or, to state it more truly, not controllable and therefore not fully conscious.... The form of the perceptual abduction is:
>
> A well-recognized kind of object, M, has for its ordinary predicates P_1, P_2, P_3, etc., indistinctly recognized.
>
> The suggesting object, S, has these same predicates, P_1, P_2, P_3, etc.
>
> Hence, S is of the kind M.
>
> The first premiss is not actually thought, though it is in the mind habitually. This, of itself, would not make the inference unconscious. But it is so because it is not recognized as an inference; the conclusion is accepted without our knowing how.

In other words, the typical competent adult human brain does this sort of thing automatically, "by itself," whether or not we are consciously focusing our attention on the underlying judgments.

Peirce always emphasized not only that logic was the key to understanding but that it was capable of being mastered by human beings. For instance, here, from an 1908

---

[7] This axiom does not extend to the applications of quantification outside of the exact sciences, such as those in psychology, wherever constructs (such as attitudes and other reports of internal status) are not thought subject to lawful perseveration in this way. But as Peirce notes in this same source, "my principles absolutely debar me from making the least use of psychology in logic."

[8] In Section E4.5 we hear Watson saying, "Suddenly I became aware" (of the structure that the double helix had to have).

article curiously entitled "A neglected argument for the reality of God," is a more extensive analysis emphasizing that origin:

> Every inquiry whatsoever takes its rise in the observation... of some surprising phenomenon, some experience which either disappoints an expectation, or breaks in upon some habit of expectation of the [investigator]; and each apparent exception to this rule only confirms it.... The inquiry begins with pondering these phenomena in all their aspects, in the search of some point of view whence the wonder shall be resolved. At length a conjecture arises that furnishes a possible Explanation, by which I mean a syllogism exhibiting the surprising fact as necessarily consequent upon the circumstances of its occurrence together with the truth of the credible conjecture, as premises. On account of this Explanation, the inquirer is led to regard his conjecture, or hypothesis, with favour. As I phrase it, he provisionally holds it to be "Plausible"; this acceptance ranges in different cases – and reasonably so – from a mere expression of it in the interrogative mood, as a question meriting attention and reply, up through all appraisals of Plausibility, to uncontrollable inclination to believe. The whole series of mental performances between the notice of the wonderful phenomenon and the acceptance of the hypothesis, during which the usually docile understanding seems to hold the bit between its teeth and to have us at its mercy – the search for pertinent circumstances and the laying hold of them, sometimes without our cognisance, the scrutiny of them, the dark labouring, the bursting out of the startling conjecture, the remarking of its smooth fitting to the anomaly, as it is turned back and forth like a key in a lock, and the final estimation of its Plausibility, I reckon as composing the First Stage of Inquiry. Its characteristic formula of reasoning I term Retroduction, i.e. reasoning from consequent to antecedent.[9] In one respect the designation seems inappropriate; for in most instances where conjecture mounts the high peaks of Plausibility – and is really most worthy of confidence – the inquirer is unable definitely to formulate just what the explained wonder is; or can only do so in the light of the hypothesis. (Peirce, 1908, pp. 320–321)

Peirce's argument is thus that our ability to do this – to guess at riddles correctly – is inexplicable, and this appearance of the Divine within our quotidian selves is actually that "neglected argument" to which the title of his essay refers. The argument is thus a transcendental one, exactly akin to Wigner's comment (already discussed in Chapter 2) about the miracle of applied mathematics or, for that matter, the existence of mathematical proofs involving hundreds or thousands of steps. Peirce continues,

> What sort of validity can be attributed to the first stage of inquiry? Observe that neither deduction nor induction contributes the smallest positive item to the final conclusion of the inquiry. They render the indefinite definite; deduction explicates; induction evaluates; that is all.... We are building a cantilever bridge of induction, held together by scientific struts and ties. Yet every plank of its advance is first laid by retroduction [abduction] alone, that is to say, by the spontaneous conjectures of instinctive reason; and neither deduction nor induction contributes a single new concept to the structure. (Peirce, 1908, pp. 323–324)

---

[9] That is, if A implies B, then the usual syllogism reasons from the truth of A to the truth of B; but abduction goes "backwards" (hence the prefix "retro-"), from the truth of B to the increased plausibility or likelihood of A.

3.2 That Hypothesis Wins That Is the Only One That Fits the Data

This psychological *aperçu* is not unique with Peirce. Kuhn, in a 1961 essay to which I am referring quite often over the course of this book, notes on pages 178–179:

> Current scientific practice always embraces countless discrepancies between theory and experiment. During the course of his career, every natural scientist again and again notices *and passes by* qualitative and quantitative anomalies that just conceivably might, if pursued, have resulted in fundamental discovery. Isolated discrepancies with this potential occur so regularly that no scientist could bring his research problems to a conclusion if he paused for many of them. In any case, experience has repeatedly shown that, in overwhelming proportion, these discrepancies disappear upon closer scrutiny. They may prove to be instrumental effects, or they may result from previously unnoticed approximations in the theory, or they may, simply and mysteriously, cease to occur when the experiment is repeated under slightly different conditions. [But] if the effect is particularly large when compared with well-established measures of 'reasonable agreement' applicable to similar problems, or if it seems to resemble other difficulties encountered repeatedly before, or if, for personal reasons, it intrigues the experimenter, then a special research project is likely to be dedicated to it. At this point the discrepancy will probably vanish through an adjustment of theory or apparatus.... But it may resist, and [scientists] may be forced to recognize that something has gone wrong, and their behavior as scientists will change accordingly.... Often crises are resolved by the discovery of a new natural phenomenon; occasionally their resolution demands a fundamental revision of existing theory.... It is probably the ability to recognize a significant anomaly against the background of current theory that most distinguishes the successful victim of an 'accident' from those of his contemporaries who passed the same phenomenon by. (Italics Kuhn's.)

This is the same argument, with a twist: that the facts "resist," implying that reality is *acting* to thwart a potentially false or misleading insight. See also Pickering (1995). This attitude, in turn, echoes Ludwik Fleck's great *Genesis and Development of a Scientific Fact*, the pioneering work of modern science studies. There Fleck explains how the construal as a fact can be usefully pursued as a *sociological* notion. Epigraph 14 offers a précis of this argument in the form of an aphorism: "A scientific fact is a socially imposed constraint on speculative thought." The cognitive mechanism by which that constraint is imposed Fleck named the "thought-collective" *(Denkkollektiv)*, a concept that has turned into the contemporary notion of the frontier scientific community (Bauer, 1994). When he wrote this, Fleck was referring to the way that a prequantitative science, the study of syphilis, turned semiquantitative as a result of the transition to modern serology (the study of chemicals in the blood) including the Wassermann reaction for antibodies to the bacterium that transmits the disease. (And indeed Kuhn acknowledges Fleck as having anticipated most of the contribution that is generally considered Kuhn's own, the idea of scientific revolutions as changes of "paradigm." For the numerical version of paradigm shifts, see Section 3.3.) For another example, from a completely different domain, consider the puzzlement of Rosencrantz and Guildenstern, in Tom Stoppard's play *Rosencrantz and Guildenstern Are Dead*, when a series of 92 consecutive flips of a coin come up heads. There *must* be an explanation for this; the drive to abduce is taken for granted as part of the contract between the characters, which is to say, between playwright and audience.

3.3 Numerical Inferences Are the Natural Home of Abductive Reasoning

In the same way that this book's notion of consilience often simplifies into an unusually fruitful special case, the agreement of numerical estimates based on fundamentally different types of data or levels of measurement, likewise the notion of abduction simplifies here. In the context of numerical inference, the choice of that "best" hypothesis is made a whole lot simpler by the possibility of a scale of measurement against which they can all be evaluated at once. You may already have seen this approach in the context of "the general linear model," which is actually a whole class of statistical hypotheses, one for each subset of terms deemed statistically significant. But it applies much more broadly than that, to the comparison of multiple models on any quantitative data resource whatever. Some algebraic details of this approach, based on considerations of information theory, are postponed until Chapter 5; but we can sketch the argument here informally.

Numbers can serve either of the initial clauses of Peirce's version, the "surprising fact" part and the "matter of course" part. Regarding the "surprising fact" part, Thomas Kuhn says it well in that 1961 essay, pages 180–182:

> When measurement departs from theory, it is likely to yield mere numbers, and their very neutrality makes them particularly sterile as a source of remedial suggestions. But numbers register the departure from theory with an authority and finesse that no qualitative technique can duplicate, and that departure is often enough to start a search. . . . Once the relevant measurements have been stabilized and the theoretical approximations fully investigated, a quantitative discrepancy proves persistently obtrusive to a degree that few qualitative anomalies can match. By their very nature, qualitative anomalies usually suggest *ad hoc* modifications of theory that will disguise them, and once these modifications have been suggested there is little way of telling whether they are "good enough." An established quantitative anomaly, in contrast, usually suggests nothing except trouble, but at its best it provides a razor-sharp instrument for judging the adequacy of proposed solutions.

Regarding the "matter of course" part, Lipton (2004) has a shrewd comment about one part of this, the difference between "accommodation" (explanation of previously known facts) and "prediction" of measurements that have not yet been made. How is it that the same facts support the insight of the discoverer with two different saliences this way? The answer is a psychological one: the "accommodation" mode always entertains at least one additional competing explanation, namely, that the current candidate was produced specifically for the purpose of explaining all those quantitative anomalies, and is thus, as the statistician would say, "overfitted"; but in the predictive context this hypothesis is no longer available. Such an accommodation would necessarily have characterized the IPCC's report if, for instance, the slope of the Keeling curve had shown some nonmonotone pattern of change after Keeling first announced its persistence in 1960. But that slope appears to be gently and smoothly accelerating upward, and so no accommodation is necessary.

Indeed the IPCC's story as retold earlier in this chapter illustrated a surprising variety of abductions with virtually no need for accommodations of the sort Lipton was apologizing for. We saw surprise in the clear mismatch of numerical patterns to the observed data in the lower panel of Figure 3.4, and also in very improbable trends such as the long run of very warm years right up through the present. We saw reassurance in many aspects of the authors' own preferred hypothesis: stationarity of residuals around the model arrived at; consistency over short-term forecasting; confirmation region by region, without exception. Other chapters of the IPCC report that we do not have the space to review here show similar surprise-cum-consilience for the several other data flows that underlie today's climate science: data such as sea level, extreme events, paleoclimate, snow and ice accumulations, and chronobiogeography.

In a sense, the entire rest of this book is an attempt to reassure readers that this form of reasoning, albeit difficult to manage (these are very demanding norms, after all, not mere slogans!), characterizes the entire range of natural and social sciences that use properly grounded inferences from numerical data. In the next few pages I would like to review a wide variety of these examples, each very briefly. Some are discussed at greater length later in the book, while for the others, interested readers might pursue on-line encyclopedias and other linked resources for further information.

Mathematics. Pólya's volumes of 1954 include extensive discussions of abduction within mathematics, including the central role of "guessing" about the possible extensions of pattern. One classic site of this technique is the technique of mathematical induction, which is an explicit, formal variant of abduction from a set of examples. Consider, for instance, the problem of finding the formula for the sum of the first n odd integers, $1 + 3 + \cdots + (2n - 1)$. One classic approach would be to rearrange the sum. If n is even, we have $1 + 3 + \cdots + (2n - 1) = (1 + (2n - 1)) + (3 + (2n - 3)) + \cdots + ((n - 1) + (n + 1)) = 2n + 2n + \cdots + 2n$, a total of $n/2$ terms, making n^2, while if n is odd, we have $1 + 3 + \cdots + (2n - 1) = (1 + (2n - 1)) + (3 + (2n - 3)) + \cdots + ((n - 2) + (n + 2)) + n$, which amounts to $(n - 1)/2$ terms each equal to $2n$ (hence a total of $n^2 - n$)) plus one final term of n there at the end, totaling, again, n^2. This is a *constructive* proof. But there is also an abductive proof, one along more empirical lines. Suppose some 10-year-old among us notices, with mild surprise, that $1 = 1^2$, $1 + 3 = 4 = 2^2$, and $1 + 3 + 5 = 9 = 3^2$. Given this pattern, let us suppose that we have proved it for the first n odd numbers – assume $1 + 3 + \cdots + (2n - 1) = n^2$, and then we check that on that assumption $1 + 3 + \cdots + (2n - 1) + (2n + 1) = (1 + 3 + \cdots + (2n - 1)) + 2n + 1 = n^2 + 2n + 1 = (n + 1)^2$. This is, of course, an example of the celebrated technique called *mathematical induction*, but it is actually an abduction instead, beginning with the surprise of the observed pattern prior to the formal mathematical proof.

Physics. There are many examples in physics in which a quantity has been found, quite surprisingly, to be a constant, and its constancy is later converted into some much deeper form of understanding. One of the most famous of these cases is the constancy of the speed of light, noted experimentally by the Americans Albert Michelson and Edward Morley in 1887. A measurement of the speed of light in two perpendicular directions, one along the earth's orbit and one perpendicular to it, led to the same numerical outcome, an invariance that was surely surprising (indeed, inexplicable) in terms of the physics of the time. After several flawed attempts at an explanation in terms of conventional approaches (for instance, the "Lorentz contraction" of measured lengths in the direction of motion), Einstein's great 1905 paper on relativity[10] introduced the geometry in which the invariance of the measured speed of light is guaranteed: the geometry, not three-dimensional but four-dimensional, that is now taught to every undergraduate physics major. Martin Krieger's intriguing short book *Doing Physics* of 1992 emphasizes the ways that good physicists assume the possibility of abduction in every experimental setup they explore. His fifth principle, for instance, is that "we find out about the world by poking at it," with a probe that, if good, affects only one degree of freedom of the system, leaving the others unchanged. "The idea, always, is to be able to probe the world yet not affect it, at least in the end.... Probing makes for an objective world, a world we may eventually consider as out there, independent of us" (Krieger, 1992, p. 114) That "objective world" is the abduction of which we are speaking, the inferred reality of the black box that behaves according to our expectations once the experimental setting is sufficiently elegant (once it has sufficiently isolated the degrees of freedom we are interested in probing, by such stratagems as vacuums, constancy of temperature, and isolation from vibrations).

Public health. My oldest readers may remember advertisements from the 1950s that showed a man in a white coat, stethoscope in pocket, cigarette in hand, diplomas on the wall, over a caption along the lines of "9 out of 10 doctors who smoke prefer Camels." Over that decade, evidence accrued that rates of lung cancer death were inexplicably (i.e., surprisingly) rising all over the Western world. The incidence of this cancer was rising with what appeared to be a fixed time lag of some 30 years behind the rise in per-capita consumption of cigarettes by smokers, and furthermore the cancers were found overwhelmingly in smokers. This would make sense only if either (1) smoking caused cancer, or (2) some factor, such as a genetic condition, caused the desire to smoke and also, separately, caused the lung cancers. This second possibility, which would convert the observed ratio of death rates into an "accommodation" of the Lipton type, was supported by no less an authority than Sir R. A. Fisher, the greatest living statistician of the time (who was also a heavy smoker; see Stolley, 1991). See, for instance, the arguments in Fisher (1958), which are cautious, authoritative, and dead wrong. The abduction represented by hypothesis (1) was first confirmed by the famous Doll-Hill (1956) "doctors study." The 1964 report of the U.S. Surgeon General summarized the evidence to that date consilient with this hypothesis, principally because its conclusions needed to be robust against a politico-economic assault on

---

[10] Not to be confused with the great 1905 Einstein paper on the photoelectric effect, confirmed by Millikan in the paper I excerpt in Section 4.2, or the great 1905 Einstein paper on Brownian motion, confirmed by Perrin in the work I excerpt in Section 4.3. Calling 1905 Einstein's *annus mirabilis* is no exaggeration.

them by the tobacco industry. Rather as the IPCC defenses of its main anthropogenic hypothesis included a good deal of ancillary data in addition to the main chart, the Surgeon General's defense of the main carcinogenetic finding included a good deal of supplementary data about dose–response dependencies (the more you smoke, the higher your risk of cancer; the longer the period after you ceased smoking, the lower your risk), and the bench scientists quickly came up with the animal and cell experimental strategies that confirmed the mechanism(s) of damage. But the abduction began, epidemiologically speaking, with the doctors themselves, the data from whom were among the most precise at the time when this truly horrifying possibility was first suspected.

There is a similarly ironic finding regarding the fetal alcohol diseases. It was certainly "surprising" back in 1973 that as ancient and honorable a beverage as alcohol, one often recommended by obstetricians to their more nervous patients "to help them relax," might actually be causing birth defects. But, as a celebrated paper of 1973 by Kenneth Lyons Jones and colleagues reported, a run of babies born to alcoholic mothers in a Seattle hospital ward all had aberrant (funny-looking) faces and small size and showed failure to thrive. The abduction that the damage to the babies was from a common cause, the mothers' alcoholism, was obvious, and once this hypothesis was enunciated (abduced), its repeated confirmation followed quickly. We now consider fetal alcohol exposure to be the most common known cause of *any* avoidable birth defect, and wonder that the syndrome was not discovered until the amazingly late date of 1973. See Streissguth (1997).

For another example of the role of abduction in public health, this time involving the persuasion of the public regarding a hypothesis that the investigator had already arrived at by "shoe leather" (that is, by getting out of the laboratory and walking to all the places at which additional data needed to be gathered afresh), see the discussion of John Snow's demonstration of the mode of "communication" (propagation) of cholera in Section 4.1.

Scorpion. In my retelling of the reconstruction of the sinking of this submarine (Section 1.2), the investigator, John Craven, searched through piles of charted evidence until he fastened on a remarkably surprising finding indeed, the fact that the blasts he reconstructed along the intended path of the lost submarine indicated she was traveling *east* at the time of the catastrophe. This is inexplicable for a submarine on a westward assignment except under one circumstance, the 180° reversal that would follow some serious malfunction in the torpedo room. In simulations, this hypothesis also accounted for the observed temporal spacing of the blasts over the 91 seconds of the sequence. Craven (but nobody else) adopted the abductive inference as his own narrative and proceeded to work out the two investigations that might most directly confirm it: a predicted *location* of the wreckage, given that hypothesis, and a predicted *form* of the wreckage (namely, one having a torpedo room that had not been crushed by water pressure). These were both confirmed, eventually, although not without administrative resistance (the sort of resistance that is common whenever an abduction is bruited in the presence of believers in earlier or politically more convenient theories). Were it not for the surprising pattern of easterly movement,

Craven would not have had his abduction, and we would not have found the wreckage confirming his reconstruction of the event.

Seafloor spreading. This was our example in Chapter 2. Alfred Wegener came up with the correct hypothesis, the drift of the continents, back during World War I, just by looking at patterns. He was "guessing" in the same sense that Peirce was talking about – there was no mechanism available to which his hypothesis could be reduced, a lacuna that remained patent for nearly 50 years. Among Wegener's semiquantitative arguments (matches of paleobiology, matches of mountain ranges) there was one that was genuinely quantitative (the match of the outline of the western continental shelf of Africa with the eastern shelf of South America). Other quantitative matches began coming to light in the 1950s (the match of paleomagnetic data to one unique pattern of polar wandering, the match of most earthquake centers to a set of curves drawn over a negligible area of Earth's surface). When the equally surprising data about the symmetry of magnetic reversal traces over mid-ocean ridges came to light in the 1960s (recall the "Zebra Diagram," Figure 2.3), the two abductions resolved jointly in the hypothesis of seafloor spreading at the ridges, at 2 cm per year on either side, as the mechanism accounting for both Wegener's observations and those of the 1950s. By now the total weight of this consilience is overwhelming (see, e.g., Figure 2.6); but figures like these were constructed well after the underlying abduction was already generally accepted.

Earthquakes off the Washington state coastline. Though seafloor spreading is a serenely measured process, its consequence of the greatest relevance for humans, the subduction earthquake, is not. Seattle, one of the cities I call home, is uncomfortably close to one of the seismologically most active subduction zones in the world, the Cascadia subduction zone (named after the Cascade Mountains of the American Pacific Northwest, with their chain of volcanoes). Here the Juan de Fuca plate, or what's left of it, is driving under the North American plate along an arc from Eureka, California, north past Vancouver Island, British Columbia. Earthquakes that attend on this sort of process (Sataki and Atwater, 2007), such as the Tōhoku earthquake of March 2011, typically involve abrupt vertical shifts of offshore ocean bottom. These are the shifts that generate tsunamis like the one that demolished the Fukushima Nuclear Power Plants on the shore of the Japanese island of Honshu a few minutes after that earthquake. How large a quake might we expect in Seattle? A few score years of written history is not enough for a sound estimate.

In 1986, Brian Atwater, a geologist working at the University of Washington, noted a striking pattern in tidal lands to my southwest. Some event had caused intertidal mud, often accompanied by anomalous sand layers, to bury vegetated lowlands six different times over the last 7000 years. Using a phrasing explicitly echoing the classic language of abduction, Atwater (1987, pp. 942, 943) concludes that "these events may have been great earthquakes from the subduction zone between the Juan de Fuca and North American plates.... Nothing other than rapid tectonic subsidence readily explains the burial of the peaty layers." Ten years later, by consideration of tree rings from red cedar stumps and roots, the date of the most recent of these events was assigned to the interval between the 1699 and 1700 growing seasons (Yamaguchi et al., 1997). In as fine an interdisciplinary consilience as this book has to offer, this timing proves to agree beyond a doubt with reports of an "orphan tsunami," a tsunami without a premonitory earthshaking, mentioned widely across the island of Honshu in diaries and other records dated the night of January 26–27, 1700 (Atwater et al., 2005). Given this identification, in turn, by inverse estimation of the physics of the earthquake from the physics of the tsunami and the shoreline subsidence, the strength of the 1700 Cascadia earthquake can be estimated as between 8.7 and 9.2 on the M_w (moment magnitude) scale. Such a consilience, matching paleoecological records at a scale of centimeters (Washington shoreline) to human autobiographical records at a scale of hundreds of kilometers, leads directly to a strengthening of the local building codes that, presumably, protect the home in which I am currently typing these comments from collapsing the next time the Juan de Fuca plate slips underneath me.

Solar system. Alvarez et al. (1980) noted a huge anomaly in the density of the element iridium at a particular level of an earth cut in Gubbio, Italy: it was 30-fold its density at neighboring levels above and below. A systematic series of abductions, alternating surprise (from additional instrument readings) with regularization (from revised hypotheses), resulted in the scenario of an asteroid of radius about 10 kilometers striking Earth catastrophically at the time of extinction of the dinosaurs. "We would like to find the crater produced by the impacting object," the authors concluded, and 12 years later the abduction was closed when that crater was, most unexpectedly, discovered off the coast of the Yucatán. Section E4.8 reviews this extraordinarily brilliant and closely interwoven series of numerical inferences.

Challenger, Columbia. Two American space shuttles have exploded, the *Challenger* in 1986 and the *Columbia* in 2003. In both cases, a tremendous amount of forensic reconstruction had to be extrapolated from necessarily truncated streams of telemetry in order to abduce the probable ultimate cause of the disaster (for *Challenger*, a cracking O-ring seal on one of the booster rockets, as famously demonstrated using his glass of ice water by Richard Feynman in one of his last public appearances; for *Columbia*, damage to the leading edge of one of the wings during takeoff) and thereby to reconstruct in gruelingly detailed simulations the actual stream of instrument readings observed prior to failure of the machine. These can be thought of as engineering-world examples of the same logic that drives the "police procedural" style of detective story mentioned in Chapter 1: the reconstruction of an initially astounding event as the ineluctable consequence of a hitherto-overlooked singular cause. Such a reconstruction will almost always come under the heading of the abductive inference we are discussing here. One has to guess at the cause before one can demonstrate that the observed data do, indeed, match the effects one would expect were it the case. This should apply wherever a catastrophe eradicates the traces of its own developmental sequence.

Biology. One of the most famous abductions in all of biology is James Watson's "sudden awareness" that a hydrogen-bonded pairing of adenine and thymine has the same form as the analogous pairing of guanine and cytosine, and thus supplies a perfect explanation of Chargaff's rules that the counts of these nucleotides are always "nearly the same" in any sample of DNA. The constancy of shape also permits the reconstruction of a double helix with geometric characteristics exactly matching the measurable parameters of Rosalind Franklin's celebrated Mode-B crystallogram. We review these and other topics in Section E4.5.

A completely different sort of abduction, for which there is no room elsewhere in this book, is abduction from the specific values of regression coefficients. In the standard treatment of regression, there are only a few particularly interesting values of the computed slope, the parameter usually notated with the Greek letter β: slopes of 0, corresponding to statistical null hypotheses; slopes of 1, corresponding to simple hypotheses of persistence; and slopes matching published values of earlier slopes. In biology, however, there can be other privileged values of computed slopes. A slope of $1/2$ characterizes "regression to the mean" in simple random walk (see the discussion of Galton's quincunx at Section L4.4), and other slopes correspond to exact path models after the fashion of Sewall Wright's path diagrams (Section L4.5). Perhaps the neatest application of this approach is to identify regression slopes that are simple fractions, like $2/3$ or $3/4$, with distinctive models of distributed evolutionary-developmental control mechanisms. The slope of $2/3$ is central to many discussions of encephalization in the course of human evolution (see, e.g., Jerison, 1973), while the slope of $3/4$ approximated in regressions of metabolic rate (or net branch number) on body mass was argued by West et al. (1997, 1999) to be crucial to their theory of the ontogeny of the branching processes entailed in the anatomy of the cardiovascular and respiratory systems across the animal kingdom.

3.4 Strong Inference

Pragmatism, Peirce said, is the logic of abduction (Collected Papers 5, p. 121). The claim is not that abduced propositions are true but that they are potentially useful: "What we think is to be interpreted in terms of what we are prepared to do." Abductive reasoning thus counts as a species of American pragmatism, which, after all, Peirce founded. As we already saw in the case of climate change, the culmination of abductive reasoning best lies in rational reflections about alternative futures (see, e.g., the discussion at Figure 3.6). The purpose of numerical inferences, according to Peirce, is not to get at truth; it is to get at a rational basis for human action.

This insight applies not only to the so-called "exact sciences" – physics and chemistry – but also the historical natural sciences such as paleontology, geology, or genetics. In fact, some of these models seem to apply far more widely, into the domains of social behaviors and even historical trends. There is an extensive social-science literature under the heading of "complexity theory," such as Mitchell (2009), and many publications out of the Santa Fe (New Mexico) Institute for the study of complex systems that as often as not deal with sciences of collective behavior or other social sciences. John Gaddis (2002) argues at length for the relevance of the same models in historiography.

Kuhn, in the essay from which I have quoted several times already, continues on to note on pages 183–184 (see also the discussion on page 40):

3.4 Strong Inference

Measurement can be an immensely powerful weapon in the battle between two theories, and that, I think, is its second particularly significant function. [The first was the construction of the awareness of the anomaly itself, its robustness against instrumental improvements and small tweaks of the rules of correspondence.] Furthermore, it is for this function – aid in the choice between theories – and for it alone, that we must reserve the word "confirmation."... In scientific practice the real confirmation questions always involve the comparison of two theories with each other and with the world, not the comparison of a single theory with the world.

Kuhn's comment introduces my final topic in this chapter, a method that has become the core method of several quantitative sciences, both basic (e.g., molecular biology) and applied (e.g., ecology). In these contexts it is usually given the name of the *method of multiple hypotheses* and attributed to the geologist T. C. Chamberlin writing in 1890. Chamberlin's original, in turn, inspired the biochemist John Platt (1928–1992) to promulgate his equally influential polemic "Strong inference" of 1964, the essay that motivated *Science* to reprint that 19th-century lecture in 1965.

Both Chamberlin and Platt, in language suited to their different centuries, concentrated on the "method of multiple working hypotheses," the complete opposite of the 20th-century method of null-hypothesis statistical significance testing, which allows only one hypothesis, and that one usually not very plausible (see at Section L4.3.4). These ideas, now 120 years old, continue to feature in any serious examination of how hypotheses are formed in actual biological studies. See, for instance, Elliot and Brook (2007), who note that Chamberlin "espouses a view that values lateral thinking and multiple possibilities," which "can only be a positive development."

I will speak here mainly about Platt's views, which were formed in the context of mid-20th-century molecular biology, the period that Gunther Stent (1969) was about to call, with Hesiodic pessimism, the "Golden Age." This was the source of Platt's examples of the systematic methods of scientific thinking that characterized "some fields of science that are moving forward much more rapidly than others." Rejecting explanations in terms of the quality of the researchers or the "tractability" of the subject, Platt highlights a method that "consists of applying the following steps to every problem in science, formally and explicitly and regularly":

(1) Devising alternative hypotheses
(2) Devising a crucial experiment (or several of them), with alternative possible outcomes, each of of which will, as nearly as possible, exclude one or more of the hypotheses
(3) Carrying out the experiment so as to get a clean result
(1′) Repeating this sequence of three steps *ad libitum*.

"It is clear why this makes for rapid and powerful progress," he notes on page 347. "For exploring the unknown, there is no faster method; this is the minimum sequence of steps. Any conclusion that is not an exclusion is insecure and must be checked." He cites examples of great researchers whose notebooks explicitly show that they have worked this way: Faraday, Röntgen, Pasteur.

Although Platt nowhere mentions Peirce, pragmatism, or the logic of abduction (indeed, the only two philosophers he credits are Karl Popper and Francis Bacon), the relation of his logic to the subject of this chapter is clear. The combination of Platt's items (1) and (2) is equivalent to the combination of clauses (1) and (2) of Peirce's original abduction, Epigraph 9. Platt has combined the surprise and the "following as a routine matter" in a parallel empirical examination of a wide range (in practice, the widest feasible range) of simultaneous explanations. In a well-designed experiment, only one of these potential explanations will come anywhere near the data. The same observations that abduce it serve to reject all the others at the same time. The scientist's effort goes equally into the construction of those hypotheses and into the care with which the experiment or investigation provides the information necessary to assess the credibility of each of them at the same time on the basis of the selfsame data.

Platt selects molecular biology as one of the fields where this method has become "widespread and effective," and and reviews a good many sites where it is applied, including papers of Monod and Jacob, Benzer, Meselson (all Nobel Prize laureates) and, of course, Watson and Crick themselves (see my Section E4.5). He suggests the style of journals such as *Journal of Molecular Biology* in the 1960s, where articles had paragraph after paragraph of linked inductions, text of the form "Our conclusions... might be invalid if... (i)... (ii)... or (iii).... We shall describe experiments which eliminate these alternatives."

The susceptibility of a field to strong inference is partly due, Platt concedes, to the nature of the fields themselves. In the high-information field of biology, he notes, "years and decades can easily be wasted on the usual type of 'low-information' observations or experiments if one does not think carefully in advance about what the most important and conclusive experiments would be." The contrast between this view of 1964 and the current obsession with genomic, proteomic and other -omic databases, within which the average information content of any byte of data is absurdly low, is hugely ironic.

The solution to the ordinary human frailty of fondness for your children (attachment to your own hypotheses) is to always have several of them in your field of attention: this is Chamberlin's "method of multiple hypotheses." Conflicts that would otherwise be between individuals become purely conflicts of ideas, which can be resolved with much less "reluctance or combativeness." "When multiple hypotheses become coupled to strong inference, the scientific search becomes an emotional powerhouse as well as an intellectual one."

Perhaps because of Platt's focus on molecular biology, he departs from my views at one important juncture, the role of quantification in scientific advances. His theoretical entities are all binary, structures or phenomena that either exist or do not, without any essential parameters. In his world, no arguments depend on the values of numerical quantities, not even the arguments by Watson and Crick that actually established the validity of the double helix model that made for such rapid progress just as Platt was writing the 1964 paper. Contradicting Kuhn (whose work he actually cites, footnote 15, though claiming that it "does not... invalidate any of these conclusions"), he notes on pages 351–352:

Measurements and equations are supposed to sharpen thinking, but, in my observation, they more often tend to make the thinking noncausal and fuzzy. They tend to become the object of scientific manipulation instead of auxiliary tests of crucial inferences. Many – perhaps most – of the great issues of science are qualitative, not quantitative, even in physics and chemistry. Equations and measurements are useful when and only when they are related to proof; but proof or disproof comes first and is in fact strongest when it is absolutely convincing without any quantitative measurement. Or to say it another way, you can catch phenomena in a logical box or in a mathematical box. The logical box is coarse but strong. The mathematical box is fine-grained but flimsy. The mathematical box is a beautiful way of wrapping up a problem, but it will not hold the phenomena unless they have been caught in a logical box to begin with. What I am saying is that, in numerous areas that we call science, we have come to like our habitual ways, and our studies that can be continued indefinitely. We measure, we define, we compute, we analyze, but we do not exclude. And this is not the way to use our minds most effectively or to make the fastest progress in solving scientific questions.

This passage, in my opinion, is the most serious objection to the main themes of this book that I have yet seen in my review of others' philosophies of quantitative science, and so I feel I should respond to it at least briefly. What Platt gets wrong, among other matters, are these:

- The role of surprise, that is, failure of expectation, in calling attention to possible explanations (i.e., in focusing the community's attention on where new explanations are most likely to prove exciting and interesting)
- The role of quantitative anomalies, sensu Kuhn, in focusing the attention of a scientific community (what Ludwik Fleck called the "thought-collective") on these surprises
- The quantitative bases of appeals to mechanism and to reduction in most of the biosciences today, including the -omic sciences that are the direct descendants of the Golden Age studies in molecular biology that Platt was apotheosizing
- The fact that strong inferences can go forward just as easily in reference to quantitative data summaries as via the qualitative experiments (something happens or it does not, something has three structural variants or only one) that Platt was accustomed to
- The far greater power of statistical summaries of data when those data are quantitative
- The great inconvenience of pursuing hypotheses about magnitudes of effects – regressions, trends, multivariate patterns – in qualitative methodology

Even in 1964, Platt was ignoring the great mass of research in genetics and in evolutionary and developmental biology, which have always dealt with quantitative assessments of causal processes. I wonder what Sewall Wright, inventor of path analysis and as brilliant a scientist as the 20th century ever produced, would have had to say about Platt's ideas.

Indeed this same John Platt, writing just a bit later, was involved in a quite different exploration, the role of quantification in the *social sciences*. Deutsch, Platt, and Senghaas, 1971, emphasize the social determinants of important social science research along entirely different lines than the methods of strong inferences and multiple hypotheses just reviewed. (See also Deutsch, Markovits, and Platt, 1986, a book-length work on the same themes.) Writing in 1971, Deutsch et al. (1971, p. 457) argue, in gross contradiction to what Platt had just published on his own for the biological sciences:

> Quantitative problems or findings (or both) characterized two-thirds of all [major advances in the social sciences], and five-sixths of those were made after 1930. Completely nonquantitative contributions – the recognition of new patterns without any clear implication of quantitative problems – were rare throughout the period and extremely rare since 1930. Nonetheless, they include such substantial contributions as psychoanalysis, Rorschach tests, and the work on personality and culture; thus, their potential for major contributions in the future should not be underrated. Certainly both types of scientific personalities, the quantifiers and the pattern-recognizers – the "counters" and the "poets" – will continue to be needed in the social sciences.[11]

His comments about that "flimsy mathematical box" aside, Platt's ideas about strong inference and multiple hypotheses are first-rate, and serve as an excellent antidote to the flimsy and shallow notions of null-hypothesis statistical significance testing that I deprecate over and over in these pages. Section 5.3, returning to the consideration of multiple hypotheses, introduces a formal method, based on the celebrated *Akaike Information Criterion* (AIC), for describing the support afforded by a given quantitative data set for any number of alternate hypotheses at the same time. By exploiting the AIC one can decide to declare that one hypothesis is far more credible than the others or else, when no single hypothesis is left standing in this way, to proceed rationally with the ensemble of all the surviving alternatives.

My argument in this short but dense chapter is now complete. Beginning from a third canonical example, the IPCC's conclusion of 2007 about the likelihood of the role anthropogenic greenhouse gases are playing in the rise of global average temperature, I reviewed the underlying principle of inference, the method of *abduction* first formalized by C. S. Peirce beginning in the 1870s. Whereas induction and deduction address hypotheses already enunciated, abduction is the process by which hypotheses are generated in the first place. The corresponding cognitive style seems congenial to the deep structure of the human mind. I briefly mentioned many additional examples that illustrate this rhetorical style. Finally, I called attention to

---

[11] *Historical note.* The categorization of quantification that the authors were using divided their 62 "basic innovations" into QFE (quantitative findings explicit), QPE (quantitative problems explicit), QPI (quantitative problems implied), and non-Q (predominantly nonquantitative). For instance, the two advances attributed to V. I. Lenin, "Theory of one-party organization and revolution" and "Soviet type of one-party state," are classified as QPI, as is the method of general systems analysis of Bertalanffy, Rashevsky, Gerard, and Boulding that constituted my own fundamental training in philosophy of science at the University of Michigan in the 1960s. Examples of the QFE type include the foundations of econometrics that were being set all over the world beginning in the 1930s, and the work by Anatol Rapoport on Prisoner's Dilemma and other conflict games that was going forward at the University of Michigan at the same time that Boulding and I were there, work that continues in the hands of Robert Axelrod and others.

a correlated cognitive style, strong inference or the method of multiple hypotheses, that is a normative method of inference often encountered across a variety of the natural and the historical sciences. Taken as a whole, these forms of argumentation place numerical inference squarely in the tradition of American pragmatism, the epistemology whereby human thought is given meaning only via human actions, as when the IPCC aimed its presentation mainly at the "policymakers" who control not the science but the social systems responsible for stimulating or slowing the rise in the levels of those greenhouse gases.

3.5 Summary of Part I

And so we arrive at the end of Part I. The argument so far has been grounded in detailed discussions of three apparently unrelated examples: the fate of the submarine *Scorpion*, the anthropogenic nature of global warming, and the role of seafloor spreading as the explanation of continental drift. Though these three are as diverse in their scientific contexts as anyone could wish, and though the actual intellectual history intertwines original arguments and their ironies with aspects of real or imagined scientific diagrams that are quite different (compare, e.g., Figure 2.5 with Figure 3.4), nevertheless they all rely crucially on both of the fundamental strategies according to which arithmetic leads to scientific inferences. These are the principles of consilience (inference from numerical agreement, Chapter 2) and abduction (inference to the best explanation, this chapter). Both principles were used in Chapter 1 prior to being introduced properly. Each was a focus of the example in its own chapter, and the second and third were then expanded and discussed each in its own separate philosophical context.

Still, if this is a book about numerical inference, where is the actual arithmetic, and, more to the point, where is the algebra that explains how the arithmetic is actually working to persuade (the match of those magnetic reversals in Figure 2.5, the separation of the orange curves from the black line in Figure 3.4)? The algebra we've seen so far (mostly in connection with Newton's Law of Gravitation or the rule about sums of consecutive odd numbers) hasn't had anything to do with the actual examples. In other words, to this point the notion of what constitutes a "numerical pattern," or a match of such patterns between two data sets, has necessarily been superficial.

To do this topic justice requires a return to the concerns of geometry (Chapter 2), but this time with a deeper consideration of the nature of patterns on paper, their analogues in our state-of-the-art data representations, and especially the combinations of algebra and geometry whereby the implications of these pattern representations can be explored quantitatively. The treatment of this theme begins in the first few sections of Chapter 4 with the algebra of single variables and broadens to some extent later in that chapter in the discussion of predicting one variable from others. The rest of the necessary sophistication is deferred to Part III, where we import a whole additional geometric language, that of patterns among points. The book then cycles through its paired main themes of abduction and consilience once again from first principles, now in a context of measuring more than one quantity at the same time: the domain of multivariable data and multivariate statistics.

Appendix 3.A: Update on the Rhetoric of Climate Change

Most of the discussion of climate change in the main text of this chapter was written in 2010 using resources dated a few years earlier. But of course this topic – whether the climate is changing, and if so whether human activity is among the causes – continues to be of great interest politically, economically, and geographically. As I've already mentioned, the next IPCC report (the *Fifth Assessment Report*, analogue of the Fourth Report, 2007, from which all the extracts in this chapter have been taken) arrives too late for review in this book. But on the Web and in the pages of "newspapers of record" such as the *New York Times* there are some early signs of how the rhetoric of consilience and abduction is managing to disseminate itself, or not, within the larger community at which the "summary for policymakers" of the IPCC was aimed. In this Appendix I sketch the import of three activities of that sort.

Contradicting IPCC's answer to FAQ 9.1. Section 3.2 mentioned FAQ 9.1 of the IPCC 2007 report, the box explaining that individual episodes of extreme weather should not be attributed to global warming as their "cause." This conservative stance is now coming under challenge from researchers who focus on such episodes over very large geographical scales. In calculations reported, to considerable fanfare, in Hansen et al. (2012), the climatologist James Hansen and his team at the NASA Goddard Institute have tabulated 60 years' worth of departures from mean summer temperatures over a grid of 250-kilometer square cells covering the continental United States. The frequency distributions of these "anomalies" (a word used here in its technical sense, not its judgmental one) has clearly shifted over the time period studied here, as graphed quite evocatively in Figure 3.7. Hansen et al. emphasize "the emergence of a category of summertime extremely hot outliers, more than three standard deviations warmer than the climatology of the 1961–1980 base period," that now, according to a different chart that I have not reproduced here, "typically cover more than 10% of the land area." Even without reasoning from any global climate models, they claim, it follows, "with a high degree of confidence," that anomalies in the range of four standard deviations or higher should be regarded as consequences of global warming "because their likelihood in the absence of global warming was exceedingly small." Notice the bell curve in the figure, which "the lay public may appreciate."

The causal pathway of interest here is not from humans to warming but from warming to weather. Because large areas of extreme warming "have come into existence only with large global warming," the authors infer, "today's extreme anomalies occur as a result of simultaneous contributions of specific weather patterns and global warming." The version of causation here is essentially statistical: given a particular particularly hot month somewhere, is it likelier to have come from the 1950s or the present century? For temperatures four or more standard deviations above baseline, the answer is "the present, beyond any reasonable doubt." This interpretation is

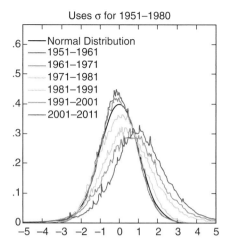

Figure 3.7. Distribution of temperature anomalies, by decade, over 250 km² square gridboxes for the continental United States. Horizontal, anomaly in units of 1951–1980 standard deviation; vertical, frequency, normalized to unit area under each curve. Both the mean and the standard deviation of this distribution have increased over time, so that the probability of a deviation from the 1951–1980 mean of three 1951–1980 standard deviations, which was 0.004, is now approximately 0.16, or 40 times as likely. (From Hansen et al. [2012], Figure 9 [left].)

clarified in an alternate diagram style (Figure 3.8) from a slightly different data source, the National Climatic Data Center's chart showing that July 2012 was the hottest month across the continental United States since systematic record-keeping by county began in 1895. Hansen's point about spatial extent is made far more helpfully by a map than by the graph that his group actually used. Is it a coincidence that the desk where this book is being written is in one of the very few counties that were actually cool over that period?

Stronger data, weaker models. The Berkeley Earth Land Temperature project (www.berkeleyearth.org) was set up with two purposes: to assemble the largest feasible public data resource of land temperatures for the entire community of climate researchers to share, and to instruct in modes of statistical inference (as distinct from AOGCM modeling per se) suitable to drawing conclusions from that corpus. In contrast to the IPCC presentation I recounted in Section 3.1, a presentation that shows the fits of a great many models of two different classes, the Berkeley Earth summary figure shows only a single summary curve (Figure 3.9), their best estimate of "Earth's average land surface temperature, 1753–2011" (Rohde et al., 2012), and a statistical fit of this average to identified volcanic events together with "the log of atmospheric CO_2."

The figure caption accompanying this graphic, from a press release by the same group, indicates the nature of the rhetoric here. Readers are supposed to be reassured by the very large number (36,000) of temperature stations aggregated in the underlying data, whereas the meaning of the phrase "well fit" and the scientific details of the

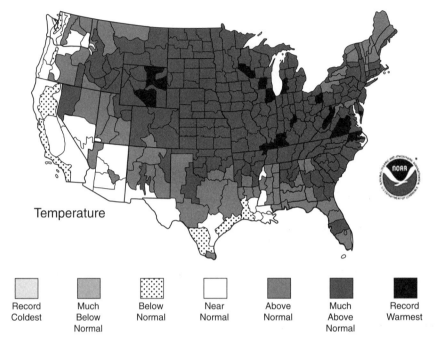

Figure 3.8. Map of ranks of July 2012 average temperatures by county, continental United States, demonstrating the extent of heat for this "hottest month ever." Dark gray: in the top 10% of July mean temperatures, 1895 to present. Black: single highest mean temperature. (From http://www.ncdc.noaa.gov/sotc/service/national/Divisionaltrank/201207-201207.gif, courtesy of the National Oceanic and Atmospheric Administration.)

"simple model" are shielded from view by the caption. In particular, there is no comment about the appearance of the measured CO_2 in the form of its logarithm or the relevance of that transform to the detailed dynamic climate models summarized by the IPCC. Notice, for example, the obvious, but unexplained, discrepancy between data and fitted model around 1940. The ultimate phrasing of their conclusion is thus much weaker than the IPCC's that I quoted in Section 3.1: "Berkeley Earth has now found that the best explanation for the warming seen over the past 250 years is human greenhouse gas emissions. While this does not prove that global warming is caused by greenhouse gas emissions, it does set the bar for alternative explanations." In other words, Berkeley Earth is saying that the observed curve looks like a version of CO_2 with notches after every volcano. But the IPCC's reasoning is far deeper; the models it summarizes show how CO_2 *produces* this increase in temperature. What is a regression coefficient in the Berkeley Earth approach is the integral of a coefficient in the IPCC approach that varies according to the values of other predictors. By virtue of this complexity, the IPCC claims have much more consilience, more authority.

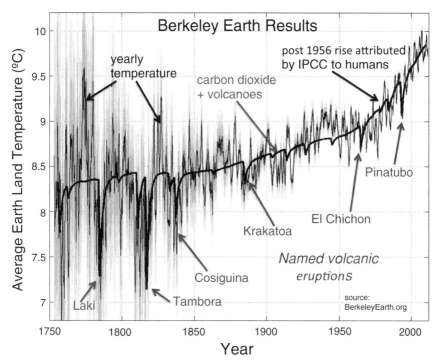

Figure 3.9. Summary graphic from the Berkeley Earth Land Temperature project, press release of July 29, 2012, emphasizing a simplified report declaring two causes for global mean temperature shifts to date: atmospheric CO_2 and volcanoes. Light jagged line: the temperature of the Earth's land surface, as determined from over 36,000 temperature stations around the globe. According to Berkeley Earth, "the data is well fit by a simple model (heavy line) containing only known volcanic eruptions and carbon dioxide (dark line). No contribution from solar variability was necessary to make a good match. The rapid but short (decadal) variations are believed to be due to changes in ocean flows, such as El Nino and Gulf Stream." By permission of Berkeley Earth (www.berkeleyearth.org).

In short, by privileging the physical modeling of these spatiotemporal series over the statistical approach the Berkeley group uses, the IPCC was able to proceed from weaker (less systematic) temperature data to a rhetorically stronger inference.

Both of these changes of rhetoric are intended to simplify the message for the public, but, in doing so, both fail to align with the IPCC's actual scientific purpose, which, you will recall, was to forecast future temperature regimes under six different political-economic scenarios (Figure 3.6). Both the Rohde et al. presentation and the Hansen presentation are oriented retrospectively – each is explaining something that has already been observed (rise of global mean temperature, incidents of extreme heat over relatively large areas). Both thereby substitute statistical reasoning (about probabilities) for scientific reasoning (about consiliences or surprises). Hansen et al. offer no explanation of the magnitude of the shifts of mean or standard deviation of the bell curves in Figure 3.7 – why, indeed, should that standard error have increased at all, anyway? – nor do Rohde et al. explain why it doesn't matter that the fit of the curve in Figure 3.9 is to the logarithm of CO_2, even though the two scales differ by a substantial factor in extrapolating the effect of the doubling of this concentration that can reasonably be forecast.

Neither Hansen's group nor Rohde's group is guilty to any unusual extent of derogation of the science in this connection. The problem is systemic, where the system is the knowledge transfer system that feeds scientific conclusions to citizens and policymakers alike. The articulation between the science of global warming and the corresponding public understanding is fraught with miscues and failures on the part of all parties, including the distinction between the two types of rhetoric I am distinguishing here. The possibility of a consilient abduction is lost when investigators revert only to the simple systems that they are accustomed to reporting (in both Hansen et al. and Rohde et al., trends but not mechanisms), and, with the loss of consilience, the draining of the possibility of forecasting, which is the social purpose that drives all of this computing. See, in general, Collins (2007) or Knutti (2008).

This insight has not been lost on the statisticians. I am aware of at least one large-scale statistical project that, *beginning* from the consilient models (the AOGCM models), attempts to bring conceptions of uncertainty to bear with uncommonly punctilious statistical rigor. Peter Challenor, National Oceanography Centre, United Kingdom, heads the RAPIT project (http://www.noc.soton.ac.uk/rapid/rapit), which is investigating the probability of "the Gulf Stream stopping" (a cessation or substantial weakening of the Atlantic Meridional Overturning Circulation [AMOC]) by running a great many global climate simulations at different parameter settings. Even in today's era of laptops with gigabytes of main memory, the weight of computations he needs is too great for any conventional academic computing context. Rather, his simulations are carried out in a "crowd-sourced" way, after the fashion of the search for alien intelligences or some protein-folding problems: by downloading the task onto thousands of PC's all over the world that have spare computing cycles to offer. See the overview at http://climateprediction.net/content/rapid-rapit-what-risk-thermohaline-circulation-collapse (accessed 8/7/2012).

Unlike the IPCC, which typically reports the results of many intrinsically different approaches to climate modeling, this project involves only a few basic models, but covers a wide range of uncertainty about the values of the crucial underlying physical parameters. If scientific evidence is to inform mitigation, adaptation, or geoengineering, then the statisticians and the physical scientists must accommodate each other's inferential styles much more graciously than they do at present. The languages of abduction and consilience currently serve them separately, and the example of the IPCC report of 2007 shows that it can serve them together as well. But a language of uncertainty of *computations* is not enough. There must also be a language of the uncertainty of model *assumptions*, both value uncertainty and structural uncertainty, and likewise a language for the uncertainty of predicted *events*. Far more extensive collaborations between statisticians and physical scientists will be necessary if the public is to be properly informed about the implications of the science for regional changes of climate and the extremes of weather. See, in general, Frame et al. (2007) or Challenor et al. (2010). Back in 2006, Challenor et al. wrote (page 60), "Substantial weakening of the overturning circulation is generally assumed to be a 'low probability, high impact' event,... [but] our results show that the probability is in the range 0.30–0.46, depending on the scenario [of the IPCC, Figure 3.6]...; this could not reasonably be described as 'low.'" It is likely no accident that most of the research on the AMOC is being carried out in Europe, where the climate is directly moderated by the Gulf Stream. Perhaps there will be a brief exhortation in the forthcoming Fifth Assessment Report of the IPCC on the necessity that statisticians and specialists in climate dynamics work together to craft a new, better rhetoric by which their findings can be reported more authoritatively.

Part II

A Sampler of Strategies

4

The Undergraduate Course

The principles of abduction and consilience (Part I) combine in a scientific methodology according to which one first searches for surprising numerical patterns in data, "surprising" in the sense that they are not consistent with anybody else's hypothesis, and then constructs and further tests the hypotheses that render these agreements unsurprising. Using these two principles as a scaffolding, most of the important practical themes of the quantitative sciences can be construed as a superstructure built atop them. This chapter surveys the "undergraduate side" of such a syllabus: applications for which the complexity of data or instrumentation resources is relatively restricted. Included are several sets of "lecture notes" that more formally review the algebraic or logical bases of the inferences about to be demonstrated. Collectively these notes constitute a pedagogical experiment in communicating those ideas not by cookbook recipes (if the question on the test is X, use formula Y) but in light of the notions of numerical inference that are this book's principal theme. The examples, in turn, set the stage for the three chapters of Part III, "The Graduate Course," which will delve deeper into methods and metaphors for problems where the complexity of both data and explanations is more demanding.

The present chapter begins with an example (E4.1) requiring no preamble at all – no statistical algebra, only tables and a truly compelling abductive argument. This is John Snow's famous 1855 demonstration that in two consecutive London cholera epidemics the disease was spread via the city's drinking water. For one of the epidemics, the proximal cause was the contamination of one company's water supply; for the other, it was the contamination of the water from a single pump on Broad Street. Following Snow's work, the syllabus turns to a classic type of argument in physics, the inference from the appearance of a straight line in measured data. The example here (E4.2) is Robert Millikan's demonstration (1916) of the validity of Einstein's law of the photoelectric effect. As for several other examples in this chapter, the narrative is preceded by a lecture (L4.1) on the corresponding elementary statistical methods: here, the elementary aspects of the principle of least squares that lead to its applications in averaging and the calibration of linear laws (which reduce to the same thing). Millikan received the Nobel Prize for the work in E4.2, as did Einstein for the theory that preceded it, one of his three great contributions of 1905.

To explicate Millikan we need to know only about straight lines. But the next example (E4.3) of consilient abduction, Jean Perrin's demonstration that atoms exist (some 70 years before any instrument could actually render any image of them), requires considerably more of the elementary algebra and also a bit of calculus, mathematics that makes up the topic of section L4.2. The focus here, the celebrated *Normal distribution* (Gaussian distribution, "bell curve"), is explored from several points of view, in order to hint at the remarkable robustness of the mathematical truths it embodies. Perrin's work, which earned *him* a Nobel Prize, demonstrated the validity of Einstein's model of Brownian motion (another of his projects of 1905).

Our next scientific example reverts to a trope of Snow's, abduction from a *dis*agreement (rather than an agreement) between quantities. This example, E4.4, will be another Nobel Prize–winning project, but in this case the prize is quite recent, from 2005: the demonstration that most stomach ulcers arise as an infectious disease. That demonstration, by Barry Marshall and Robin Warren, had to climb a steep wall of preexisting dogma about causation of ulcers by stress, according to which their claim was preposterous. The lecture preceding this history (L4.3) centers on an introduction to the arithmetic according to which data can be claimed to support hypotheses, a topic usually called "likelihood" or "inverse probability." I hope to persuade you that this approach deserves to completely supersede the vastly inferior alternative too commonly taught in the undergraduate courses required for students who are intending to be something other than statisticians, the approach via *null-hypothesis statistical significance tests* (NHSSTs). Section L4.3 has a great deal to say about NHSSTs, none of it favorable.

Section E4.5 springs from the most celebrated biology journal article of all time: the announcement by James Watson and Francis Crick (1953) that the DNA molecule has the structure of a double helix. The book James Watson wrote in 1968 explaining the process that led to that article and *his* subsequent Nobel Prize is one of the masterpieces of history of science; it explains the discovery as one big abduction between two consilient channels of quantitative data. Preceding this retelling, in keeping with the spirit of this book, there is an interlude (Section L4.4) exploring the characteristics of pathological reasoning in science when a scientist is insufficiently skeptical about his interpretations of his own data.

The abductions of the examples so far close down on the reader's opinions like a bear trap on a hunter's leg. Other numerical inferences, equally sound, are not nearly so obvious. Our next example (E4.6) is a dual one, consecutive articles in the *British Medical Journal* on two aspects of the health damage caused by passive smoking (tobacco smoke in the environment). Both of them illustrate abduction by consilience – by agreement with other data – but they embody different degrees of subtlety about exactly what quantification it is that is supposed to be consilient. As both of these articles involve *regression coefficients* and their *adjusted values*, it is necessary to precede the example by a lecture (L4.5) about just what that arithmetical

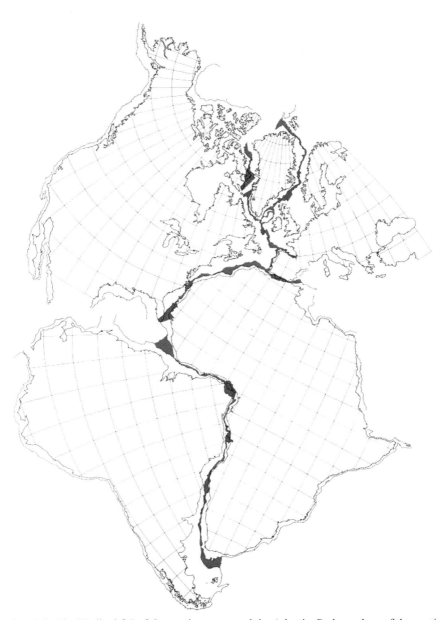

Plate 2.1. The "Bullard fit" of the continents around the Atlantic. Red: overlaps of the continental shelves at a depth of 3000 feet. Blue: gaps between the shelves. (From Bullard, Everett, and Smith, 1965, Figure 8. Both this original publication and the Raff and Mason article from which Figure 2.3 is extracted include an additional version of the crucial figure in enormously more detail than the downsampled versions I have reproduced here.)

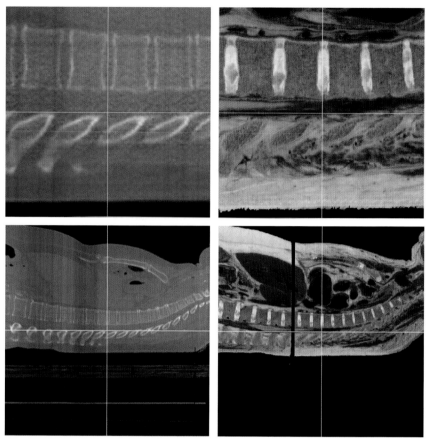

Plate 2.8. Example of a hidden consilience: CT geometry, left, matches physical tomographic geometry, right, of a solid organism (here, the Visible Female, Eve). (Data from the Visible Female project, National Library of Medicine. The images in the right-hand column are sections computed in Bill Green's software package Edgewarp. The scale of the lower image pair is one-fourth of that for the upper pair.) The panels in the right column are actually in true anatomical colors.

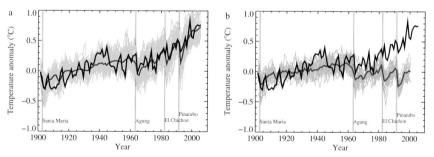

Plate 3.4. Central figure of the central IPCC abduction: demonstration of the perfect separation between two ensembles of partial differential equation computations corresponding to two classes of large-scale models, those including (left panel) or not including (right panel) the anthropogenic forcing functions of the greenhouse gases. (From IPCC, 2007, p. 684, with permission.)

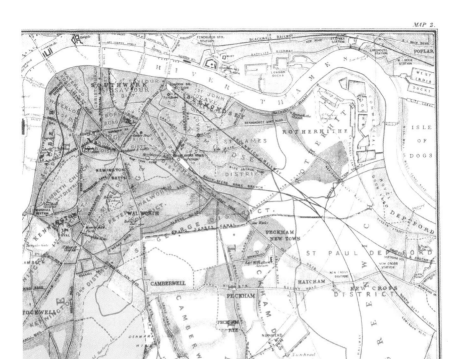

Plate 4.3. Detail of the second map from Snow (1855) – a full-color foldout in the original publication – indicating the neighborhoods served by the two water companies of interest, Lambeth and Southwark & Vauxhall, and the region of their overlap making Snow's inference possible.

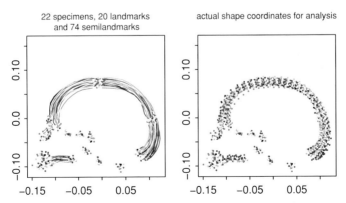

Plate 7.10. Semilandmarks represent curves in a scene for purposes of Procrustes analysis. Each specimen is in a randomly selected color. (After Weber and Bookstein, 2011.)

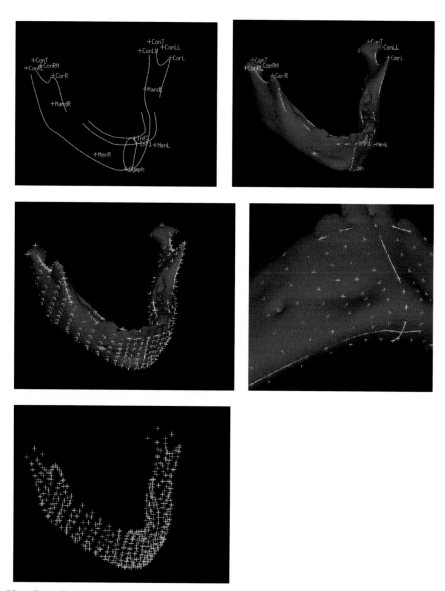

Plate 7.14. Example of an extended semilandmark scheme. (Top left) Curves and landmark points. Plane P1 of the text is implied by its closed curve of intersection with the symphysis; plane P2, spanning the tops of coronoids and condyles, is not shown. (Top right) Curves on the anatomical surface. (Middle left) Adding surface semilandmarks. (Middle right) Detail at Menton. (Lower left) The data as received by the Generalized Procrustes algorithm that would follow. (Figure from Bookstein, 2008; its data courtesy of Michael Coquerelle, see Coquerelle et al., 2010.)

maneuver might mean. It turns out to have a fairly complicated epistemology and formulary, with built-in abduction tools of its own.

Recall from Chapter 3 that there are three clauses in Peirce's abduction syllogism. We have not yet inspected the major premise closely, the assertion that a fact encountered is "surprising," that is, not to be expected on anybody else's theory. For that premise to be meaningful there must be some discipline overseeing the search for other theories. This matter is the topic of the last lecture in this chapter (L4.6), a review of one good systematic attempt to generate an appropriate range of such alternative hypotheses in a particular social science setting (the policy intervention).

The remaining pair of examples in this chapter can each be understood via this rubric of "plausible rival hypotheses," as each in its own way proceeds to demolish the complete list of those other hypotheses one by one. From a great many possible choices of example I have selected two to illustrate the diversity of potential application areas. In Section E4.7, Stanley Milgram succeeded in arguing the causal primacy of a scripted psychological setting (as opposed to the individual subject's personality) in his study of obedience. By manipulating the script of these experimental sessions, Milgram could induce any outcome behavior between nearly 100% obedience and 100% defiance (for definitions, see the main discussion). This was thus an abduction following from actual numerical control of a psychological process. The final example (E4.8) précises an article from 1980 in the weekly journal *Science*, abducing that the cause of the Cretaceous–Tertiary extinction must have been the collision with Earth of an asteroid about 10 kilometers across. The logic here is a startlingly long chain of abductions first rejecting everybody else's hypothesis and then showing a consilience of multiple data streams with respect to their own. The example is particularly striking inasmuch as after the death of the senior author the "smoking gun" of this theory, the crater of the actual asteroid responsible for the mayhem, was discovered off the Yucatán peninsula of Mexico at just the size and just the age it was supposed to have. This is remarkably like the denouement of the *Scorpion* story, when the wreckage was discovered just where it was supposed to be – in either case, closing the abduction by proof positive.

E4.1 John Snow on the Origin of Cholera

John Snow's *On the Mode of Communication of Cholera* is generally regarded as the first great work of epidemiology and surely the finest ever carried out by a lone investigator. It is here that the idea of the "natural experiment" arose: a type of *rhetoric* corresponding to an observational study (hence *not* an experiment) in which, owing to mixing or other chaotic processes, differences in groups can be argued to support solely the hypothesis proferred by the investigator. Any reader who wants to follow the argument here in more detail should study Snow's original text. A mere 139 pages long, it is available on the Web at http://www.ph.ucla.edu/epi/snow/snowbook.html, the same website from which the illustrations in Section E4.1.1 have been drawn.

The study Snow is describing here was his second on the subject of London cholera epidemics. Already in 1853, based on carefully scrutinized anecdotal evidence, Snow had reason to suspect that cholera was a communicable disease caused by inadvertently ingesting "morbid matter" from the "evacuations" of earlier victims, and the most plausible mechanism of such ingestion was drinking water. In one of its early weeks the 1854 epidemic seemed to have a concentration in the vicinity of Broad Street, and Snow came to suspect the pump there (the water from which was indeed funny-looking). For the 89 cholera deaths in the vicinity in the week ending September 2, 1854, for instance,

> ...there were sixty-one instances in which I was informed that the deceased persons used to drink the pump-water from Broad Street, either constantly or occasionally. In six instances I could get no information, owing to the death or departure of every one connected with the deceased individuals; and in six cases I was informed that the deceased persons did not drink the pump-water before their illness.
>
> The result of the inquiry then was, that there had been no particular outbreak or increase of cholera, in this part of London, except among the persons who were in the habit of drinking the water of the above-mentioned pump-well.... It will be observed [on the famous map of 616 cases, over the full run of the epidemic, my Figure 4.1] that the deaths either very much diminished, or ceased altogether, at every point where it becomes decidedly nearer to send to another pump than to the one in Broad Street.

(I have omitted the charming text passage explaining away many populations that did *not* have an elevated risk of cholera, for instance, workers at a brewery, who drank water from their own well.)

The 61-to-6 ratio certainly meets Peirce's requirements for an abduction as they would be set out 20 years later, in that the intrusion of so large a numerical value into a discussion previously the domain of merely anecdotal evidence is indeed quite surprising. The hypothesis that the disease was waterborne thereby instantly became a great deal more plausible, and this new map-based evidence persuaded the "Board of Guardians of St. James's parish" to remove the pump handle on September 7. Even though the epidemic was already fading anyway, Snow had no doubt that his hypothesis was correct: "Whilst the presumed contamination of the water of the Broad Street pump with the evacuations of cholera patients affords an exact explanation of the fearful outbreak of cholera in St. James's parish, there is no other circumstance which offers any explanation at all, whatever hypothesis of the nature and cause of the malady be adopted." Note the explicit logic of the abduction here, and also the odd usage of the word "exact" to mean "qualitatively correct" instead of "quantitatively correct." We will see this usage again in Perrin's text, but since then it has become more fugitive.

E4.1.1 The Basic Abduction

The intervention Snow adopted, removing pump handles, was capable only (at best) of truncating epidemics, not of preventing them. But water companies had the power of more extensive interventions. Over the period from 1849 to 1853, while London

Figure 4.1. Detail of John Snow's map of cases of cholera in the vicinity of the Broad Street pump, epidemic of 1854. This and Figures 4.2 to 4.5 are from Snow's 1855 publication. (Courtesy of School of Public Health, University of California at Los Angeles. Available at: http://www.ph.ucla.edu/EPI/snow/snowbook.html.)

was free of cholera, one water company, the Lambeth Company, moved its intake upstream, but not the the Southwark and Vauxhall Company. When the cholera returned in 1853, there was now a pattern of differential prevalence that had not been there in 1849 (see his Table 3, my Figure 4.2). "The districts partially supplied with the improved water suffered much less than the others, although, in 1849, when the Lambeth Company obtained their supply opposite Hungerford Market, these same districts suffered quite as much as those supplied entirely by the Southwark and Vauxhall Company." The districts referred to in this table are also named on Snow's own map of the coverage of London by these two water companies. I reproduce a detail of that map here as Figure 4.3.

Snow's observation about the emerging pattern of differential cholera risk in 1853 comes after a long narrative on the history of clean water in London, a history that makes pretty depressing reading even today. Specifically, after the epidemic of 1832, some water companies moved their intakes out of the Thames,

TABLE VI.

Sub-Districts.	Population in 1851.	Deaths from Cholera in 1853.	Deaths by Cholera in each 100,000 living.	Water Supply.
St. Saviour, Southwark	19,709	45	227	
St. Olave . . .	8,015	19	237	
St. John, Horsleydown	11,360	7	61	
St. James, Bermondsey	18,899	21	111	
St. Mary Magdalen .	13,934	27	193	
Leather Market .	15,295	23	153	Southwark and
Rotherhithe* . .	17,805	20	112	Vauxhall Water
Wandsworth . .	9,611	3	31	Company only.
Battersea . . .	10,560	11	104	
Putney . . .	5,280	–	–	
Camberwell . .	17,742	9	50	
Peckham . . .	19,444	7	36	
Christchurch, Southwk.	16,022	7	43	
Kent Road . .	18,126	37	204	
Borough Road . .	15,862	26	163	
London Road . .	17,836	9	50	
Trinity, Newington .	20,922	11	52	
St. Peter, Walworth .	29,861	23	77	
St. Mary, Newington .	14,033	5	35	Lambeth Water
Waterloo (1st part) .	14,088	1	7	Company, and
Waterloo (2nd part) .	18,348	7	38	Southwark and
Lambeth Church (1st part) .	18,409	9	48	Vauxhall Company.
Lambeth Church (2nd part)	26,784	11	41	
Kennington (1st part) .	24,261	12	49	
Kennington (2nd part) .	18,848	6	31	
Brixton . . .	14,610	2	13	
Clapham . .	16,290	10	61	
St. George, Camberwell .	15,849	6	37	
Norwood . . .	3,977	–	–	Lambeth Water
Streatham . .	9,023	–	–	Company only.
Dulwich . . .	1,632	–	–	
First 12 sub-districts .	167,654	192	114	Southwk. & Vaux.
Next 16 sub-districts .	301,149	182	60	Both Companies.
Last 3 sub-districts .	14,632	–	–	Lambeth Comp.

Figure 4.2. Table VI of Snow, 1855: deaths from cholera by neighborhood, epidemic of 1853, in the regions served by one or the other of the water companies in Figure 4.3.

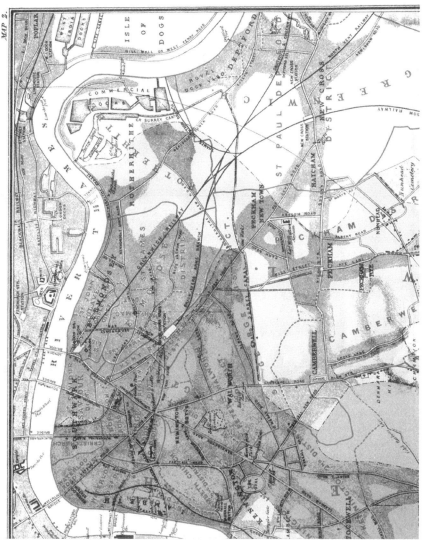

Figure 4.3. Detail of the second map from Snow (1855) – a full-color foldout in the original publication (see Plate 4.3) – indicating the neighborhoods served by the two water companies of interest, Lambeth and Southwark & Vauxhall, and the region of their overlap making Snow's inference possible.

and others began to filter their water, but not all, and so ecological correlations (to use the modern jargon) began to emerge between water service and death rates from cholera. Nevertheless, neighborhoods that differ in water supply would differ as well in a tremendous number of other potentially relevant factors. Snow determined to expend an enormous amount of "shoe leather," in David Freedman's (1991) felicitous phrase, to close the issue: to *prove* that it was the water supply that was responsible for differences in cholera prevalence.

At this point Snow explains the reasoning of the natural experiment, as I have already hinted.

> The intermixing of the water supply of the Southwark and Vauxhall Company with that of the Lambeth Company, over an extensive part of London, admitted of the subject being sifted in such a way as to yield the most incontrovertible proof on one side or the other. In the sub-districts enumerated in the above table as being supplied by both Companies, the mixing of the supply is of the most intimate kind. The pipes of each Company go down all the streets, and into nearly all the courts and alleys. A few houses are supplied by one Company and a few by the other, according to the decision of the owner or occupier at that time when the Water Companies were in active competition. In many cases a single house has a supply different from that on either side. Each company supplies both rich and poor, both large houses and small; there is no difference either in the condition or occupation of the persons receiving the water of the different Companies. As there is no difference whatever, either in the houses or the people receiving the supply of the two Water Companies, or in any of the physical conditions with which they are surrounded, it is obvious that no experiment could have been devised which would more thoroughly test the effect of water supply on the progress of cholera than this, which circumstances placed ready made before the observer.
>
> The experiment, too, was on the grandest scale. No fewer than three hundred thousand people of both sexes, of every age and occupation, and of every rank and station, from gentlefolks down to the very poor, were divided into two groups without their choice, and, in most cases, without their knowledge; one group being supplied with water containing the sewage of London, and, amongst it, whatever might have come from the cholera patients, the other group having water quite free from such impurity.
>
> To turn this grand experiment to account, all that was required was to learn the supply of water to each individual house where a fatal attack of cholera might occur. I regret that, in the short days at the latter part of last year [1853], I could not spare the time to make the inquiry; and, indeed, I was not fully aware, at that time, of the very intimate mixture of the supply of the two Water Companies, and the consequently important nature of the desired inquiry.
>
> When the cholera returned to London in July of the present year [1854], however, I resolved to spare no exertion which might be necessary to ascertain the exact effect of the water supply on the progress of the epidemic, in the places where all the circumstances were so happily adapted for the inquiry.

With extraordinary care, Snow ascertained the identity of the water supply household by household for every case of cholera in the neighborhoods served by both companies. It turned out that in addition to checking by payment records, he could verify by salt content: the downstream intake was so far downstream that in a dry summer it took on salt from sea water! (Remarkably enough, he was also able to locate the initial case setting off the epidemic, that of a sailor dying along the Thames in July whose bedlinens were thereafter washed in the river.)

Snow presents his results in an extraordinary three-way table, district by water supply by temporal collection window (see his Tables 7 and 8, my Figures 4.4 and 4.5).

> As 286 fatal attacks of cholera took place, in the first four weeks of the epidemic, in houses supplied by the former Company, and only 14 in houses supplied by the latter, the proportion of fatal attacks to each 10,000 houses was as follows. Southwark and Vauxhall 71. Lambeth 5. The cholera was therefore fourteen times as fatal at this period, amongst persons having the impure water of the Southwark and Vauxhall Company, as amongst those having the purer water from Thames Ditton.

Something that would interest a modern epidemiologist happens thereafter in this quantification: the relative prevalence rate *falls* from 14:1 at four weeks to 8.5:1 at seven weeks and, finally, 5.8:1 for the full run of the epidemic. Snow explicitly explains this away:

> As the epidemic advanced, the disproportion between the number of cases in houses supplied by the Southwark and Vauxhall Company and those supplied by the Lambeth Company, became not quite so great, although it continued very striking. In the beginning of the epidemic the cases appear to have been almost altogether produced through the agency of the Thames water obtained amongst the sewers; and the small number of cases occurring in houses not so supplied, might be accounted for by the fact of persons not keeping always at home and taking all their meals in the houses in which they live; but as the epidemic advanced it would necessarily spread amongst the customers of the Lambeth Company, as in parts of London where the water was not in fault, by all the usual means of its communication.

In these extensive tables, Snow takes great pains to show that the effect of water supply dominates any ecological effect that may remain. The tables cross-classify 1514 cholera cases by water supply and subdistrict. (The matter of subdistricts, which were highly segregated by geography, social class, and mode of labor, was a central concern of many of the competing explanations of cholera.) In the text corresponding to this tabulation Snow touches on the ecological arguments from historical data sources with which he began; but the argument from prevalences in 1854 alone is the finest, strongest, most effective epidemiological argument ever made, in view of the natural experimental design.

> It may, indeed, be confidently asserted, that if the Southwark and Vauxhall Water Company had been able to use the same expedition as the Lambeth Company in completing their new works, and obtaining water free from the contents of sewers, the late epidemic of cholera would have been confined in a great measure to persons employed among the shipping, and to poor people who get water by pailsful direct from the Thames or tidal ditches.

TABLE VII.
The mortality from Cholera in the four weeks ending 5th August.

Sub-Districts.	Population in 1851.	Deaths from Cholera in the four wks. ending 5th August.	Water Supply.				
			Southwark & Vauxhall.	Lambeth.	Pump-wells.	River Thames, ditches, etc.	Unascertained.
St. Saviour, Southwark	19,709	26	24	–	–	2	–
St. Olave, Southwark	8,015	19	15	–	–	2	2
St. John, Horsleydown	11,360	18	17	–	–	1	–
St. James, Bermondsey	18,899	29	23	–	–	6	–
St. Mary Magdalen	13,934	20	19	–	–	1	–
Leather Market	15,295	23	23	–	–	–	–
Rotherhithe	17,805	26	17	–	–	9	–
Battersea	10,560	13	10	–	1	2	–
Wandsworth	9,611	2	–	–	–	2	–
Putney	5,280	1	–	–	1	–	–
Camberwell	17,742	19	19	–	–	–	–
Peckham	19,444	4	4	–	–	–	–
Christchurch, Southwk.	16,022	3	2	1	–	–	–
Kent Road	18,126	8	7	1	–	–	–
Borough Road	15,862	21	20	1	–	–	–
London Road	17,836	9	5	4	–	–	–
Trinity, Newington	20,922	14	14	–	–	–	–
St. Peter, Walworth	29,861	20	20	–	–	–	–
St. Mary, Newington	14,033	5	5	–	–	–	–
Waterloo Road (1st)	14,088	5	5	–	–	–	–
Waterloo Road (2nd)	18,348	5	5	–	–	–	–
Lambeth Church (1st)	18,409	5	2	1	–	1	1
Lambeth Church (2nd)	26,784	10	7	2	–	–	1
Kennington (1st)	24,261	11	9	1	1	–	–
Kennington (2nd)	18,848	3	3	–	–	–	–
Brixton	14,610	1	–	1	–	–	–
Clapham	16,290	5	4	–	1	–	–
St. George, Camberwell	15,849	9	7	2	–	–	–
Norwood	3,977	–	–	–	–	–	–
Streatham	9,023	–	–	–	–	–	–
Dulwich	1,632	–	–	–	–	–	–
Sydenham	4,501	–	–	–	–	–	–
	486,936	334	286	14	4	26	4

Figure 4.4. Table VII of Snow, 1855, the table that heralded the invention of modern epidemiology. Cholera deaths by source of drinking water, epidemic of 1854: first four weeks. The relative risk Snow reports is the cross-ratio of column totals for Lambeth and Southwark & Vauxhall to the populations served.

TABLE VIII.
Mortality from Cholera in the seven weeks ending 26th August.

Sub-Districts.	Population in 1851.	Deaths from Cholera in the seven weeks ending 26th August.	Water Supply. Southwark & Vauxhall.	Lambeth.	Pump-wells.	River Thames and ditches.	Unascertained.
*St. Saviour, Southwark	19,709	125	115	–	–	10	–
*St. Olave, Southwark	8,015	53	43	–	–	5	5
*St. John, Horsleydown	11,360	51	48	–	–	3	–
*St. James, Bermondsey	18,899	123	102	–	–	21	–
*St. Mary Magdalen	13,934	87	83	–	–	4	–
*Leather Market	15,295	81	81	–	–	–	–
*Rotherhithe	17,805	103	68	–	–	35	–
*Battersea	10,560	54	42	–	4	8	–
Wandsworth	9,611	11	1	–	2	8	–
Putney	5,280	1	–	–	1	–	–
*Camberwell	17,742	96	96	–	–	–	–
*Peckham	19,444	59	59	–	–	–	–
Christchurch, Southwk.	16,022	25	11	13	–	–	1
Kent Road	18,126	57	52	5	–	–	–
Borough Road	15,862	71	61	7	–	–	3
London Road	17,836	29	21	8	–	–	–
Trinity, Newington	20,922	58	52	6	–	–	–
St. Peter, Walworth	29,861	90	84	4	–	–	2
St. Mary, Newington	14,033	21	19	1	1	–	–
Waterloo Road (1st)	14,088	10	9	1	–	–	–
Waterloo Road (2nd)	18,348	36	25	8	1	2	–
Lambeth Church (1st)	18,409	18	6	9	–	1	2
Lambeth Church (2nd)	26,748	53	34	13	1	–	5
Kennington (1st)	24,261	71	63	5	3	–	–
Kennington (2nd)	18,848	38	34	3	1	–	–
Brixton	14,610	9	5	2	–	–	2
*Clapham	16,290	24	19	–	5	–	–
St. George, Camberwell	15,849	42	30	9	2	–	1
Norwood	3,977	8	–	2	1	5	–
Streatham	9,023	6	–	1	5	–	–
Dulwich	1,632	–	–	–	–	–	–
Sydenham	4,501	4	–	1	2	–	1
	486,936	1514	1263	98	29	102	22

Figure 4.5. Table VIII of Snow, 1855: The same after seven weeks of the epidemic. The relative risk has dropped, but Snow's hypothesis continues to be strongly supported.

(In its rhetoric this is the classic counterfactual interpretation of a causal relationship. See Section L4.5.2.)

What makes the persuasion so crushing, as in my other favorite examples of coercive numerical rhetoric (Perrin, Watson, Marshall), is not the numerical values of the relative risks per se, but the survival of one sole hypothesis (one the researcher knows in advance to be true) at the end of the project. These studies are not tests of hypotheses, but serious play with rhetoric designed to refute possible objections. The methodology may be quantitative, but the conclusion is entirely a qualitative one: the water supply accounts for the epidemic of cholera (or atoms exist, or ulcers are infectious, or DNA is a double helix). That is, the process is a numerical inference, from an arithmetic summary to a verbal proposition it demonstrates. Snow himself admits that his motivation was in no way limited to the interest of the scientist (a word that had just been coined by his countryman William Whewell):

> I was desirous of making the investigation myself, in order that I might have the most satisfactory proof of the truth or fallacy of the doctrine which I had been advocating for five years. I had no reason to doubt the correctness of the conclusions I had drawn from the great number of facts already in my possession, but I felt that the circumstance of the cholera-poison passing down the sewers into a great river, and being distributed through miles of pipes, and yet producing its specific effects, was a fact of so startling a nature, and of so vast importance to the community, that it could not be too rigidly examined, or established on too firm a basis.

That is, his purpose in carrying out this huge study was a *rhetorical* purpose: to be as persuasive as possible regarding what he knew had to be done.

Snow has no theory specifying a relative risk of 14:1 for the onset of the epidemic, or for the decline of that ratio by a little more half over the next 10 weeks. Instead his essay emphasizes two facts that follow from the hypothesis of waterborne cholera transmission "as a matter of course," in Peirce's phrase, but that, as Snow notes, are incomprehensible on any other explanation whatsoever: the huge magnitude of these relative risks, and the nearly universal polarity of the prevalence ratio over the 32 subdistricts over all time intervals. It was "incontrovertible proof" that Snow was after – likewise Watson, Perrin, and Marshall. Attempts at "controverting" are a function of the nature of the scientific discipline involved, but the way in which quantitative argument closes off controversy, in all these canonical examples of mine, is aligned with Peirce's mode of abduction: maximize the surprise of a "surprising fact C," then show that it follows as a matter of course from hypothesis A. If C is surprising enough, and no other hypothesis is left standing, it is reasonable for the reader to consider hypothesis A, and proceed accordingly (to move water intakes, study DNA, treat ulcers with antibiotics, or smash atoms).

E4.1.2 Two Modern Analogues

Snow's narrative has the arc of a theatrical drama, with the closing of the abduction as its climax. The inference is powerful beyond all reasonable doubt, as otherwise an incomprehensible coincidence must have occurred in a fantastic number of separate cases. Such an inference, so similar to that of a good police procedural novel or sound criminal conviction, is of course not restricted to nineteenth-century London. Here are two further examples likewise from the domain of epidemiology, one concerned with an infectious disease, the other not.

The 1979 outbreak of anthrax in Sverdlovsk. This story is the topic of a full-length book (Guillemin, 1999). Unlike the situation for John Snow, the epidemic was traced to its cause not in real time but only 20 years later, after major political changes made investigation possible (although immensely labor-intensive).

In April 1979 there seems to have been a great increase in the rate of deaths from anthrax in Sverdlovsk, Soviet Union (now Russia). The issue was not the existence of the epidemic, but its cause. Anthrax is relatively common in animals but rare in humans except when contracted from animals, and the Soviet government had insisted that this was such a meat-related outbreak. For the purpose of a political analogy with another anthropogenic epidemic, related to defoliants in Vietnam, Guillemin wished to assess the cause of this particular outbreak more precisely. In a project at least as dependent on "shoe leather" – household-by-household interviews – as Snow's, she examined the lives of a number of persons whose headstones in the Sverdlovsk cemetery indicated their date of death as April 1979 at an inappropriately young age. By interviews with their surviving family members and friends, she placed most of them either within a secret factory or downwind of that factory early in the afternoon of April 2. The reported deaths of farm animals arose along the same direction line, but further downwind. This is the consilience step highlighted in Chapter 2: in this case, a numerical agreement between a human infection pattern, an animal infection pattern, and the historical records of wind direction. Like the case for *Scorpion*, there are other facts in evidence that, although not numerical, are consistent with this: a rich series of interview reports that something strange happened at that factory early on the afternoon of April 2, resulting in emission of odd clouds of vapor.

The abduction here corresponds to that in Snow's 1853 study, an abduction from a map. To Snow, consilience came from household billing information, or water chloride content, or both; for Meselson and Guillemin, it came from meteorology. In the three maps from Guillemin, 1999 (Maps 1, 2, 3 between pages 238 and 239, which I have collected in Figure 4.6), we see (from left to right) an (uninformative) pattern of the residences of those additional deaths, then a map of where these citizens were likely to have been located during the workday on Monday, April 2, 1979, and finally a map of the reported animal deaths. The ellipses drawn through the center of that factory in Map 2 are aligned with the reported wind direction that afternoon to visualize the surviving theory here: they are plausible contours of density of the anthrax spores that participated in those releases of "vapor." The inference seems ineluctable: the animals were not a cause of the human deaths but another consequence of the same cause, a cause located somewhere near the uppermost cluster of circles in Map 2, where there happened to be a biowarfare factory whose existence was denied until well after the fall of the Soviet Union 10 years later. See Meselson et al. (1994) or Guillemin (1999).

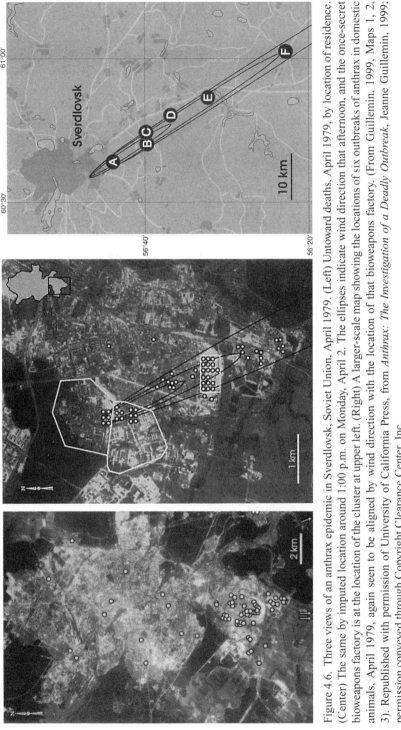

Figure 4.6. Three views of an anthrax epidemic in Sverdlovsk, Soviet Union, April 1979. (Left) Untoward deaths, April 1979, by location of residence. (Center) The same by imputed location around 1:00 p.m. on Monday, April 2. The ellipses indicate wind direction that afternoon, and the once-secret bioweapons factory is at the location of the cluster at upper left. (Right) A larger-scale map showing the locations of six outbreaks of anthrax in domestic animals, April 1979, again seen to be aligned by wind direction with the location of that bioweapons factory. (From Guillemin, 1999, Maps 1, 2, 3). Republished with permission of University of California Press, from *Anthrax: The Investigation of a Deadly Outbreak*, Jeanne Guillemin, 1999; permission conveyed through Copyright Clearance Center, Inc.

A pattern of asthma in Barcelona. The discussion here is from Goldstein and Goldstein (2002). In the early 1980s Barcelona, Spain, was experiencing a series of sporadic outbreaks of asthma as detected by tabulations of hospital admission reports. In a section entitled "Barcelona: A detective story," the Goldsteins write, "One of the epidemics involved a forty-fold abrupt increase in asthma visits at several hospital clinics within a four-hour period." There is then a meaningless significance test: "The probability that this could have happened by chance ... was 1.5×10^{-17}." A map recording the locations where the attacks took place (analogous to Guillemin's map of "where they were on the afternoon of Monday, April 2") showed that most of the attacks occurred near the port area of Barcelona (Figure 4.7, map). A records search of twelve other asthma epidemics showed that their victims, too, had been either in the port area or downwind of it. Following a range of other guesses, two physicians thought to check on what was being unloaded in the port, resulting in the table on the right in the figure. The comparison of interest is the 13-to-0 row for epidemic days versus soybean unloading. Further investigation showed that most asthma victims were indeed sensitive to soybeans. Once filters against soybean dust were installed throughout the harbor, the epidemics ceased.

We see here all of the logical elements that we reviewed in Chapters 1, 2, and 3: the surprise of a pattern, the sytematic examination of hypotheses, and the survival of one of them by explicit numerical consilience with other channels of quantitative evidence. In the cholera and the asthma examples, as for *Scorpion*, verification was by a direct intervention. In the anthrax matter, verification was by more standard historical techniques, interviews with people who actually knew what had been going on and who were no longer sworn to secrecy. In all these cases, we have met the "interocular trauma test" (see Epigraph 1) of Edwards et al.: the combination of abduction and consilience results in an important conclusion that "hits you between the eyes."

L4.1 The Logic and Algebra of Least Squares

There is a small core of statistical ideas that every educated person, scientist or not, ought to understand: least squares, the Normal or Gaussian distribution, inverse probability (probability of hypotheses), and regression. Without these ideas, one cannot appreciate the effective summaries of quantitative data that drive our technology and our society.

L4.1.1 Sums of Squares

Whenever a ship is in distress, and the master deliberately makes a sacrifice of any part of the lading, or of the ship's furniture, masts, spars, rigging, &c. for the preservation of the rest, all the property on board, which is saved by the sacrifice, must contribute toward the value of what is thus sacrificed. The contribution is called a GENERAL AVERAGE; and the property sacrificed is called the jettison.

– Emerson (1850, p. 124)

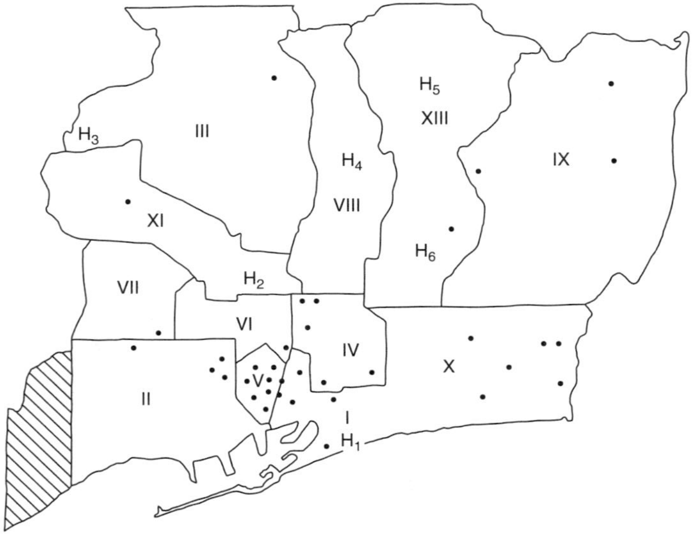

TABLE 9-1 Cargoes Unloaded and Asthma Epidemic Days

	Number of Days Each Cargo Was Unloaded		Number of Days Each Cargo Was not Unloading	
Cargo	Number of Days of No Epidemics	Number of Epidemic Days	Number of Days of No Epidemics	Number of Epidemic Days
Coal	196	4	521	9
Fuel oil	150	3	567	10
Gasoline	180	2	537	11
Cotton	399	7	318	6
Coffee	300	5	417	8
Corn	135	1	582	12
Soybeans	249	13	468	0
Butane	140	1	577	12

Figure 4.7. Convincing analysis of a pattern of epidemics of asthma, Barcelona, Spain, 1984. (Left) Map of cases, showing a concentration in the area of the port. (Right) Table showing asthma epidemic occurrence by nature of cargoes being unloaded, two years' data. Inference is from the absence of asthma epidemics on days when soybeans are not being unloaded. (From Goldstein and Goldstein, 2002.) By permission of Oxford University Press.

Take any collection x_i of N numbers. You are familiar with their **average** or **mean** $\bar{x} = (x_1 + x_2 + \cdots + x_N)/N = (\Sigma_{i=1}^{N} x_i)/N$, their total divided by their count. (The symbol that is the letter x with a bar over it is read, reasonably enough, as "x bar," and Σ, Greek capital sigma, is the "summation symbol." This is a nice example of the power of notation to condense. In fact, we immediately condense further: the $_{i=1}^{N}$ will be dropped.) Sometimes this sort of computation seems intuitively to be nonsense (for instance, the average number of hockey goals scored per citizen of the United States); sometimes it seems to be perfectly reasonable (average of the radii of the gamboge spheres that we will encounter in Perrin's experiment, Section E4.3). In a sense, *all* of statistics is a systematic philosophicoscientific exploration of the properties and meanings of averages. But this is a course in numerical inferences: the criterion of "meaning" refers to the *inferences* in the scientific or social context in which an average is intended to be used.

The **sum of squares** of N x's around a value y is just what it says: the quantity $\Sigma(x_i - y)^2$. As a function of y, this sum of squares is smallest when $y = \bar{x}$: the sum of squares is smallest around the average. This is easy to show by taking derivatives ($\frac{d}{dy} \Sigma(x_i - y)^2 = \Sigma \frac{d}{dy}(x_i - y)^2 = \Sigma(-2)(x_i - y) = -2(\Sigma x_i - Ny)$; this is zero when $\Sigma x_i = Ny$ or $y = \bar{x}$), but it's just as convenient to use elementary algebra:

$$\begin{aligned} \Sigma(x_i - y)^2 &= \Sigma((x_i - \bar{x}) + (\bar{x} - y))^2 \\ &= \Sigma(x_i - \bar{x})^2 + \Sigma(\bar{x} - y)^2 + 2\Sigma(x_i - \bar{x})(\bar{x} - y) \\ &= \Sigma(x_i - \bar{x})^2 + N(\bar{x} - y)^2 + 2(\bar{x} - y)\Sigma(x_i - \bar{x}) \\ &= \Sigma(x_i - \bar{x})^2 + N(\bar{x} - y)^2, \end{aligned} \quad (4.1)$$

since $\Sigma(x_i - \bar{x}) = \Sigma x_i - N\bar{x} = 0$. Only the second term involves y, and it is clearly smallest when $y = \bar{x}$.

Mathematical note 4.1. On extended equalities. Let me digress (in the eyes of many of my readers, anyway) to have a closer look at what is going on in the preceding few lines. We are manipulating letters on a page or a blackboard. At the same time, we are stating that two ways of doing the arithmetic will amount to the same outcome, and so the "meaning" of the computation can go with whichever interpretation, the first line or the last, is the more congenial. This is mathematics in its role of producing identities. At many other spots in this chapter and the next we will encounter chains of equalities or near-equalities like this, and one of the central cultural facts about mathematics is that competent mathematicians can all be brought to agree on the manipulation across each equality separately and agree also on the inference that the left and right ends of the chain, *no matter how long it is* (Wigner refers to "thousands" of elementary steps like these), must thereby be the same quantity in every application to real quantities. For more on the culture of mathematics, see, for instance, Hersh (1997) or Byers (2007).

Only a mathematician would even *think* of launching herself down such a chain, because the very first step, the subtracting and then adding $\bar{x}$ in getting from the first expression to the second, is so obviously arbitrary *in this role*. It is justified only by the simplification at which we arrive by the end, which is the cancellation $\Sigma(x_i - \bar{x}) = 0$ that accompanies the last equality (a justification appearing in the text that immediately followed).

Both of these quantities, $\bar{x}$ and $\Sigma(x_i - \bar{x})^2$, are "physical," meaning, in this case, **kinematical**, expressions of the formalized geometry of the (nonrelativistic, pre-Einsteinian) real world. If we imagine a collection of particles of mass 1 located at each value x_i on a "massless rod," then the value $\bar{x}$ is the point at which the rod will balance (say, on your finger), and the quantity $\Sigma(x_i - \bar{x})^2$ is proportional to the energy it will take to spin the rod around that point of support. (This is a typical way that physicists talk about idealized experiments – for instance, there is no mention of whatever armature would be required to hold those particles in place, only the assumption that it wouldn't weigh anything.) Sums of squares are like energies in lots of ways; we make extensive use of this analogy in Part III of this book.

It will be useful to talk about a second derived quantity, that actual minimizing sum of squares per original item. This is $(\Sigma(x_i - \bar{x})^2)/N$, the **variance** of the set of N x's. (It is also sometimes called the **mean square**, as it likewise takes the form of an average.) It is often more useful to talk about a quantity that has the same units as the x's that were originally measured; one such quantity is the **standard deviation** (s.d.), square-root of the variance. The 19th-century scientist would use the *probable error*, the median of $|x_i - \bar{x}|$. When data are Normally distributed, this is about two-thirds of the standard deviation. By definition, 50% of the x's are within one probable error of the mean, whereas about 68% of the x's can be expected to be within 1 s.d. of their average (the segment inside the "waist" of the curve, Figure 4.12).... But I am getting much too far ahead of the argument at this early point in the lecture series.

We'll need one more simple property of these descriptors. Suppose we change all the x's by multiplication by a constant. For instance, the new x's might be the old ones divided by 10. Then, of course, the average is changed by the same factor of 10. The variance, because it is a sum of *squares*, will change by the *square* of the factor (here, a division by $10^2 = 100$); the standard deviation, scaling as the square-root of the variance, will be altered by the same factor as the original data. This is good, since it's supposed to represent a sort of "width" on the physical bar we're talking about.

L4.1.2 Precision of Averages, and Weighted Averages

For convenience, let's take a very simple "measurement," the propensity of an ordinary coin to come up heads. We measure this by flipping the coin and observing what comes up: heads or tails, each with probability one-half.[1] The example of the coin-flip will occupy us extensively later on, in connection with the "bell curve" (the Normal or

---

[1] The "flip of a fair coin" is the epitome of this fully randomized random process for our teaching purpose here. However, it is a poor actual physical prototype. Diaconis et al. (2007) show that for tosses of realistic magnitude (torque and height), the chance of an actually tossed coin flipping over before being caught is less than 0.5. Gas molecules, as in Section E4.3, are much more effective; even better are the computational random-number generators built into most statistical software.

L4.1 The Logic and Algebra of Least Squares

Gaussian distribution). For now, there is no need to go to limits, as all we need is the logic of sums of squares.

We can imagine measuring the propensity of a coin to come up heads as the average of a large number of coin flips each coded 1 for heads, 0 for tails. It will be very interesting to understand how the variance of this average (or its standard deviation, which conveys the same information) changes as a function of the number K of times we have flipped the same coin.

Begin as simply as possible. One coin flip yields 0 half the time, and 1 half the time. The average is $1/2$ and the the deviation from the mean is always $1/2$ one way or the other, so that its square is always $1/4$, and so must average $1/4$ – this is the variance. We can think of this as "two cases," the two outcome possibilities of the one coin.

Flip the coin again, and total the heads. You get no heads 25% of the time, two heads 25% of the time, and just one head the other half of the time. The average is 1, as it must be $(1/2 + 1/2)$; the variance of the total is what is interesting. We are 1 away from the average half the time (all heads or all tails), and 0 away the other half the time (1 of each), for a total sum of squares around the mean of 2; divided by $N = 2^K = 4$ cases, that gives us a mean-square of $1/2$, which is $1/4 + 1/4$ – the sum of the variances of the two coins considered separately.

This fact is perfectly general – actually, it is just another version of the Pythagorean theorem: **variances of independent processes add when the scores are totaled**. As a theorem it rests upon the definition of "independent," which is an important part of probability theory, more than on the algebra of sums of squares, so I won't go into it further here.

If you don't want to believe the general statement I just set out (and this is, after all, a course in skeptical reasoning about numbers), you can compute the same result for a fair coin just as easily from first principles. We want to know the variance of a collection of x's that number, in total, $N = 2^K$ – the number of ways of flipping K fair coins is $2 \times 2 \times \cdots \times 2$ – where each x is the count of heads for its sequence of coins. We'll proceed by mathematical induction (Section 3.3). Supposing the variance of the count of heads is $K/4$ for K coins (which we have seen is true for $K = 1$ or 2), here's how to show that the variance is exactly $1/4$ more for the same computation but with $K + 1$ coins.

If we divide equation (4.1) by N on both sides, we arrive at

$$\Sigma(x_i - y)^2/N = \Sigma(x_i - \bar{x})^2/N + (\bar{x} - y)^2,$$

meaning that the mean square of a list of x's around any number y is equal to their variance plus the square of the distance from the average to y. Let's use that for all the flips of $K + 1$ coins by dividing those series of flips into two subsets: those that ended with a head, and those that ended with a tail.

For those 2^K sequences that ended with a head, by assumption, the sum of squares around the mean is $K/4$. The mean around which they have that average, though, is $(K/2) + 1$. That mean is not the correct mean for the $(K + 1)$-coin problem; we have displaced the correct mean $(K + 1)/2$ by half a head, and so we need to add the term $(\bar{x} - y)^2$, which is another $(1/2)^2 = 1/4$, to the mean square, which therefore becomes $K/4 + 1/4 = (K + 1)/4$. Similarly, the sum of squares of the second half of the flips, those that ended in a tail, is $K/4$ around a mean of $K/2$, so we add in another mean square of $(1/2)^2 = 1/4$ corresponding to the shift of *that* mean to $(K + 1)/2$.

Hence: however the last coin came up, heads or tails, its effect on the whole distribution of all the series of flips that preceded is to add just $1/4$ to the expected mean-square: the square of the $1/2$ by which the new coin *must* fail to shift the mean by precisely the $1/2$ of its average. The first coin is just like the others in this regard.

Mathematical note 4.2. On coin models. Is this an argument about the real world (the effect of counting one more coin flip) or the properties of mathematical symbols? Both, at the same time. The passage from k coins to $k + 1$, which is a physical act, is isomorphic with the passage from the sum of k events to the sum of $k + 1$ that rearranges symbols on the blackboard. All we are assuming is that the physical act of counting is properly interpreted as adding 1 over and over: that the integers match the counting numbers.

We've shown, then, that the variance of the total number of heads in a flip of K fair coins is $K/4$. The average number of heads is this count divided by K, so the **variance of the average**, the measuring process we're really interested in, is diminished as the *square* of this factor; it becomes $(K/4)/K^2$, or $1/4K$. The standard deviation of this average (which is called **the standard error of the mean** in this simple model) varies as the square-root of the variance, or $1/2\sqrt{K}$. This is a precision of the mean, the plus-or-minus around it for the frequency of heads. For 100 flips of a fair coin, for instance, we expect the frequency of heads to be 0.50 ± 0.05.

The precision of an average increases as the inverse square root of the number of observations that contributed to the average. We've shown this for coin flips, but it is generally true of averages of any measurement that *has* a variance. (Not every physically realistic process has a variance in this sense: one very simple kinematic system that does not is the location of the spot on the wall pointed at by a flashlight spinning around its axis that is sampled at a random moment.)

Suppose we flip a coin 25 times, and then 100 times, and then 400 times, and we get an average "propensity of heads" for each set of flips. Clearly we would multiply the 25-flip mean by 25 to get its total of heads, and the 100-flip mean by 100 to get its total of heads, and the 400-flip mean by 400 to get its total of heads, and add up all the resulting totals, then divide by the total number of flips. The 25-flip mean has a standard error of $1/2\sqrt{25}$, however, the 100-flip mean of $1/2\sqrt{100}$, and the 400-flip mean a standard error of $1/2\sqrt{400}$.

So in converting back from averages to totals, for these coins, we have in effect **divided by the variances of the individual means**. The weight of each contribution to a pooled estimate is proportional to the reciprocal of the **variance** of that contribution. Again, it is easy enough to prove, using only a little more algebra, that this statement is just as true of the general estimate of almost *anything* that arises by combination of independent observations.

Here's how this works. It's enough to talk about combining two different observations of the same underlying physical quantity (though this is itself a hypothesis, not an axiom). Let A and B be independent observations of the same mensurand (e.g., a measurement twice by the same instrument, or by two instruments each calibrated at the factory) with variances V_A and V_B. For instance, A might be an average of $0.25/V_A$ coin flips, and B an average of $0.25/V_B$; or A might be a slope estimate made at a moment $1/\sqrt{V_A}$ units away from the mean time in one of the imaginary experiments with automobiles we will model in connection with Figure 4.8, and B a slope estimate made at a moment $1/\sqrt{V_B}$ units from the mean time, etc.

A weighted average of the measurements A and B is a new quantity $aA + bB$ where $a + b = 1$. Assuming the measurements are independent, the variance of this weighted average is $a^2 V_A + b^2 V_B$. This expression is easy enough to minimize – we have $\frac{d}{da}(a^2 V_A + b^2 V_B) = 2aV_A - 2(1-a)V_B$, which is zero where $a(V_A + V_B) = V_B$ or $a = \frac{V_B}{V_A + V_B}$, whereupon $b = \frac{V_A}{V_A + V_B}$ and the minimum value is $\frac{V_A V_B}{V_A + V_B} = 1/(\frac{1}{V_A} + \frac{1}{V_B})$. The weights are in precisely inverse proportion to the variances.

We have not yet justified **why** it might be a good idea to use this simple least-squares machinery for estimating quantities that are observed indirectly, or with noise. That argument requires a criterion for what it means to behave rationally in estimating quantities of this sort, and we need to observe a contingent aspect of the real world according to which the least-squares answer is the "right" one in a specific logical sense. The next subsection goes more deeply into all these issues.

L4.1.3 Least-Squares Fits to Linear Laws

Now consider an apparently more complicated situation: a physical system governed by a "law," which, for the purposes of these notes, will mean an exact empirical regularity – an exact equation – relating two or more items you've repeatedly measured. Assume, furthermore, that that regularity takes the mathematical form of a straight line. Theories tend to talk about the coefficients in the formulas for these lines – why the slopes have the values they do, why the intercepts are what they are. For now, just consider those "constants" – those slopes or intercepts (and their extensions for more complicated laws, like planetary orbits) – to be numbers we are interested in. Can we use the observed data that are supposed to be lawful as if collectively they supplied some sort of measurement of the underlying quantities that theory argues should be seen as the more real?

Yes, we can. Suppose, for instance, a vehicle is moving straight away from you at a constant speed. (How do you know that the speed is constant? What does that mean, anyway?) You have a stopwatch, and the car has an odometer, maybe the kind with digital read-out. (But how do you know if the readings of the odometer are evenly spaced? What does that mean, anyway?) At some moment when the car is already moving "at constant speed," you start the stopwatch. Then at N different moments, the driver of the car calls out the numbers on the odometer, and you note the reading of the stopwatch each time you hear him yell. **How fast is the car traveling?** (Ignore the speed of sound, the speed of neural processing of acoustic stimuli, etc.)

(But notice how those parenthetical phrases seem to pile up after a while. They are all worrying about the "evenness" (really, the lawfulness) of the phenomenon, the linearity of the measuring instruments (with respect to what? each other, it turns out), and the other real physical factors being treated as "negligible." The problems with evenness and linearity are ordinarily answered, at least in kinematic problems, by a gently circular argument centered around a modern version of Newton's First Law – "there exist inertial reference frames." The question of what to ignore is never closed.)

The crucial features of this setup, then, are the following:

- We can measure distance [position] and time.
- Our measurements always incorporate small errors.
- The laws in which we believe do not use these measurements directly. Raw data depend far too much on accidental factors such as where one happens to be standing when observing a mass on Earth or in the heavens. The successful laws are those that are independent of all these arbitrary choices, by specifying constant coefficients ("invariants") for relations among them. For instance, theories in astronomy do not deal with the direction in which the telescope is set to point, but reduce these observations first to a stable reference coordinate frame oriented on the plane of Earth's orbit (the "ecliptic").
- **The laws are exactly true**, in the sense that the apparent errors can be made as small as we please by incorporating instruments of steadily greater precision and adjusting again and again for small but measurably relevant contingent factors.

As laws go, the one about the motion of this car has unusually small scope: it will be expected to apply only when observers are on the same line as that on which the car is traveling, and only when the whole propulsive mechanism of the car is in a thermodynamic and aerodynamic and rheodynamic steady state with respect to its surroundings. Level road, warm engine, constant throttle setting, and so forth: all these are "calibrations."

Let's write s_1 for the first odometer reading called out by the driver, and t_1 for the time we read on the stopwatch when we hear him doing so. The pair of numbers (t_1, s_1) is the *first pair* of observations. Similarly, (t_2, s_2) is the second pair, et cetera, up through (t_N, s_N), for a total of N pairs. The typical intermediate term is (t_i, s_i), the ith pair.

We **know** a priori that these quantities, had they been measured perfectly, would satisfy some equation

$$s_i = s_0 + rt_i,$$

where s_0 is the odometer reading at the moment you started the stopwatch, and r is the speed of the car, the number we are seeking. (This "knowledge" is what we mean by saying that the car is traveling "at constant speed." Operationally, this is roughly the same as saying that there is a class of objects that we know to be *all* traveling at "constant velocity," constant direction as well as constant speed.) We are considerably more interested in r than in s_0: different observers with different stopwatches will

L4.1 The Logic and Algebra of Least Squares

have unrelated values for s_0, after all, but ought to arrive at nearly the same values for r if we are really "measuring" something. Nevertheless we will end up computing s_0 and r together, in a surprisingly symmetrical way.

Recall, now, that we are measuring with error. There is error in reading the stopwatch, and in its running; there is error in the mechanism of the odometer, and in the calling-out of its values. This means that the "law" really is

$$s_i - \epsilon_{s_i} = s_0 + r(t_i - \epsilon_{t_i})$$

where ϵ_{s_i} is the error of the ith odometer reading and ϵ_{t_i} that of the ith stopwatch reading.

The observer cannot separate the values of the two ϵ's given only the data of this simplistic measurement scheme. In principle, they could be estimated from additional data all their own, repeated measurement of the same (the ith) shout by the driver, or of the same stopwatch click. The 19th century made many important improvements in the understanding of this sort of measurement. But the car was assuredly *somewhere* when the stopwatch read "3 seconds," for instance, so we can rewrite the error equation for data pair i as

$$s_i = s_0 + rt_i - (r\epsilon_{t_i} - \epsilon_{s_i})$$
$$= s_0 + rt_i + v_i, \text{ say,}$$

where v_i is a new "composite" error – *not* a "measurement error" – referring to the ith equation as a whole. More explicitly, we have

$$v_i = s_i - s_0 - rt_i.$$

By rearranging terms in this way, diverse errors that were once physically distinguishable (here, two: ϵ_{s_i} and ϵ_{t_i}, one error of measuring position and one of measuring time) have been combined into a single term, an "error in the equation." A cost is incurred thereby. For instance, whereas the ϵ's are functions of one observation at a time, the value of v_i, error in the equation for the ith observation, will turn out to depend on *all* the observations, not just the ith pair. If these errors v are unacceptably large, you can't tell if their magnitude would be more effectively reduced by improvements in the error ϵ_{s_i} of the odometer, in the error ϵ_{t_i} of the stopwatch, or in the calibration of the vehicle to constant speed.

These are the errors we wish to make "small": errors in the *law* we are expecting to find. In the usual graphic (Figure 4.8), a line drawn on the plane of the points (t, s), this "error" looks like the vertical distance of the point from the line of the law. But there is a better way to understand what is being computed – it is that something reasonable is being averaged, as we shall see in a moment.

You can imagine many criteria for smallness, many desiderata for the v's that would each generate values for s_0 and for r, the speed we wish to have "measured" in this indirect fashion. It turns out to be most useful to measure "smallness" by the mean square of the v's, the same quantity we were calling "variance" above as it applied to collections of single measurements.

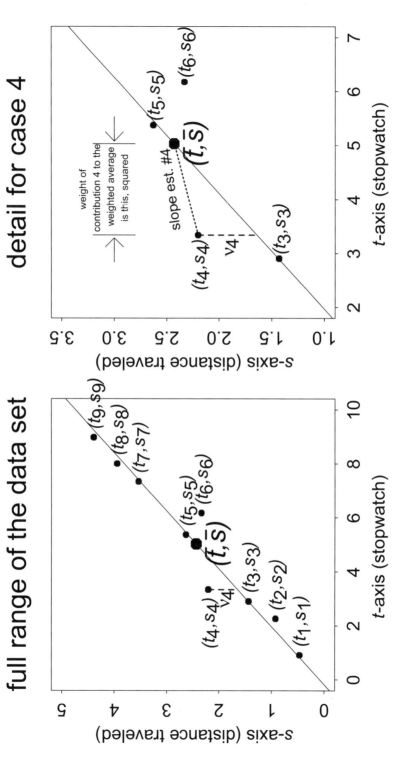

Figure 4.8. The simple regression line as a weighted average. (Left) The data and the fitted line. Horizontal, the predictor (t, time read from the stopwatch); vertical, the quantity predicted (s, distance traveled according to the car's odometer). The regression line has to go through the point $(\bar{t}, \bar{s})$ right at the middle of the data scatter. (Right) Enhancement of the previous plot showing the contribution of one single data point to the weighted average expression for the slope r, equation (4.4).

L4.1 The Logic and Algebra of Least Squares

What makes the least-squares estimates of s_0 and r worth considering is that under some circumstances they are the most *reasonable* estimates of the corresponding "underlying quantities." They are maximum-likelihood (ML) (in the same sense that the ordinary average will prove ML) whenever the probability distribution of v takes the form e^{-cv^2}. If this is the case,[2] the arithmetical values s_0 and r that minimize Σv^2 are the most reasonable estimates of those two quantities.

In that case, as Legendre argued in 1805 and as any manual for a classroom calculator could inform you today, the estimation of the true speed r, along with the true zero point s_0, comes from straightforward algebra. The formula turns out to involve simple averages in an instructive way.

Begin with the formula for the derivative of a squared simple binomial:

$$\frac{d}{dx}(a+bx)^2 = 2b(a+bx).$$

We want the values of s_0 and r that minimize $\Sigma v_i^2 = \Sigma(s_i - s_0 - rt_i)^2$. At the minimum, we must have

$$\frac{\partial}{\partial s_0}\Sigma v_i^2 = \frac{\partial}{\partial r}\Sigma v_i^2 = 0.$$

But, from the formula for derivatives of squared binomials and the definition of the v's,

$$\frac{\partial}{\partial s_0}\Sigma v_i^2 = 2\Sigma(-1)(s_i - s_0 - rt_i)$$

and

$$\frac{\partial}{\partial r}\Sigma v_i^2 = 2\Sigma(-t_i)(s_i - s_0 - rt_i).$$

If we set both of these to zero, we produce two new equations in the two unknowns s_0 and r from the N original equations

$$v_i = s_i - s_0 - rt_i = 0$$

stating that each error-in-equation is 0 in turn. We can't solve the whole set of N, but we can solve the two, which are formed by adding up the N after multiplying by two different sets of coefficients. In the first equation, for the $\partial/\partial s_0$ criterion, the expression for each error term is multiplied by -1, which is the coefficient of s_0, the first unknown parameter; the sum of all these is then set to 0. In the second equation, that for $\partial/\partial r$, each original term v is multiplied by the value $-t_i$ that is the coefficient of the other parameter r that is sought; then, again, their sum is set to 0. In all of the equations we are free to drop the minus sign multiplying the whole.

---

[2] Although this is the conventional assumption, the scenario here is actually more consistent with the "errors-in-variables model," the assumption that primary psychophysical measures like ϵ_{s_i} and ϵ_{t_i} are the entities that are Normally distributed. This assumption is reasonable to the extent that the underlying processes are imagined to be cascades of innumerable elementary steps themselves having independent errors. This is "Hagen's Hypothesis," which we encounter in Section L4.2.4.

The first equation is thus

$$\Sigma v_i = \Sigma(s_i - s_0 - rt_i) = 0$$

– the v's themselves average 0. Rewrite this as a single linear equation in the two unknowns:

$$N s_0 + (\Sigma t_i) r = \Sigma s_i.$$

(The situation is more symmetric than it looks, since $N = \Sigma 1$, the sum of all the coefficients of s_0.) Likewise, from the second equation,

$$(\Sigma t_i) s_0 + \left(\Sigma t_i^2\right) r = \Sigma s_i t_i.$$

This is the method of **normal equations**, which took about 90 years to develop, from 1716 through 1805 (Stigler, 1986). This use of the word "normal" is unfortunately not related to the use of the identical word for the "bell curve," the Normal distribution; I'll try to tell them apart here by preserving the capital N for the bell curve. We will encounter the normal equations again in Section L4.5.3.3, where they arise in the totally different context of *explanation*.

We **solve** the normal equations by the determinantal rule: the solution of $ax + by = c$, $dx + ey = f$ is $x = \frac{ce - fb}{ae - bd}$, $y = \frac{af - dc}{ae - bd}$. There results

$$r = \frac{N \Sigma s_i t_i - \Sigma s_i \Sigma t_i}{N \Sigma t_i^2 - (\Sigma t_i)^2,} \tag{4.2}$$

after which we simply evaluate s_0 from the simpler equation:

$$s_0 = \frac{\Sigma s_i}{N} - r \frac{\Sigma t_i}{N},$$

so as to guarantee that the errors v average 0.

In this way we have converted raw data – those pairs (t_i, s_i) – to a pair of new quantities: an estimate of s_0, which is the location of the car at time $t = 0$ on the stopwatch, and an estimate of the speed r at which the car is moving. I prefer to speak of these quantities, the coefficients in the law $s = s_0 + rt$, as having been **measured**, through this modest extent of algebra, by the application of the law to the data set (t, s). The quantities s_0 and r have been measured, so to speak, right along with the separate s's and t's, as averages of certain modifications of those observations. Their measurement has been **indirect**, mediated via that ostensible law of constant speed rather than the simpler laws of stopwatches or odometers (or, for that matter, internal combustion engines) separately. We have, in passing, underlined one of the principal points of Kuhn's great essay on measurement in the physical sciences (Section 2.3): there can be no effective measurement without equivalently effective theory. Both of the examples from physicists in this chapter, Millikan's (E4.2) and Perrin's (E4.3), rely on this interpretation of a regression slope as the indirect measurement of a real quantity; so does Sewall Wright, the biologist, when he interprets them as path coefficients (Section L4.5.2.2).

L4.1.4 Regression as Weighted Averaging

The numerator of r, equation (4.2), can be rearranged as

$$N(\Sigma(s_i - \bar{s})(t_i - \bar{t})).$$

The denominator can be rearranged as $N(\Sigma(t_i - \bar{t})^2)$. This is N^2 times a number we have already seen, the **variance** of t. The numerator is N^2 times another useful quantity, the **covariance**

$$(\Sigma(s_i - \bar{s})(t_i - \bar{t}))/N \qquad (4.3)$$

of s and t. (That is, the regression slope is the covariance of the predictor by the measure being predicted, divided by the variance of the variable being predicted. We exploit this in much more detail in Section L4.3, in the discussion of Pearson's mother–daughter height data.) We can further rearrange the whole expression for r in a very interesting way:

$$r = \frac{\Sigma(s_i - \bar{s})(t_i - \bar{t})}{\Sigma(t_i - \bar{t})^2}$$
$$= \frac{\Sigma\left(\frac{s_i - \bar{s}}{t_i - \bar{t}}\right)(t_i - \bar{t})^2}{\Sigma(t_i - \bar{t})^2}. \qquad (4.4)$$

This takes the form of a **weighted average**: a collection of raw ratios $\frac{s_i - \bar{s}}{t_i - \bar{t}}$ are each multiplied by a factor $(t_i - \bar{t})^2$, the products added up, and their total divided by the sum of those factors. This simple manipulation was apparently first noticed by F. Y. Edgeworth in 1892, nearly a century after the formula itself was first derived (see Stigler, 1986). (The "rearrangement" is another of those informative no-ops [null operations] already discussed in Mathematical Note 4.1. There we added and subtracted $\bar{x}$; here we multiply and simultaneously divide by $(t_i - \bar{t})^2$.)

Both of the factors in this rearrangement, the quantity being averaged and the quantity doing the weighting, are meaningful separately, and add strength to the interpretation of this process of line-fitting as a form of (indirect) measurement. The quantity being averaged, $\frac{s_i - \bar{s}}{t_i - \bar{t}}$, is an elementary slope estimate for r: a distance divided by a time. The distance is distance from the **mean**, however, not from our arbitrary zero, and the time is likewise time from the mean of the stopwatch readings, not from zero. That is, they are **mean-centered** measurements.

(If you **knew** the stopwatch was started just as the car ran over it (between the tires, so as not to crush it), you would modify this formula: you would use $r = \Sigma s_i t_i / \Sigma t_i^2$, the same formula "referred to (0,0)." In this form it is clearer that we are averaging separate estimates s_i/t_i of speed using squared weighting factors t_i^2. Either way, this is the same operation of squaring that went into the analogy of the variance of a balanced rod with its rotational kinetic energy.)

Supposing now that all values of s were measured with the same standard error, then the associated guess $\frac{s_i - \bar{s}}{t_i - \bar{t}}$ for the speed r has standard error equal to the standard error of s divided by $(t_i - \bar{t})$. The smaller this quantity – the closer t_i is to the mean

$\bar{t}$ – the more error-prone is this estimate of the common slope r. The weight of the "observation" $\frac{s_i - \bar{s}}{t_i - \bar{t}}$ of r is the reciprocal of this error variance, and so should be proportional to $(t_i - \bar{t})^2$, just as the formula says. The closer t_i is to the mean $\bar{t}$, the lower the weight of this observation of s_i; the farther t_i is from $\bar{t}$, the greater the weight of the observation.[3]

So the formula (4.4) for the speed of the car,

$$r = \frac{\Sigma \left(\frac{s_i - \bar{s}}{t_i - \bar{t}} \right) (t_i - \bar{t})^2}{\Sigma (t_i - \bar{t})^2},$$

is ultimately no more mysterious than any other kind of average. We thought we were asking about a least-squares fit of the "law of constant speed" to time, but what we actually got was much more relevant to *measurement* than we expected: a (weighted) least-squares estimate, that is, an average, of the slope we were really after, the specific object of the indirect measurement. What is now being averaged is not the original measurements, but the component ($\frac{s_i - \bar{s}}{t_i - \bar{t}}$) of equation (4.4), the individually derived estimates of that common slope. (For instance, if the car's velocity were not presumed to be constant, this whole procedure would be meaningless.) The least-squares logic of kinematics dictates that the line we are looking for has to go through the center of mass of the whole set of data as graphed – the centroid $(\bar{t}, \bar{s})$ of the set of points (t_i, s_i) considered as particles of identical mass on the plane. From the lines through that point we end up selecting the one that has the weighted least-squares estimate of average slope. That selection is equivalent to reweighting the slopes alone, each by the square of its separation from the centroid (Figure 4.8). The effect of one point at separation 2 seconds is equivalent to the effect of four points at separation 1 second from the mean time $\bar{t}$ read off the stopwatch, even though in this particular setup we don't have access to the requisite number of replicate stopwatches.

For problems where there is a natural zero, the slope is the weighted average of the quantities s_i/t_i and the intercept is, analogously, the (unweighted) average of the *adjusted differences* $s_i - rt_i$, the apparent "starting odometer readings" of the observations corresponding to a time $t = 0$ for each, all on that assumption of constant speed. We see an example of this in Section E4.6.1.

Averages of elements that are not numbers. In this chapter, the entities we are averaging are typically single numbers. In many contemporary applications, however, what are being averaged are considerably more complicated (but still instrument-derived) patterns: maps, or pictures, or profiles, or spectra, for example. The most common way of establishing analogues to all of the univariate operations – not just averaging and least-squares fits, as in this lecture, but also the inference engines of Part III – is by recourse to equation (4.1), the identity that $\bar{x}$ was the minimizer of $\Sigma(x_i - y)^2$ as a function of y. Averages of the more complex patterns are *defined*

---

[3] In other words, on average the regression line fits the distant points better than the points in the vicinity of the average. Errors around the line at the distant points have been *attenuated* relative to those near the mean. See the discussion of the hat matrix in Section L4.5.2.3.

as minimizers of the corresponding "sums of squares" over the data. Of course the meaning of an expression like "$(x_i - y)^2$" will itself require some adaptation when y and the x's are no longer numbers but pictures or other patterns. In fact, we have to redefine the minus sign (the subtraction) as well as the squaring. The combination is replaced by a more general sort of mathematical notion, a "squared distance function," and a more general notion of an average, the Fréchet mean. We encounter one example of this in the course of our discussion of morphometrics in Chapter 7, but in its full generality the topic is too advanced for this book.

E4.2 Millikan and the Photoelectric Effect

Sometimes the success of an experiment in physics takes the form of an event. In 1845 you pointed your telescope at a location in the sky whose coordinates were computed from anomalies in the orbit of Uranus ... and there, right there, was a new planet, shortly to be named Neptune. Or perhaps you were there when they triggered the first atomic bomb at Alamogordo, New Mexico, on July 16, 1945. More commonly, though, the success is communicated via a quantity. You may be Newton, confirming that the acceleration of an apple toward the ground is 1/60th of 1/60th of the acceleration of the moon toward Earth, corresponding to the fact that the moon is just about 60 times as far away from the center of Earth. Or you are monitoring the atomic pile under Stagg Field at the University of Chicago in 1942, and as you pull out the carbon rods, the rate of clicks on your Geiger counter rises and then begins the slowest possible exponential increase (it would explode in about a week), confirming the theory of chain reaction decay of the ^{238}U nucleus. Or you observe that the deflection of light that just grazes the sun in the solar eclipse of 1919 is exactly what Einstein predicted and twice what a 19th-century physicist would have predicted, confirming general relativity even more strongly than when Einstein had earlier resolved a 19th-century anomaly, the precession [rotation of the axes of the elliptical orbit] of the planet Mercury. As a special case of this estimation of a constant, one has the sort of experiment in which the dependence of a measurement on a machine control or a material property is investigated and found to be null. The physicist James Clerk Maxwell (he of Maxwell's equations) predicted that the viscosity of a gas was not a function of its pressure, a law that "appeared very remarkable at its first announcement," so that its verification in 1866 "constituted one of the first important successes of the kinetic theory." (The quotations are from Perrin, 1923, a resource we have already encountered in Chapter 3.) Others may have been surprised, but presumably not Maxwell himself. Albert Michelson and Edward Morley, on the other hand, were not prepared for the invariance that they found in 1887 when they measured the speed of light as a function of direction; it was not explained until Einstein came along 18 years later.

These correspond to one stereotype of quantitative reasoning in physics: the production of a simple numerical quantity that, zero or not, has a constant value. We will see, however, first in the Millikan experiment in this section and subsequently in the

Perrin narrative, that insofar as physics is based on equations, assertions of "physical law," it is possible to be far more aggressive about checking those equations directly. In Section E4.3, we see how Jean Perrin, who already knows that Brownian motion is real, will be checking for the slope of an exponential curve and then for the shoulder width of a bell curve. Robert Millikan checked instead for a straight line, and confirmed it, even though he believed there was no experimental evidence whatever suggesting that such a line should exist, only Einstein's "reckless hypothesis" (see below).

Indeed, it is startling to modern readers how skeptical a tone Millikan (1916) takes at the beginning of the article under review here, his "A direct photoelectric determination of Planck's h." (In retrospect, of course, that tone served ultimately to strengthen the reader's appreciation of abduction that resulted.) He begins as follows (pages 355–356):

> Quantum theory was not originally developed for the sake of interpreting photoelectric phenomena.... Up to this day the form of the theory developed by its author [Planck] has not been able to account satisfactorily for the photoelectric effect presented herewith. We are confronted, however, by the astonishing situation that these facts were correctly and exactly predicted nine years ago by a form of quantum theory which has now been pretty generally abandoned.

(Note that word "astonishing.")

> It was in 1905 that Einstein ... [brought] forward the bold, not to say the reckless, hypothesis of an electromagnetic light corpuscle of energy $h\nu$, which energy was transferred upon absorption to an electron. This hypothesis may well be called reckless first because an electromagnetic disturbance which remains localized in space seems a violation of the very conception of an electromagnetic disturbance, and second because it flies in the face of the thoroughly established facts of interference [owing to the wave nature of light]. The hypothesis was apparently solely because it furnished a ready explanation of one of the most remarkable facts brought to light by recent investigations, namely, that the energy with which an electron is thrown out of a metal by ultraviolet light or X-rays is independent of the intensity of the light while it depends on its frequency.

> This assumption enabled him at once to predict that the maximum energy of emission of corpuscles under the influence of light would be governed by the equation $\frac{1}{2}mv^2 = h\nu - p$, where $h\nu$ is the energy absorbed by the electron from the light wave, ... p is the work necessary to get the electron out of the metal, and $\frac{1}{2}mv^2$ is the [kinetic] energy with which it leaves the surface, an energy evidently measured by the product of its charge e by the potential difference [V] against which it is just able to drive itself before being brought to rest. At the time at which it was made, this prediction was as bold as the hypothesis which suggested it, for at the time there were available no experiments whatever for determining anything about how potential difference varies with ν.

Millikan reported that he confirmed *five separate experimental relationships* implied by Einstein's equation – they are "rigorously correct." And he was baffled: "No one of these points except the first had been tested even roughly when Einstein made his prediction, and the correctness of this one has recently been vigorously denied by Ramsauer...." Nevertheless, Einstein's equation deserved to be checked, as if it is valid we have a way of estimating the value of Planck's constant h (because Millikan himself had just finished the experiment estimating the value of the other constant, e, the charge on the electron – this was the other justification for his Nobel Prize).

So Millikan's job was to get a *precise* line, from which slope and intercept could *both* be quantified. At the outset, while some aspects of Einstein's equation appear to be valid (e.g., emission energies are independent of temperature, and do not appear to depend on the metal), there is not yet any experimental reason to believe even that the relation between the frequency v and kinetic energy is linear.

E4.2.1 *The Role of Lines; the Role of Certainty*

To rigorously examine Einstein's claim, Millikan was required more or less to "build a machine shop in vacuo," which, he noted dryly, was not at all easy. Metal cylinders had to be alternately exposed to an electron beam and to a knife scraping their surfaces clean again. The vacuum, the purity of the metal cylinders, the monochromaticity of the incident light – all had to be managed with "the utmost possible accuracy" in order to test Einstein's equation to a precision adequate for acquiring information (in this case, Planck's constant, h) from the value of an experimental slope.

The line claimed by Einstein's equation takes the form of a limiting voltage against frequency. Millikan's raw data, my Figure 4.9, took the form of individual parabolic curves, of which these in particular were from "the most reliable single set of photocurrent potential curves yet taken." From curves like these the intersections with the zero axis had to be found by inspection. The zeroes, copied over onto a new chart, yield the triumphant Figure 6 reproduced here as my Figure 4.10.

Millikan (1916, p. 372) wrote:

> The results of plotting the intercepts on the potential axis against the frequencies are given in Fig. 6. The first result is to strikingly confirm the conclusion ... as to the correctness of the predicted linear relationship between potential difference and v, no point missing the line by more than about a hundredth of a volt.

Interestingly, what Millikan used to fit the line he needs was not the least-squares approach of equation (4.2), which respects the input data points equally, but a version of the slope-averaging approach, equation (4.4), that accommodated his intimate knowledge of the measurement apparatus he had built and its ineluctable remaining uncertainties. He knows from experience how the equipment functions: "The slope of this line was fixed primarily by a consideration of the five [leftmost] points, of which the first, the fourth and the fifth are particularly reliable." So

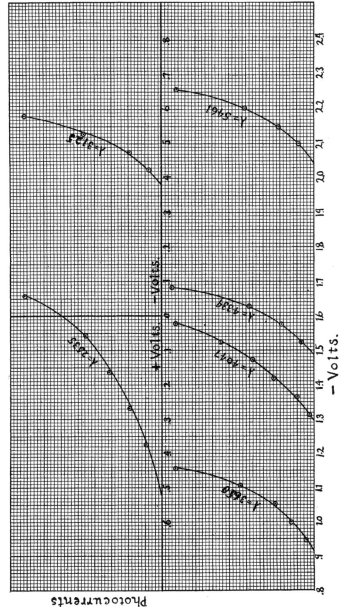

Figure 4.9. Plot of raw data from Millikan (1916). The quantities here are not those subject to the linear law Einstein predicted. By permission of Oxford University Press.

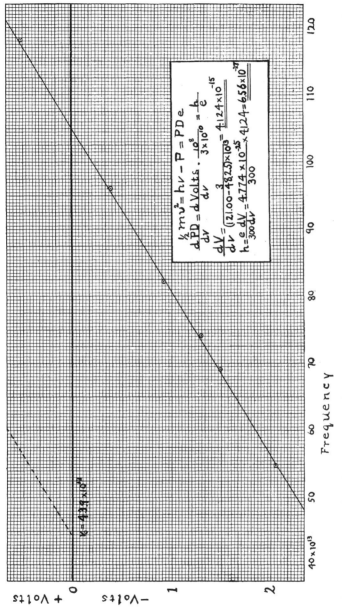

Figure 4.10. Conversion of Figure 4.9 to a format for which Einstein's law appears as a straight line: frequency against limiting voltage (the intercept of each little curve with the horizontal axis in Figure 4.9).

it is reasonable for Millikan to overrule the least-squares machinery explicitly, by manually selecting the particular slopes he will average. "Having placed beyond a doubt the fact of the linear relation between V and v, it was possible to determine the slope of the line in the highest available vacuum by locating two distant points on it in rapid succession. Wavelength 5461 [in units of Å, Ångstroms] was always chosen for one of these points because its great intensity and large wavelength ... adapt it admirably to an accurate determination of the intercept on the potential axis" (pp. 374–375). So he measured the voltages at 5461, "a few minutes after shaving," and then at each of the two further wavelengths, and then at 5461 again (to correct for short-term trends in the purity of the metal surface). The two other wavelengths were the two experimentally stable points at the farthest remove from 5461 that yielded the most precise slope estimates (4.124 and 4.131). These two estimates were then averaged ("without weighting") to provide the value of 4.128 on which his publication ultimately settles. Hence the analysis is not of the pool of his data, but of these two specific runs, the pair in which he had the most trust. Millikan has become notorious for selecting only the "good" data in this way. For his estimate of e, the charge on the electron, he did exactly the same thing. (For a spirited defense of this style of calculation, see Goodstein [2000]. It is not that Millikan was selecting the data points that best confirmed his hypothesis; it seems rather that he was selecting those for which the experimentally derived value was the most precise.)

From this line, finally, we need a number, Planck's constant h. To derive its value, one must multiply the slope of the line in the graph by e, the charge on the electron; the arithmetic for this is written out by hand, for one of the slopes (the value of 4.124), on his Figure 6 (my Figure 4.10) *as published*. What results is an estimate of $h \approx 6.569 \times 10^{-27}$ erg-seconds. The accepted modern value is 6.626×10^{-27} erg-seconds; Millikan came within 1% of the current consensus value on this very first try. He also carried out the same experiment for lithium, and got the same value. Planck's estimate, Millikan noted, differs from Millikan's own by about 3.5%.

Measurement now completed, Millikan returns to speculative philosophy toward the end of his article. He notes (page 379, in italics in the original), "Planck's h appears then to stand out in connection with photoelectric measurements more sharply, more exactly and more certainly than in connection with any other type of measurements thus far made." Even though "the semicorpuscular theory by which Einstein arrived at his equation seems at present to be wholly untenable," still "it must be admitted that the present experiments constitute very much better justification for such an assertion than has hitherto been found, and if that equation be of general validity, then it must certainly be regarded as one of the most fundamental and far reaching of the equations of physics." In short, "Einstein's photoelectric equation has been subjected to very searching tests and it appears in every case to predict exactly the observed results" (pp. 383–384). The existence of an apparently exact mathematical model (a straight line) comprises sufficient evidence that this is a promising direction of empirical research even though theoretical justification of this line or indeed any straight line was totally lacking at the time Millikan was writing.

E4.2.2 The Unreasonable Effectiveness of Mathematics in the Natural Sciences

That italicized mantra is of course the title of the great essay in natural philosophy by Eugene Wigner (1960), an essay that has its own entry in Wikipedia. Wigner's essay, which we have already encountered in Chapter 2, was not an ordinary journal article, but the written text of a named lecture, the Richard Courant lecture, delivered at New York University in 1959. Venues like this afford the opportunity to consider foundational issues in a speculative, interdisciplinary manner. There are many good exegeses of this article, and I have no desire to add to the bulk of this literature of commentary. But for present purposes it is appropriate to note four points of Wigner's that we have already confirmed in the course of the Millikan history just reviewed:

- Although the world around us is of "baffling complexity," nevertheless, as Schrödinger commented, certain regularities in the events can be discovered. The laws of nature are concerned with such regularities, which are surprising for three reasons: they are true everywhere and for all time (the *invariance* property); they are independent of many conditions that could have an effect, but do not; and they are expressed in the language of mathematics.
- We are willing to conjecture that equations (such as Einstein's linear equation for kinetic energy as a function of wavelength) might be *exactly* true, meaning, true even beyond the limits of current instruments to confirm them. Physical laws written in a simple mathematical notation are accurate "beyond all reasonable expectation" – are of "almost fantastic accuracy" – . . .
- . . . but they are also of strictly limited scope. Newton's laws deal with second derivatives. About the boundary conditions, such as "the existence, the present positions, or the velocities" of bodies, they give no information at all. Nevertheless, we get far more out of the equations than we ever put in by way of experimental data.
- The mathematics can be there first, or the physics; it doesn't seem to matter which. The languages *match*; that is the principal tenet of the physicist. "The enormous usefulness of mathematics in the natural sciences is something bordering on the mysterious, and there is no rational explanation for it. . . . It is not at all natural that 'laws of nature' exist, much less that man is able to discover them. . . . The miracle of the appropriateness of mathematics for the formulation of the laws of physics is a wonderful gift which we neither understand nor deserve." (This is my Epigraph 5.)

Millikan illustrates his faith in these propositions in several ways.

- He is willing (as his predecessors would have been) to check an equation of Einstein's for which there was, at the time, no persuasive logical justification.
- He relies on graph paper, a ruler, and all the other tools of data geometry. Although it is astonishing that there should be an equation about this phenomenon, given that there *is* an equation it is no longer astonishing that it should be a linear

one, a matter of simple subtraction and multiplication, and we trust our laboratory instruments, when properly calibrated, to obey the law of conservation of energy and otherwise to get matters of addition and subtraction of energy right. Put another way, we trust our machines, which instantiate laws of physics themselves: especially the sensors, such as spectroscopes and vacuum gauges.
- He is confident that by sheer experimental craftsmanship he can bring the precision of his measurements under sufficient control to see, by direct inspection, whether Einstein's equation looked "reasonably accurate" in Kuhn's sense (Section 2.3).
- He derives essentially the same value of Planck's constant regardless of the experimental details (here, for different metals inside his vacuum tube). Of course a constant of Nature must show behavior like that.
- Finally, like Wigner, he is willing to accept the empirical pattern as true, and as a justified ground for future work, even in the complete absence of any a-priori logical reason for doing so. He is willing, in other words, to accept yet another example of mathematics' "unreasonable" application in domains like this. But this faith itself has already required serious work: the design of experiments that permit all the irrelevancies (the parts of reality that are not lawful) to be stripped away – what Krieger (1992) calls the production of the "good enough vacuum" – and the insistence on *numerical* consilience, not just vague agreement, as the coin of intellectual currency. If Einstein's equation is valid, then different estimates of Planck's constant must yield the same value. That is what Millikan has managed to demonstrate, and the implication, a classical abduction, is that the equation is now much likelier than before to be true.

Some have said that God is a Geometer. From the arguments I've just been reviewing, it would appear that His profession might better be described as Logician. See C. S. Peirce, "A Guess at the Riddle," unpublished in his lifetime, which can be found on pages 245–279 of Houser and Kloeser (1992), or, more recently, Stephen Wolfram's (2002) great work of speculative natural philosophy. For a specific elaboration on Wigner's argument, see Hamming (1980).

L4.2 Galton's Machine

L4.2.1 The Large-Scale Pattern of Binomial Coefficients

We return to the topic of coin tosses, regarding which section L4.2.2 derived the inverse-square-root law of the precision of averages. We now want to learn about more than the decrease of variance of those averages with sample size – we want some details about how those averages vary around the true mean. It will turn out that for "sufficiently many coins" (in practice, more than a couple of dozen), the distribution of fractions of heads varies about its mean of $1/2$ according to one underlying formula that requires only a conjoint scaling of width and height to describe any mass coin-flipping operation. This derivation is very old; it is usually attributed to Abraham de Moivre, writing in the 1730s.

We begin even earlier than that, with a formula from Blaise Pascal published posthumously in 1665. Write $n!$, read "n factorial," for the product of the integers from 1 through n, the value $1 \times 2 \times \cdots \times n$. The total number of flips of n coins that come up with k heads is $\frac{n!}{k!(n-k)!}$. (The proof is based on the identity ${}_nC_k = {}_{n-1}C_k + {}_{n-1}C_{k-1}$, the same formula that underlies the Binomial Theorem, Pascal's Triangle, etc., or you can just argue that there are n ways of choosing the first coin to come up heads, times $n-1$ ways of choosing the second, ..., times $n-(k-1)$ ways of choosing the kth – this product is $\frac{n!}{(n-k)!}$ – but we get the same set of heads whatever the order we decided that one was first, one second, ..., and so we have to divide this ratio by a further factor of $k!$.) This quantity is written ${}_nC_k$, pronounced "enn choose kay," the number of ways of "choosing" k heads out of the n coins; that symbol labels a button on most scientific calculators. If that coin was fair, all 2^n sequences of heads and tails are equally likely – that is actually what the statistician *means* by fairness – and so the probability of getting k heads (or k tails, for that matter) when flipping n coins is ${}_nC_k/2^n$. The peak is at $k = \frac{n}{2}$, if n is even, and otherwise is tied between $k = \frac{n-1}{2}$ and $k = \frac{n+1}{2}$. The minimum, 1, is always at $k=0$ and $k=n$: there is only one way of getting all-heads or all-tails. If you interchange k and $n-k$ in the formula, you get the same expression for ${}_nC_k$, meaning that the chance of k Heads and the chance of k Tails are the same. (This formula applies for a fair coin only. For an unfair coin, or any other two-outcome process for which each trial has a constant probability p of a "head," the formula becomes ${}_nC_k p^k (1-p)^{n-k}$. Now the chances of k heads and the chance of k tails are no longer the same. We will need this formula in connection with the introduction to Bayesian reasoning at Section L4.3 below.) Can we get the mean and the variance of counts of heads from unfair coins like these?

First, a useful recursion. From the definition, we have $\frac{{}_nC_k}{{}_{n-1}C_{k-1}} = \frac{(n!/k!(n-k)!)}{((n-1)!/(k-1)!(n-k)!)} = n/k$. Then

$$k \, {}_nC_k = n \, {}_{n-1}C_{k-1}. \qquad (4.5)$$

The binomial distribution for number of heads for n tosses of a coin with probability q of heads is $p(k) = {}_nC_k q^k (1-q)^{n-k}$, from the identity

$$1 = 1^n = (q + (1-q))^n = \Sigma_{k=0}^{n} {}_nC_k q^k (1-q)^{n-k}. \qquad (4.6)$$

Then the average number of heads for n tosses of that biased coin is

$$\Sigma_{k=0}^{n} k p(k) = \Sigma_{k=0}^{n} k \, {}_nC_k q^k (1-q)^{n-k} = n \Sigma_{k=1}^{n} {}_{n-1}C_{k-1} q^k (1-q)^{n-k}$$
$$= nq \, \Sigma_{k-1=0}^{n-1} {}_{n-1}C_{k-1} q^{k-1} (1-q)^{(n-1)-(k-1)} = nq.$$

Mathematical note 4.3. On what a "step" in a derivation is. In this equation chain and in the next two, the mathematical style is to have every step be explicable as the application of some simple rule that is left implicit for the reader or a teacher to reconstruct. The first equality is by plugging in the definition of $p(k)$; the second uses the recursive identity just introduced and also pulls the n out of the summation sign; the third drops the zeroth term in the summation and then pulls one factor q outside the summation sign; and the fourth recognizes the expression under the summation as another version of the identity (4.6). Part of the art of being a good math student is to figure out all of these elementary steps oneself.

To get the variance of this number of heads, first iterate the identity (4.5): if $k\,_nC_k = n\,_{n-1}C_{k-1}$, then $k(k-1)\,_nC_k = n(n-1)\,_{n-2}C_{k-2}$. Apply this to the expected square of the number of heads:

$$\Sigma_{k=0}^n k^2 p(k) = \Sigma_{k=0}^n k(k-1) p(k) + \Sigma_{k=0}^n k p(k)$$
$$= nq + n(n-1)\Sigma_{k=2}^n\,_{n-2}C_{k-2} q^k (1-q)^{n-k} = nq + n(n-1)q^2.$$

The variance is the expected value of k^2 minus the square of the mean of k; this is $nq + n(n-1)q^2 - (nq)^2$, which is $nq(1-q)$. For $n=1$ this is $q(1-q)$, the variance of a binomial of frequency q (or $1-q$).

We will eventually need an approximation for the factorial function included in this definition, and one has been available for about 300 years via the famous *Stirling's formula* for the factorials in the numerator and denominator: $n! \approx n^n e^{-n}\sqrt{2\pi n}$, where e is the base $2.71828\cdots$ of natural logarithms. (Thus the whole theory of the Normal distribution, which [see later] really does describe some aspects of the physical world, rests on large-scale properties of the integers: a lovely Pythagorean epiphany.) We see shortly (Section L4.2.3) how the number π gets in there, and based on that formalism Section L4.2.7 will present one modern derivation of the factorial formula itself, but the pedagogy goes best if we just assume its validity for now.

From this approximation one begins right away to learn some things about the behavior of the frequency of heads as the number of coins flipped becomes larger and larger. For instance, assume we've flipped $k = 2n$ coins, an even integer; how often do we get exactly n heads (i.e., precisely a 50:50 outcome)? The count of these cases is exactly $(2n)!/n!n!$. By Stirling's formula, this is approximately

$$\frac{(2n)^{2n} e^{-2n} \sqrt{2\pi(2n)}}{(n^n e^{-n}\sqrt{2\pi n})^2} = \frac{2^{2n} n^{2n} e^{-2n}\sqrt{2\pi(2n)}}{n^{2n} e^{-2n}(2\pi n)} = 2^{2n}\sqrt{2}/\sqrt{2\pi n}.$$

(Remember that "canceling" is another of those symbolic manipulations – inscriptions on a blackboard – that the mathematical community assures you are valid.) Converting this from the count of patterns with n heads to the chance of n heads is to divide by 2^{2n}; so the probability we want is $\sqrt{2/2\pi n}$. Rewriting to refer to the total of k coins, this is $\sqrt{2/\pi k}$. Hence: the chance of getting a fraction of heads of *exactly* $1/2$ falls off as the square root of sample size. This makes sense from Section L4.1.2: the variance of the count of heads is proportional to the number n of coins, so the standard deviation of the count of heads increases as $\sqrt{n}$. Then the chance of getting the one specific outcome right in the middle of the expected range should fall by the inverse, the factor $1/\sqrt{n}$.

The total number of heads is not a function of their order, so this probability is the same as the chance of ending in a tie in a coin-flipping game in which you gain a point for heads but lose a point for tails. For $n = 100$ flips, this chance is about 0.08.

L4.2.2 Ratio of Chances of Heads to Its Maximum

Decades before de Moivre's work, Bernoulli had noticed that the counts $_{2n}C_k$ fall off from their maximum at $k = n$ faster and faster (in terms of their ratios to their upstream neighbor) with distance from n in either direction. Continuing with what deMoivre discovered, we ask how fast the terms $_{2n}C_{n\pm l}/2^{2n}$, representing the probabilities of $n \pm 1$, $n \pm 2, \ldots, n \pm n$ heads, fall as we depart from the peak. The answer, again, is a symbolic manipulation, not a computation. One constructs the formula for $_{2n}C_{n\pm l} = (2n)!/(n-l)!(n+l)!$ from the formula for $_{2n}C_n = (2n)!/n!n!$ by dropping factors n, $n - 1, \ldots, n - (l - 1)$ from the first factorial in the denominator and adding factors $n + 1$, $n + 2, \ldots, n + l$ to the second factorial in the denominator. We have thus increased the denominator, as a whole, by a factor that we can rearrange helpfully as

$$\frac{n+1}{n} \cdot \frac{n+2}{n-1} \cdot \ldots \cdot \frac{n+l}{n-(l-1)}.$$

We need two approximations from elementary calculus: for a, b both small, $(1 + a)/(1 - b) \approx 1 + (a + b)$, and $\log(1 + a) \approx a$. (Part of the craft of the mathematician consists in remembering things like these that you were taught years or even decades ago. Note this is the *natural* logarithm of $1 + a$, not the log to base 10 but the log to base e.) We apply the first of these with $a = 1/n$, $b = 0$, then with $a = 2/n$, $b = 1/n, \ldots$, up through $a = l/n$, $b = (l - 1)/n$. The relevant a's and b's are small if n is large enough – they will be of the order of $1/\sqrt{n}$ or less – so that our factor by which the denominator of $_{2n}C_{n\pm l}$ has increased over the denominator of $_{2n}C_n$ becomes, approximately,

$$\left(1 + \frac{1}{n}\right) \cdot \left(1 + \frac{3}{n}\right) \cdot \ldots \cdot \left(1 + \frac{2l-1}{n}\right).$$

Logarithms reduce multiplication to addition, which is easier. The logarithm of this product is the sum of the logs of the factors, each of which is approximated via a formula like $\log(1 + \frac{1}{n}) \approx 1/n$. Thus the log of $_{2n}C_{n\pm l}/_{2n}C_n$ is very nearly $(1/n) + (3/n) + (5/n) + \cdots + (2l - 1)/n$. But $1 + 3 = 4$, $1 + 3 + 5 = 9, \ldots$, and in general the sum of the first l odd integers is l^2 (we proved this in Chapter 3). Hence the logarithm of this increase in the denominator is just l^2/n.

Now undo the logging and remember that we were working in the *denominator* of the formula for $_{2n}C_{n\pm l}$. It turns out that **the terms $_{2n}C_{n\pm l}$ fall off from their maximum $_{2n}C_n$ as** $e^{-l^2/n}$. Again we see a square-root rule: for equal ratios l^2/n, quadrupling n only doubles the value of l for which the fall-off achieves any particular threshold of negligibility. (This is the scaling that authorizes us to treat a and b as small in the preceding approximations.)

If we rewrite to refer to the flip of n coins instead of $2n$, the exponent becomes $-l^2/(n/2)$, or $-2l^2/n$. So for a count of heads $n/2 \pm l$ that differs by l from the maximum ($n/2$ heads) in either direction, the peak frequency is multiplied by $e^{-2l^2/n}$. This is what de Moivre wrote out in 1733, although his derivation is much more circuitous. A simple calculation of the second derivative of $e^{-2l^2/n}$ shows that the *waist* of this curve, the segment between the inflections where the curvature changes sign, is at $l = \pm\sqrt{n}/2$, which happens to equal the standard deviation of the count of heads. The height of the curve there is $1/\sqrt{e}$, about 0.61, of its peak height. These curves are shown for various values of n in Figure 4.11, along with the formula in the exponent. For any reasonable value of n that exponent, the logarithm of the bell curve probability, is indistinguishable from a parabola opening downward.

L4.2.3 Notation: The Normal Distribution

For n large, the distribution of l can be imagined as continuous rather than discrete if we convert from the count $\frac{n}{2} + l$ of heads to the fraction $\frac{1}{2} + \frac{l}{n}$ of heads. There arises the famous "bell curve," the **Normal distribution**. Remember that the negative exponential was multiplying a maximum probability of $\sqrt{2/\pi n}$ corresponding to the likeliest total, $n/2$ heads. As a fraction, this corresponds to the value 0 for l/n. But that maximum probability applies now to an interval of width only $1/n$, the increment of fraction-of-heads between $\frac{1}{2} + \frac{l}{n}$ and $\frac{1}{2} + \frac{l+1}{n}$ for $l = 0$. The probability *density* at the maximum is thus actually n times that point probability $\sqrt{2/\pi n}$; this product is $\sqrt{2n/\pi}$. For the fraction of heads, then, the probability goes as this new, properly normalized peak height times that factor of fall-off: the product

$$\sqrt{2n/\pi}\ e^{-2l^2/n}.$$

It is customary to rewrite this formula so as to show the standard deviation explicitly in both the premultiplying factor and the exponential:

$$\sqrt{2n/\pi}\ e^{-2l^2/n} = \frac{1}{\sqrt{2\pi}\sigma} e^{-\frac{1}{2}\left(\frac{x}{\sigma}\right)^2}, \tag{4.7}$$

where, traditionally, the Greek letter σ, lowercase "sigma," denotes the standard deviation – here, $1/2\sqrt{n}$, corresponding to the variance $1/4n$ in the fraction of heads as we found in Section L4.1 – and x renames l/n, the deviation of what we observe from the value (in this case, $\frac{1}{2}$) that we "expect." The factors in front look different but are the same.

We began all this algebra with an approximate formula for the factorial function, but now that we are at the formal level of infinitesimals, the result is exact:

$$\int_{-\infty}^{\infty} \frac{1}{\sqrt{2\pi}\sigma} e^{-\frac{1}{2}\left(\frac{x}{\sigma}\right)^2} = 1,$$

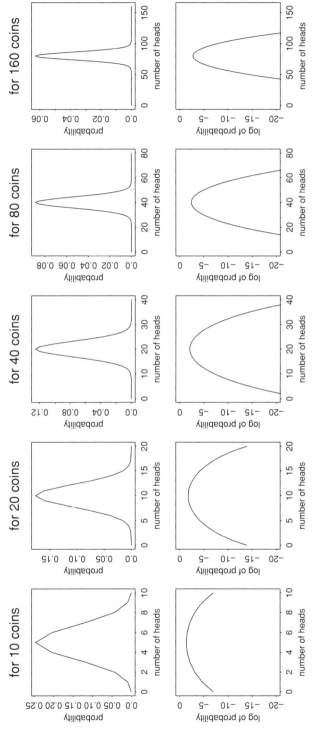

Figure 4.11. Exact representations, five binomial distributions (counts of heads in 10, 20, 40, 80, or 160 flips of a fair coin). (Upper row) Raw probabilities for each count of heads. (Lower row) Logarithm of the curves in the upper row, all to the same vertical scale; each is a parabola opening downward. A logarithm of −10 corresponds to a probability of about 0.000005.

as it must be in order to be a probability distribution. It is sufficient to prove this for $\sigma = 1$. But (in a very old trick) the square of $\int_{-\infty}^{\infty} e^{-x^2/2} dx$ is

$$\int_{-\infty}^{\infty} e^{-x^2/2} dx \int_{-\infty}^{\infty} e^{-y^2/2} dy = \iint_{\mathbf{R}^2} e^{-(x^2+y^2)/2} dx\, dy = \iint_{\mathbf{R}^2} e^{-r^2/2} r\, dr\, d\theta$$

$$= \int_0^{2\pi} d\theta \int_0^{\infty} e^{-r^2/2} r\, dr = 2\pi \left[e^{-r^2/2} \right]_0^{\infty} = 2\pi,$$

as was to be shown. (This is also where the π in Stirling's formula comes from, as shown in Section L4.2.7.)

> *Mathematical note 4.4. On "clever tricks."* In the course of my own mathematical education, the appreciation of this style of proof was one of the signs that one had achieved an appropriate mathematical imagination. By calling this a "trick," one implies that there is nothing particularly natural about the evaluation. Instead of applying any particular set of rules for how you work an integral (change the variable, try integration by parts as in Section L4.2.7, etc.), you simply treat the whole expression as an unknown and see if you can evaluate some function of it, then invert the function. The particular trick here is based on the change-of-variables theorem for the Cartesian plane. An integral with respect to *two* dummy variables, say x and y, that cover a plane, is treated as if it were an integral with respect to a *different* set of dummy variables, in this case, polar coordinates, that cover the same plane. We can multiply the integrands, which means adding the exponents of the exponential; but each is a pure quadratic, so the sum of the two, $(x^2 + y^2)/2$, is the $r^2/2$ of the new polar coordinate system. The Jacobian changes according to the identity $dx\, dy = r\, dr\, d\theta$, which is about the first thing one learns about polar coordinates in a calculus course. The trick works because with that additional factor r in there, the integral of $re^{-r^2/2}$ is exact – it has an *in*definite integral, unlike the original integrand $e^{-x^2/2}$. The genius of the anonymous inventor of this proof consisted in linking three separate steps – the product of two copies of e^{-x^2} as an expression $e^{-x^2} \times e^{-y^2} = e^{-(x^2+y^2)}$, the identification of $x^2 + y^2$ with a new dummy variable r^2, and the observation that the Jacobian of the change of variables make the integrand exact, the differential of a finite function e^{-r^2} – in a way that is obvious once written down. Culturally, part of being a good applied mathematician is to have a retentive memory for definite integrals like these and a reasonable facility with the manipulations necessary to reconstruct them.

Regardless of the value of σ, half the probability is contained within $\pm 0.674\sigma$ of the peak; 68% within $\pm\sigma = \pm 1/2\sqrt{n}$, the "waist"; 95.4% within $\pm 2\sigma$ (here, $\pm 1/\sqrt{n}$); 99.7% within $\pm 3\sigma$, here $\pm 3/2\sqrt{n}$. All these percentages are independent of n as long as n is large enough.

In passing, we have come to a much deeper understanding of that precision of the average fraction of heads over n coins. In Section L4.1, just by summing squares, we found it to improve as the square root of sample size. We have now found that the distribution of that variation approaches one specific distribution, having a remarkably simple formula that involves the two main constants of applied mathematics (e and π), and so might have broader relevance than mere games of chance.

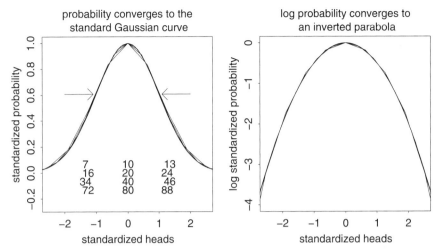

Figure 4.12. (Left) The rightmost four curves from Figure 4.11 after scaling to constant height and to constant "waist" (width between the inflection points left and right, arrows). Convergence to the underlying bell shape is evidently quite rapid. (Right) Logarithms of the curves at left. The common parabolic form is obvious.

It is very interesting to examine just how quickly this approximation comes to resemble its limiting form. Figure 4.11 shows, in the upper row, a set of five graphs of the exact binomial probabilities of various numbers of heads for $n = 10, 20, 40, 80, 160$ coins, all scaled to the same horizontal range (0 to 1 fraction of heads) and the same peak height (the probability $2/\sqrt{\pi n}$ for $n/2$ Heads). As n increases, the scaled probability distribution becomes steadily sharper. The maximum height of the curve and the scaled width of the curve at its waist (61% of the maximum) both decrease as $\sqrt{n}$: for instance, the curve for $n = 20$ appears twice as tall and also twice as wide as that for $n = 80$.

Figure 4.12 shows how well the curves line up once this horizontal scaling is corrected to go as $1/\sqrt{n}$, rather than $1/n$, around an axis of symmetry that is now set at 0. (Vertical scaling remains at $\sqrt{n}$, correcting for the fall in peak height noted before.) The location of the waist at $\pm 1\sigma$ and about 61% of the full height is quite clear. Placing this waist too low on the page and too far out from the center is the commonest error in hand-drawn sketches of this curve.

Underneath the left panel I have labeled some of these normalized curves in terms of raw counts of heads. On this peculiar axis are plotted the original abscissas $n/2$ and $n/2 \pm \sqrt{n/2}$ (this is about 1.4 standard deviations) for $n = 20, 40, 80, 160$ so that you can see the square-root scaling in action. For instance, the interval 72–88 around 80 heads out of 160 spans the same range of the curve, and hence the same total probability, as the range 16–24 around 20 heads out of 40. Four times as many coins, four times the variance, but only twice the standard deviation: $88 - 72 = 16$ is double $24 - 16 = 8$.

L4.2.4 Other Origins of the Normal Distribution

If gambling games were the only context in which the Normal distribution arose, there would be no reason to pay it much attention. But in fact it is woven deep into the foundations of the physical world and of the averages that represent the social world as well. Following C. R. Rao (1973), let us review these physical origins.

The derivation I find most uncanny is James Clerk Maxwell's, from the kinetic theory of gases. (This argument lies not too far from the explicit reasoning of Perrin's you will be shown shortly.) Let us imagine a gas at thermal equilibrium, aerodynamically still and not showing much effect of gravity: for instance, the atmosphere of the location where you are reading this. Our concern is for the probability distribution of the velocity of the individual molecule as it bounces off other molecules, the walls, you, other people and so forth. We make two hypotheses (sensu Perrin), neither of which involves any algebraic notation.

- We assume that the distribution of velocities is the same in all directions (or else thermometers inside the room could tell us which way was east).
- We also assume that the distributions of the *components* of velocity in perpendicular (orthogonal) directions are independent. (That is, it makes no difference for the east–west speed of the molecule how fast it is going in the north–south or up–down directions.)

These are hypotheses about *symmetry*, not about formulas. Ultimately they arise from kinetic theory (hypotheses about the mechanical interactions of molecules): they express the mechanics of kinetic energy and its partitions, not the geometry of space.

Then:

The velocities of gas molecules must be Normally distributed as vectors about the zero vector, and the distribution has spherical symmetry.

That is, the *complete* description of these molecular velocities as a collective involves only one parameter, the standard deviation of that velocity (which turns out to be proportional to the square root of temperature).

It seems absurd for so specific a formula to emerge from assumptions that appear so abstract and generalized. But the derivation is very short and not subtle at all. The independence of the components arises from Newton's laws; the isotropy of space is something we observe to be true (in volumes small enough that gravity may be ignored). The backbone of the proof is an interplay between the Pythagorean formula and the definition of independence. The probability density for velocities such as $(u, u, 0)$ (speed $u\sqrt{2}$ to the northeast, say) must be the product of the densities for its northerly and easterly components separately. That is, the probability for a velocity component of magnitude $u\sqrt{2}$ must be the square of the probability for a component of u. It can be shown to follow that the distribution law must be loglinear in u^2 – that is, it must take a form $e^{-u^2/c}$ for some quantity c that will turn out to be measured on a scale of temperature. (We take advantage of a similar argument in Section 5.3.1 on why a function for which $f(xy) = f(x) + f(y)$ can only be a multiple of the log function.) The mathematical physicists have had over a century to make this reasoning rigorous, and have done so. Figure 4.22, part of Section E4.3, is a relatively recent experimental confirmation of this theorem. (How does an experiment "confirm" a theorem? Recall our discussion of Wigner earlier in this lecture – we are testing the match of the mathematics to the underlying physical reality.)

It will follow, from some arguments of Perrin's to be reviewed in Section E4.3, that whether particles are observed in the true three-dimensional space of their suspension or in the two-dimensional image of a microscope objective, **the steps between equally spaced successive observations of a Brownian motion are Normally distributed around zero.** Variance of the step length is linear in the time interval between observations.

Another version of this deals with shots (arrows, missiles) fired at a point target. The target is a point on a plane, on a barn door, perhaps, or maybe on a map of some military enemy (Mackenzie, 1990). Suppose the probability of an error of magnitude v depends only on v, independent of direction, and suppose that the errors in any two perpendicular directions are statistically independent (representing flips of separate coins, perturbations of separate aiming motors, etc.). These are in some sense "the same" assumptions of isotropy and independence as Maxwell posited for gases. Then **the distribution of the errors in any direction is Normally distributed around zero, and these distributions all have the same standard deviation.** That is, the distribution of the shots around the target has the same complexion, statistically speaking, as the distribution of movements of those spheres of gamboge in Perrin's microscope, or the velocities of the molecules you are breathing.

Other realizations of the Normal distribution arise from generalizations of the coin-flipping model we have explored in detail. Successive flips of a coin are one instance of averages of independent, identically distributed (i.i.d.) events; as n increases, the distribution of the average of n elementary events of this sort approaches the Normal distribution as long as the elementary event itself has a mean and a finite variance. This is the celebrated *Central Limit Theorem*, which actually holds under rather more general conditions than these.[4] The commonest ways of proving this theorem, including the very earliest (due to Laplace), have recourse to complex analysis, the calculus of terms involving $\sqrt{-1}$, which is traditionally denoted i. (There's the *third* main constant of applied mathematics, right up alongside e and π.)

So the Gaussian distribution has nothing essential to do with the facts that coins have two faces, coin flips can be described using only one bit of information, and

---

[4] It is important to remember that the Central Limit Theorem is phrased in terms of "approaching" a Normal distribution. The rate at which that limit is approached, although quite satisfactory for coin flips and for gas molecules, can be unacceptably slow for other sorts of applications, including "flips" of coins that come up heads very rarely (so that the fraction of heads is the relative frequency of a rare event). In the phrase "central limit theorem," the adjective modifies the word "theorem," not the word "limit." This name, incidentally, was coined by George Pólya, whom we met in Chapter 3.

so forth. The distribution arises from the sheer accumulation of repeated (replicated) independent events, not from any detail of the underlying experiment being replicated. (This is already implied in Maxwell's derivation for gas molecules, which certainly do not move like coins.) In *survey research* (estimates of propensities for populations of citizens by random sampling), the role of the "fairness" of the coin is played by the true total frequency of "yes" answers (or whatever) as it might be found by instantaneous census. Instead of flipping one coin many times, one averages a sample of i.i.d. events each of which is the answer supplied by one randomly selected respondent. The "one respondent at a time" supplies the independence, whereas the "identical distributions" are guaranteed by the randomness of sample selection. (Respondents are not identical one by one. The assertion is not about that, only about the identity of the successive sampling protocols.)

The situation in biology is more complex than in these models from the physical sciences, and the appeal to Normal distributions thus much more delicate. Chapter 2 has already mentioned the argument of Williams (1956) that the sheer fact of individuality as a continuing separateness over a life cycle causes enormous problems for theories of meaningful measurement by *any* finite list of parameters, let alone as few as two, and those a "mean" and a "variance." In population genetics, the "coins" might be individual chromosomes, which are "flipped" to one side or the other during meiosis. But there is far more disorder to biology than that, as alleles are not in any one-to-one relation with chromosomes over populations of any size, and measurements of the developmental trajectory or of the adult form, physiology, or fitness cannot be reduced to the mutual actions of genes in any logical way, and thus not to any sort of Central Limit Theorem for sums of biological factors.

Pertinent here as well are the views of Elsasser that I have already introduced in Chapter 2. We have no methodology (i.e., no reliable rules) for determining just *which* aspects of an organic form can be expected to be Normally distributed. In his discussion of "analysis of actual frequency distributions," Wright (1968, Chapter 11) emphasizes the importance of one specific "hidden heterogeneity" (hidden, that is, from least-squares statistics such as means and variances), the presence within a population of a subset in which a measure is zero (a structure is missing). Yablokov (1974) is likewise a remarkable compendium of pitfalls. The problem is that organisms are organized complex systems, whereas our Gaussian models arise explicitly by reduction to *dis*organized systems. In general, then, our discussion of distributions of measures of organisms must await the more sophisticated criteria of Part III, which emphasize inspection of diagrams of similarity ("ordinations") over any modeling of covariation as Gaussian. Until that discussion, we probably should rely mostly on the analysis of experimental interventions, which are automatically more straightforward than analysis of natural variability insofar as experimental interventions take systems outside of their usual homeostatic neighborhoods.

The justification of Normal distributions in psychophysics is sometimes called *Hagen's Hypothesis*. This models errors in perceptual processes as sums of many very small "elementary errors," all with mean zero and constant variance. The assumption is that of fundamental parallelism of processing at some level, perhaps multiple effectors (such as muscle cells), perhaps multiple sensors integrated by some summative mechanism. In this case, the elementary "coin flips" contributing to the final distribution are truly hypothetical, incapable in principle of being observed. Given that we can't observe them, obviously we can't tell if they're being "summed" or being combined in another way: perhaps added to unity and then multiplied – the so-called *lognormal* distributions.

It is particularly convenient that **sums of variables that are each Normally distributed are again Normally distributed**. The mean of the sum is the sum of the separate means, and if the variables being summed are independent then the variance of the sum is the sum of the separate variances.

An intriguingly different derivation, somewhat more mysterious than those from kinetic theory, is couched in another construct from thermodynamics, the entropy

$$-\int p(x) \log p(x)\, dx$$

of a probability distribution $p(x)$. The Normal distribution has the *greatest entropy* of any distribution with fixed mean and variance. In a specific mathematical sense, it is the *most disorderly* distribution – a tribute to the kinetics of mixing gas molecules in a room, which destroys all the order it possibly can consistent with the laws of conservation of energy and momentum. We will have a great deal more to say about *this* mathematical coincidence when we take up the matter of information-based statistical inference in Section 5.3.

L4.2.5 The Role of the Normal Distribution in Statistics

The importance of the Normal distribution for statistical practice corresponds to the two different ways in which it can arise: from physical or biophysical processes involving diffusion or other kinetics of disorder, and from conscientious, orderly human arithmetic following on the one chaotic act of controlled random sampling. The quantities produced in the course of social measurement are generally averages of rather large numbers of elementary observations. They can be expected to be approximately Normally distributed, then, to the extent that the elementary quantities being averaged are independent and identically distributed and have a reasonable variance.

In survey practice, for instance, the first two of these criteria are satisfied by very carefully overseen random sampling[5] of a specified population. The third is most often managed by truncating variables like income, which offers no real equivalents for "mean" or "variance," into categories such as "below $100,000/above $100,000" that do have means (which, in turn, as proportions, are much better-behaved).

---

[5] Under this innocuous phrase is concealed much subtlety, grief, and embarrassment. A proper randomization protocol includes guarantees that each selection is independent of all the others and that each element of a population had the same probability of being selected. Such procedures rely on algorithms for random production of Gaussian deviates (which rely in turn on algorithms for random selections uniformly over the interval from 0 to 1), random permutations of the order of a list of elements, and so on.

There is another scientific use for these surveys. In fields such as econometrics or demography, what matters for the patterns of predictable behavior are not averages as much as the actual *totals* (from which averages are constructed by division). Surveys serve as a sampling of the elements being totaled, and the uncertainty of the totals combines the uncertainty of the sampling with the uncertainty of the measurements of each entity sampled. In such fields, the requirement of consilience supplies an often useful critique of the internal validity of such data records, the applicability of "conservation equations." For instance, in economics, it would be reasonable to enforce the requirement that the world's total of exports matches the world's total of imports. In demography, the equivalent consistency would be between censuses: in any particular geographical region, the number of 5-year-olds enumerated in year $X + 5$ should approximately equal the number of children born in the region in year X, plus the number of children born somewhere else in year X and immigrating to the region by age 5, minus the number of children born in year X who emigrated from the region, minus the number of children born in year X who died there before age 5. In economics, the figures do not match very well (see Morgenstern, 1950); in demography, regional and international migration flows are notoriously unreliable, for understandable reasons (politics of resource allocation, fears of bureaucracy). This book will not not assume that such data resources qualify as quantitative for the purposes of abduction or consilience that are its central themes.

In other forms of statistical practice, such as the biometrics that will concern us later, theoretical quantities such as path coefficients and factor loadings will appear as averages of sample quantities. If these arise from properly randomized samples, they are typically distributed close enough to Normally for the Gaussian probabilities above (95.4% within 2 s.d. of the mean, etc.) to apply.

But it is a very great fallacy – let us call it the Queteletian fallacy, after Adolphe Quetelet, who first fell into it around 1840 – to assume that a variable observed to be more or less Normally distributed in one's data has arisen as an average of some sort or by the summation of many small, independent, identically distributed errors about a true mean value. For 130 years, since Galton's discovery of correlation, we have known that there are mechanisms other than averaging and diffusion that can produce Normally distributed measurements.

Most important is what the statistician and zoologist Lancelot Hogben (1950) calls **the Umpire-bonus model:** a Normal distribution arises whenever to the sum of a very large number of very small "errors" you add one great big perturbation that is already Normally distributed. Mathematically, this is just a special case of something we already noted – the sum of any number of Normals still must be Normal – but in any practical context of prediction or explanation it represents the most important logical confound. For instance, outputs of many biological processes are approximately Normally distributed – at least, in "healthy" populations – even as their values are strongly determined by inheritance from ancestors (one great big "bonus" from the "Umpire"). This model, and the rhetoric by which evidence is argued to support it in real studies of biometric and psychometric data, are the topic of Sections L4.5.3 and E4.6.

If you do not *know* that a distribution arose from averages over a random sample, or that it is rooted in some physical process analogous to Brownian motion, it is unwise to presume its origin in either of these ways. A "Normally distributed" regression error, for instance, may still very well conceal other causes; it is necessary to continue exploring. On the other hand, no modern statistical method requires that you assume Normality; if a method seems to *demand* that your data be distributed Normally, pick another method.

L4.2.6 Galton's Quincunx; Regression Revisited

We need another way of generating Normal distributions, one that is not quite as random as coin flips or gas molecules but instead permits us to observe *cause–effect relationships* in data.

This new line of thought began with an actual physical device invented by Francis Galton, whom we have already met in Chapter 2. Galton, cousin of Charles Darwin and surely the most diverse scientist that Victorian England ever produced, worked, among other problems, on weather maps, fingerprints, fidgeting behavior, the efficacy of prayer, "hereditary genius," averaged photography, and eugenics. In 1877, in a letter to a Darwin cousin (published in facsimile in Stigler [1986]), Galton devised a physical device that could produce normal variability while still permitting there to be "causes" of differences in the final outcome. Our attention is drawn to a clever contraption, the *quincunx* (the word properly refers to the X-shape of dots on the five-face of a die, or the equivalent in horticulture), for the physical production of Normal distributions (Figure 4.13). His little sketches constitute my Figure 4.15, but I've also supplied a clearer diagram in a newly drawn Figure 4.14. In any of the figures, the top is where lead shot enter a pattern of successive, independent left-or-right choices just like flips of a coin. Their accumulation at the bottom precisely delimits the Normal distribution governing such processes. One describes position at the bottom by the accumulated number of lefts versus rights, or heads versus tails, so the horizontal separation between every neighboring pair of dots in the same row corresponds to the switch of one left bounce for a right bounce, or vice versa: it is 2 units.

At every row past the tenth or so, these distributions are approximately Normal, as we've seen in connection with Figure 4.12. There is no standardizing of width such as applied to show the convergence to one shared bell-shaped curve; rather, each row has a different standard deviation (or waist, or variance). Remember (from Section L4.1) that the variance of the number of heads in flips of n coins is $n/4$, and heads minus tails goes as double the number of heads, minus n. So the variance of heads minus tails, which is the quincunx bin variable we are looking at, is four times the variance of heads, which happens to equal the row number of the quincunx we're looking at.

This description, net excess of heads over tails in k flips of a fair coin, can be applied at any stage of the shot's fall, not just in reference to the point where it comes

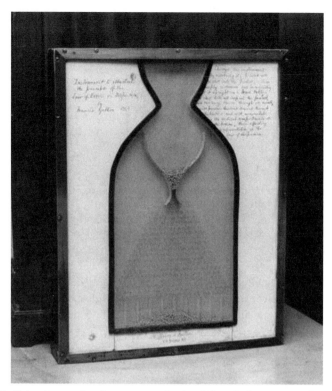

Figure 4.13. Galton's actual quincunx, now in the Galton Laboratory, University College London. That is Galton's own handwriting on the back panel.

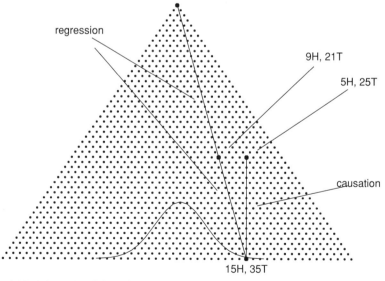

Figure 4.14. Schematic of the quincunx: a machine for producing the distribution of excess of heads over tails in multiple trials with a fair coin as the distribution of left bounces minus right bounces for a sphere falling downward over pins; hence, a machine for producing Gaussian (Normal) distributions.

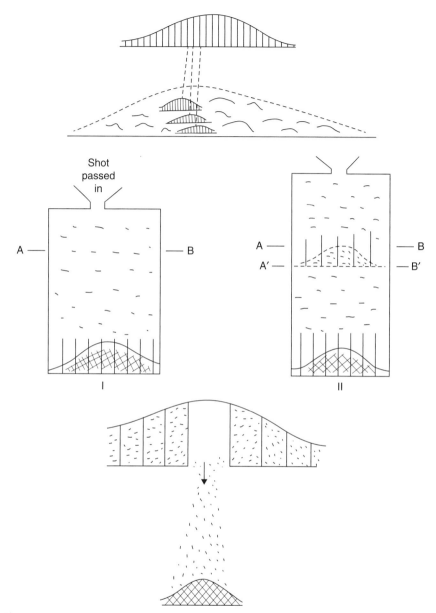

Figure 4.15. Sketches from Galton's letter to his cousin George Darwin, explaining the reproducing principle of the quincunx: Normal distributions added to Normal distributions remain Normal. (From Pearson, 1914–1930, vol. III, pp. 465–466.)

to rest. Galton's realization reduces to imagining the process of falling through the quincunx as the concatenation of two processes – say, falling through the first half, and then falling through the second. You can imagine this, if you like, as a modification of the quincunx by placing a card across its width partway down, perhaps at the 30th row of the 50-row quincunx here, and then removing the card. Falling through the first half results in a normal distribution of standard deviation $\sqrt{30} \approx 5.5$. When the card is removed and each of those cells of accumulation is permitted to fall the remaining 20 rows, it generates its own normal distribution, now of s.d. $\sqrt{20} \approx 4.5$, around its own starting position. **The composite of all of those Normal distributions of variance 20, weighted by the distribution of variance 30, must be identical to the actual Normal distribution of variance 50 (s.d. 7, waist 14) drawn in the bottom row of the figure.**

Before this insight of Galton's, Normal distributions had been viewed as *homogeneous*, necessarily arising from the replication of processes (flips of a coin) themselves identical over the cases of a series. Immanent in this sketch is a brilliant advance in applied probability theory: a model by which a Normal distribution can arise from a set of initial values that are not identical but themselves variable. That is, a Normally distributed pattern of outcomes – say, the weights of a batch of seeds, or the heights of Britons – can be viewed as the result of a process by which each seed or youth began at a *different* setting, the parents' seed weight or average height, and then proceeded to diffuse outward by coin flips of its or his own. (In a happy convergence of model with fact, these "coins" would later be identified – though not by Galton or Pearson – with Mendelian genes.) The overwhelming temptation, which need not be resisted, is to refer to the starting value, the location intermediate along the quincunx, as the *cause* of the value observed at the bottom. It accounts for that value *to some rigorously quantifiable extent*, by influencing the mean about which it varies by its own Normal noise.

For an application to human heredity, which is what mainly concerned Galton, this is half of what we need – a model, free of the "hereditary particles" whose existence would not be known (outside of Gregor Mendel's monastery in Brno) for decades, for the passage from parents' scores to children's scores on any quantitative variate. The other half of the required methodology follows from a study of *reversion*, nowadays called *regression*, using this same chart.

A specific score at the bottom, at the 50-coin level – say, 35 heads, 15 tails – could have come from, say, 25 heads 5 tails at level 30, or from 20 heads 10 tails, or from 15 heads 15 tails, and so on. As Galton did, take the model of causation as "falling down the quincunx": the sample of shot that enter at the top are distributed Normally with variance k at the kth row downward. Let us score position in the quincunx by distance left or right from the spot directly under the entrance point: count of heads minus count of tails for the equivalent series of coin flips. After k of these binary choices, a particle is found in the yth bin of its row. Where is it likeliest to have been in the lth row, $(k-l)$ rows above?

We use, in advance, the principles of inverse probability from the next lecture. The shot were distributed Normally in the kth row of the quincunx, and are distributed Normally again in the remainder of their fall. More specifically, the probability of having been in position x in the lth row is proportional to $e^{-x^2/2l}$, and the probability of having fallen from position x to position y over $(k-l)$ rows is proportional to

$e^{-(y-x)^2/2(k-l)}$. These events, the fall through the first l rows and the fall through the next $(k-l)$, are independent; so the likelihood of having arrived at the yth bin of the kth row via the xth bin of the lth row is proportional (leaving out those factors of $\sqrt{\pi}$) to

$$e^{-x^2/2l} \times e^{-(y-x)^2/2(k-l)} = e^{-(x^2/2l+(y-x)^2/2(k-l))}. \tag{4.8}$$

We maximize this by minimizing the negative of its logarithm, which is, after multiplying out by $2l(k-l)$, the formula $(k-l)x^2 + l(y-x)^2$. This is maximized where its derivative $2(k-l)x - 2l(y-x)$ with respect to the unknown x is 0; but this is obviously for $\frac{x}{y-x} = \frac{l}{k-l}$, or $\frac{x}{y} = \frac{l}{k}$, or $x = \frac{l}{k}y$, meaning that x is just the point in its row (l) on the line between the point y of the kth row and the origin of the quincunx. This line is the **regression line.**

With y fixed (it was the observed data, after all), the likelihood expression in equation (4.8) takes the form of just another Normal distribution, where the coefficient of $-x^2$ in the exponent is $\frac{1}{2}(\frac{1}{l} + \frac{1}{k-l}) = \frac{1}{2}\frac{k}{l(k-l)}$. From formula (4.7) in Section L4.2.3, twice this coefficient must be the reciprocal of the variance of that x-distribution. Then over repeated observations (the whole set of lead shot falling down that ended in the yth bin of the kth row), the originating value x in the $(k-l)$th row is distributed around this likeliest location with effective variance $l(1 - \frac{l}{k})$, the variance of x reduced by the same factor $\frac{l}{k}$. (This would be an easy experiment to do using Galton's "card": after pausing the process at the $(k-l)$th row, you paint each shot there with its slot number, then remove the card.)

So, although on the average gravity causes the lead shot to fall "straight down," from an excess of 20 heads over tails at the 30th row (25H, 5T) to an average excess of 20 at the 50th row (35H, 15T), if you actually *observed* an excess of 20 heads at the 50th row, it most likely did not arise from an excess of 20 in earlier rows – such an excess (indeed, *any* specific excess of heads over tails) is less probable as the row number drops – but from an excess proportionally "regressed" toward 0. At the 30th row, this would be an estimated excess of $20 \cdot \frac{30}{50} = 12$, the point (21H, 9T) that is marked on the figure.

Hence: linear regression of causes on effects follows from the design of the quincunx. This was Galton's great intuitive discovery.

But, also, linear regression of effects on causes follows directly, because beginning from any bin x in the lth row, the expected position in any row underneath is the same bin x, a prediction just about as linear as you can get, and the variance around this prediction must be just $k - l$, the number of additional coin flips required to pass from the lth row to the kth.

Now imagine that your quincunx is turned sideways, so that variation of the final shot position is now read "vertically." The lines in the quincunx figure, which evidently are serving the function of maximum-likelihood *prediction* or *explanation*, are now also serving the function of the lines in Section L4.1, least-squares fits to a "law." (This is the geometry of Pearson's own figure of a eugenic application, my Figure 7.34.) A remarkably impoverished law, indeed, with one point (the observed data) fitted without error, and every other point having the same value of the "stopwatch," but a law nevertheless. *Hence the formulas of that section must still apply here.*

To use them, however, we need a substitute for the summation operator Σ over "all the observed data." The substitution, actually, is for $\frac{1}{N}\Sigma$, the *averaging* over the observations. Instead of $\frac{1}{N}\Sigma$ we invoke an operator E, called **expectation**, that stands for the value we would obtain in the limit of indefinitely many repeated samples from distributions whose probability laws we know. The equivalent formulas for the quincunx are easy enough to compute just from its symmetries, but to simplify the notation let's take the regression formulas in the form

$$\Sigma xy / \Sigma x^2, \qquad (4.9)$$

the form when we know the prediction is already centered. For x the "cause" – known position in the quincunx at any level – and y the "effect" – position any number of rows lower down – the expectation $E(xy)$ is the expected value of x times a quantity $y = x + \epsilon$ where ϵ has mean (= expectation) zero. For each value of x, the expectation of $x\epsilon$ is still zero, and so $E(xy) = E(xx) = \text{var}(x)$. Then the regression of y on x is $E(xy)/\text{var}(x) = \text{var}(x)/\text{var}(x) = 1$, and that of x on y is $E(xy)/\text{var}(y) = \text{var}(x)/\text{var}(y) = 1/k$, exactly as we've computed already from first principles of inverse probability.

Using this same identity between quincunx and linear "laws," we can go on to compute the regression relating *consequences of the same cause*. (This was actually the question with which Galton began: the regression relating the heights of pairs of brothers, modeled as if they were generated by falling down a quincunx from a point corresponding to midparent height at some upper row.) On the quincunx, these are pairs of shot on lower rows (not necessarily the same lower row) that arise from the same starting position some rows above. In effect, a pair of shot fall from the top of the quincunx while tied together up to the kth row, at which time the tie is broken and they thereafter fall separately, one l_1 rows further and one l_2 rows further. Let x denote the position at the kth row. Then the regression relating the position of the two shot at the lower rows has numerator $E(x + \epsilon_1)(x + \epsilon_2)$, and since the expectations for either ϵ multiplying anything, including the other one, are 0, this is just the variance of x, the same as the row number k we've been using all along. The denominator is still the variance of the $(k + l_2)$th row, the predictor variance in the regression; this is the same as its row number. Hence the regression coefficient for the l_1-row descendant on the l_2-row descendant is thus $k/(k + l_2)$, independent of the value of l_1. The earlier result for the pure inverse inference, predicting the cause from the effect, is the case of this formula for $l_1 = 0$.

There exists a geometric interpretation of this sort of algebra that proves remarkably helpful – Galton himself, who was not much of a mathematician, originally discovered the possible realism of regression in this way. To introduce it, it will be helpful to have an example,[6] and here is a particularly convenient one, from some

---

[6] Galton's own example is easy to track down – Stigler (1986) conveniently reproduces it on pages 286–287 – but it is not as helpful a teaching device as the version I am offering here.

data published by Karl Pearson more than a hundred years ago (Pearson and Lee, 1903). The example is the first in Weisberg's popular textbook of applied regression (Weisberg, 2005), and as a courtesy to his student readers the data are posted for anybody to download from http://www.stat.umn.edu/alr/data/heights.txt. (In spite of its age, the original Pearson–Lee publication is also easily obtained – the complete run of *Biometrika* is available from the scholarly Web service JSTOR.) These data actually arise as random small perturbations ("jittering"), one per count per cell, of the centers of the tabulated bins counted by Pearson and Lee as in their Table 31 (my Figure 4.16). Pearson and Lee measured these heights to the nearest quarter inch; I believe the fractional counts in their tabulations are an attempt to accommodate measurements that are exact inches (and thus lie on the boundary between two bins).

By ignoring the facts of sexual reproduction, we can think of the daughter's height as a function of the mother's height alone, plus some noise. This is not as unreasonable an assumption as it might initially seem, as the mothers' and fathers' heights were substantially correlated in the sample Pearson and Lee were using: see Table 13, page 408 of their text.[7] And the Gaussianity of that noise is likewise not unreasonable (for most of the measurements in the Pearson–Lee paper, not just this pair). Reasonable or not, we can ask if there is a regression that relates the mother's height to the daughter's height, and vice versa.

Figure 4.17 charts these expected heights as solid circles, one per unit interval of mother's height. (These are essentially the same as the column-by-column means of the Table in Figure 4.16.) In a manner that would have been very gratifying to Pearson, you can see that they lie very nearly along the line whose slope formula is the one already given in Section L4.2, the ratio of covariance to variance, which is likewise the version "$\Sigma xy/\Sigma x^2$" of equation (4.9).

But there is also a regression of mother's height on daughter's height – the regression of cause on effect – *and this regression is likewise linear* as shown by the open circles in the same figure, which vary meaninglessly around the regression line in *their* orientation. This illustrates a handy theorem that is not as well-known as it ought to be: if both of these regressions are linear, then the distribution of the two variables together must be multivariate Gaussian. Pearson already suspected this, and argues, both in this paper and in others of the same period, that linearity of all the regressions like these is strong evidence for the assumption of a Gaussian distribution for the underlying variables. Furthermore, the slope of the second line, *taken in the correct coordinate system* (i.e., mother's height vertical, daughter's height horizontal), is very nearly the same as that of the first line – both are just about 0.5. We already noted, just as Galton did 135 years ago, that the shape of this scatterplot (or of the underlying table) is approximately that of an ellipse.

---

[7] This illustrates a point dear to economic historians: in Europe and the United States, the major determinant of adult size, which is also the value of manual labor, is the quality of nutrition, which in turn is heavily determined by social class. Marriage is assortative by height for developmental, not just cultural, reasons. See Floud et al. (2011).

TABLE XXXI.

Mother's Stature and Daughter's Stature.

Mother's Stature.

Daughter's Stature.	*52—53*	*53—54*	*54—55*	*55—56*	*56—57*	*57—58*	*58—59*	*59—60*	*60—61*	*61—62*	*62—63*	*63—64*	*64—65*	*65—66*	*66—67*	*67—68*	*68—69*	*69—70*	*70—71*	Totals
52—53	—	—	—	—	—	—	—	·5	—	—	—	—	—	—	—	—	—	—	—	·5
53—54	—	—	—	—	—	—	—	·5	—	—	—	—	—	—	—	—	—	—	—	·5
54—55	—	—	—	—	—	—	—	—	—	—	—	—	—	—	—	—	—	—	—	—
55—56	—	—	—	—	—	—	—	1	—	—	—	—	—	—	—	—	—	—	—	1
56—57	—	—	—	—	—	—	1	·25	1·5	1·25	·25	·25	—	—	—	—	—	—	—	4·5
57—58	—	—	—	1·5	1·25	1·5	1·25	1·5	3·25	1·25	1·25	·75	1	—	—	—	—	—	—	14·5
58—59	—	—	—	·5	2·25	1·5	2·25	4·75	2·5	·25	·5	—	—	·25	·75	—	—	—	—	15·5
59—60	—	—	—	—	—	3·5	5·75	12	11·25	9·25	3	1·5	·75	·25	1·25	—	—	—	—	48·5
60—61	·25	·25	—	—	2·75	4·75	9·25	15	19·5	15·75	17	7·75	3·25	1·75	1·5	·25	1	—	—	100
61—62	·25	·25	·25	·25	·75	2	8	18·5	18·25	34	27·25	15	8·75	6	3·25	·25	—	—	—	143
62—63	·5	—	·25	·25	·25	4	11·5	19·75	33·25	34·75	34·75	24	12·75	10·25	2·75	1	·5	—	—	190·5
63—64	·5	—	—	—	·75	3·75	6·25	17·5	29	41·5	46·5	33·5	18·25	10·25	3·75	·5	·5	—	—	212·5
64—65	—	—	—	—	—	2·5	4	6·25	23·75	32	38·5	40	27·75	18·75	3·5	1·5	·5	—	—	199
65—66	—	—	—	—	—	·25	1	5·75	10·5	18	27·5	35	26	16·5	12·25	3·25	—	—	·5	156·5
66—67	—	—	—	—	—	·25	1·25	3·25	10·25	14·75	23	29·25	21·75	16·25	15·75	2·75	2	·5	·5	141·5
67—68	—	—	—	—	—	—	—	1·5	5·5	7·75	7·75	15	14·25	11·25	8·25	5	1·25	·5	1	79
68—69	—	—	—	—	—	—	—	·5	—	1·5	5·75	6·5	4·75	5·75	3·25	4·75	1·25	1	—	35
69—70	—	—	—	—	—	—	—	—	—	—	1	2·5	5·25	3·25	3·25	·75	1	1	·5	18·5
70—71	—	—	—	—	—	—	—	—	—	—	1	1	1	·25	1·25	2	1	1·5	·5	9·5
71—72	—	—	—	—	—	—	—	—	—	—	—	·5	—	·75	·75	·5	—	·5	1	4
72—73	—	—	—	—	—	—	—	—	—	—	—	—	—	—	—	—	—	—	—	—
73—74	—	—	—	—	—	—	—	—	—	—	—	1	—	—	—	—	—	—	—	1
Totals	1·5	·5	·5	2·5	8	24	51·5	108·5	168·5	212	235	213·5	145·5	101·5	61·5	22·5	9	5	4	1375

Figure 4.16. Table 31 of Pearson and Lee (1903), with the original data underlying Figure 4.17. The ellipse of Figure 4.18 can just be glimpsed as the boundary of the region characterized by empty cells. (Notice that the values of the daughter's heights in the table here run downward – of course the correlation displayed in the table must be positive.)

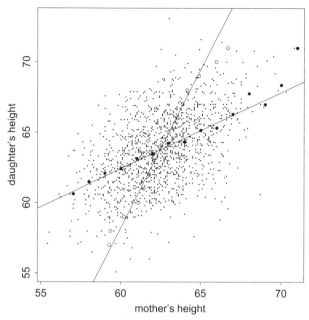

Figure 4.17. The equivalent as a scatterplot, after resampling with jitter as explained in the text. Solid circles, means of the daughter's height by width-1-inch bins of mother's height; open circles, means of mother's height by width-1-inch bins of daughter's height. Both regressions appear linear. (Data from Weisberg [2005]; online resource `heights.txt`.)

If the one regression slope is $\Sigma xy/\Sigma x^2$ and the other is $\Sigma xy/\Sigma y^2$, then the product of the two is $(\Sigma xy)^2/(\Sigma x^2 \Sigma y^2)$, a formula that is symmetric in the two centered variables x and y under study. This is the celebrated index r^2, square of the *correlation coefficient* $r = \Sigma xy/\sqrt{\Sigma x^2 \Sigma y^2}$ between two centered variables. We discuss it at length later in this chapter.

Mathematical Interlude

Taking our leave of Victorian England, let us put all this together from the point of view of an undergraduate statistics class of the present era. Today the ellipse we just glimpsed as the edge of the occupied region of the table is *modeled* as an *equiprobability contour* of a bivariate Gaussian scatterplot, which is modeled, in turn, as the rotated version of a prior bivariate Gaussian of two variables (the principal components, see Section 6.3) that were not themselves directly measured. The observed variables, in other words, are treated as the sum $x_1 + x_2$ and the difference $x_1 - x_2$ of two new constructs x_1 and x_2, each Gaussian, that are independent of one another (i.e., uncorrelated) and that have different variances σ_1, σ_2. The distribution of the point (x_1, x_2) is a cross-product of two bell curves. We know that the density

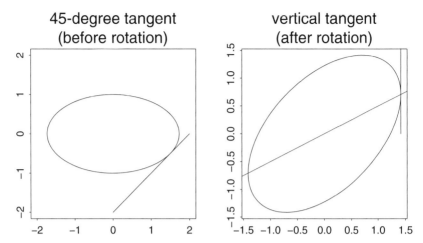

Figure 4.18. Mathematical interlude: why regression lines hit ellipses of constant probability at the points of tangents horizontal or vertical. The locus of the 45° tangent is produced in the left-hand diagram and interpreted as the touching point of the regression line after the 45° rotation into the right-hand diagram. See text.

of each of the x's separately is proportional to $e^{-x_i^2/\sigma_i^2}$, so the density of their bivariate scatter is proportional to the product $e^{-x_1^2/\sigma_1^2} e^{-x_2^2/\sigma_2^2} = e^{-((x_1^2/\sigma_1^2)+(x_2^2/\sigma_2^2))}$. So the densities are constant on the curves where $(x_1^2/\sigma_1^2) + (x_2^2/\sigma_2^2)$ is constant – but these are just ellipses around the origin with axes of symmetry along the coordinate axes.

The situation is as in the left-hand panel of Figure 4.18, with isoprobability curves (loci of equal probability) for the combination (x_1, x_2) ellipses lying horizontally around $(0, 0)$. To shift to the coordinates $x_1 + x_2$ (horizontal) and $x_1 - x_2$ (vertical) is just to rotate the picture by 45° counterclockwise, arriving at the version at right in the figure.

Let us now renotate equation (4.9) for the present context. We can compute the regression of $x_1 - x_2$ on $x_1 + x_2$ as their covariance divided by the variance of the predictor. Because x_1 and x_2 are uncorrelated, by assumption or by inspection of the left-hand panel in Figure 4.18, the covariance of $x_1 + x_2$ with $x_1 - x_2$ is $\sigma_1^2 - \sigma_2^2$, and the variance of $x_1 + x_2$ is $\sigma_1^2 + \sigma_2^2$, so the regression slope is just their ratio, which is

$$\frac{\sigma_1^2 - \sigma_2^2}{\sigma_1^2 + \sigma_2^2}. \tag{4.10}$$

But this quantity has a geometric meaning! Let us go back to the left-hand panel in the figure and ask, seemingly out of nowhere, where the point is at which the tangent to this ellipse makes an angle of 45° with the axes. If the equation of this stretched ellipse is $x_1^2/\sigma_1^2 + x_2^2/\sigma_2^2 = c$, where it won't matter what c is, then the tangent direction is for $2x_1 dx_1/\sigma_1^2 + 2x_2 dx_2/\sigma_2^2 = 0$. If this is to be in the direction $(dx_1, dx_2) = (1, 1)$, which is the tangent direction we have specified, we have to have $x_2/x_1 = -\sigma_2^2/\sigma_1^2$.

Our picture of correlated variation was constructed by rotating the left-hand panel of Figure 4.18 by 45°. The locus of tangent at 45° to the horizontal *before* the rotation will be the locus of tangent vertical *after* the rotation. We can work out that direction by the usual formula: the rotation of the direction $(\sigma_1^2, -\sigma_2^2)$ by 45° is the direction $(\sigma_1^2 + \sigma_2^2, \sigma_1^2 - \sigma_2^2)$.

But this ratio is just the slope of the regression of $x_1 - x_2$ on $x_1 + x_2$ set out in equation (4.10).

Therefore:

When bivariate data are Gaussianly distributed (i.e. when they arise as the sum and difference of uncorrelated Gaussians of different variance, as in the construction here), *then the regression line of* **either** *on the other intercepts ellipses of constant bivariate probability at the points of tangent horizontal or vertical.*[8]

You can draw regressions by hand this way, by sketching that curve by eye – it is through the isolines of the cells of the tabular approximation, like we already saw in Pearson's paper – and then drawing the regressions by connecting the pairs of places where your approximating curve, adequately smoothed, appears to be running vertically and horizontally. (These lines must intersect right at the centroid of the scatter.)

This locus, the vertical tangent, is always below the diagonal of the ellipse: hence "regression to [meaning: toward] the mean." But it is also "below" that diagonal when we reflect the whole picture, *including that regression line*, in the main diagonal: so whether we regress x on y or y on x, there is a regression toward the mean. In Pearson's example, tall mothers have daughters whose average height is shorter than that of their mothers (after correcting for the shift in the average between the generations), and at the same time, tall daughters have mothers whose average height is shorter than that of their daughters (again after correcting for the shift of means). This comes from the geometry of ellipses, and has nothing whatever to do with causation – it is just built into what we mean by *correlation*, which is thus confirmed to be no sort of causal concept at all.

L4.2.7 *Appendix to section L4.2: Stirling's Formula*

The Gaussian formula comes from Stirling's formula, which is nearly 300 years old. But you should never accept mathematics without proof, as long as the proof is at a level you are capable of following and finding instructive, so here is a very neat and relatively modern demonstration of that identity.

---

[8] I just illustrated one of Pólya's points from the discussion in Chapter 3. Having guessed the role of that vertical tangent, perhaps by looking at Pearson's figure (or Galton's) in one particular situation, we proceed to prove it for *every* case; and the proof here consists of "working backward" – of starting with the vertical tangent and showing that the slope of the line connecting it to the center of the ellipse is the same as the slope of the regression line. And *then* we get the slope of that line of vertical tangent by rotating into a different coordinate system, solving the problem there, and rotating back to the correlated case. All told, this is a very nice example of how a mathematician approaches a statistical question. The 45° rotation has nothing much to do with the scientific question here; its sole function is to make the proof transparent.

To begin, recall from Section L4.2.3 the normalizing constant for the error integral: by a truly crafty mathematical trick we found that $\int_{-\infty}^{\infty} e^{-x^2/2} dx = \sqrt{2\pi}$. Using that identity, we can arrive at Stirling's approximation by essentially elementary methods. Remember the familiar formula for *integration by parts:* from $d(uv) = u\, dv + v\, du$ one gets $\int u\, dv = uv - \int v\, du$. Then (the formal proof is the obvious induction)

$$\int_0^\infty e^{-x} dx = -e^{-x}\Big|_0^\infty = 1,$$

$$\int_0^\infty xe^{-x} dx = -xe^{-x}\Big|_0^\infty + \int_0^\infty e^{-x} dx = \int_0^\infty e^{-x} dx = 1,$$

$$\int_0^\infty x^2 e^{-x} dx = -x^2 e^{-x}\Big|_0^\infty + 2\int_0^\infty xe^{-x} dx = 2\int_0^\infty xe^{-x} dx = 2,$$

$$\int_0^\infty x^3 e^{-x} dx = -x^3 e^{-x}\Big|_0^\infty + 3\int_0^\infty x^2 e^{-x} dx = 3\int_0^\infty x^2 e^{-x} dx = 6, \ldots$$

$$\int_0^\infty x^n e^{-x} dx = -x^n e^{-x}\Big|_0^\infty + n\int_0^\infty x^{n-1} e^{-x} dx = n\int_0^\infty x^{n-1} e^{-x} dx = n!.$$

(We used the fact that $x^n e^{-x}$ is zero for $x = 0$ and also tends to 0 as x tends to infinity for any positive integer n.)

Also, the maximum of $x^n e^{-x}$ is located where $nx^{n-1} e^{-x} - x^n e^{-x} = 0$, or $x = n$.

Stirling's formula then follows from expanding the integral for $n!$ around this maximum, using the substitution $x = n + \sqrt{n}u$. Because $\log(1 + u/\sqrt{n}) \approx u/\sqrt{n} - u^2/2n$, the first two terms of the Taylor series, one has

$$n! = \int_0^\infty x^n e^{-x} dx = \sqrt{n} \int_{-\sqrt{n}}^\infty (n + u\sqrt{n})^n e^{-(n+u\sqrt{n})} du$$

$$= n^n \sqrt{n} \int_{-\sqrt{n}}^\infty e^{n \log(1+u/\sqrt{n})} e^{-(n+u\sqrt{n})} du$$

$$\approx n^n \sqrt{n} \int_{-\sqrt{n}}^\infty e^{n(u/\sqrt{n} - u^2/2n)} e^{-(n+u\sqrt{n})} du$$

$$= n^n e^{-n} \sqrt{n} \int_{-\sqrt{n}}^\infty e^{-u^2/2} du \to n^n e^{-n} \sqrt{2\pi n}.$$

(This demonstration is from Lindsay, 1941.)

> *Mathematical note 4.5. On the interplay between integers and continua.* Unlike the evaluation of the error function integral, this proof comes fairly close to a standard method used by physicists and other people who need approximations. The fundamental tactic would occur only to a mathematician – to take a formula about integers, the definition of $n!$, and expand it as an integral. Historically, this certainly went the other way: any classical mathematician would want to know what the integral of polynomials times the exponential function is, and whoever first noticed that one particular patterned integral reduced to a simple integer notation for a complicated function would certainly announce it as widely as possible. An applied mathematician (or mathematical physicist)

is also very accustomed to the Taylor series approximation, which replaces a function by an approximating polynomial (usually, a line or, as here, a quadratic).

Notice that the first three transformations of the Lindsay proof are all exact equalities, as are the last two. There is only one step at which something essential has changed, and that is the fourth, at which the log (in the exponent) is replaced by that quadratic (the second-order Taylor expansion). The *reason* for this maneuver is apparent almost instantly. It is that the n (the power of x) and the $\sqrt{n}$ (from the variable substitution) combine to cancel the e^{-x} of the original integrand, leaving only the second-order term from the Taylor series, which is the same term as in the error function integral itself. (This cancellation is different in its flavor from those we saw earlier that involved simultaneously adding and subtracting the same symbol or simultaneously multiplying and dividing. The intention here is to cancel part of a previous symbol string against part of a new one produced for this specific purpose. Foreseeing the possibility that something along these lines might "work" [i.e., might simplify the expression] is a specific mathematical skill quite analogous, I would imagine, to the skill of being able to plan combinations in chess. I didn't have it, and so I am not, in fact, a mathematician: only a statistician, which is a less imaginative occupation by far.) The substitution then converges because of the effect on the lower limit of the integral. The idea of taking a Taylor series precisely at the maximum of a function in order to make the first term vanish is an extremely old idea indeed – it occurred to Taylor himself in the 1600s.

Retelling the mathematics this way is a lot like retelling a joke – just like the joke immediately ceases to be funny, the mathematics immediately ceases to look brilliant. But it *is* brilliant, because it was seen as a whole, by somebody guessing that this should work, trying the substitution a few different ways (probably first with $x = n + u$ or $x = n + nu$, neither of which "works," and then with $x = n + \sqrt{nu}$, which does).

I passed over the first set of equations above with a pat phrase ("by the obvious induction"). Being comfortable with this, likewise, is one of the things that separates the mathematicians from the rest of us. We want to establish an identity for all n? It is just not obvious to break this into two parts, a proof for $n = 1$ and then a proof for $n + 1$ *assuming the truth of the proposition for n already*. Isn't this induction the same as assuming what you're trying to prove? Well, no, it isn't, the mathematicians all agree; it is a valid interpretation of the meaning of a phrase like "and so on." Training in proof techniques like these is actually the core of an education in pure mathematics. They are different in algebra, in geometry, and in analysis; they are different in probability theory and statistics; and they represent a very specific cognitive skill, one apparently localized to a relatively small part of the brain.

E4.3 Jean Perrin and the Reality of Atoms

In my judgment this is the most powerful numerical inference of all time in any natural science. The story begins around 1873, with a clever manipulation originally due to van der Waals. Avogadro's Law is that equal numbers of gas molecules (or gas atoms, for monatomic gases like argon), regardless of species, enclosed in the same volume v at the same temperature, exert equal pressure. So let N denote Avogadro's number, "the number of molecules in a gram-molecule" (e.g., 2 grams of gaseous hydrogen). Write L for the mean free path of a gas, measurable from its viscosity (by an equation of Maxwell's); let D be the "diameter of impact" of a gas molecule represented as a sphere; and let B be the van der Waals constant for the liquefied gas, measured by fitting adjustments to the gas law to account for dense gases or fluids. We then have

$$\pi N D^2 = \frac{v}{L\sqrt{2}}, \quad \pi N D^3 = 6B$$

for the "total surface area of spheres of impact" and the total volume of these molecules. These are two equations in two unknowns; from the pair of them we extract an estimate of N as $\frac{1}{\pi}(\pi N D^2)^3 / (\pi N D^3)^2$. From the best data available in 1908, for the monatomic noble gas argon, Jean Perrin estimated $N \approx 62 \cdot 10^{22}$ with a probable error of about 30%.

Perrin tries, not too successfully, to put this into human terms. (Here and elsewhere in these notes I quote from the second English edition (1923) of Perrin's book of 1911, *Les Atomes* – astonishingly, in 2013 *both* are still in print.)

> The mass of the hydrogen atom will be
>
> $$\frac{1.6}{1,000,000,000,000,000,000,000,000} \text{ gramme.}$$
>
> ... Each molecule of the air we breathe is moving with the velocity of a rifle bullet; travels in a straight line between two impacts for a distance of nearly .0001 millimetre; is deflected from its course 5,000,000,000 times per second, and would be able, if stopped, to raise a spherical drop of water 1μ in diameter to a height of nearly 1μ. [The unit μ is a *micron*, 0.001 millimeter or 0.000001 meter.] There are $3 \cdot 10^{19}$ molecules in a cubic centimeter of air, under normal conditions. Three thousand of them placed side by side in a straight line would be required to make up one millimetre....
>
> The Kinetic Theory justly excites our admiration. It fails to carry complete conviction, because of the many hypotheses it involves. If by entirely independent routes we are led to the same values for the molecular magnitudes, we shall certainly find our faith in the theory considerably strengthened. (Perrin, 1923, pp. 81–82)

In other words, the human scaling I have just quoted is not persuasive; the scale is, in the end, not yet credibly human. The subject of Perrin's book, summarizing several years of work that earned its author the Nobel Prize in physics (1926), is a splendidly original project to bridge these two domains of scaling: to make "molecular reality" visible. For a recounting of Perrin's life work at greater length, see Nye (1972).

Brownian motion had been known since the 1820s, when it was discovered (by an English botanist, of all people). Over the rest of the 19th century, many of its most puzzling properties had been demonstrated:

> The movements of any two particles are completely independent, even when they approach one another to a distance less than their diameter; ... The Brownian movement is produced on a firmly fixed support, at night and in the country, just as clearly as in the daytime, in town and on a table constantly shaken by the passage of heavy vehicles.... The agitation is more active the smaller the grains. The nature of the grains

appears to exert little influence. The Brownian movement never ceases. *It is eternal and spontaneous.*

If the agitation of the molecules is really the cause, and if that phenomenon constitutes an accessible connecting link between our dimensions and those of the molecules, we might expect to find therein some means for getting at these latter dimensions. This is indeed the case. (Perrin, 1923, pp. 84–85; italics in original)

Now Perrin reminds the reader of two hints in the literature that for some laws molecular size may not matter: Raoult's Law, the proportionality of freezing-point depression in liquids to the molecular concentration of a solute, and van't Hoff's demonstration that dilute solutions can be modeled by the gas laws by substituting osmotic pressure for gas pressure.

Is it not conceivable that there may be no limit to the size of the atomic assemblages that obey these laws? Is it not conceivable that even visible particles might still obey them accurately? I have sought in this direction for crucial experiments that should provide a solid experimental basis from which to attack or defend the Kinetic Theory. (Perrin, 1923, p. 89)

From here on, the demonstration is simply a matter of extremely painstaking measurement. Perrin begins by reminding us, via a dimensional analysis, that the rate of rarefaction by height in a gas (or a dilute solution) – the equilibrium between pressure and gravity – is exponential with a rate constant proportional to pressure gradient, which is proportional to numbers of molecules in unit volume and therefore inverse to gram-molecular weight. In pure oxygen at $0\,°C$, 5 kilometres of rise would halve the pressure. "But if we find that we have only to rise $\frac{1}{20}$ of a millimetre, that is, 10^8 times less than in oxygen, before the concentration of the particles becomes halved, we must conclude that the effective weight of each particle is 10^8 times greater than that of an oxygen molecule. *We shall thus be able to use the weight of the particle, which is measureable, as an intermediary or connecting link between masses on our usual scale of magnitude and the masses of the molecules*" (Perrin, 1923, pp. 93–94; italics in the original). Because Earth's atmosphere is adequately calibrated, the analysis reduces to one extremely careful measurement of a contrast with an "artificial atmosphere" as different from Earth's as possible.

E4.3.1 Multiple Routes to the Same Quantity

There followed years of determined work with an emulsion of *gamboge* (a yellow pigment) and mastic (a resin) in water. From a kilogram of gamboge, "several months" of centrifuging yielded a few decigrams of spherules of an appropriate size.

The **density** of these little balls was measured in three different ways, yielding 1.1942, 1.194, and 1.195 g/cc.

The **volume** of the grains was measured in three different ways: by effective *radius* when aggregated into long rows under desiccation; from *weighing* the desiccated result from a known volume of solution at known dilution; and from Stokes' Law for the *rate of fall* of the "edge" of the distribution under gravity. "The three methods give concordant results . . . ": for instance, average radii of 0.371, 0.3667, and 0.3675 μ in one analysis of about ten thousand grains.

The **concentration** of the emulsion as a function of height was measured by mounting a microscope objective horizontally, viewing the emulsion through a small pinhole, and, in modern jargon, *subitizing* the small integer count of spherules in the field of view, over and over, thousands of times. After verifying that the column was in statistical equilibrium, Perrin counted a total of 13,000 spherules on four equally spaced horizontal planes.

> The readings gave concentrations proportional to the numbers 100, 47, 22.6, 12, which are approximately equal to the numbers 100, 48, 23, 11.1, which are in geometric progression. Another series [using larger grains but more closely spaced imaging planes] show respectively 1880, 940, 530, 305 images of grains; these differ but little from 1880, 995, 528, 280, which decrease in geometric progression. (Perrin, 1923, p. 103)

(In Chapter 2 we already noted these interesting phrases "approximately equal' and "differ but little," which are a remarkable anticipation of Kuhn's observation that the role of tables of data in physics texts is mainly to show what "reasonable agreement" is.)

The ratios of half-heights for air and for these spherules can thus be converted into estimates of Avogadro's number. In a section entitled "A Decisive Proof," Perrin snaps the trap shut:

> It now only remains to be seen whether numbers obtained by this method are the same as those deduced from the kinetic theory. . . . It was with the liveliest emotion that I found, at the first attempt, the very numbers that had been obtained from the widely different point of view of the kinetic theory. In addition, I have varied widely the conditions of experiment. The volumes of the grains have had values [over a 50:1 range]. I have also varied the nature of the grains . . . I have varied the intergranular liquid [over a viscosity range of 125:1]. I have varied the apparent density of the grains [over a 1:5 range; in glycerine, in fact, they floated *upward*]. Finally, I have studied the influence of temperature [from $-9°$ to $+60°$].
>
> In spite of all these variations, the value found for Avogadro's number N remains approximately constant, varying irregularly between $65 \cdot 10^{22}$ and $72 \cdot 10^{22}$. Even if no other information were available, *such constant results would justify the very suggestive hypotheses that have guided us.* . . . But the number found agrees with that given by the kinetic theory from the consideration of the viscosity of gases. *Such decisive agreement can leave no doubt as to the origin of the Brownian movement.* To appreciate how particularly striking the agreement is, it must be remembered that before these experiments were carried out we should certainly not have been in a position either to deny that the fall in concentration through the minute height of a few microns would be negligible, in which case an infinitely small value for N would be indicated, or, on the other hand, to assert that all the grains do not ultimately collect in the immediate vicinity of the bottom, which would indicate an infinitely large value for N. It cannot be supposed that, out of the enormous number of values a priori possible, values so near to

the predicted number have been obtained by chance for every emulsion and under the most varied experimental conditions. The objective reality of the molecules therefore becomes hard to deny. As we might have supposed, an emulsion is an atmosphere of colossal molecules which are actually visible. (Perrin, 1923, pp. 104–105; italics in the original)

(In fact, the height difference for rarefaction by half in these series was of the order of 30μ, about $1/150{,}000{,}000$ of the 5 km characterizing molecular oxygen; so the "molecular weight" of these gamboge balls was about $4.8 \cdot 10^9$.)

Perrin wants one single number, and so from a "more accurate series" of 17,000 grains he reports a "probable mean value" of $68.2 \cdot 10^{22}$. The actual value we assign this constant today is $60.22798 \cdot 10^{22}$ (see Section E4.3.2). But it makes no difference that he was off by 13% or so in his final estimate, for **he has proved his case: yes, there are atoms.**

The embrace of the language of abduction here is obvious, in spite of the absence of the word or the formal syllogism. Two separate arguments are woven together, each an abduction of its own. The experiments involving solutions varied all of the accessible physical parameters over their full feasible ranges, and yet the estimates of N varied hardly at all (by $\sim 10\%$, as physical parameters varied by factors up to hundreds). Such an invariance is incomprehensible except on the assumption that some aspect of the true structure of the world constrains the interrelationships of the measurements a priori.

Separately, we see an arithmetical agreement between two approaches that are completely independent conceptually: one from viscosity of gases and one from rarefaction of liquid solutions. **The outcome is thus doubly surprising.** By the reference in his section title to "decisive proof" Perrin means not to suggest but to *declare* that there is no other explanation possible. Only the atomic formalism, in combination with the physicist's expertise at manipulating the symbolic contents of his equations, can justify the convergence of so many experimental measurements on the same arithmetic evaluation of N over and over.

E4.3.2 More Detail about Brownian Motion

The original publication of 1909 on which Perrin's book is based was entitled *Mouvement Brownien et réalité moléculaire*. This paper begins with actual observations of single particles executing a Brownian motion, as in Figure 4.19. The actual challenge up for verification was not the existence of atoms – this had been conceded, more or less, by everyone except Mach and Ostwald – but the considerably more fraught claim of Einstein (1905) to have found the actual coefficients of Brownian diffusion from first principles. Writing x for diffusion in position and A for rotational degrees of freedom, Einstein derived, for observations over unit time,

$$\overline{x^2} = \frac{RT}{N}\frac{1}{6\pi a \zeta},$$

FIG. 9.

Figure 4.19. Among the first published random walks: observations of a grain every 30 seconds over a long period of time. "Diagrams of this sort," Perrin comments (1923, pp. 115–116), "give only a very meagre idea of the extraordinary discontinuity of the actual trajectory. For if the positions were to be marked at intervals of time 100 times shorter, each segment would be replaced by a polygonal contour relatively just as complicated as the whole figure, and so on. Obviously it becomes meaningless to speak of a tangent to the trajectory of this kind." This is thus also a pioneering published figure of a two-dimensional statistical fractal (compare Figure 2.12, which is simulated, not observed). (From Perrin, 1923, Figure 9.)

where ζ is viscosity, T is the absolute temperature, R is the Boltzmann constant, and a is the radius of the grain, and

$$\overline{A^2} = \frac{RT}{N} \frac{1}{4\pi a^3 \zeta}.$$

That such is actually the case appeared at the time to be far from certain [note that one scales as diameter and the other as volume]. An attempt by V. Henri to settle the question... had just led to results distinctly unfavourable to Einstein's theory. I draw attention to this fact because I have been very much struck by the readiness with which at that time it was assumed that the theory rested on some unsupported hypothesis. I am convinced by this of how limited at bottom is our faith in theories; we regard them as instruments useful in discovery rather than actual demonstrations of fact. (Perrin, 1923, p. 121)

Displacement between:-	P for each ring	n Calculated	a Found
0 and $\frac{e}{4}$	.063	82	34
$\frac{e}{4}$,, $2\frac{e}{4}$	.167	83	78
$2\frac{e}{4}$,, $3\frac{e}{4}$	.214	107	106
$3\frac{e}{4}$,, $4\frac{e}{4}$	.210	105	103
$4\frac{e}{4}$,, $5\frac{e}{4}$	.150	75	75
$5\frac{e}{4}$,, $6\frac{e}{4}$	.100	50	40
$6\frac{e}{4}$,, $7\frac{e}{4}$	.054	27	30
$7\frac{e}{4}$,, $8\frac{e}{4}$	.028	14	17
$8\frac{e}{4}$,, $9\frac{e}{4}$	.014	7	9

A third verification is to be found in the agreement established between the values calculated and found for the quotient $\frac{d}{e}$ of the mean horizontal displacement d by the mean quadratic displacement d. By a line of reasoning quite analogous to that (p. 55) which gives the mean speed G in terms of the mean square U^2 of the molecular speed, it is shown that d is very nearly equal to $\frac{8}{9}e$. As a matter of fact, for 360 displacements of grains of radius .53 μ, I found $\frac{d}{e}$ equal to .886 instead of .894 required by the theory.

Further verification of the same kind might still be quoted, but to do so would serve no useful purpose. In short, the irregular nature of the movement is quantitatively rigorous. Incidently we have in this one of the most striking applications of the laws of chance.

Figure 4.20. Confirmation of Einstein's law of Brownian motion, I: the correct functional of a Normal distribution (a χ^2 on two degrees of freedom) for the distribution of displacements of a particle over time intervals of constant length. Note that Perrin calls this a "verification" without resorting to Pearson's formula (another use of the same χ^2 probability law) for comparing calculated and observed distributions like these. (From Perrin, 1923, p. 119.)

Remarkably enough, Perrin goes on to observe the elementary distributions underlying the expected mean square of displacement (Figure 4.20), shows that it fits the χ_2^2 distribution well enough (by tabulation, not by test) – another eruption of Kuhnian pragmatism – and verifies Einstein's equation, in the sense that the value of N it supplies is actually in the right range. In fact an entire additional table of estimates of N follows, over a total of 4200 particle displacements at intervals of 30 seconds, with spherule masses ranging in a ratio of 15000:1, viscosities by 125:1, and so on. The estimates range only from 55 to 80 times 10^{22}, "as in the vertical distribution experiments. This remarkable agreement proves the rigorous accuracy of Einstein's formula and in a striking manner confirms the molecular theory" (Perrin, 1923, p. 123).

In these pages Perrin shows his mastery of what Kuhn describes half a century later as the "limits of reasonable agreement." Now it is not only the estimated value of Avogadro's constant that is astonishingly stable against experimental variations of such an extent that, were it not lawfully invariant, it could not possibly remain unchanged; it is also that the detailed construction of *variation* is consistent with Einstein's suggested law. Perrin, like Millikan, was able to estimate the slope of a line, and find it invariant over experimental preparations. But, unlike Millikan, Perrin was also able to model the variation of his experimentally measured points

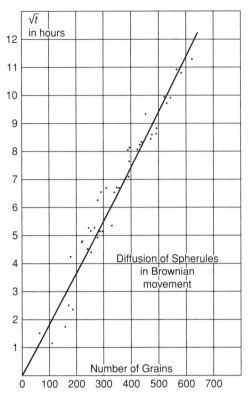

Figure 4.21. Confirmation of Einstein's law of Brownian motion, II: Straight-line fit of Brownian motion in a plot with square root of time on the vertical axis. (From Perrin, 1923, Fig. 11.)

from that line, and show that an appropriate function of the variance of *those* was lawfully invariant over experimental preparations. Finally, and most magisterially, Perrin could show that the two sets of models, those for the slope and those for the variance, themselves result in substantially the same estimate of N.

Perrin even investigates Einstein's translational/rotational equipartition claim, by preparing spherules so large that they rotate so slowly that the rotation itself can be measured by speed of apparent movement of "inclusions." He finds the root mean square of A to be about $14.5°$ per minute, versus the value of about $14°$ that would be expected on the estimate $N \approx 69 \cdot 10^{22}$.

A final verification in this series considers the rate of accumulation of spherules on an absorbing barrier instead of their displacements in a microscope objective. The slope of a straight line fitted to the scatter of $\sqrt{t}$ by count of particles absorbed (Figure 4.21) leads, again, to $N \approx 69 \cdot 10^{22}$. This is the only actual line to be fit in the entire project, and Perrin does not even pause to tell us how it was computed. It is sufficient that, in a figure based on "several thousand grains" absorbed over about six days, the relationship with $\sqrt{t}$ is obviously linear.

The Gaussian distribution is absolutely central to Perrin's ultimate message mainly because it has a crucial additional parameter that can supply *independent* information about the same question asked of Nature. The Gaussian distribution matches Perrin's data (the images of random walk [Figure 4.19] and the corresponding quantitative summaries like that in Figure 4.20) not only in its essential disorder but also in the orderliness that underlies it: its scaling as a function of time, and the relative frequency of the more extreme excursions with respect to the typical range. In other words, Perrin demonstrates not merely that the error around his lines of the speed of diffusion is Gaussian, but also that it is Gaussian with a scaling in terms of time whose coefficient yields the selfsame estimate of N. To argue for a Normal model for this variation is thus to argue, along with Boltzmann (or Elsasser, for that matter), that the experimentalist *knows how to measure what matters*, and that the rest is indeed completely uninformative noise. The Boltzmann distribution is as real as the atom. After Avogadro's number has been measured, there is no further information in these data. Knowing that – knowing how to design an experiment that measures only one thing, and that in a context of exact lawfulness – is the key to understanding how physicists judge each other's experiments (see Krieger, 1992, or Galison, 1987).

Perrin pauses for a summary:

> The laws of perfect gases are thus applicable in all their details to emulsions. This fact provides us with a solid experimental foundation upon which to base the molecular theories. The field wherein verification has been achieved will certainly appear sufficiently wide when we remember: That the nature of the grains has been varied ...; That the nature of the intergranular liquid has been varied ...; That the temperature varied ...; That the apparent density of the grains varied ...; That the viscosity of the intergranular liquid varied ...; That the masses of the grains varied (in the enormous ratio of 1 to 70000) as well as their volume (in the ratio of 1 to 90000).
>
> From the study of emulsions the following values have been obtained for $N/10^{22}$ – 68.2 deduced from the vertical distribution of grains. 64 deduced from their translatory displacements. 65 deduced from observations on their rotation. 69 deduced from diffusion measurements. (Perrin, 1923, p. 132)

Remarkably enough, we are still not finished. Perrin goes on for another 80 pages, extracting another *eleven* completely independent estimates of N from density fluctuations in emulsions, opalescence, the blueness of the sky, the diffusion of light in argon, the black body spectrum, Millikan's measurement of the charge on the electron, and four different aspects of experimental radioactivity (e.g., by counting helium atoms emitted by radioactive decay of a known weight of radium).

The 16 separate summary estimates of Avogadro's number, based on 11 fundamentally different experimental investigations, instruments, and phenomena, range from $60 \cdot 10^{22}$ to $75 \cdot 10^{22}$. His experiments completed, Perrin steps all the way back for an apotheosis:

> Our wonder is aroused at the very remarkable agreement found between values derived from the consideration of such widely different phenomena. Seeing that not only is the

same magnitude obtained by each method when the conditions under which it is applied are varied as much as possible, but that the numbers thus established also agree among themselves, without discrepancy, for all the methods employed, the real existence of the molecule is given a probability bordering on certainty.... These equations express fundamental connections between the phenomena, at first sight completely independent, of gaseous viscosity, the Brownian movement, the blueness of the sky, black body spectra, and radioactivity.

The atomic theory has triumphed. Its opponents, which until recently were numerous, have been convinced and have abandoned one after the other the sceptical position that was for a long time legitimate and no doubt useful. Equilibrium between the instincts towards caution and towards boldness is necessary to the slow progress of human science; the conflict between them will henceforth be waged in other realms of thought. (Perrin, 1923, pp. 215–216)

One can take this conclusion in several different directions. Perrin himself, hearkening back to the preface to his book, emphasizes the interplay between continuity and discontinuity in modeling physical phenomena. (In that preface he actually introduced the problem of representing "the coast line of Brittany," parent of the comparable question for Britain that introduced Mandelbrot's first popularizations of fractals 70 years later.)

But in achieving this victory we see that all the definiteness and finality of the original theory has vanished. Atoms are no longer eternal indivisible entities, setting a limit to the possible by their irreducible simplicity; inconceivably minute though they be, we are beginning to see in them a vast host of new worlds. In the same way the astronomer is discovering, beyond the familiar skies, dark abysses that the light from dim star clouds lost in space takes æons to span. The feeble light from Milky Ways immeasurably distant tells of the fiery life of a million giant stars. Nature reveals the same wide grandeur in the atom and the nebula, and each new aid to knowledge shows her vaster and more diverse, more fruitful and more unexpected, and, above all, unfathomably immense. (Perrin, 1923, p. 217)

Other interpretations are equally possible. I prefer the one in the spirit of Eugene Wigner's great essay: in the Gaussian model for Brownian motion we surely "have got something out of the equations that we did not put in," in Wigner's phrase. Perrin was satisfied to fit his models to a few percent only, and that linear verification of Einstein's law of diffusion was fitted by hand. In the century since Perrin's work, we have grown accustomed to the extension into statistical physics of the trust in precision instruments that previously characterized only celestial mechanics. If a law is exactly true, even a law about distributions, eventually a way will be found to verify both the conditions on which it does *not* depend and the shape of the distribution on which it does.

At the same time, we see powerful confirmation of another two themes of Wigner's. First, the laws of physics are of universal application, but of strictly limited scope. In Millikan's hands, it was not the measured currents (Figure 4.9) that satisfied Einstein's equation, but the line plotting voltage against the vanishing of any current at all (Figure 4.10). For Perrin, Avogadro's number was not directly observed, the way it is today (by counting; see later), but instead was extracted from a variety of arithmetical maneuvers that combine slopes (physical densities over space, or over time, or over the square root of time) with standardizations for a wide variety of scaling factors such as molecular weight. Second, the universe is so constructed that the rules of algebra (canceling common factors, isolating the symbol N on one side of an equation or the other, taking logarithms) are justified physically, no matter how many such symbolic steps have been taken. The cultural agreement among mathematicians about which such *symbolic* steps are justified is confirmed here by the (presumably noncognitive) behavior of the little balls of gamboge and mastic in Perrin's laboratory.

Perrin exploits throughout his book the extraordinary power of scientific hypotheses that explain otherwise preposterous coincidences or other outrageously improbable patterns in data. Psychologically, the intensity of the elegant surprise driving step 1 of the abduction is converted into an equivalently strong confidence about the plausibility of the hypothesis that has reduced the finding to "a matter of course." For 16 different algebraic manipulations of data from a dozen wholly different instruments and experiments, to arrive at numbers all in the range of $60 \cdot 10^{22}$ to $80 \cdot 10^{22}$ – all having logarithms, in other words, between 54.75 and 55.04 – would be an absurd coincidence, unless atoms existed, in which case the approximate agreement of all these counts of the same thing (molecules in a gram-mole) would be "a matter of course." With this abduction from overwhelming numerical agreement (consilience), Perrin simply closes the case.

But in fact this was not post-hoc reasoning. Perrin *knew* that atoms existed, and set out to demonstrate the fact in the way that most expeditiously blocked any possibility of counterargument. In short, his task was to convince a community to get on with more fruitful debates. Note those graceful emotional terms: "It was with the liveliest emotion" that the first of the rarefaction studies closely matched the classic van der Waals value from kinetic theory; even though "at bottom we have limited faith in theories," by the end of the argument, "our wonder is aroused" at the remarkable agreement among these algebraic derivations from superficially entirely unrelated phenomena; "Nature reveals the same wide grandeur in the atom and the nebula." One is reminded of Darwin's equally elegant summation, "There is grandeur in this view of life," even though it took somewhat longer for the underlying proposition in that domain to come to be taken as fact.

Recall Fleck's observation (Epigraph 14) that the factuality of a scientific fact can be interpreted as a socially imposed constraint on creative thought. Perrin set out to create a fact in this specifically *social* sense – to justify "faith in the kinetic theory" by explicit apodictic evidence – and did just that. When, around 1980, the scanning tunneling microscope first allowed us to "actually see" atoms (that is, to image individual atoms: von Baeyer, 1992), the result, however elegant, was epistemologically entirely an anticlimax. We already knew for certain that they had to be there. Imaging them standing still, in the way that cloud chambers imaged them in motion, was merely a matter of the application of sufficient cleverness (and the earning of another Nobel Prize).

I noted that it did not matter that Perrin's estimate of N was off by about 15% from the accepted value today, because he had closed the argument he was trying to close. However, we have already seen from the Millikan case that there are other purposes to measurements of this style than closure. One is simply to get the most precise possible value of a constant so as to test *other* physical laws that claim constancy of certain ratios and especially the invariance of the constants over human time. From a late 20th-century textbook of metrology (Petley, 1985) we learn of a representative current way of measuring N that has nothing to do with *any* of Perrin's twelve but that results in an estimate $60.228 \cdot 10^{22}$ having a precision of about one part in ten million, which is a million times better, more or less, than anything Perrin could do. This modern approach is entirely different: the way to count a bunch of atoms is to make them sit relatively still, inside unit cells of a crystal, and then count the number of crystal cells, which is inverse to their spacing. We then have only to figure out how many of these cells make up a weight equal to the atomic weight of the atom in the crystal (in this case, silicon), and there we have N. The precision is a function of the measure of the spacing of the crystal lattice (which comes from X-ray diffraction just like the image that drove the discovery of the structure of DNA, my Figure 4.27) and the accuracy of the density estimate of the silicon in the crystal (together with a knowledge of the relative abundances of the three different isotopes of silicon [^{28}Si, ^{29}Si, ^{30}Si] in the crystal). But this is also, itself, a massive demonstration of the consilience underlying physical science and more specifically of that "trust in consilience" of which Wilson spoke: the trust that entirely different experiments can, in sufficiently competent hands, lead to *exactly* the same number.

E4.3.3 The Maxwell Distribution Is Exact

In Section L4.2 we noted that a handful of symmetry assumptions let the mathematician *deduce* a unique form for the velocity distribution of the molecules of a gas. As Ruhla (1992, pp. 76–78) dryly notes, "A good deal of work has been devoted to testing the predicted velocity distribution." He quotes one particular experiment, reported in 1959, "whose results are particularly precise and convincing." The layout (Figure 4.22) is nearly as complicated as Millikan's in the endeavor to control all the relevant physical parameters at the same time. It involves a molecular beam of single potassium atoms traveling in a straight line at extremely low ambient pressure. The straight line is interrupted by two disks from each of which a notch has been cut out. The notched circles are separated by some distance and rotated relative to one another by an angle of 4°. We spin the pair of notched circles at various speeds and count the number of atoms that make it past both notches as a function of rotational speed. Then we can recover the Maxwell–Boltzmann velocity distribution by the speed distribution of these atoms in the direction of the axis of rotation of the wheels, and that distribution, in turn, from the dependence of the density of arriving atoms (counted by their impacts on a detector) on the rotational velocity of the pair of notched circles. There results the chart in Figure 4.23, where the data points (dots) lie precisely on the fitted curve to the accuracy of the printed medium – accuracy many orders of magnitude better than what Perrin could achieve half a century earlier. Nevertheless,

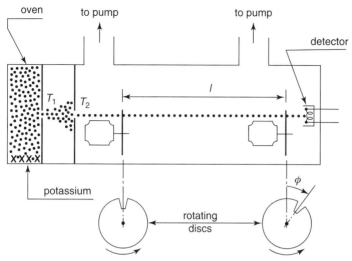

Figure 4.22. The Marcus–McFee apparatus for verifying the Maxwell–Boltzmann distribution. Atomic potassium is heated (far left) and collimated into one single line (left to right) of particles whose speed distribution is assessed by cutoffs that pass only a narrow range of molecular speeds as a function of their rotational speeds. (This figure and the next are from Marcus and McFee, 1959, via Ruhla, 1992.) Republished with permission of Elsevier Books, from *Recent Research in Molecular Beams*, P.M. Marcus and J.H. McFee 1959; permission conveyed through Copyright Clearance Center, Inc.

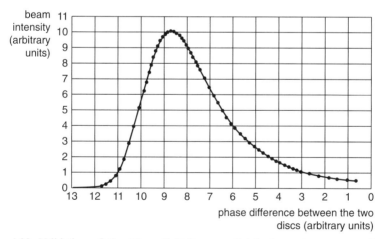

Figure 4.23. Validation of the Maxwell–Boltzmann distribution using enormously more molecules than Perrin could. Dots, data; curve, theoretical distribution (density of the reciprocal of molecular speed). The curve is the detector performance from Figure 4.22 as a function of the angular spacing between the notches of the two disks. Republished with permission of Elsevier Books, from *Recent Research in Molecular Beams*, P.M. Marcus and J.H. McFee 1959; permission conveyed through Copyright Clearance Center, Inc.

the numerical inference is precisely the same: "The agreement between theory and experiment is quite remarkable and well worth stressing," Ruhla concludes.

L4.3 Likelihood-Based Statistical Inference

L4.3.1 Bayes' Theorem

The ultimate justification of the lectures so far – the least-squares properties of averages (Section L4.1) and the origins of the Gaussian distribution in deviations from averages (Section L4.2) – is to construct machinery for reasoning from effect to cause. The initial statement of this setup comes from a posthumously published paper (1764) by the reverend Thomas Bayes:

> *Given* the number of times in which an unknown event has happened and failed: *Required* the chance that the probability of its happening in a single trial lies somewhere between any two degrees of probability that can be named [i.e., the probability that the true long-term fraction of Heads lies in the interval (.45,.55) or the like]. (Stigler, 1986, p. 123)

(For the intellectual context that helped to shape the phrasing of this inquiry, see Chapter 1 of McGrayne, 2011.)

The thrust of this question is evidently different from that of our hitherto customary discourse about coins. Thus far, a coin has had a "propensity for heads," which is some physically fixed characterization it is in our interest to measure, and (along with Bernoulli and de Moivre and many others) we asked the chances of particular numbers of heads given this fixed propensity. Bayes has changed the question. Now we have fixed *data* – counts of heads and tails – and ask about probabilities of propensities of the coin. Bayes argues with the aid of a familiar vernacular image, a billiards table, as in Figure 4.24. On the table a black ball is thrown first; its position sets the "propensity of heads." Subsequently, any number of white balls are thrown one by one (none of them jarring the black ball) and we count the number of white balls that landed closer to the left edge of the table than the black one did – that fell to the left of the black ball. Thus on the one substrate *both* processes could go forward: the setting of the original propensity, and the occurrence of those "unknown events" given that propensity. We count 32 white balls out of 50 falling to the left of the black. Where was that black ball, and what precision should we attribute to our estimate of that location?

The philosophical crux of all modern statistical inference lies in that innocuous assumption about the first ball, the black one, the "hypothesis." *Is* that the way Nature selects true values? Asked in this form, the question is nearly meaningless. I prefer to construe probabilities as relative frequencies of classes of outcomes, whereas there is no evidence that Nature throws most billiard balls (i.e., for the speed of light) more than once. (For a contrary speculation, see Smolin, 1997.) But there is another approach to probabilities that, when information about relative frequency is unavailable, grounds them in *degrees of belief* calibrated by rules (also applying to rational gambling) that enforce consistency with the rules of probability. These

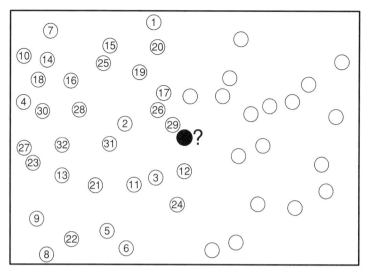

Figure 4.24. Schematic of Thomas Bayes' 1764 explanation of the probability of hypotheses. Black ball: Nature's value. White balls: individual trials that do or do not exceed the threshold set by the black ball. Bayes' question: Given the count of white balls that lie beyond the black ball, what is the distribution of the location of the black ball?

lecture notes are not the place to go into this literature, the doctrine of "personal probability." Let us simply agree, if only temporarily, that hypotheses about true values may be treated as if they were subject to such probability distributions, and see what consequences arise.

In all this talk about billiard balls we have not yet stated the theorem that permitted Bayes to answer his question about "the probability of [the unknown event] happening in a single trial." In fact, he stated it geometrically, in a form difficult to encapsulate. Laplace, playing the role of the city slicker to this country reverend, publicized Bayes' principle in essentially its modern form:

> If an event can be produced by a number n of different [and equiprobable] causes, then the probabilities of these causes given these events are to each other as the probabilities of the event given the causes.

In contemporary language, the **likelihood** of each **hypothesis**, the factor by which we multiply its "prior probability" to arrive at a quantity proportional to "posterior probability," is proportional to the probability of the data given the hypothesis. Here and throughout modern statistics, "prior" means "before the data are gathered," and "posterior" means "after the data are gathered." "Equiprobable" will turn out to refer to the little baize patches of the billiard table, the locus of symmetry of the underlying stochastic assumption.

Laplace stated his principle as a theorem, but in the modern approach to probability it is very nearly a tautology. The probabilist *defines* the occurrence of a hypothetical Cause – call it C – and the occurrence of a particular Outcome – call it O – as *events*, subsets (identified howsoever) of a space of all possibilities, the *sample space*. The causes and the outcomes can be indexed by integers – $C_1, C_2, \ldots$ and $O_1, O_2, \ldots$ – or by continuous measurements $C = x$, $O = y$. In the billiard-ball example, the causes lie in the (continuous) space of left–right positions for one black ball; the outcomes, in the discrete space of binary sequences for the n white ones.

Any event like C or O has a probability; write these $P(C)$, $P(O)$, and so on. These P's may have, but need not have, originated in "frequentist" observations (percentages of a sample actually encountered). Write $C \& O$ for the occurrence of *both* C and O; this is a compound event like "black ball halfway up the table and 5 out of 7 white balls to its left." Then, by definition, the "probability $P(C|O)$" (read "probability of C given O," here the probability of the cause C given the effect O) is *defined* as the ratio $P(C \& O)/P(O)$. If we had access to frequencies, this would be the fraction of instances of O having C for cause out of the total number of cases of O. Back on the billiard table, we are talking about the frequency distribution of locations of the black ball given that some *fixed* fraction $O = r/n$ of the white balls were found to its left. Events C and O are *independent* if $P(C|O) = P(C)$, or, equivalently, if $P(C \& O) = P(C)P(O)$ – which is not the case here. (Independence would imply that the chance of having, say, five out of seven white balls past the black ball would be independent of the location of the black ball. Although that sort of independence of past and future characterizes sequences of flips of fair coins, it wouldn't allow us to learn much about our experiment with the billiards table.)

As in one of the derivations of the physical Normal distribution, at the core of the mathematics is an argument from symmetry: The event $C \& O$ is the same as the event $O \& C$. Then the same conditional probability $P(C|O)$ can be written

$$P(C|O) = P(C \& O)/P(O)$$
$$= \bigl(P(C \& O)P(C)\bigr) / \bigl(P(O)P(C)\bigr)$$
$$= \bigl(P(O \& C)/P(C)\bigr)\bigl(P(C)/P(O)\bigr)$$
$$= P(O|C)\bigl(P(C)/P(O)\bigr).$$

Now consider this identity for fixed O (the "effect") while varying C (the "cause"). The term $P(O)$ is a constant and so drops out of comparisons; and assuming $P(C)$ to be a constant for all C (by judicious relabeling of categories, if necessary), Laplace's statement follows, as it were, by definition: $P(C|O) \propto P(O|C)$. (The symbol $\propto$ is read "is proportional to.") But these quantities $P(O|C)$ are just the probabilities of outcomes given causes – number of heads given the propensities of coins, for instance – that we've been reasoning with all along. It is rational, then, to choose as likeliest hypothesis given the data the hypothesis according to which the *data* were likeliest: the celebrated method of *maximum likelihood*. (This is the "maximum a-posteriori probability" variant. There are other choices – a-posteriori mean, minimum expected loss, etc. – that make no difference under the Normal assumptions to follow.)

For an application later in this subsection, we have to fix that constant of proportionality. Suppose we have a range of causes $C = x$ (like Bayes' black billiard ball), of which precisely one must be true (the black ball must have come to rest *somewhere*). Then on any outcome O the integral of the probabilities of all of those causes $C = x$ must be unity. No matter what data $O = r/n$ we actually encountered, we must have $\int_x P(C = x|O)\,dx = 1$. From this, we derive the constant of proportionality we need:

$$P(C = x|O) = P(O|C = x) \Big/ \int_x P(O|C = x)\,dx.$$

In Bayes' specific problem, the equiprobability assumption applies to the left–right position of the black ball: given $O = r/n$ (that is, r white balls out of n to the left of the black), the probability distribution of the black is $x^r(1-x)^{n-r} \big/ \int_0^1 p^r(1-p)^{n-r} dp$.

L4.3.2 The Gauss–Laplace Synthesis

We have now assembled quite a diversity of logical machinery: the characterization of the average as least squares, the formula for Normal distributions, and Bayes' rule for reasoning from probabilities of data given hypotheses to probabilities of hypotheses given data. These three themes combine in one of the core techniques of numerical inference, putting it on a rigorously principled basis.

Let us return to the problem of combining multiple observations of "the same quantity" into an average, Section L4.1. It turns out that there could have been a hypothesis involved in this maneuver. Specifically, conceive the data x_i we actually measured (cf. Section L4.2) as having arisen from perturbation of a **true value** μ by independent measurement errors $\Delta_i = x_i - \mu$. Suppose, further, we had a function ϕ giving the actual probability distribution of these errors. This means, precisely, that given μ (the *hypothesis*), the probability of a datum in the range $(x_i, x_i + dx)$ will be $\phi(x_i - \mu)dx = \phi(\Delta_i)\,dx$, where ϕ is the same distribution for each case i.

Because we have assumed independence (= multiplication of probabilities) among the separate acts of measurement, the net probability of the whole data set (all the x_i taken as a list) on the hypothesis that the true value is μ is the product of the ϕ's for each datum separately. By Laplace's principle, the value of μ we wish to take as our (indirect) measurement of its value – the "most likely" value for μ – is that for which this net probability $\Omega = \phi(\Delta_1)\phi(\Delta_2)\cdots\phi(\Delta_n)$ is a maximum. For given observational data x_i, we could find this maximum as a function of μ, because $\Delta_i = x_i - \mu$, if only we knew the function ϕ.

But, for a great many practical problems, *we* **can** *know the function ϕ*. Often it is reasonable to take it as having the form $e^{-c\Delta^2}$, a Normal distribution having some error variance $1/2c$ around the mean μ. (You'll see the reason for this funny form of the variance in a moment.) Under this somewhat demanding assumption, we have

$$\begin{aligned}\Omega &= e^{-c\Delta_1^2}\,e^{-c\Delta_2^2}\cdots e^{-c\Delta_n^2} \\ &= e^{-c\Sigma_i \Delta_i^2}.\end{aligned}$$

We maximize Ω by minimizing that exponent. But we *already* know how to minimize this exponent. It is a multiple of the sum of squares of the data x_i around μ, so it is guaranteed to be minimized when μ is taken as the average of the data!

Hence:

For data that are known to be distributed around a true mean by Normally distributed errors of any fixed variance, the average of the data is the likeliest candidate for that mean. An easy extension, known already by 1810: if those variances aren't equal, the correct answer is still an average, the one weighted inversely by those error variances, as in Section L4.1.2.

This is the justification of the operation of averaging, and by extension the other least-squares methods, that I have been building up to. Regardless of the variance c, as long as error is Normally distributed the average is the choice of central tendency having the maximum of likelihood for given data. That maximum of likelihood is the justification for choosing it.

Mathematical note 4.6. Symbols or reality? This maneuver entails a brilliant notational pun, the interpretation of

$$\Omega = e^{-c(x_1-\bar{x})^2} \times e^{-c(x_2-\bar{x})^2} \times \cdots \times e^{-c(x_n-\bar{x})^2}$$

as the cth power of

$$e^{-\Sigma_i(x_i-\bar{x})^2} = e^{-\left((x_1-\bar{x})^2+\cdots+(x_n-\bar{x})^2\right)},$$

the logarithm of which we already know how to maximize (by minimizing the sum of squares). The algebra is so slick because the value of c, the precision of the Normal distribution involved, is irrelevant to the maximization of likelihood by which we are selecting the best estimate of the true mean, μ. But this double mathematical meaning is combined with the knowledge that such error distributions truly exist in Nature or else truly arise from conscientious random sampling.

It is worth pausing here to recount the assumptions that justify the average of a set of data as the likeliest, most reasonable estimate of its central tendency. We must have been persuaded (perhaps by some cogent theory of physical, physiological, or social randomness) that the measured data arose from a common "true value" by independent, Normally distributed errors of mean zero and the same variance. Where this compound assumption is *not* reasonable – for instance, where the "true value" is obviously different for different subgroups of the data – the average remains the bare formula it has been all along, the total of the data divided by their count (e.g., the average number of hockey goals per citizen of the United States, or the average number of leaves per North American plant). **The meaning of the simplest of all statistics, the ordinary average, is inextricable from the validity of the (equally simple) theories in which its value is implicated.**

We seem to have assumed nothing at all about the value of μ. All of its values were "equiprobable" prior to the inspection of data. (This is a so-called *improper prior*, because it's not a valid probability distribution; it doesn't integrate to 1. That doesn't matter for the applicability of Laplace's advice. For instance, Perrin was implicitly invoking an analogous prior, the Jeffreys prior $p(N) \propto 1/N$, $p(\log N) \propto 1$, a different but likewise improper prior, when he referred to the full range of possible values of N in the text from his page 105 I have extracted in Section E4.3.) We needn't purport to be so ignorant. We can modify averages, etc., to accommodate arbitrarily strong hunches, earlier attempts at measurement, and so forth. The algebra is simplest if this prior knowledge is encoded in the so-called *conjugate prior* form, which, for Normal errors, means a Normal distribution of its own. In effect, the prior contributes one more term to the weighted average. Sometimes this is written out in a somewhat more general notation whereby the probabilities of the data on the hypotheses multiply the "prior probabilities," wherever they come from, to yield the "posterior probabilities." We have already seen an argument along exactly these lines involving the quincunx, where we derived the exactly linear form of the regression phenomenon.

The argument here can actually be run backward: if the general preference for averages as a representative summary of samples of numbers is rational, Gauss showed, it follows that errors must be Normally distributed. This discussion is deferred to Section 5.2, where I return to the characterization of this distribution at greater mathematical depth.

L4.3.3 Significance Testing: Decisions about Theories

References specific to this section: Jeffreys (1939); Edwards et al. (1963).

The developments of the preceding section have, at last, justified the method of least squares as applied to the problem of indirect measurement that occupied us in Section L4.1. Under the assumption of independent "errors," of whatever origin, each Normally distributed around 0, the weighted average of a sample of values, or of a sample of slopes, is the likeliest estimate of the true value or slope around which the manifest measurements were distributed. Starting from the standard deviation of those measurements, the familiar factor of $1/\sqrt{n}$ gets us to standard errors and thereby to the probabilities (at least, approximate probabilities) of ranges of true values.

But there are more challenging scientific activities than the "scholium" underlying Bayes' question, the position of a black billiard ball. Some explorations require more than the estimation of single parameters and their standard errors. For the confrontation of theory with theory, which is when scientific debates are at their most exciting, we need to know the extent to which data support one theory over another. This lecture deals with one ritualized version of this confrontation: the problem of significance testing.

The context of significance testing is a special case of the choice of hypotheses reviewed in Section 3.4. In this scenario, the confrontation of two theories is formalized as a dispute about the value of one parameter (or, sometimes, the values of a few). Typically, one theory will entail a special value of a parameter (usually 0), whereas an alternative specifies perhaps a different specific value, perhaps a range of values within which some hitherto uncertain constant is to be located. The machinery we've already reviewed is adequate, in principle, to guide you through this problem. As we just showed, it is rational to compute the likelihoods of the data on the hypotheses severally, and then select the hypothesis for which the data were most probable (adjusted for prior information if available) as the one that has the maximum of likelihood given the data. But the confidence that should be placed in such a decision is not a function only of that maximum likelihood; it is affected by the likelihoods of the competing hypotheses as well. That is, significance testing is a **decision driven by a numerical comparison of likelihoods** in some way.

Following Jeffreys, once the information borne by the data about the hypotheses is encoded in the likelihoods, we can compute a single summary statistic – call it K, the "decisiveness of the evidence" – that indicates the weight of evidence for the null over that for the hypothesis involving the additional parameter. (The polarity of K faces the simpler alternative because Jeffreys was a physicist: to accept a null hypothesis is the overwhelmingly more desirable outcome.) The report of the decision would then go cogently in terms of the **likelihood ratio** – the ratio of likelihoods, that for the null hypothesis over that for the alternative, given the data:

$K > 1$. Data support the null hypothesis.
$1 > K > .3$. "Evidence against the null, but not worth more than a bare comment."
$.3 > K > .1$. Evidence against the null "substantial."
$.1 > K > .03$. Evidence against the null "strong."
$.03 > K > .01$. Evidence against the null "very strong."
$.01 > K$. Evidence against the null "decisive." (As a physicist, Jeffreys does not associate error with death or dread.)

Our problem, then, is to generate these ratios K from data, at least in the algebraically tractable case of coin flips, and explore some of their intuitive properties. The formulas that emerge are likely to surprise you.

Generally, small values of the parameter we are estimating will support the null hypothesis, large values the alternative. Jeffreys explains this charmingly on page 248, using a notation in which α is the true value of the parameter in question, a is the estimate of the parameter on the alternative hypothesis, and s is its standard error.

> The possibility of getting actual support for the null hypothesis from the observations really comes from the fact that the value of α indicated by it is unique. [The alternative hypothesis] indicates only a range of possible values, and if we select the one that happens to fit the observations best we must allow for the fact that it is a selected value. If $|a|$ is less than s, this is what we should expect on the hypothesis that α is 0, but if α was equally likely to be anywhere in a range of length m it requires that an event with a probability $2s/m$ shall have come off. If $|a|$ is much larger than s, however, a would be a very unlikely value to occur if α was 0, but no more unlikely than any other if α was not 0. In each case we adopt the less remarkable coincidence.

In the most straightforward situation, the "alternate" hypothesis has the same formal structure as the null: it consists of a single wholly specified model from which likelihoods can be computed without any fuss. Suppose, for instance, that you knew that a coin was either fair or biased to come up 64% heads, each with probability 1/2 (for instance, that it had been picked randomly from a bag of two coins, one of which was fair, one biased in precisely the manner specified). You flip the coin 50 times, and find 32 heads; now what are the likelihoods of the two hypotheses? Because the prior probabilities of the two hypotheses (the two coins) are the same, by assumption, the likelihoods of the hypotheses given the data are proportional to the probabilities of the data given the hypotheses. The chance of 32 heads on a fair coin is $_{50}C_{32}/2^{50} = .016$. [Recall from Section L4.1.3 that the symbol $_nC_r$ means the number of ways of choosing r objects out of n – here, the number of distinct sequences of r heads out of n; each sequence has the same probability $(1/2)^n$.] The chance of the same outcome on the biased coin is $_{50}C_{32}(0.64)^{32}(0.36)^{18} = .117$, for a likelihood ratio of about 7.3:1. The evidence against the fair coin is "substantial" but not "strong," in Jeffreys' categorization.

But notice how much weaker is even this inference than the ordinary significance test beloved of undergraduate statistics courses, the logic that calibrates the likelihood of the alternative hypothesis by the probability of the data given the null alone. On the null hypothesis of a fair coin, the standard deviation of the expected number of heads is $\sqrt{n}/2 = 3.5$, and we are thus 2 full standard deviations away from the mean, an occurrence that is fairly rare. (We already knew that because the probability of exactly 32 heads is only 1.6%; in fact, with a fair coin the probability of 32 heads *or more* is less than 3%.) But the issue is not the improbability of that null hypothesis per se; rather, we are concerned with its likelihood relative to the alternatives. Here, there is only one alternative, a coin that happens to be perfectly matched to the actual percentage of heads encountered. For that coin, the likelihood for the data actually encountered (the total of exactly 32 heads) is little more than one-ninth – remember, nearly fair coins do *exactly* what they're expected to do only about a fraction $\sqrt{2/\pi n}$ of the time – so the strength with which the alternative hypothesis is supported (7:1 odds) is substantially weaker than the "$p < .05$ level" (2 s.d. out from the mean, "odds of 19:1") "suggested" by the probability on the null hypothesis alone.

Now suppose you did not have such a simple alternative, but instead just the suspicion that "the coin might not be fair." From experience, or for lack of any better choice, you might reasonably say that the alternative hypothesis is a coin of any bias at all toward heads or tails, that is, a coin the propensity of which for heads is distributed over the interval (0,1) with uniform prior probability (Bayes' billiard table). (If you knew how biased coins tend to come out, you could adjust the following formulas appropriately. After enough flips of the coin at hand, that prior knowledge won't matter much anyway.)

Bayes' theorem continues to apply in this case to manage the likelihoods of all the competing hypotheses – in a manner, furthermore, that is fairly insensitive to one's assumptions about how the unfairness of coins might be distributed: see the discussion of "stable estimation" in Edwards et al. (1963). We saw that the likelihood of the alternative hypothesis on the assumption of a coin producing 64% heads was 0.117. Similarly, for *each* of the hypotheses included under the "alternate" (i.e., for *each* biased coin) there is a probability of having produced that frequency of 32 heads out of 50. In fact, for a coin with a propensity p for heads, that probability is just $_{50}C_{32}p^{32}(1-p)^{18}$, by a formula from Section L4.1.

By Bayes' theorem, the likelihood to use for the alternative (the hypothesis of a biased coin) is the integral of these partial likelihoods $P(C|O)$, each a probability $P(O|C)$, over all the ways (all the hypotheses) that compete with the null, each weighted by its prior probability. Assuming a uniform distribution of this possibility of bias (the case of "no information," the argument originally due to Laplace), the quantity we need is the integral from 0 to 1 of all these elementary likelihoods:

$$_{50}C_{32} \int_0^1 p^{32}(1-p)^{18} dp.$$

This is another of those integrals one learns in a second course in calculus. One verifies by mathematical induction that

$$\int_0^1 p^r (1-p)^{n-r} dp = r!(n-r)!/(n+1)!.$$

To convert from an integral to a likelihood we need to multiply by $_nC_r = n!/r!(n-r)!$, of which the two terms in the denominator cancel. What remains is just $n!/(n+1)! = 1/(n+1)$.

Then the likelihood of 32 heads out of 50, on an alternative hypothesis of a coin with propensity of heads uniform over the range of possible values, is $1/51 \approx 0.020$ – in effect, the "chance" of getting exactly the number of heads you got, knowing nothing whatever except that the number had to be in the range $[0, n]$. (This line of reasoning seems quite plausible even without the integral.) But 0.020 hardly differs at all from the likelihood of 0.016 that we computed for the "null" hypothesis that the coin was fair. Hence: **a finding of 32 heads in 50 flips of a coin about which nothing further is known is almost equivocal regarding the fairness of the coin**, even though the chance of a finding so extreme was less than 3% on that same null hypothesis. **Probability on the null and the likelihood ratio in support of that null are not directly related.** The difference between the 0.020 and the 0.117 is the informational value of knowing *precisely* the propensity for heads of the biased coin; this degree of prior knowledge is unrealistic in most contexts.

For r heads in n flips, when r and n are both large, one can apply Stirling's formula to get, approximately,

$$K \approx \sqrt{\frac{2n}{\pi}} e^{-2l^2/n} \quad \text{with } l = r - n/2.$$

(Recall that K is the odds-ratio **for** the null hypothesis of a fair coin.) You have seen this formula before: it is the height of the probability density for the Normal approximation to the fraction of heads in n tosses of a fair coin. (There, of course, it is neither a probability nor a ratio of probabilities, but a probability density.) In this combination of terms we see a dependence on $\sqrt{n}$ in the factor at the front and also a $1/n$ in the negative exponent; we need to understand the way these two dependencies combine.

The argument continues in imitation of a neat turn from Edwards et al. (1963). That factor $1/(n+1)$ decreases as we increase the number of coins flipped. Suppose we consider steadily increasing numbers of coin flips and for each a count of heads that is exactly 1.96 standard deviations from fairness (the "2.5% one-sided tail") under the assumption of a fair coin: flips of n coins that produce $r = n/2 + .98\sqrt{n}$ heads. Although I omit the details here, the previous calculations go through pretty much unchanged, with the likelihood on the null hypothesis being $_nC_r/2^n$ and the likelihood on the diffuse alternative hypothesis being $1/(n+1)$ as earlier. From approximations set out earlier in this chapter, using $l = .98\sqrt{n}$ and hence $-2l^2/n = -1.92$, we can write

$$_nC_r/2^n \approx \sqrt{2/\pi n}\, e^{-1.92}.$$

The ratio of likelihoods is thus a constant – $\sqrt{2/\pi}\, e^{-1.92}$ – times $(1/\sqrt{n})/(1/n+1)$, or about $0.117\sqrt{n}$.

This paradox is very troublesome. Evidence that, on its face, appears under the null hypothesis (the fair coin) to have a fairly constant implausibility with respect to that null, actually ends up *supporting* that hypothesis, against the most reasonable alternatives, *more and more strongly* as sample size increases. This paradox is usually called the *Jeffreys paradox*, after Harold Jeffreys, who first pointed out the correct formulas quite matter-of-factly, as if he couldn't imagine anyone ever thinking otherwise, in Jeffreys (1939).

The reason for this paradoxical outcome (and it *is* paradoxical: many applied statisticians never embrace it) is not fundamentally difficult to grasp, as long as one accepts the notion that hypotheses have probabilities (the ontology of inductive statistical inference introduced earlier). The alternative hypothesis to the null does not merely assert that the coin is "not fair": it asserts that the coin has indeed a propensity for heads, a propensity that is not 50%. That being so, the chances that the unfair coin will be found to supply a fraction of heads that is close by the expected 50:50 ratio – say, between 49% and 51%, ±2 s.d. for the flip of 10,000 coins – is itself quite unlikely if that frequency of heads were a priori to be found anywhere in the range from 0 to 1. This improbability of a count of heads so close to 50% on the alternative competes with, and for large samples eventually swamps, the improbability of a count of heads so far from 50% on the null. As n increases, the likelihood on the alternative drops as n, and that on the null drops as $\sqrt{n}$ times an exponential function. The place where they cross, the number of standard errors that is actually neutral for the null hypothesis, rises with sample size as $\sqrt{\log n}$, a rise that is slow but sure. In other words, for samples of substantial size (say, thousands), you have to be a lot farther from the null than you thought for that distance to actually imply support for the alternative in a situation like this where you have no real idea of exactly where the alternative ought to be.

Exactly the same logic applies, albeit with a constant different from 0.117, to tests of means of Normally distributed variables, of correlation coefficients, and the like. **The classical tail probabilities for these distributions on the assumption of the null hypothesis are no guide to the support that data afford for those hypotheses against reasonable alternatives.** That support is a function of the alternatives, and their prior likelihoods – and, most notoriously, of the sample size. The likelihood under the alternative, and thus the rational person's decision about the relative strength of the two hypotheses, is a function of a parameter, sample size, that is inadequately incorporated in most authors' formulas. The larger the sample, the more strongly a tail probability of 0.01, say, supports the **null**. (At $n = 1000$ the odds are even for that tail; at $n = 10,000$, for the 0.001 tail.)

In Section 5.3.2 we return to a consideration of the likelihood methods in the more complex context of multivariate data and multiparameter models.

L4.3.4 The Contemporary Critique

What the use of P implies, therefore, is that a hypothesis that may be true may be rejected because it has not predicted observable results that have not occurred. This seems a remarkable procedure.

– Jeffreys (1939, p. 385)

What bearing, if any, has the rarity of an observable occurrence as prescribed by an appropriate stochastic hypothesis on our legitimate grounds for accepting or rejecting the latter when we have already witnessed the former?

– Hogben (1957, p. 319)

Readings for this section: Cohen (1994); Ziliak and McCloskey (2008); Edwards et al. (1963); Morrison and Henkel (1970); Hogben (1957).

Recall that a **null hypothesis** or **nil hypothesis** is the hypothesis that some potentially meaningful or scientifically intriguing numerical parameter of a true statistical description of your population is "nil," that is to say, exactly 0. **Null hypothesis statistical significance testing**, or NHSST, is a common procedure for commenting on such claims with respect to your particular data set. In NHSST, you compute the actual group difference (or regression slope, or difference of two regression slopes, or whatever analogous number stands for the claim you intend to be making), and then the probability of a value at least that large (or a value of either sign with absolute magnitude at least that large, if your claim does not come with a preferred direction like positive or negative) on the assumption that its value was actually 0 on the average. This probability is computed by formula or by random sampling while other quantities that are not formally part of the comparison under inspection (for instance, the sample distributions as pooled over the groups that the hypothesis claims to be differentiating) are held unchanging. The 0 is tentatively a "true value," before sampling, in the sense of Section L4.3.1. What NHSST is computing is the *distribution* of the *computed* numerical summary on the assumption of pure random sampling of your specimens under some assumed model of how they vary *around* this true value, for instance, by uncorrelated Normal noise. From that probability, called the "*p*-value," you "infer" whether it is reasonable to believe that the null hypothesis is true.

Why are there scare quotes around the word "infer" in the last sentence? In 1994 the psychometrist Jacob Cohen published a wry lecture entitled "The earth is round ($p < 0.05$)." The fallacy about which he is writing had already been the topic of 40 years of polemics, including a complete book a quarter of a century earlier (Morrison and Henkel, 1970); yet the malign practice persisted (and persists to this day). The issue is the invalidity of the following syllogism:

- If the null hypothesis is correct, then these data are highly unlikely.
- These data have occurred.
- Therefore, the null hypothesis is highly unlikely.

Or, just as invalidly:

- If H_0 is true, then this result (statistical significance) would probably not occur. ["H_0", read "aitch-zero" or "aitch-naught," is the most common notation for the null hypothesis.]
- This result has occurred.
- Then H_0 is probably not true.

Consider the following example (after Cohen): "Most Americans are not members of Congress. This person is a Congressman. Therefore, he is probably not an American."

In other words, *the p-value on the null is* **not** *the probability that the null hypothesis is true*. Except sometimes in physics, the scientific question is *never* about the probability of the given data assuming that the null is true.

> About that exception: There is a style of inference in the physical sciences where you check that quantities that should be equal are indeed equal – the "strong form" of null hypothesis testing (Meehl, 1967). We see an example of this in Section E4.8, but many historically prior examples of the same logic come to mind. These include the Michelson–Morley experiment of 1887 showing equality of the speed of light in perpendicular directions, the studies in Section E4.3 showing the exact validity of the Maxwell velocity distribution, and Millikan's verification (Section E4.2) that the slopes h/e between results from metal cylinders made from different elements are numerically indistinguishable (as Einstein insisted they must be). *In this context only*, the typical NHSST procedure is valid. If the null model is really true, the method of abduction requires that it must almost always appear to be consistent with the data, *no matter how massive the extent of that data*. Even tiny discrepancies, if they persist over series of improvements in instrumentation, represent serious challenges to reigning theories – see the discussion of Kuhn's ideas at Section 2.3. Otherwise, whenever the null model is literally false – which is *always*, in any science outside of the physical sciences – the question is really of the hypothesis that is likeliest given the data. This is part of the abduction *against* the null that we are pursuing (namely, whether the data are surprising on *that* hypothesis, which is, presumably, one of a great many that we are considering in addition to our own).

And in practice there are nearly always several auxiliary assumptions associated with the attachment of any data set to a hypothesis – about instrumentation, sampling representativeness, omitted conditioning variables – that interfere with inferences from data. These are the sorts of concerns we have already considered several times as aspects of the process of abduction. To infer from an NHSST, a probability on the null, that "a group difference [or a regression slope] is probably not zero" (1) is invalid, but more importantly (2) is not to assert any scientific theory, nor even to support one. The statement is not consistent with any process of consilience; **you cannot confirm a quantitative theory by rejecting null hypotheses.**

This problem is only worsened by the recent turn to especially large data sets, for which even small associations become statistically significant. As a rule of thumb, a correlation great than about $2/\sqrt{N}$ in absolute value is statistically significant at $p \sim 0.05$ or better, where N is the sample size; for example, for $N \sim 10000$ correlations become significant at a value of about ± 0.02. But an indefinitely large number of causes, or none at all, can produce correlations of that magnitude in any real data set. The psychometrician Paul Meehl (1990) refers to the "crud factor" [crud, American slang: dirt, filth, or refuse] according to which almost any pair of psychological measurements are correlated at levels up to ± 0.30 or so in realistic samples. It is pointless, then, to test correlations of this magnitude in samples larger than about $N = 100$ except in properly randomized experimental studies. This is "ambient correlational noise," produced by who knows what network of background genetic and epigenetic and environmental factors. *Only* a randomized experimental design (or, like Snow's, an arguably "naturally randomized" one) can break the links between these prior factors and the cause in which your theory is interested, and *only* the methodology of abductive consilience can protect you from pursuing false leads, red herrings, and will-o'-the-wisps when data are dredged for "signals" of this particularly unreliable type.

The p-value on H_0 is **not** the probability that the null hypothesis H_0 is true. Likewise, $1 - p$ is **not** the probability of a successful replication (meaning, a significant finding in the next study). (If a numerical effect size, such as a difference, is statistically significant at the $p \sim 0.05$ level, and if that effect size is truly the correct value, then replications that involve samples of the same size and measurements of the same precision will repeat the finding of statistical significance only half the time.) A ludicrously small p-value, such as the 1.5×10^{-17} reported in the course of the Barcelona asthma analysis (Section E4.1), means only that the null hypothesis is irrelevant to your data; it bears no implications whatever for the reasonableness of your alternative hypothesis with respect to the same data. In this connection, see the discussion of the meaning of small p-values in the classic book about the authorship of the Federalist Papers (Mosteller and Wallace, 1964/1984).

One may summarize as follows. NHSST is not a rational procedure for scientific inference, in that (1) it makes no reference to the hypothesis that you actually hold, (2) in the limit of large samples, it rejects the null no matter how unlikely the alternative (the Jeffreys paradox, Section L4.3.3), and (3) the actual decimal values we use for significance testing, such as 0.05 or 0.001, come out of nowhere and have no scientific purpose.

> *Exception:* Some bureaucracies *require* a suitable NHSST as part of a decision protocol: regarding, for instance, health claims for pharmaceuticals. These would include claims of *equipoise* or *noninferiority*, meaning the demonstration that a new (presumably cheaper) technique is "not worse" than an established method.

If your goal is numerical inference, there is no alternative to attending to numerical effect sizes and their consilience via independent lines of evidence. This combines with the exquisite care you are investing in both premises of the corresponding abduction, the surprisal your data afford with respect to other people's theories and the complete reasonableness of the same data on your theory. Tukey (1969) puts it nicely:

> Data analysis needs to be both exploratory [abductive] and confirmatory [consilient]. In exploratory data analysis there can be no substitute for flexibility, for adapting what is calculated – and, we hope, plotted – both to the needs of the situation and the clues that the data have already provided. In this mode, data analysis is detective work – almost an ideal example of seeking what might be relevant. Confirmatory data analysis has its place, too. Well used, its importance may even equal that of exploratory data analysis. We dare not, however, let it be an imprimatur or a testimony of infallibility. "Not a high priestess but a handmaiden" must be our demand. Confirmatory data analysis must be the means by which we adjust optimism and pessimism, not only ours but those of our readers. . . .
>
> The Roman Catholic Church is a long-lived and careful institution. It has long held that sanctification was only for the dead. I believe that sanctification of data is equally only for dead data. (Turkey, 1969, p. 90)

Statistical significance testing is a form of sanctification of a data analysis. In connection with nearly any valid numerical inference outside of the physical sciences, it is dead, intellectually speaking, and should stay dead. If a colleague, a mentor, or an editor requires you to carry out a significance test as the culmination or indeed the only inferential step in some process of rhetorical persuasion, find another colleague, mentor, or editor. As argued most passionately by Ziliak and McCloskey (2008), significance testing offers no actual rhetorical force at all; its functions seem to be purely those of professional boundary maintenance, not scientific discovery. **NHSST makes no sense** in nearly all of its scientific applications. For the purposes of numerical inference it needs to be replaced by abductive consilience, the method foregrounded in this book.

E4.4 Ulcers Are Infectious

E4.4.1 Chronicle of a Preposterous Proposition

Barry Marshall and Robin Warren (Nobel Prize for Physiology or Medicine, 2005) were faced with an unexpected pattern (Figure 4.25, top) in a 1982 series of

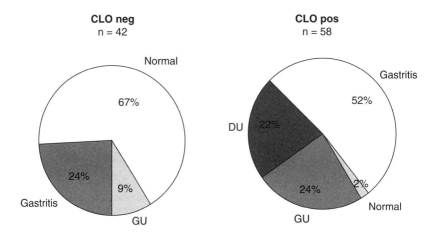

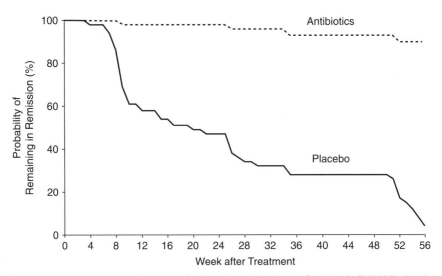

Figure 4.25. Two versions of the same finding. (Top) Pie charts after Marshall (1988) showing that out of 100 consecutive patients encountered in an Australian gastrointestinal clinic in 1982, 13 out of 13 with duodenal ulcers were found to be hosting a particular bacterium as well and only 1 hosting the bacterium was symptomatically normal. (Bottom) Survival analysis redrawn from Hentschel et al. (1993), showing that to remain ulcer-free, subjects cured of an ulcer needed to be receiving maintenance doses of a specific antibiotic regimen. The upper chart had no persuasive power, but the lower chart, outcome of a randomized experiment, signaled an official change in the rules of ulcer treatment barely 10 years later.

gastroenterological examinations from an Australian clinic. There was a *highly* statistically significant difference, one having too many zeroes in the *p*-value to be worth printing, between the clinical presentations in two subgroups of their patients, those who stained positive for the particular bacillus now called *Helicobacter pylori* versus those who stained negative.

In a famous episode, this result was declared to be absurd, as everybody knew that ulcers were caused by stress, not by an infection. In other words, the pattern was classified by its initial peer reviewers as pathological science in the sense of the discussion of the next section: a conclusion that was not fraudulent but that nevertheless expressed mostly fuzzy thinking and self-delusion. That was not the case – the hypothesis triumphed in less than ten years. The sociological-cognitive process by which opinions changed is nicely summarized by the cognitive psychologist and psychologist of science Paul Thagard (1999). Here is a short synopsis of his 80-page narrative about what happened:

1. Chronic gastritis (of which the main symptom is burping) had long been known to be associated with peptic ulcer.
2. A letter to *Lancet* (Marshall and Warren, 1983) notes that "half the patients coming to gastroscopy show colonisation of their stomachs," usually by one "unidentified curved bacillus" that was "almost always" present in patients with active chronic gastritis (1983). In 100 consecutive patients presenting at an Australian gastroenterological facility, 58 proved to have this particular bacterium; of the 58, 27 have ulcers (13 duodenal, 14 gastric). Of the 42 patients without the bacterium, 4 have ulcers, all gastric. Put another way, of the 58 subjects positive for the bacterium, only 1 was normal; the others all had gastritis or a gastric (GU) or duodenal (DU) ulcer. By comparison, the subjects who were negative for the bacterium were normal two-thirds of the time, and none had duodenal ulcers.

   The finding is considered preposterous, and hence unacceptable for presentation at a gastroenterological conference, but nevertheless a further note to this effect also eventually appears as a letter in *Lancet* (Marshall and Warren, 1984).
3. The bacterium is named *Campylobacter pylori*, a nomenclature that will be changed later, and, on the chance it may have something to do with ulcers, studies begin that add antibiotics to the usual ulcer therapies.
4. In 1988, a prospective double-blind trial of patients with duodenal ulcer and *H. pylori* infection, some treated with antibiotics and some not, found that ulcers healed in 92% of patients free of *H. pylori* at ten weeks, but in only 61% of patients with persistent *H. pylori*. After twelve months, ulcers recurred in 84% of the patients with persistent *H. pylori* but only 21% of the patients without persistent *H. pylori*.
5. Antibiotic regimens continued to improve, and the evidence about the efficacy of antibiotic treatment eventually became overwhelmingly consilient.

In a comparison of two regimens for treatment of ulcers accompanied by *H. pylori*, all 51 of the patients treated with antibiotics healed, along with 49 of the 51 treated with placebo. But of the 53 patients in whom *H. pylori* persisted, 45 had a recurrence of the duodenal ulcer, versus 1 of the 46 patients in whom *H. pylori* was eradicated (Hentschel, 1993). See Figure 4.25, bottom.

6. Hence in 1994, barely a decade after the original empirical hint was uncovered, an NIH consensus conference concludes that ulcer patients with *H. pylori* infections should be treated with antibiotics. The question is no longer whether bacteria cause ulcers (i.e., the causal issue is closed); it is of methods for best treating the bacteria that cause ulcers.

In terms of abductions, it is clear that the 1982 significance test per se persuaded nobody. A surprising fact, yes – that pair of pie charts was certainly not a product of mere random sampling from the same underlying population – but until a tenable theory could be formulated for which some analogous data would be predicted "as a matter of course," it would not be accepted as any sort of evidence regarding an explanation. It wasn't consilient with *anything*. The contents of the pie charts looked a bit like numerical information, but no numerical inference was available – because bacteria obviously can't live at the pH levels of the human stomach, something is wrong from the very outset (compare "*Scorpion* is missing"). The significance level of the clinical trial result, the 1993 figure, is indistinguishable from that of the pie charts, the 1982 figure; but this resemblance is highly misleading. It is not that the findings are "equally surprising," because the meaning of the null hypothesis has changely utterly from 1982 to 1993. In 1982, the alternative was "random chance" in the match of bacterial infection to reported symptoms; but by 1993 it had become the assertion of *clinical inutility* of a particular treatment strategy, and inutility has an opposite, utility. So no abduction was possible in 1982: the particular finding suggested by the pie chart did not follow "as a matter of course" from anything (because the percentages remained unexplained). By 1993, though, the lower finding in Figure 4.25 *does* follow as a matter of course, in explicit quantitative detail, from a practical hypothesis that ulcers should be treated with antibiotics. Antibiotic treatment should reduce the rate of reinfection to zero, which is more or less what the data here show; the rate of reinfection under placebo is not and never was the explanandum.

In essence this argument has the same structure as John Snow's of 140 years earlier. The numerical finding is of a ratio between two incidences (of cholera, for Snow; of ulcer recurrence, for the later study), but what follows as a "matter of course" is the approximate zeroing of one term of the ratio, not the value of the ratio itself. Snow could argue for the "natural experiment," the quasirandomization of the water supply across households that intercepted every explanation from geography or social class. As we saw, and Snow himself argued, he could find any sufficiently large ratio of death rates – whether 14:1, 8:1, or 5:1 did not really make any difference – and that they appeared in that calendrical order did not weaken the support of his data for his hypothesis but actually strengthened it. And Marshall and Warren won the argument over *Helicobacter* only when an experimental zero could be produced in a role for which the explanation predicted, indeed, just about zero. No quasiexperiment was available that has a logical force comparable to that from this experimental production of a value so close to the value predicted.

E4.4.2 Evolution of Koch's Postulates

Marshall and Warren's numerical inference succeeded – the physicians quickly agreed among themselves that this bacillus causes (most) ulcers – in keeping with a change in the language of causal reasoning over the century and a quarter from the dawn of modern bacteriology to the present.

In the 1880s, the German pathologist Robert Koch codified the following four *postulates* to guide research on the identification of particular infectious agents (with initial applications to the etiologies of anthrax and tuberculosis):

1. The microorganism must be found in abundance in all organisms suffering from the disease, but should not be found in healthy animals.
2. The microorganism must be isolated from a diseased organism and grown in pure culture.
3. The cultured microorganism should cause disease when introduced into a healthy organism.
4. The microorganism must be reisolated from the inoculated, diseased experimental host and identified as being identical to the original specific causative agent.

Over the interval since then, limited exceptions to all of these have accumulated in the lore of microbiology, but that list of exceptions is not our point here. Rather, there has been a sea change in the language of the postulates themselves. When Barry Marshall himself tried to invoke these postulates in their 19th-century form – one day in 1983, he mixed himself a cocktail of *H. pylori*, drank the stuff, and in a couple of days developed a bad case of gastritis (including the burping) – it did not constitute evidence of any particular power about his main hypothesis, the infectious origin of ulcers.

One set of current criteria, for instance, from Ewald (1994), could read roughly as follows:

1. *Prevalence* of the disease should be significantly higher in those exposed to the putative cause than in matched control subjects not so exposed.
2. *Exposure* to the putative cause should be present more commonly in those with the disease than in control subjects without the disease when all risk factors are held constant.
3. *Incidence* of the disease should be significantly higher in those exposed to the putative cause than in those not so exposed, as can be shown in prospective studies.

4. *Temporally*, the disease should follow exposure to the putative agent with a distribution of incubation periods on a bell-shaped curve.
5. A *spectrum* of host responses should follow exposure to the putative agent along a logical biological gradient from mild to severe.
6. A *measurable host response* after exposure to the putative cause should *regularly* appear in those lacking this response before exposure or should increase in magnitude if present before exposure; this pattern should not occur in people not so exposed.
7. *Experimental reproduction* of the disease should occur in higher incidence in animals or humans appropriately exposed to the putative cause than in those not so exposed.
8. *Elimination or modification* of the putative cause or of the vector carrying it should decrease the incidence of the disease.
9. *Prevention or modification* of the host's response on exposure to the putative cause should decrease or eliminate the disease.
10. The whole story should make biological and epidemiological sense.

Note, in comparison to the 19th-century version, all of those references to numerical confirmation: rates should increase or decrease, incubation times should show a plausible distribution, and magnitudes should generally follow monotonic relationships (the more X, the more Y) as confirmed not by regressions but by actual analysis of curves. Besides that incubation time, other aspects of the situation need to be quantified as well, such as early host response, the gradient of host response, or the changes in the disease vector. [Note also in (2) the jargon "when all [other] risk factors are held constant," which doesn't actually mean what it says – nothing is actually "held constant" in studies like these. "Held constant" is a term of art from the multiple-regression approach to covariance adjustment to be discussed at some length in Sections L4.5.2–L4.5.4.] All these changes go in the same logical direction, that of supplying greater consilience between the available data and a hypothesis of biological causation. They can also affect the strength of the abduction when competing hypotheses are inconsistent with some or all of these same numerical patterns. Grammatically, the auxiliary verb "must" in most of the original postulates has been replaced by a construction involving "should" that is amenable to degrees of validity, and in all of these items except the last there is an explicit reference to some other quantification, versus only one such reference in the Koch originals, the phrase "in abundance" in his first. The Marshall–Warren hypothesis was accepted mainly owing to the consiliences implied in Evans postulates 1, 2, 7, 8, and 10.

L4.4 Avoiding Pathologies and Pitfalls

L4.4.1 The Langmuir–Rousseau Argument

Abductive numerical inference is no automated algorithm, nor is it any sort of panacea assuring scientific success in general. Over the course of the 20 century, it has suffered its full share of erroneous or even foolish arguments and mistaken announcements. There is considerable value, both historical and prophylactic, in dissecting the most common errors that users of this methodology are risking. Among the most engaging essays on these problems are Feynman (1974), Feinstein (1988), Langmuir (1989), and Rousseau (1992). The phrase "pathological science" seems have to caught on for a type of scientific claim that is false owing to the all-too-human frailties of scientists. Episodes of pathological science are not due to "fabrication, falsification, or plagiarism," the three principal symptoms of research misconduct (Office of Research Integrity, 2009) – in pathological science, there is no such dishonesty involved. Instead, experts are found to have misled themselves by insufficient skepticism about what turn out to have been evanescent, irreproducible, or ambiguous patterns. The result is an episodic series of false discovery claims, often accompanied by immense publicity and sometimes, as in the case of parapsychology or vaccines as a cause of autism, quite difficult to extirpate from public awareness. Characteristics shared by many papers later found to be pathological in this sense involve aspects of abduction, aspects of consilience, or both. Here is Langmuir's list, from his 1953 lecture:

- The maximum effect that is observed is produced by a causative agent of barely detectable intensity, and the magnitude of the effect is substantially independent of the intensity of the cause.
- The effect is of a magnitude that remains close to the limit of detectability, or many measurements are necessary because of the very low statistical significance of the results.
- There are claims of great accuracy.
- Fantastic theories contrary to experiments are suggested.
- Criticisms are met by ad hoc excuses thought up on the spur of the moment.

Rousseau (1992) adds one more shared characteristic:

- The investigator finds it impossible to do critical experiments that allow strong inference comparing the purported theory to its competitors.

The list of episodes that count as pathological science is not, alas, diminishing. Within the last 25 years, we have, among others, cold fusion (Huizenga, 1993), infinite dilution (to be discussed in a moment), and the vaccine theory of autism as thimerosal poisoning (Offit, 2008).[9] We have seen antidotes to most of these forms of predictable illogic in the examples preceding, and we will see more in the examples to come. A poignant current episode is the parapsychological research reported, perhaps with tongue in cheek, by Bem (2011). A rebuttal, Wagenmakers (2011),

---

[9] It was not until 2010 that the instigating article of this pathology, Wakefield et al. (1998), was marked as "retracted" by the journal in which it was originally published, mainly because Andrew Wakefield, the physician responsible for this tragic episode, still refuses to concede any bad faith on his part.

argues that the problem lies squarely with the principle of NHSST and the complete failure to consider consilience that pervade the field of social psychology within the conventions of which Bem was operating.

For a book like this one, centered on numerical inference, perhaps the most instructive of the many available examples is the controversy that arose in 1988 regarding a paper published that year in *Nature*. In Davenas et al. (1988) a team headed by the immunologist Jacques Benveniste (1935–2004) claimed that an aqueous solution of an allergen, anti-immunoglobulin E, has a biological effect at a dilution of 1 part in 10^{120}, at which dilution, obviously, no molecules of the allergen remain in the solution. The principal finding was communicated in their Figure 1 reproduced below as my Figure 4.26. What is plotted is the "measured" response to this allergen by human basophil cells over 60 successive cycles of dilution of the allergen solution by factors of 10 or 100. "Transmission of the biological information could be related to the molecular organization of water," the authors concluded, a possibility that is of course completely incompatible with all existing knowledge about the actual structure of water (which is, after all, a familiar substance). "Water could act as a 'template' for the molecule, for example by an infinite hydrogen-bonded network or electric and magnetic fields; . . . the precise nature of this phenomenon remains unexplained" (Davenas et al., 1988, pp. 816, 818). Evidently there is no consilience to be had. There is also no explanation of the obvious quasiperiodic cycling of the graphs in Figure 4.26, and hence no abduction possible, either. Although the results are certainly startling, there is no theory that renders these quasiperiodicities "a matter of course" in keeping with Peirce's formulation.

The published article was accompanied by a peculiar "editorial reservation" written by the journal's editor, John Maddox. "Readers of this article may share the incredulity of the many referees who have commented on several versions of it during the past several months," he noted. "There is no physical basis for [the finding here]. With the kind collaboration of Professor Benveniste, *Nature* has therefore arranged for independent investigators to observe repetitions of the experiments. A report of this investigation will appear shortly." That report, duly appearing a month later, is headlined "High-dilution experiments a delusion," exactly in keeping with our language of pathological science (Maddox et al., 1998). The visiting group, an "oddly constituted team," comprised *Nature's* editor Maddox, the magician James Randi, and a scientist from the (American) National Institutes of Health, Walter Stewart, who specialized in studies of scientific misconduct. They conclude that "the claims [of the original paper] are not to be believed," for four reasons:

- The phenomena described are not reproducible.
- The data lack sampling errors of the magnitude that would be expected.
- No serious attempt has been made to eliminate systematic errors, including observer bias.
- "The climate of the laboratory is inimical to an objective evaluation of the exceptional data."

Fig. 1 Human basophil degranulation induced either by anti-IgE anti-serum (●) diluted tenfold from 1×10^2 down to 1×10^{60} (a) or hundredfold down to 1×10^{120} (b) or by anti-IgG antiserum (○) diluted hundredfold from 1×10^2 down to 1×10^{120} (representatives of at least 10 experiments for anti-IgE and 4 experiments for anti-IgG). The significant ($P < 0.05$) percentage of degranulation was 15% (a) and 20% (b). (....) relation to the number of counted basophils from control wells[15].

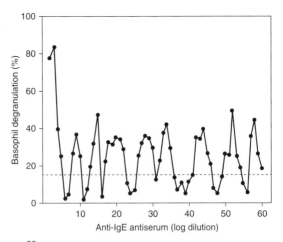

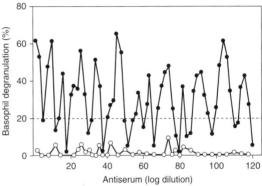

Methods Goat anti-human IgE (Fc) antiserum or as a control, goat anti-human IgG (Fc) antiserum (Nordic Immunology, The Netherlands) was serially diluted as indicated above in HEPES-buffered Tyrode's solution (in g l^{-1}: NaCl, 8; KCl, 0.195; HEPES, 2.6; EDTA-Na$_4$, 1.040; glucose, 1 human serum albumin (HSA), 1.0; heparin, 5000 U per l; pH 7.4). Between each dilution, the solution was thoroughly mixed for 10 s using a Vortex. Given the molecular weight of IgG molecules (150,000), the 1×10^{60} and 1×10^{120} dilutions correspond in the assay to 2.2×10^{-66} M (th) and 2.2×10^{-126} M (th) respectively. Venous blood (20 ml) from healthy donors was collected using heparin (1 U per ml) and a mixture of 2.5mM EDTA-Na$_4$/2.5 mM EDTA-Na$_2$ (final concentrations) as anticoagulants and allowed to sediment. The leukocyte-rich plasma was recovered, twice washed by centrifugation (400g, 10 min) and finally resuspended in an aliquot of HEPES-buffered Tyrode's solution. The cell suspension (10 μl) was deposited on the bottom of each well of a microtitre plate containing 10 μl CaCl$_2$ (5 mM final) and 10 μl of either of anti-IgE or anti-IgG antiserum dilutions. To a control well were added 10 μl CaCl$_2$ and 10 μl Tyrode's but no anti-IgE or anti-IgG antiserum. Plates were then incubated at 37°C for 30 min. Staining solution (90 ml; 100 mg toluidine blue and 280 μl glacial acetic acid in 100 ml 25% ethanol, pH 3.2 — 3.4) was added to each well and the suspension thoroughly mixed. Specifically redstained basophils (non-degranulated basophils) were counted under a microscope using a Fuchs-Rosenthal haemocytometer. The percentage of basophil degranulation was calculated using the following formula: Basophil no. in control − basophil no. in sample/ basophil no. in control × 100. Between 60 and 120 basophils were counted in cell suspensions from control wells after incubation either in the absence of anti-IgE antiserum, or in the presence of anti-IgG antiserum.

Figure 4.26. Ostensible evidence of the persistence of a biochemical effect of antiserum over 60 successive cycles of dilution. The authors offer no theory to explain the apparent quasiperiodicities shown here. (Above) Tenfold dilutions. (Below) Hundredfold dilutions. Note the truly enormous figure caption from the original publication. (From Davenas et al., 1988).

James Randi put the point best in a quote that has become famous:

> Look, if I told you that I keep a goat in the backyard of my house in Florida, and if you happened to have a man nearby, you might ask him to look over my garden fence, when he'd say, 'That man keeps a goat.' But what would you do if I said, 'I keep a unicorn in my backyard'? (Maddox et al., 1988, p. 299)

Maddox (1988) later explained that *Nature*, as a matter of policy, "consciously published a regular sprinkling of heterodoxy," as "people may find it instructive to know what is happening on the fringes of their interests," and furthermore that he had terrible trouble in reviewing this paper: "There are circumstances in which it is difficult to tell from the inspection of a manuscript, and of proferred supplementary material, whether some kinds of claims are valid." He concludes that even though "Dr Benveniste was (and, perhaps, still is) convinced of the reality of the phenomena reported in his article," nevertheless "my own conviction is that it remains to be shown that there is a phenomenon to be explained" (Maddox, 1988, pp. 761, 762, 763).

In the language of this book, everything has come out fine here except that the process should have concluded well before publication. The four criticisms of the visiting committee are all in keeping with our logic of abductive consilience. They checked against the requisites of abduction, the implausibility of alternate explanations, but find them to be quite plausible indeed. Separately, consilience is simply impossible – what is claimed is incompatible with any known theory of the physical structure of water, and no single experiment by fallible Frenchmen could possibly have sufficient leverage to open up such a well-explored area for fundamental rethinking. The other examples of pathological science I listed at the outset of this chapter likewise would have been uncovered by the honest application of the rules of Chapters 2 and 3.

A surprisingly close match between unrelated measurement domains is highly persuasive (the anthrax example, earlier, or the post–Luis-Alvarez match of the date of the Chicxulub crater to the Cretaceous–Tertiary boundary, later). Consiliences that are too vague ("I predicted something positive, didn't I!") don't count. The match must be claiming some specific value that is genuinely unlikely by chance (e.g., the exact spacing of the nucleotides in the Watson–Crick double helix model, and also the exact fit of the A–T and G–C pairs, or the full exploitation of the outcome scale in the Milgram experiment to be reviewed at Section E4.7). Competing theories must be treated with the greatest respect (the Alvarez example, or Snow's). It is somebody else's job, not ours, to argue like advocates, reporting only the evidence that supports our side. Ours is rather to reason as scientists, humbly emphasizing the evidence against our position. An abduction must be based on *genuine* surprise at the fact or pattern observed, and its *genuine* incompatibility with competing theories; and consilience must be based on "reasonable agreement" between numerical claims or other quantitative patterns arrived at as independently as possible. Consilience was blocked for the pathological traditions I just listed: for cold fusion, incompatibility with every known law of nuclear physics *and* with the known risks of radiation poisoning ("if the experiment had worked as reported, everybody in the room would be dead"); for infinite dilution, no known mechanism and no explanation at all for the observed periodicity of 10 dilutions; for the thimerosal theory of autism, no confirmation whatsoever over a great many large-scale epidemiological studies.

Still, as we have seen, often arguments are not closed until data have been generated at multiple levels of instrumentation. We cannot demand that all hidden variables be made manifest before agreeing with a claim, or else Perrin would not have carried the argument about atoms 70 years before they were actually seen. The argument about ulcers was settled teleologically (if you treat ulcers as if they are infectious, they don't recur) but now there are breath tests for the metabolic products of the specific causal agent, the bacillus. The arguments about cholera and asthma were settled for good when the interventions were found to work. The argument from seafloor spreading, as transformed via those diagrams of magnetic reversal symmetry at the midocean ridges, was settled by nearly unanimous consent within a couple of years of its announcement – it provided a mechanism for continental drift that no mass of Wegener-like arguments from coincidence (shapes of the continental shelves, fossil flora and fauna, etc.) could supply. The machinery of abduction plus consilience that I am recommending in this essay should be effective in other cases, not only these, by pushing the practitioner into attending to the many additional channels of persuasion, when they are available, and into becoming aware of their absence, thus tempering one's otherwise natural optimism, whenever they are not.

Much of this is craft knowledge. As Perrin or Millikan announced the persuasive power of a curve-fitting diagram he would have to have his own craft sense of what kinds of patterns seem to fit data even when they are meaningless, and arguments, even if not fully written out, that *this* case was not one of *those*. To Perrin, the invariance of all those estimates of N against all those changes of parameters *within* experiments plus instruments and equations between theories amounted to far more persuasion than the community required. For Millikan, the fit was extraordinarily close (by 1916 standards) of those limiting voltages to straight-line functions of wavelength for multiple metals.

In none of these settings does statistical significance have *anything* to do with the outcome of the argument. As physical scientists and engineers have wisecracked since I was a child, "If you need a statistician to make your point, you designed your experiment wrong," at least whenever a strong inference is the desired outcome. That is because significance testing was not actually about your real hypothesis at all, as explained in the previous lecture. If your explanation is correct, you *must* be able to design an argument that makes it obvious in light of data that, *at the same time*, make competing arguments look inadequate. A significance test cannot make *your* argument stronger, because the test does not refer to your argument at all. Consilience deals with actual signal magnitudes on *your* hypothesis, not probabilities on somebody else's hypothesis; that is just one component of the necessary accompanying abductions.

The great physicist Richard Feynman, a most plainspoken man indeed in his general intellectual habitus, presented a splendid send-up of these ideas in a commencement address at California Institute of Technology in 1974. Feynman argued, exactly in keeping with the preceding ideas, that what is missing from "Cargo Cult science" (activities that "follow all the apparent precepts and forms of scientific investigation, but ... the planes don't land") is

> ... a kind of scientific integrity, and utter honesty. For example, if you're doing an experiment, you should report everything that you think might make it invalid – other causes that could possibly explain your results, and things you thought of that you've eliminated by some other experiment, and how they worked. Details that could throw doubt on your interpretation must be given, if you know them. If you make a theory, for example, and advertise it, or put it out, then you must also put down all the facts that disagree with it. . . . You want to make sure, when explaining what it fits, that those things it fits are not just the things that gave you the idea for the theory, but that the finished theory makes something else come out right, in addition.
>
> It's this type of integrity, this kind of care not to fool yourself, that is missing to a large extent in much of the research in Cargo Cult science. (Feynman, 1974, pp. 11–12)

The disciplined combination of abduction and consilience prevents you from turning your science into advocacy, and this seems like a very fine consequence indeed.

L4.4.2 An Unflawed Example from Frank Livingstone

In hindsight it is often possible to find evidence of strong inference and consilience in arguments made well before the machinery of numerical inference for the pertinent data domain had been invented. One enlightening example of this is Frank Livingstone's (1958) great paper on the geography of the sickle cell gene in West Africa. Livingstone was studying a quantitative trait, the frequency of the sickle cell trait,[10] which was already known to be highly deleterious when homozygous; some theory was required to explain what had to be the relative fitness of the heterozygote (Neel, 1957). The dominant theory was one of resistance to malaria, but that was not sufficiently strongly supported by any single data set.

As Livingstone explains his reasoning, the distribution of the sickle cell trait in West Africa is in equilibrium with the distribution of falciparum malaria in some regions but not in others. He will divide the map of West Africa into two parts, one in which the frequency of the trait is in equilibrium with the selective advantage of the heterozygote and the other where it isn't, and he will infer an explanation of its prevalence in this second region from the distribution of languages and "certain domesticated plants." His Table 1 presents the prevalence of the sickle-cell trait in a hundred different tribes, which are plotted on a map and compared to the distribution

---

[10] The "trait" represents either one or two copies of the sickle cell allele, which codes for a variant form of hemoglobin. These were hard to tell apart in the 1950s, but the homozygotes would hardly ever live to produce offspring, so populations were adequately summarized by the frequency with which their members bore *any* count of this allele.

of language families, and Livingstone notices a *partial* consilience, whereby the frequency of the trait is is in accord with the incidence of malaria in some high-malaria areas but not in others. (The accord here is a quantitative one, not correlational or metaphorical. All areas with 15% or more incidence of the sickle-cell trait have endemic malaria – a Boolean zero of the sort we have already seen in the Barcelona asthma example, Fig. 4.7.) In an argument that could not remain numerical, as the necessary methods for regressions over maps had not yet been invented, Livingstone traces the implications of two positive hypotheses:

1. The sickle cell trait has been present in some parts of West Africa for a considerable time but, owing to the comparative isolation of the low-frequency populations in Portuguese Guinea and Eastern Liberia, is only now being introduced to them.
2. The environmental conditions responsible for the high frequencies of the sickle cell trait have been present for a relatively short time among these populations, so that the spread of the trait is only now following the spread of the selective advantage of the allele.

Livingstone then demonstrates these propositions by detailed analysis of West African language families and by archeological evidence of the history of the area's cultures, concluding that the cline of the frequency of the sickle cell trait coincides with the spread of yam cultivation in a manner complicated by the complicated epidemiology of malaria itself. "It is only when man cuts down the forest [for agriculture] that breeding places for *Anopheles gambiae* become almost infinite." Livingstone concludes, in as pretty an example of abductive consilience as any in this book, that this trait "is the first known genetic response to a very important event in man's evolution [the invention of agriculture ca. 7000 years ago] when disease became a major factor determining the direction of that evolution" (Livingstone, 1958, pp. 554, 557). For similar examples of ongoing human evolution, though perhaps not so carefully argued, see Harpending and Cochran (2009).

We will see most clearly in our discussion of Milgram's experiment, Section E4.7, how effective this discipline of consilient abduction is in comparison with the less common alternative rhetorics. To the extent that the method is out of reach for some sciences, their conclusions should be appropriately tempered. Along the boundaries of biology I have in mind such fields as evolutionary psychology or paleoanthropology, fields in which the sorts of numerical inferences I am praising in this book seem not to be possible. The appropriate response is to acknowledge that constraint, not to renormalize one's rhetoric like a college professor grading on a curve. There is no substitute for strength of a numerical inference, and if sufficient rhetorical force is not available, practitioners in the nonconsilient disciplines should be willing to say so. Of course they should continue with their work anyway, on grounds of its importance to man's conception of himself, or its intrinsic fascination, or a regulatory bureaucracy's desire to behave rationally in the course of regulating something. I will have more to say on the virtues of methodological self-awareness in Chapter 6.

E4.5 Consilience and the Double Helix

James Watson's and Francis Crick's paper "A structure for deoxyribose nucleic acid" (Watson and Crick, 1953) is perhaps the most powerful and effective one-page publication in all of scientific history. It is also one of the most misleading, nearly all the intellectual steps leading to its argument having been artfully redirected within its actual text.

In particular, the published paper systematically downplays the centrality of quantitative reasoning in this fundamental discovery. Fortunately, we also have Watson's *The Double Helix* (1968), his bestselling autobiographical account of this same episode. The interested reader might also have a look at Crick's version of the same events, Chapter 3 of his autobiographical comments published two decades later as Crick (1988), or the intriguing synthesis of the two published two decades after *that* as Ogle (2007), Chapter 2. For a detailed scientific history, see Olby (1974).

Reading the later Watson, it is possible to make sense of the earlier publication (W–C) as an exemplar of quantitative reasoning that is, again, strongly reminiscent of the mode of "interrogatory" scientific inference we have been calling abduction. Surprising facts are observed, which, if hypothesis A (this particular double helix) were true, would arise as a matter of course; hence there is reason to believe that hypothesis A is true. The story of the double helix is in fact a narrative of the systematic attempt to construct some such single hypothesis A. The task was to arrange the atoms "in positions which satisfied both the X-ray data and the laws of stereochemistry" (Watson, 1968, p. 200) – to produce *some* plausible model consistent with the data. Unfortunately, that required that the data be available first: "We had to know the correct DNA structure before the right models could be made" (loc. cit., p. 85). The published paper completely conceals this abduction. Yet, in a paper this short, the nature of the substitution by a fiction can be followed in some detail.

E4.5.1 The Nature Version

My point is not that Watson and Crick denied the role of this rhetorical standard, the match of model to data. Their paper begins with a demonstration of how somebody else's otherwise plausible model crumbles against this standard of confrontation. "A structure for nucleic acid has already been proposed by Pauling and Corey.,... In our opinion, this structure is unsatisfactory for two reasons: ... (1) It is not clear what forces would hold the structure together. (2) Some of the van der Waals distances appear to be too small." Watson's book (1968, pp. 160 ff.) is much less gentle on this point: "Pauling's nucleic acid was not an acid at all. Without the hydrogen atoms, the chains would immediately fly apart. If a student had made a similar mistake, he would be thought unfit to benefit from Cal Tech's chemistry faculty [where Pauling was professor]." Later, they learned that Pauling's manuscript had been published "before his collaborator could accurately measure the interatomic distances. When this was finally done, they found several unacceptable contacts that could not be overcome by minor jiggling. The model was thus impossible on straightforward stereochemical grounds." Ironically, the authority for this impossibility was Pauling's own master monograph, *The Nature of the Chemical Bond.*

"We wish to put forward a radically different structure," W–C goes on. They introduce the double helix and then some "assumptions": "There is a residue on each chain every 3.4 A[ngstroms].... We have assumed an angle of 36° between adjacent residues in the same chain, so that the structure repeats after ten residues on each chain, that is, after 34 A. The distance of a phosphorus atom from the fiber axis is 10 A.... The previously published X-ray data on deoxyribose nucleic acid is insufficient for a rigorous test of our structure.... It must be regarded as unproved until it has been checked against more exact results. Some of these are presented in the following communications [including Franklin and Gosling, 1953]. We were not aware of these results when we derived our structure, which rests mainly though not entirely on published experimental data and stereochemical arguments."

E4.5.2 The True Version

But the text of the Watson–Crick article is, alas, seriously misleading. It *is* the case that the "assumed" geometry of the helix matches the data in the "following communications," but that is because the helix was constructed specifically to match that data: an abduction, not a predictive verification. In his book, Watson explains that (in a possible ethical breach that has concerned historians and ethicists of science ever since) in fact he had seen the diffraction image (my Figure 4.27) in the Franklin–Gosling communication long before, without Franklin's having been aware of his seeing it. He recalls, in terms that (to this practicing statistician, anyway) are not necessarily an overdramatization:

> The instant I saw the picture [of what is now called 'the B structure'] my mouth fell open and my pulse began to race. The pattern was unbelievably simpler than those obtained previously. Moreover, the black cross of reflections which dominated the picture could arise only from a helical structure.... Mere inspection of its X-ray picture gave several of the vital helical parameters. (Watson, 1968, p. 167)

The numbers 3.4 A., 34 A., and 10 A. of the published article thus were neither assumed nor predicted but instead actually *measured.* The article in the same issue of *Nature* by Rosalind Franklin and R. G. Gosling explains this in the stereotyped, anonymous voice of calm authority:

> The X-ray diagram of structure B shows in striking manner the features characteristic of helical structures, first worked out in this laboratory [by a team that included Francis Crick].... If we adopt the hypothesis of a helical structure, it is immediately possible to make certain deductions as to the nature and dimensions of the helix. In the present case the fibre-axis period is 34 A. and the very strong reflection at 3.4 A. lies on the tenth layer line.... This suggests strongly that there are exactly 10 residues per turn of the helix.... Measurements of [the distances of those bands from the center] lead to values of the radius of about 10 A.... We find that the phosphate groups or phosphorus atoms

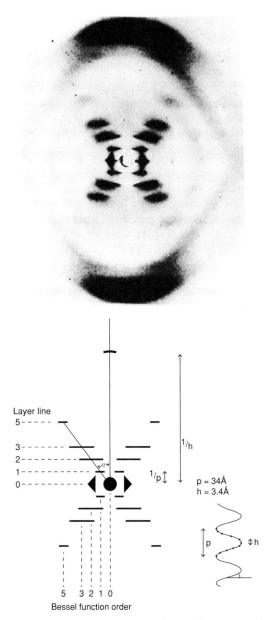

Figure 4.27. The B form of the crystallized DNA diffraction image, a version of which Watson saw on Franklin's desk. (Top) X-ray diffraction pattern from B form of DNA. (Bottom) The geometric quantities that can be measured on this picture are referred to in Watson and Crick (1953) as "assumptions." (From K. N. Trueblood and J. P. Glusker, *Crystal Structure Analysis: A Primer*, Oxford University Press, 1972, p. 137, Fig. 39b. The *actual* image that Watson saw was published in Franklin and Gosling (1953) and is available on the web as http://www-project.slac.stanford.edu/wis/images/photo_51.jpg. an original that, however widely available, cannot be freely duplicated. As another Wikipedia entry points out, the figure is copyright by Raymond Gosling and Kings College London, and reproduction is approved only for "minimal use," to wit, "only to illustrate the article on the photograph, [but not] to illustrate articles on related subjects, including articles on the photographer, Rosalind Franklin, Kings College, or DNA or the history of DNA research.") By permission of Oxford University Press.

lie on a helix of diameter about 20 A., and the sugar and base groups must accordingly be turned inward toward the helical axis.... The structural unit probably consists of two co-axial molecules which are not equally spaced along the fibre axis; ... this would account for the absence of the fourth layer line maxima and the weakness of the sixth. Thus our general ideas are not inconsistent with the model proposed by Watson and Crick in the preceding communication.

"Not inconsistent," indeed! – they *are* the model. Given this strong hint about what the geometry of the helical backbone *must* be, Watson explains, the only remaining task was to figure out how the base pairs fit inside. He was hobbled by having an incorrect textbook geometry for the base pairs. A guardian angel (Jerry Donohue) supplied him with the correct shapes, which, in passing, completely destroyed the possibility of another stereo configuration (a like-with-like pairing) that Watson had been working on up until about a week earlier. (Watson notes that the like-with-like model also required an impossible rotation angle between bases, and further that it offered no explanation for Chargaff's rules.) We continue in Watson's words:

Suddenly I became aware that an adenine-thymine pair held together by two hydrogen bonds was identical in shape to a guanine-cytosine pair held together by at least two hydrogen bonds. All the hydrogen bonds seemed to form naturally; no fudging was required.... Chargaff's rules then suddenly stood out as a consequence of a double-helical structure for DNA.... In about an hour [the next day, when the metal models finally arrived] I had arranged the atoms in positions which satisfied both the X-ray data and the laws of stereochemistry. (Watson, 1968, pp. 194, 196, 200)

This model is, of course, the structure in the photograph of the press conference, my Figure 4.28. Note how the published paper completely conceals the existence of that "Eureka!" moment, when "suddenly" Watson became aware of the correct diagrams. Concealment of this moment is the overwhelmingly common rhetorical trope in peer-reviewed science, but (as in this case) it often conceals precisely the core abductive logic that leads to confidence in a hypothesis on the part of the investigator. The awareness, however suddenly arrived at, proves almost impossible to scrutinize skeptically.

"Chargaff's rules," known since about 1949, are the experimental finding that in DNA preparations the ratios of A to T and G to C are "nearly unity." In the W–C paper, it is presented as a post-hoc explanandum; in Watson's book, as I have just quoted, it appears instead as just another part of the abduction. Surprisingly, neither source notes that the model supplies a testable prediction, namely, that the ratios should be not "nearly" but *exactly* unity. Chargaff, who lived into the present century, insisted to his dying day that he deserved partial credit for the discovery of the double helix based on this abduction, but in fact it was not used as a constraint on the model-building, merely as rhetorical support later on. (But it raises an interesting logical question: does a finding that a value is $1 \pm \epsilon$ establish a claim of priority with respect to a separate finding that ϵ is actually 0?)

Figure 4.28. James Watson and Francis Crick showing the double helix model at a press conference, 1953. The object that Crick is using as a pointer is a slide rule, a now-obsolete consilience machine for scaling up molecular dimensions to models. Appropriately scaled, the geometry of this model matches the measurements that can be made on Figure 4.27. (From A. Barrington Brown/Science Source.)

At this point, what actually *happened* is what the paper says has not yet been done. From Watson's book:

> Exact coordinates [needed to be] obtained for all the atoms. It was all too easy to fudge a successful series of atomic contacts so that, while each looked almost acceptable, the whole collection was energetically impossible.... Thus the next several days were to be spent using a plumb line and a measuring stick to obtain the relative positions of all atoms in a single nucleotide.... The final refinements of the coordinates were finished the following evening. Lacking the exact x-ray evidence, we were not confident that the configuration chosen was precisely correct. But this did not bother us, for we only wished to establish that at least one specific two-chain complementary helix was stereochemically possible. Until this was clear, the objection could be raised that, although our idea was aesthetically elegant, the shape of the sugar-phosphate backbone might not permit its existence. Happily, now we knew that this was not true; ... a structure this pretty just had to exist.[11] (Watson, 1968, pp. 201–202)

[11] This echoes the comment of Einstein's about beauty that I have quoted from Wigner in Section 3.1.

The rest of the story rushes forward in the few remaining pages of Watson's book, and in proportionately little actual calendar time. Faced with the astonishing fit of this elegant and biologically plausible model to the available data (the X-ray diffraction image, Chargaff's rules, the standard knowledge of atomic contact geometry, a mechanism for gene copying), everyone in the community, even Pauling, conceded with startling grace that Watson and Crick had got it right. In this, at least, the story of the double helix goes forward like the other histories of "coercive quantitative reasoning" that I am reviewing here. As do these other examples, it closely matches the structure of a Peircean abduction of extraordinary force: a detailed match between a hypothesis and surprising facts available prior to the enunciation of the hypothesis. Those prior facts, Franklin's X-ray crystallograph and Chargaff's rules, are *numerical*, and so this fabulous paper is in fact a numerical inference, the very topic of this book (and the best example in all of biology). As Franklin and Gosling explain things, the entire Watson–Crick model can be construed as a retroduction from this single X-ray diagram (Figure 4.27), once such a model could be invented. Watson's book explains this, but the original paper conceals it completely.

There is one more consilience than those already mentioned. Figure 4.28, a figure from Watson's book (not from the *Nature* article), shows our young heroes during a press conference at which they were explaining just what a double helix would look like if it were big enough to be seen by visible light. You see the "backbones" of the structure, the pair of helices twisting around each other, and you see the base pairs that link them together step by step up the ladder. But what is that object in the hand of Francis Crick (on the right in the figure)? It is in fact a slide rule (German, *Rechenschieber*; French, *règle à calcul*), a machine invented about 400 years ago to carry out multiplication by addition. We learn more about it in Section 6.5, but for now, think of it as an actual *consilience machine*, designed to help ascertain whether the spacing of the atoms in this physical model did indeed correspond to the facts tabulated by Pauling in his book on how near atoms liked to sit to others. The model had to match Pauling's tables as scaled up by the appropriate factor, and the slide rule greatly simplified the task of checking between the model and the published tables.

So the third consilience of the *Nature* paper, which didn't merit even one single sentence there, was the agreement of the model with Pauling's laws of atomic spacing. In the article, its only trace is in the three-word phrase "published experimental data" in the fourth-to-last paragraph. This is in addition to the match to the crystallograph (Figure 4.27) and the explanation of the hitherto perplexing rules of equality between A and T, and between C and G, from Chargaff's work. But regarding its centrality, Watson is more honest in his book. If this consilience had failed, there would be no model – it would *have* to be rejected, as incompatible with the accepted laws of chemistry (see also Rousseau's comment above dealing with the relation of established facts or laws to the claims of pathological science). Yet given its *consilience* with those earlier publications, its plausibility became very much greater. As Watson said, "A structure this pretty just had to exist." The conclusion arose from a well-founded *numerical* inference: that is the fact that the *Nature* paper almost wholly conceals.

L4.5 From Correlation to Multiple Regression

Galton's quincunx is a mechanical model for linear *causation*, just as the experiments of Millikan or Perrin that checked Einstein's equations were models for linear *laws*. The machine is realistic, in that the causation it produces (balls falling down the machine) really is linear and operates with "noise," and likewise the regressions upward. Over the last century there has been an enormous ramification of statistical methods that generalize this approach, or try to, for all the other contexts in which quantitative scientists are accustomed to use the word "cause." (Remember that the other main point of the quincunx was its cautionary note regarding the inference of "the direction of gravity" from correlational statistics based on its bin scores.) This has proved a far more difficult task than was hoped in the 1970s when I was a graduate student. The problem, in my view, arises from the circumstance that *all* the standard methods in this domain lack formalisms for either abduction or consilience. This critique substantially aligns with Pearl's (2009) theme that standard statistical data analysis has no symbol for causation at all. A similar point enspirits the late David Freedman's wonderful linear models textbook of 2009.

Section L4.5 reviews the standard methods, indicating how aspects of "comparative causal force" are encoded in linear equations, but will preach them and exemplify them only in the quincunx-like context where one single cause is dominant or otherwise of primary interest while other potential causes serve only as competing hypotheses *to be discarded*. Only in this context will abduction and consilience manage to articulate with the otherwise remarkably weak inferential formalisms collected under the heading of "multiple regression analysis." The next section, E4.6, offers two two examples where the formalism "works," but only because the surprise (the abduction) comes from consensus over multiple studies, not just one, and the consilience comes, as Wilson so earnestly hoped it would, from agreement with experiments on other levels of the systems under study.

This theme, the algebraic calibration of multiple causes of the same consequence, is one of the grand themes of 20th-century statistics. In the opening sentence of his textbook of regression, Weisberg (2005, p. xiii) expresses the current conventional wisdom quite clearly using the trope of *dependence:*

> Regression analysis answers questions about the dependence of a response variable on one or more predictors, including prediction of future values of a response, discovering which predictors are important, and estimating the impact of changing a predictor or a treatment on the value of the response.

A claim this forthright deserves an equally forthright rebuttal. From Freedman and Zeisel (1988, p. 46), a "parable":

L4.5 From Correlation to Multiple Regression   221

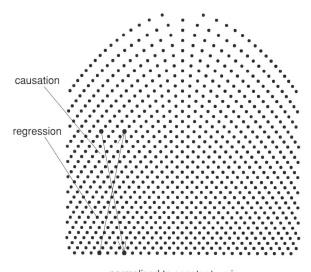

Figure 4.29. Modification of Figure 4.14 to hold constant the variance of the rows. Now the equivalence of the two regression lines (see Figure 4.17) is geometrically obvious: they have the same slope with respect to the x-axis for the two different orientations (upward or downward) of the y-axis.

> Some friends are lost in the woods and using a compass to find their way. If the compass is broken, pointing that out is a positive contribution to getting home safely.

Indeed, over the course of this section we will see that every clause of Weisberg's sentence after the word "including" is actively misleading and that the set of all three taken together is just plain false, along with the main statement itself. Regression analysis is not capable of answering questions about "dependence." Regression analysis is just arithmetic; by itself, it cannot "predict future values," "discover which predictors are important," or "estimate the impact of changing a predictor or a treatment." Scientific meaning, if there is any to be had, must be based on details of samples, conditions of observation or intervention, choice(s) of measurements to be made, and, crucially, the possibilities of abduction and consilience, the main themes of this book.

L4.5.1 From Causation to Correlation

The quincunx that concerned us in Section L4.3 was getting wider and wider (counts of heads minus tails more and more variable) with every row. This doesn't match many phenomena in the world of natural science. To circumvent this inconvenience, divide each bin index, x or y, by the standard deviation of the distribution of shot in its row, l or k, respectively, and call these *standard scores*. That is, the "standard score" z_y of y, in the kth row, is $y/\sqrt{k}$, and the "standard score" of x, in the lth row, is $x/\sqrt{l}$, and now every row has the same variance of 1. See Figure 4.29.

The linear least-squares prediction of x from y was $x \sim \frac{l}{k}y$, so the prediction of z_x is

$$z_x = x/\sqrt{l} \sim \frac{\frac{l}{k}y}{\sqrt{l}} = \frac{\frac{l}{k}(z_y\sqrt{k})}{\sqrt{l}} = \sqrt{\frac{l}{k}}\, z_y.$$

And the prediction of y from x was $y \sim x$, so the prediction of z_y is

$$z_y = y/\sqrt{k} \sim \frac{x}{\sqrt{k}} = \frac{z_x\sqrt{l}}{\sqrt{k}} = \sqrt{\frac{l}{k}}\, z_x,$$

– which is exactly the same formula but with x and y interchanged. In terms of the cross-product formulas from Section L4.3 (I use the version with Σ, but the strategy is exactly the same for the version with E), the computed slope is now

$$\Sigma z_x z_y / \Sigma z_x^2 = \Sigma z_x z_y = \Sigma(x/\sqrt{\Sigma x^2})(y/\sqrt{\Sigma y^2}) = \frac{\Sigma xy}{\sqrt{\Sigma x^2 \Sigma y^2}},$$

where now the symmetry between x and y is obvious.

For the remainder of this section, I drop the z-subnotation. What had been z_x is now written just x, and similarly y. With this change, we have *standardized* for the (increasing) variance of the position of the lead shot as we go down the quincunx. Now that that has been done, the prediction of cause by effect (x by y) and the prediction of effect by cause (y by x) have the same slope: $x \sim ry$ or $y \sim rx$, with $r = \sqrt{\frac{l}{k}}$, the **correlation** between position in the kth row of the quincunx and position in the lth row above it. In other words, a correlation is just a regression between standardized scores, in either direction. *That* is why correlation tells you nothing whatsoever about cause. The slopes for the two regressions in Figure 4.17, for the mother's height on the daughter's and the daughter's height on the mother's, are nearly the same because their variances are nearly the same. So each of those slopes was very nearly a correlation already, and we have already verified that indeed, by virtue of this symmetry, the relationship tells us nothing at all about causes.

The probability of either standardized observation is distributed about this predicted value with variance $1 - \frac{l}{k}$, the famous "$1 - r^2$" of *unexplained variance*. The complementary fraction r^2 of variance is what is "explained," namely, the variance of the standardized prediction $y \sim rx$ or $x \sim ry$.

> Because the predicted part and the residual part are uncorrelated, their variances must add to 1, the variance of either standardized variable, and so the variance of the residual must be $1 - r^2$, the difference of the variances of y and rx after the standardization. Why are they uncorrelated? This is an algebraic check: if x and y are both centered of variance 1, then the covariance of y, say, with $x - ry$ for $r = \Sigma xy/N$ is $\text{cov}(x, y) - r\,\text{var}(y) = r - r = 0$.

Before we standardized, one regression coefficient was $\frac{l}{k}$ and the other simply 1. The product of these two coefficients is $\frac{l}{k}$, and the correlation coefficient r is the square root of that product whether variables were standardized or not.

Yet the words "explained" and "unexplained" here are remarkably misleading, because the variable x finds itself "explained" to exactly the same extent as y, regardless of any actual scientific understanding achieved. Evidently explanation here is not in the sense of scientific explanation, but instead in some weaker sense beholden more to the numbers than to the phenomena. It is the *relation* between x and y that is explained in the context of this particularly powerful stochastic model, the quincunx. There are some settings, notably the agricultural experiments in which this terminology originated, within which the two uses of the word "explanation" are reasonably consistent; but the formulation for correlations and "explained" variance apply regardless of whether that phrase makes any scientific sense. We delve much more deeply into the connection between line-fitting and explanation when we discuss *multiple regression*, regression on more than one predictor, in Section L4.5.2.

This second version of the quincunx, the one in Figure 4.29 with the normalized variables, shows "the paradox of regression to the mean" quite clearly. Extreme causes produce, on the average, less extreme effects; but also, extreme effects are produced, on the average, by less extreme causes. As the figure makes clear, the slopes are in fact the same: the two lines drawn are symmetric with respect to a vertical line (not drawn) that runs between them through their point of intersection. It was this same paradox that first started Galton puzzling over the tie between causation and bell curves, when he noticed that the parents of tall children, though tall on average, were nevertheless not as tall on average as those tall children; and yet the children of tall parents, though tall on average, were likewise on average not as tall as those tall parents. The "paradox" arises because the standardized regression coefficients r are identical, not reciprocal, for prediction in the two temporal directions; both are less than 1.

For the case of the common cause (end of Section L4.2.6), after standardization the regression coefficients $\frac{k}{k+l_1}$ and $\frac{k}{k+l_2}$ relating l_2-row and l_1-row descendants of a common ancestor in the kth row both become the correlation coefficient $\frac{k}{\sqrt{(k+l_1)(k+l_2)}}$. But this is the same as the product $\sqrt{\frac{k}{k+l_1}}\sqrt{\frac{k}{k+l_2}}$ of the correlations linking the cause (the row-k value) to the two effects separately: one factor for the regression "up" and a second one for the passage back "down." Just as the correlation pays no attention to the direction of causation, so this product pays no attention to the fact that one prediction is along the direction of causation, while the other is in the opposite temporal direction. In practice, we are asking the following question, which corresponds to a good many practical questions in today's applied natural and social sciences: how can one reasonably predict the consequence of an imperfectly observed cause? If all errors are these Gaussian errors, the answer (according to the inverse-probability argument) is the simple formula I've just given you: the regression of one effect on another is by the product of their two separate correlations on their common cause, once these effects are variance-standardized themselves. If we use a different topology of causation, row k to row l to row m in steady downward order, then the product of the separate correlations is $\sqrt{\frac{k}{l}}\sqrt{\frac{l}{m}}=\sqrt{\frac{k}{m}}$: such *causal chains* behave the way they ought to.

L4.5.2 The Algebra of Multiple Regression: Three Interpretations

This was a simple version of causation indeed: only one process at work, "falling down the quincunx." The trouble starts once we imagine there to be more than one separate process at work. The experimentalist would surely choose to work on one at a time, but for most of the last century, the statistically minded researcher has attempted to argue instead from tests of multiple models at the same time. Abduction is very difficult in a context like this (see Anderson, 2008), and consilience works only one process at a time in any case. Success, in these contexts, consists in finding a data set for which only one of a set of alternate explanations applies *within the single experiment*. This section reviews the algebra by which multiple hypotheses are (rather weakly) assessed in one single computation, and the next states the criteria according to which one (at most) can be supported abductively by one data set and consiliently by its relationships to others. In Section E4.6 all this lore is applied to one persistent issue in public health, the effect of environmental tobacco smoke (passive smoking).

Just so as to avoid perseverating with natural-science examples, the initial exposition here will involve a small artificial data set (what is often called a "toy" data set) of three "psychosocial" variables – IQ (the "intelligence quotient"), SES ("socioeconomic status"), and "Success" (SCS) – in two versions. In **Version 1**, IQ and SES are uncorrelated: being high on either will not alter the average value of the other. In **Version 2**, IQ and SES are moderately correlated ($r \sim 0.39$). In **both versions**, there is a **true model** by which IQ and SES jointly predict SCS according to the following suspiciously simple equation:

$$SCS = IQ + 3.5SES + \epsilon,$$

which is to say,

$$SCS = IQ + 3.5SES + \text{a little bit of other stuff}.$$

The full data set I'm using for Version 1 is shown in the left panel of Figure 4.30, where I indicate the value of SCS for each combination of an IQ score and an SES score. There are 36 cases here, and, as promised, there is no correlation between IQ and SES: each variable covers its full range with perfect evenness regardless of the value of the other. This is called the case of **orthogonal** or **uncorrelated** predictors, and so I'll refer to this figure as *the uncorrelated-predictors figure*.

The other two panels in Figure 4.30 show the relationships between these two predictors and their joint outcome *separately*. In the middle panel, you see the scatter of SCS against IQ; in the right panel, the same outcome against SES. The "little bit of other stuff" in the equation is just a fluctuation of ± 2, not really random (as it is usually supposed to be) but instead designed to perfectly balance against variation in either IQ or SES, so that it won't affect the arithmetic to follow. You can see that the correlation of success with SES is rather lower than that with IQ – 0.82 vs. 0.57 – but that won't affect our arithmetic here.

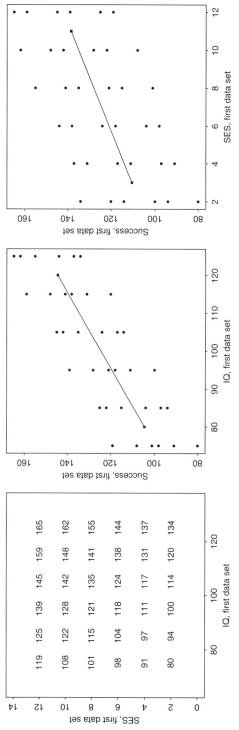

Figure 4.30. The "uncorrelated predictors figure." A multiple regression that reduces to two simple regressions: prediction of an outcome by two uncorrelated predictors. (Left) IQ by SES. (Middle) IQ predicting the outcome (Success). (Right) SES predicting the same outcome. Lines are Tukey's approximate regressions, connecting means of predictor and outcome for upper and lower thirds of the distribution of the predictor.

225

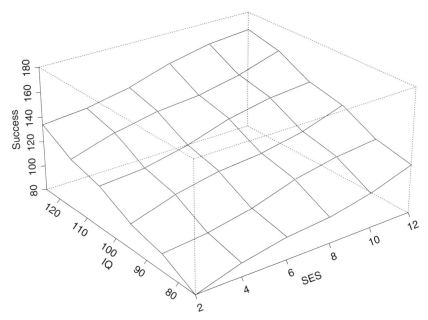

Figure 4.31. The actual data set for Version 1: Success $\sim$ IQ + 3.5 SES plus a little bit of noise.

The coefficients of the equation above can be recovered from these plots by slightly altering the formula for regression slopes – instead of taking averages of slopes $(SCS - \overline{SCS})/(IQ - \overline{IQ})$, we take the slope connecting averages:

$$(\overline{SCS}_{IQ>110} - \overline{SCS}_{IQ<90})/(120 - 80) = (144.5 - 104.5)/40 = 1,$$

and, similarly,

$$(\overline{SCS}_{SES>9} - \overline{SCS}_{SES<5})/(11 - 3) = 28/8 = 3.5.$$

If we graph the same data as a surface over the plane of combinations of values of IQ and SES (Figure 4.31), we immediately recover the equation that (by assumption) produced these data. The surface is a wobbly plane that rises 40 points when IQ rises 40 points (see the curve on the left-hand cut face, at SES equal to 2) and a bit less when SES rises 10 points (the cut face toward the right, at IQ equal to 75).

L4.5.2.1 Interpretation 1: the Counterfactual of Uncorrelated Predictors
In real applications, the geometry displayed in Figure 4.31 is not incidental to the method but an absolute demand. If the following algebraic computations are to have any scientific meaning, the surface (or hypersurface, if there are more than two predictors) representing averaged values of the outcome over clusters of jointly specified predictor values must at least approximately manifest the symmetries indicated in the diagram. Specifically: for the coefficients of a multiple regression equation to correspond to what one would reasonably expect from an actual *intervention* on the

system under study, whether intentional or hypothetical (e.g., historical), the observed effect of any incremental change in any single predictor value must be independent of the values of *all* the predictors. As Lieberson (1985) notes, it must also be the case that intervening by *raising* the value of a predictor must have precisely the opposite effect of intervening by *lowering* the predictor to the same extent, or, putting the same point another way, if you intervene to set the value of a predictor to 4, it cannot matter whether you brought it up from 3 or down from 5. Moreover, the *error* around this prediction must be independent of the values of *all* the predictors. Geometrically, any cut of the surface by a vertical plane parallel to one of the predictor axes must result in a curve of *conditional expectation* that takes the form of a straight line never changing in slope regardless of how the sectioning plane is translated. For more than two predictors, these statements about intervention and independence apply to all of the possible combinations.

These assumptions are demanding and usually highly unrealistic. Usually, too, they are hidden from the reader's scrutiny by concealment under a meaningless phrase such as the Latin *ceteris paribus*, "other things being equal." For more of the underlying epistemology, see, for instance, Freedman (2009) or Mosteller and Tukey (1977).

The composite assumption is very difficult to verify as a whole in general scientific practice, but can occasionally be confirmed in very carefully controlled observational studies in the natural sciences with very few plausible predictors. The main purpose of **multiple regression analysis** in such applications is to make possible the recovery of these same underlying coefficients (for IQ, 1; for SES, 3.5) when the competing predictors are no longer uncorrelated. To bring about this more complicated but also more realistic circumstance, I will alter the data set in the easiest way I know how: I will augment the data in the uncorrelated-predictors figure by a new data set obeying exactly the same prediction equation SCS = IQ + 3.5SES, this time without any wobble, but now IQ and SES will be quite substantially correlated.

The new data points number 35, giving us a total of 71 for *the correlated-predictors figure* (Figure 4.32). As you see in the left panel, I have cobbled them together so as to preserve the symmetry of the original construction. The new data line up right along the diagonal of the original square table, but put a good deal more weight of data there than used to be the case. (In passing, we illustrate one way of generating the "essential heterogeneity" that Wright [1968] claims to be a particularly common observation in population studies: two subsamples of the same mean but wholly different covariance structures. See Bookstein and Kowell, 2010.) No real data set is likely to have arisen in this way, but the unrealism does not affect any of the logic or arithmetic by which we are going to be disentangling "effects on success."

The figure continues with the new scatters of the outcome variable Success on the two predictors of this augmented data set. It is clear that the slopes of the partial predictions in the rightmost pair of plots have changed from those in the uncorrelated case *even though the model has not changed.* In the middle frame, for instance, we've thrown in a new block of points that lie below the earlier average when IQ is low and above the earlier average when IQ is high. The estimate that used to be $40/40 = 1$

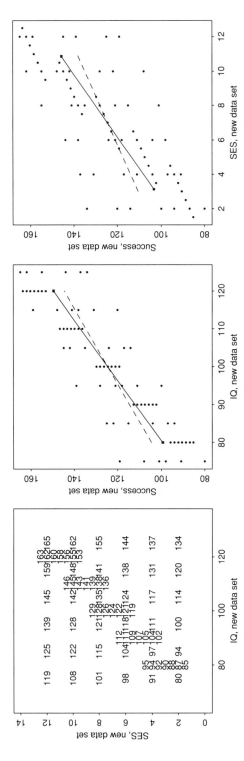

Figure 4.32. The "correlated-predictors" figure. A somewhat less unrealistic situation: prediction of an outcome by the same two predictors, now correlated. (Left) The simulation of correlated predictors, a mixture of the distribution in Figure 4.30 and an additional subgroup lying much closer to the diagonal. Again, at each point is plotted the corresponding Success score. (Middle, right) Tukey-type regression lines for the modified data (solid lines) compared to those for the original scheme (dashed lines). Both slopes have been raised above their declared values; the purpose of multiple regression is to get back to the correct coefficients from the equation.

228

success point per IQ point is now

$$(\overline{SCS}_{IQ>110} - \overline{SCS}_{IQ<90})/40 = (149.65 - 99.34)/40 = 1.258,$$

which is an overestimate; but, also, the estimated slope for the effect of SES is now

$$(\overline{SCS}_{SES>9} - \overline{SCS}_{SES<5})/(\overline{SES}_{SES>9} - \overline{SES}_{SES<5})$$
$$= (145.75 - 103.25)/(10.86 - 3.13) = 5.50,$$

overestimated by an even greater ratio.

Our task is to compensate for the double-counting of the correlated effects of IQ and SES – to reduce the estimate of each back from the panels of the correlated-predictors figure to those from the corresponding panels of the uncorrelated-predictors figure, which, as we have seen, are "correct."

The logic by which this is done has an odd history. Its equations are the same **normal equations**, lower-case n, that were developed by Gauss and Legendre (see Section L4.3) to handle the estimation of parameters in celestial mechanics (the astronomical application of Newton's laws) from fallible data. But the application to this quite different **causal** context came almost a century later, after Galton's development of the causal version of regression. The multivariable elaboration was primarily at the hands of G. U. Yule in the 1890's (see Stigler, 1986, Chapter 10), and the version sketched in the rest of this subsection was developed around 1920 by the American geneticist Sewall Wright (1889–1988).

To recover the **true values** 1 and 3.5 of our first equation from the data of the correlated-predictors Figure 4.31, we need a notation for the quantities we're trying to compute. It is customary to call them the **direct effects** d_{IQ}, d_{SES} of the variables. While this language sounds magisterially wise, it merely means that these are the coefficients that will appear in equations like the first one here, SCS = IQ + 3.5SES + ϵ, equations intended to simulate the mechanisms of ultimate scientific interest underlying the data actually encountered. The quantities we extracted from the correlated-predictors figure are the (numerically different) **total effects** t_{IQ}, t_{SES} expressing only the observed correlations among the predictors, not any deeper understanding of the situation.

In what way did the discrepancy between the d's and the t's arise from the correlation between IQ and SES? I have already given you a hint in the phrase "double-counting" that I used above. The total effect t_{IQ} of IQ exceeds the direct effect d_{IQ} we're interested in by the amount that is **indirect via SES.** To be specific, a rise of 1 unit in IQ, according to this logic, is "associated" (we are making no claim as to how, and may actually remain clueless) with a rise of some number of units $b_{SES.IQ}$ in SES, and that rise, in turn, produces the "extra" effect $t - d$ that we are trying to cancel out.

We can estimate this new abstract quantity $b_{SES.IQ}$ easily enough from the data at the left in the correlated-predictors figure (again, using slopes of averages instead of averages of slopes): it is just

$$(\overline{SES}_{IQ>110} - \overline{SES}_{IQ<90})/(\overline{IQ}_{IQ>110} - \overline{IQ}_{IQ<90}) = (8.47 - 5.53)/40 = 0.0737.$$

This quantity 0.0737 is the net effect on SES of the variation caused for IQ of all the causes of IQ that have effects on SES – it is those unknown "joint causes" that are abstractly held responsible for the correlation we see.

So we get **one equation** relating the two d's:

$$d_{IQ} + .0737 d_{SES} = 1.258.$$

This is called the first **normal equation**, the name you have seen before.

Similarly, to correct the total effect 5.5 of SES for the indirect path through IQ (even though we have no theory as to why they covary), we first extract an adjustment factor for the rise in IQ associated (inexplicably) with a rise in SES:

$$b_{IQ.SES} = (\overline{IQ}_{SES>9} - \overline{IQ}_{SES<5})/(\overline{SES}_{SES>9} - \overline{SES}_{SES<5})$$
$$= (107.7 - 92.3)/(10.86 - 3.14) = 2.$$

This value is, like the value 0.0737, assumed to express the effect on IQ of variation in whatever were the causes of SES that are ultimately going to be found accountable for its correlation with IQ (the same mechanism(s) at which we have already guessed before).

There results the other normal equation, now one telling us how to recover the direct effect d_{SES} we seek from the total effects t:

$$d_{SES} + 2 d_{IQ} = 5.5.$$

Copy the first one again:

$$d_{IQ} + .0737 d_{SES} = 1.258.$$

Just as in Section L4.1, these equations are solved by determinants:

$$d_{IQ} = (1.258 \times 1 - 5.5 \times 0.0737)/(1 \times 1 - 0.0737 \times 2) = 1,$$
$$d_{SES} = (1 \times 5.5 - 2 \times 1.258)/(1 \times 1 - 0.0737 \times 2) = 3.5.$$

In this way **we have recovered the coefficients of the uncorrelated-predictors figure from the data of the correlated-predictors figure.** (Geometrically, we have recovered the slope of the coordinate plane *sections* in the geometric figure from slopes of coordinate plane *projections* – not a bad trick. All those logical assumptions pertaining to the method were necessary for the subterfuge to succeed.) These coefficients, called **multiple regression coefficients** or **beta weights** (because many textbooks write them using the Greek letter β, beta), are usually computed using averages of slopes rather than ratios of averages, as I have here, but the exposition using ratios of averages is easier, and makes the arithmetic match the diagrams more clearly.

Pause now to try and decide what it means to "get back to the uncorrelated-predictors figure" from data that started out looking like the correlated-predictors figure. The coefficients 1 and 3.5 are usually phrased as "the effect of IQ holding everything else (in this case, SES) constant" and, symmetrically, "the effect of SES holding IQ constant." What that means is, precisely, a conversion of the explanation to the scenario of the uncorrelated-predictors figure, in which the predictors have been forcibly **un**correlated by some abstract cancellation of *all* of the common causes they actually share. Nothing is in fact "held constant" in any real sense. It is all an "as if": the fiction of uncorrelated predictors, exploited in order to guess at the effect of an intervention that was not actually carried out.

Therefore, although the coefficients 1 and 3.5 enter into a least-squares prediction of the value of SCS from the values of IQ and SES, this is an accident of algebra. What these numbers really represent is **a kind of explanation**, one that philosophers of science call a *counterfactual*. **If** SES and IQ were found to be uncorrelated, in some alternate world (or alternate data set), **then** their separate predictions of SCS would be the numbers we just computed. You could change IQ without changing SES, or vice versa, and the effect of the IQ change would then be 1 unit of SCS change per unit change in IQ, or 3.5 units of SCS change per one unit of SES. It is because we cannot guarantee the invariance of IQ over that change of SES, or the invariance of SES over that change of IQ, that these expectations are counterfactual, which is to say, unreal.

L4.5.2.2 Getting Back to Correlations

The preceding was analysis by thirds, John Tukey style, but one can rework the same argument using ordinary least-squares regression slopes of any of the variables on any of the others. For simplicity, assume, along with my hero Sewall Wright, that all variables have mean 0 and variance 1, so I can write all slopes using the same letter r instead of the letter t (for "total effect") I have been using up to this point. Let the predictors be X and Y and the dependent variable be Z, and write r_{XY}, r_{XZ}, r_{YZ} for the three correlations among these three variables. Then, as Figure 4.33 diagrams, we want to estimate coefficients $d_{Z.X}, d_{Z.Y}$, the *direct effects* of X or Y on Z, where, in words,

> the total effect r_{XZ} on Z of a unit change in X is the sum of its direct effect $d_{Z.X}$ and its indirect effect via Y, which is r_{XY}, the expected rise in Y given that same unit change in X, times the direct effect $d_{Z.Y}$ of Y on Z;

or, in an equation,

$$r_{XZ} = d_{Z.X} + r_{XY}\, d_{Z.Y}.$$

Similarly, in words (not shown) and then in a second equation,

$$r_{YZ} = d_{Z.Y} + r_{XY}\, d_{Z.X} = r_{XY}\, d_{Z.X} + d_{Z.Y}.$$

L4.5.2.3 Even Further Back, to Least Squares

Now start all over, temporarily ceasing all talk about science or reality but just sticking to the arithmetic. For whatever reason, imagine we are trying to fit a plane "$Z \sim d_{Z.X} X + d_{Z.Y} Y$" to some data about X, Y, and Z by least squares. In the subscript, the dot means "regression on." If you want you can think of this as a model

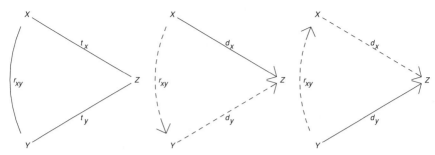

Figure 4.33. Path model of the direct and indirect effects for a bivariate regression. (Left) The correlational data: a total effect t_x of x on z equal to the correlation r_{xz} (and hence drawn without an arrowhead), a total effect t_y of y on z equal to the correlation r_{yz}, and a correlation r_{xy} between the two predictors that holds no causal interest for us, but merely complicates all our algebra. (Center) Decomposition of t_x into the sum of two processes, the direct effect d_x of x on z, solid line, and the indirect effect, dashed line, which is the product of the coefficients along its two segments: the correlation r_{xy} and the direct effect d_y of y on z. (Right) The same decomposition of t_y, reversing x and y in the previous diagram. Notice that the correlation r_{xy} participates in two indirect paths, one in each "direction." The *path equations* according to which d_x and d_y reconstitute t_x and t_y are the same as the *normal equations* generated by demanding that $d_x x + d_y y$ be the optimal (least-squares) linear predictor of z.

$Z = d_{Z.X} X + d_{Z.Y} Y + \epsilon$ for a least-squares estimated ϵ (which will be maximum-likelihood if the model is true and the real ϵ is normally distributed, etc.). For convenience, let N stand for the number of cases and assume X, Y, and Z have all been set to mean 0 and variance 1 before we started calculating.

To minimize $Q = \Sigma(Z - d_{Z.X} X - d_{Z.Y} Y)^2$ we just set to 0 its derivatives with respect to the two unknowns:

$$\frac{\partial Q}{\partial d_{Z.X}} = -2\Sigma X(Z - d_{Z.X} X - d_{Z.Y} Y) = 0,$$

and likewise

$$\frac{\partial Q}{\partial d_{Z.Y}} = -2\Sigma Y(Z - d_{Z.X} X - d_{Z.Y} Y) = 0.$$

Rearrange: these are equivalent to the pair of equations

$$d_{Z.X} \Sigma X^2 + d_{Z.Y} \Sigma XY = \Sigma XZ,$$
$$d_{Z.X} \Sigma XY + d_{Z.Y} \Sigma Y^2 = \Sigma YZ.$$

But our notation implies $\Sigma X^2 = \Sigma Y^2 = N$ and $\Sigma XY = Nr_{XY}$, $\Sigma XZ = Nr_{XZ}$, $\Sigma YZ = Nr_{YZ}$. Canceling the N in every term, then, we get

$$r_{XZ} = d_{Z.X} + r_{XY} d_{Z.Y},$$
$$r_{YZ} = r_{XY} d_{Z.X} + d_{Z.Y}.$$

But those are exactly the equations at which we arrived at the end of the path analysis. So the path analysis, which is explicitly causal, gets us to the same

coefficients as the ordinary least-squares fitting of the linear model, which, as usually set down, does not sound like it entails causality at all. To predict a third variable by a linear combination of two others – to "fit a plane" (or a hyperplane, if there are more than two predictors) – is just arithmetic; but no attempt to ascribe meaning to the coefficients of the plane, or to expect the coefficients to apply to instances not already included in those sums of squares and cross-products, can circumvent these postulates of direct and indirect causality.

If we change notation so that all the predictors X, Y, etc. are collected in one matrix X, rename Z as the letter Y freed up thereby, and introduce a vector d for the list $d_{Y.X_1}, d_{Y.X_2}, \ldots$ of all the regression coefficients together, we get to write the equations setting derivatives to 0 (the *normal equations*, all together), as $(X'X)d = X'Y$, or $d = (X'X)^{-1}X'Y$, which is where most textbooks anchor their notation. Likewise the predicted value can be written $\hat{Y} = Xd = X(X'X)^{-1}X'Y = HY$, where H is the *hat matrix* $X(X'X)^{-1}X'$, projection of the observed dependent variable Y onto the column space of the X's. One standard modern approach to regression residuals v is via the *studentized version* $v_i/\sqrt{1-h_{ii}}$ where h_{ii} is the ith diagonal element of H. For predictor values a long way from the average, h_{ii} will be relatively large, $\sqrt{1-h_{ii}}$ relatively small, and hence the observed residual v_i will be assigned a relatively larger importance than if it had arisen for a predictor right at the average of that distribution. Equations like these are at the core of a first course in linear statistical models as exposited in, for example, Weisberg (2005) or Freedman (2009).

We now have seen three different interpretations of the least-squares equation for multiple predictors X_i of a single outcome Y. There were the *normal equations* for fitting a plane (that reduced to $\beta = (X'X)^{-1}X'Y$), the *counterfactual* adjustment-to-uncorrelated-predictors, and the *path equations*, after Sewall Wright, that arise from chipping apart a joint causal explanation as sketched in Figure 4.33. Path equations decompose each "total effect" (regression of the outcome score on any predictor separately) into a *direct effect*, having coefficient equal to the corresponding component of β, and a sum of *indirect effects*, one for each of the other predictors. Each indirect effect is the product of the direct effect through that other predictor times the regression of the other predictor on the one that we are adjusting.

This ends the "epistemological" portion of the present subsection. Readers who are less than enthusiastic about aspects of linear algebra and statistical modeling may choose to skip the three subsections of Section L4.5.3 and jump ahead to Section L4.5.4, which turns to a related issue, that of using regressions in an abductively effective way. The message will have to be simpler: only one variable of interest, with the others serving as distractions from the main message. Even that reader will be helped by yet another witty quote from David Freedman, this one from page 177 of his essay "Statistical models and shoe leather" of 1991:

A crude four-point scale may be useful:

1. Regression usually works, although it is (like everything else) imperfect and may sometimes go wrong.
2. Regression sometimes works in the hands of skillful practitioners, but it isn't suitable for routine use.

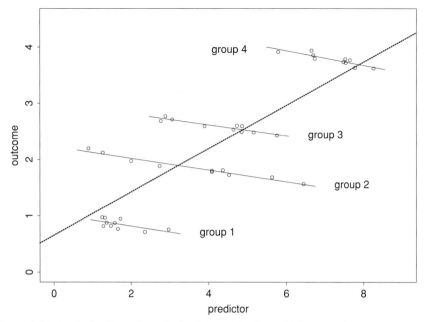

Figure 4.34. Analysis of covariance in the framework of a multiple regression.

3. Regression might work, but it hasn't yet.
4. Regression can't work.

Textbooks, courtroom testimony, and newspaper interviews seem to put regression into category 1. Category 4 seems too pessimistic. My own view is bracketed by categories 2 and 3, although good examples are quite hard to find.

What follows is an illustration of Freedman's point. Regression might work, but good examples, like the following three, were hard to find.

L4.5.3 Three Special Cases of Multiple Regression

The algebra we have just reviewed turns out to work well in more contexts than the straightforward bilinear prediction sketched in Figure 4.31. Here are three more contexts, rhetorically quite different from the preceding, where the same computations apply, but not necessarily the same interpretations.

L4.5.3.1 Special Case I: Analysis of Covariance

Not all measurements come as decimal numbers; some are categories. The same logic of direct and indirect effects, or their equivalent by least-squares fitting of linear equations, applies to the more general setup where some of the predictors are continuous variables and others are categories. It is convenient to think of the situation as analogous to the schematic in Figure 4.34 (inspired by one on page 233 of Andersen and Skovgaard [2010], a regression textbook that I like to teach from).

Now we are modeling some outcome as the sum of indicator functions for the subgroups together with one common slope for each of the continuous predictors (here, there is only the one). The indicator functions (grouping variables) specify the averages for *both* variables, the continuous predictor as well as the outcome. It does not matter that the geometry in Figure 4.31 is no longer applicable. We can still use the path equations (direct and indirect effects) or the counterfactual version (effects as if all groups had the same average) for interpretation even if the usual computer software uses the least-squares formulation to carry out the actual computations.

Let's turn to the simplest possible situation, two groups (hence one grouping dummy variable) and one single continuous predictor. Figure 4.35 shows six different realizations of a single model $Y = aG + bX$, in this case without any random errors (i.e., all residuals zero), where G is a grouping variable taking values $-\frac{1}{2}$ and $\frac{1}{2}$ half the time each. Write X_w for a continuous predictor that is being modeled simply as a uniform from -1.715 to $+1.715$ (so that it has variance 1) within groups. The second predictor X of Y, after G, is this within-group predictor X_w modified by adding some coefficient c times the grouping variable G. That is, $X = X_w + cG$.

We compute variances and covariances as follows:

$\text{var}(X_w) = 1$ (by construction),
$\text{var}(G) = \frac{1}{4}$ (flip of one fair coin),
$\text{cov}(X_w, G) = 0$ (these two are independent), hence
$\text{var}(X) = \text{var}(X_w + cG) = \text{var}(X_w) + c^2 \text{var}(G) = 1 + \frac{c^2}{4}$;
$\text{cov}(X, G) = c \text{ var}(G) = \frac{c}{4}$,
$\text{cov}(Y, G) = \text{cov}(aG + bX, G) = a \text{ var}(G) + b \text{ cov}(X, G) = \frac{a+bc}{4}$;
$\text{cov}(X, Y) = \text{cov}(X, aG + bX) = a \text{ cov}(X, G) + b \text{ var}(X) = \frac{ac}{4} + b(1 + \frac{c^2}{4})$.

The path-equations approach directs us to reconstruct the *direct effects* a and b of G and X on Y by decomposing their *total effects* using the associations among the predictors themselves. Here is how this works out in this scenario:

the total effect of G on Y, $\text{cov}(Y, G)/\text{var}(G)$, is $a + bc$, which equals a, the direct effect, plus bc, the *indirect effect*, which in turn decomposes as b, the *other* direct effect, times c, the regression of X on G; and,

the total effect of X on Y, $\text{cov}(X, Y)/\text{var}(X)$, is $(\frac{ac}{4} + b(1 + \frac{c^2}{4}))/(1 + \frac{c^2}{4})$, which breaks out as b, the direct effect, plus $(\frac{ac}{4})/(1 + \frac{c^2}{4})$, which is the product of the other direct effect a by the regression $\text{cov}(X, G)/\text{var}(X) = (\frac{c}{4})/(1 + \frac{c^2}{4})$ of G on X.

So the path equations work in this case that mixes categorical and continuous variables just as they did in the situation we were describing in the previous section, where the predictors were both continuous variables. In fact, they work out for *any* combination of predictors of *either* type. Hence the labels in Figure 4.35 match the dashed lines, which are the predictors from an ordinary linear-models software facility that knows nothing about the difference between group indicators and continuous measurements.

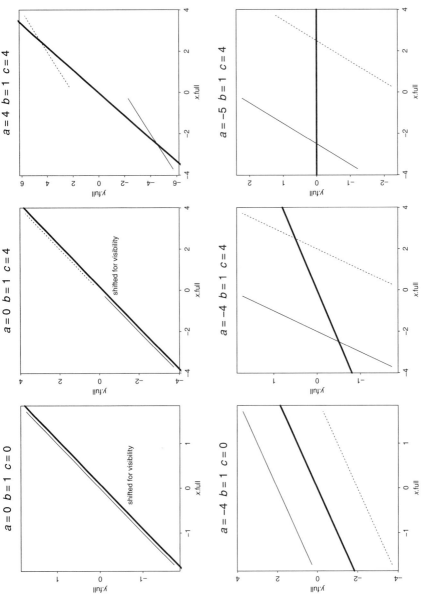

Figure 4.35. Six different versions of a two-group analysis of covariance, including most of the ecological fallacy you are likely to encounter in practice. Above each frame are written the parameters a, b, c of the models $Y = aG + bX$ and $X = X_w + cG$, where X_w is a uniform variable and G is a group indicator. There is no noise (error term) in these models. Light lines: the two-group models (group 1, solid light line; group 2, dotted light line), also the predicted values from the least-squares regression in the text. Heavy solid line: the wrong regression every time (Y on X

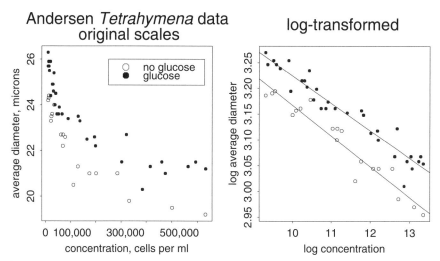

Figure 4.36. Example of an analysis of covariance: the *Tetrahymena* data set from Andersen and Skovgaard (2010), online resource, www.biostat.ku.dk/~linearpredictors.

As the figure hints, by adjusting a and c you can produce any relation you like between the within-group slope b and this pooled between-group slope for the same pair of variables X, Y as long as the grouping variable alters both of their means. The lower right panel is a particularly interesting case often called the *ecological fallacy*. I produced an exact zero for that slope by setting $a = -5, b = 1, c = 4$. We have thereby demonstrated yet another example, different from the ones that textbooks usually show, of a pair of variables that are uncorrelated but not independent.

A very modest modification of this algebra allows us to model the two lines with different slopes. That is the case in the example to follow, which is based in real data, from the Web resource associated with Andersen and Skovgaard (2010). *Tetrahymena*, a protozoan, is a particularly useful model organism for a wide range of studies, including studies of the dependence of growth rate on the environment. These data arose from an experiment, not any sort of random sample of *Tetrahymena* colonies found floating in a pond somewhere. Glucose was either added, or not, and, as a source of energy, may have had an effect on the biochemical processes that govern the life of this little creature. Also, temperature and oxygenation were varied, with consequences for concentration. Is an abduction lurking somewhere?

Examine the left panel of Figure 4.36. It is very difficult to say anything cogent about the situation here from the glucose-comparison point of view. Yes, the filled dots are usually above the general run of open circles in every concentration range, but the curvature of the trend line, and the difference of these curvatures between the subsamples, makes it impossible to say anything authoritative. When data look like these, the log-log transformation often linearizes them, because physiological processes in biology often follow power laws $y = ax^b$ that appear linear after a lon-log scaling. Here a transformation of the variables to the logarithm works well to linearize the whole situation. (In fact, we didn't need to transform average diameter, but only the concentration variable: only the abscissa, not the ordinate.) From the differently scaled plot of the same data, Figure 4.36 (right), we notice, with the help of a pair of regression lines, that –

- The regressions for each treatment group, glucose or no-glucose, are evidently *linear*, so we can summarize the associated predictions with two parameters, a slope and an intercept.
- The transformation shears the region of the plot at the upper left until we can make some visual sense of trends there, too, instead of just "points falling down a ladder."
- The slopes look parallel (and indeed do not differ significantly), so the effect of glucose is only on an intercept, making discussion even easier.
- But the interpretation of the situation here does *not* go well in any sort of causal language. What we have is a system in *homeostasis* against environmental variation. The different experimental runs here (made at different temperatures, different oxygen levels, etc.) result in different colonies that appear to be obeying an equation that, from the regressions, takes the form $\log \text{diameter} \approx -0.055 \log \text{concentration}$, or $\log \text{diameter} + 0.055 \log \text{concentration} \approx \text{constant}$, or, because the volume of a sphere goes roughly as the cube of the diameter,[12]

$$6 \log \text{volume} + \log \text{concentration} \approx \text{constant}$$

where the constant differs by about 0.06 between the glucose and no-glucose conditions (meaning, on the original [non-log-transformed] scale, about 6.2%). This effect presumably has something to do with the environmental cues this creature uses when deciding when to reproduce (which, for a protist, means just splitting in two). Because this is a field that really prefers coefficients to be integers, one might report as follows: "Volume seems to scale as concentration to the minus one-sixth power."

The regressions are not causal. Concentration does not produce diameter, or log diameter; nor vice versa. The experiment controlled temperature, nutrients (including glucose), and perhaps other stuff, and both concentration *and* diameter (and hence their correlation) depended on what the experimenter was doing.

Hence the log transformations were *extremely* fruitful, and (although I am no protistologist) potentially abductive (the parallel slopes on the log scale, the integer coefficients of the rearranged equation). The transformations converted a nonlinear model into a linear one – converted the study from something crassly empirical, in other words, into a finding that just might have some implications for how one does protist biology.[13] The rearrangement converts a regression from the misleading

---

[12] Old joke: the mathematical biologist, asked to give a talk on milk production, begins, "Consider the cow as a sphere."

[13] Were this a real analysis instead of just a demonstration, we would be obliged to point to that seventh-rightmost filled point, which appears to be plotting with the wrong glucose group, and ask whether it represents a measurement blunder of some sort.

form $y \sim -ax + b$, with all its misleading baggage of causation (wrong) and error (wrong), into the arrangement $y + ax \sim b$, with its implications of homeostasis and optimal resource management (by an organism that is not only not a bureaucrat but that in fact lacks any analogue of a nervous system). We have fitted lines by least squares, but our purpose was never to predict cell volume, the variable along the ordinate; the purpose was to understand the trivariate relation among cell volume, concentration, and glucose.

L4.5.3.2 Special Case II: Added-Variable Plots

The complexity of the path models, with the forced dissection of total effects into direct plus indirect, is necessary because predictors are correlated. Suppose that, by some preliminary stratagem, we can produce an additional predictor that is, by design, *exactly* uncorrelated with the others, so that all of the indirect paths through it are zero. Then we should be able to produce the corresponding *direct effect* as a *total effect*, that is, a correlation or simple covariance. Once the predictor is made orthogonal to the others, the regression itself doesn't need to adjust its slopes any further. In the regression textbooks, this approach is called the method of *added-variable plots* for adding one variable to a regression that has already invoked one or more predictors. The following discussion is from Mosteller and Tukey (1977), and to simplify the notation I'm imagining a situation with just two predictors x and y of some outcome z. Write $y_{.1x}$ for the residual from the prediction of the variable y by the variable x along with a constant term. (Now in the subscript, the dot means "after regression on" and that strange notation "1" stands for the constant term.) Similarly, write $z_{.1x}$ for the residual of the variable z after the prediction by x and a constant. ("Prediction by a constant" is equivalent to mean-centering; I am maintaining this somewhat clumsy notation for consistency with the Mosteller and Tukey version.)

Then if we regress $z_{.1x}$ on $y_{.1x}$, and plot the regression line (the combination of scatter and line is what is called *the added-variable plot for y after x has been entered into the regression for z*),

- this regression goes through (0,0);
- the coefficient of $y_{.1x}$ in this regression is the same as the coefficient of y in the regression of z on x, y, and a constant term; and
- the residuals from this regression are the same as the residuals from the regression of z on x, y, and a constant term.

A little notation will help the reader who is trying to relate this to an ongoing statistics class or accessible computer software package. Write out, explicitly,

$$z_{.1x} = z - a - bx, \quad y_{.1x} = y - c - dx$$

(i.e., the prediction equations for z and y as functions of x are $z \sim a + bx$, $y \sim c + dx$). Let the least-squares line in the **added-variable plot** of $z_{.1x}$ on $y_{.1x}$ be $z_{.1x} \sim e + fy_{.1x}$. In other words, we have just defined six letters a, b, c, d, e, f all in terms of one or another of the three regressions involved here. Then the least-squares fit of z to $1, x, y$ is, in fact,

$$z \sim a + bx + fy_{.1x} = (a - fc) + (b - df)x + fy,$$

and the three bulleted statements above follow after a bit of manipulation, for which consult pages 304–305 of Mosteller and Tukey.

Notice that the coefficient of x in the equation just above is the coefficient adjusted as per our path models: the value b, from the simple regression, minus the indirect effect df which is the product of the coefficient d for y on x and the coefficient f for z on y.

Sometimes the scheme in Figure 4.33 is complete nonsense. If, for instance, the variables x, y, z are the same measurements on a growing organism at three different ages – call them now x_1, x_2, x_3 – then while one can imagine an "effect" of x_1 on x_3 that is indirect through x_2, the middle observation, one cannot imagine the corresponding indirect effect of x_2 on x_3 that goes via x_1, the *earlier* observation.

Because the symmetry of Wright's derivation of the normal equations is thereby broken, we must change our approach. Instead of a least-squares fit of x_3 to any linear combination of x_1 and x_2, let the penultimate (second-last) observation "explain" (predict) all it can of the last score. Then invoke the antepenultimate (third-last) observation to "explain" whatever was left over from prediction by the penultimate, and so on. By the rules of the added-variable plot, we can replace the antepenultimate observation by its residual after its own prediction by the penultimate observation, and so on.

Again a didactic example might be helpful. I turn to another of the data sets from the textbook of Weisberg (2005), this one the *Berkeley Guidance Study* of Tuddenham and Snyder (1954). The original study analyzes observations on seven variables at up to 33 different ages on 136 children. In the data set publicly available as http://www.stat.umn.edu/alr/data/BGS.boys.txt, Weisberg has extracted five variables (of which one is height) for the 66 boys of this study at up to three ages (2 years, 9 years, and 18 years).

Figure 4.37 shows the longitudinal analysis, in three frames. At left is the complete raw data – 66 little three-point time series. In the center, we see the prediction of age-18 height on age-9 height, with its regression line. At right is the added-variable plot for the *additional* predictive power of the age-2 measure. It offers no further useful information. The residual of the prediction from age 9 to age 18 is not meaningfully correlated with the similarly adjusted age-2 value, meaning that this growth process seems to have no "memory." While this is not a particularly surprising result, it is nevertheless interesting for its implications about the corresponding biological control mechanisms, at least as long as we are not trying to make sense of the patterns during puberty, when individuals differ greatly in the applicable time scales. For a far more sophisticated analysis of these same data, the interested reader can turn to Ramsay and Silverman (2010), but the additional details are not relevant to the main themes of this book.

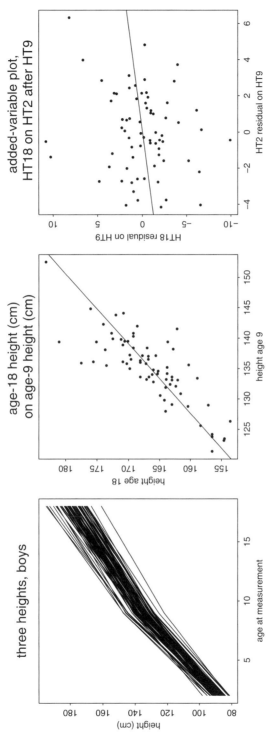

Figure 4.37. Longitudinal analysis of a little series of three heights on 66 Berkeley boys. (Left) The full data set, from Tuddenham and Snyder (1954) by way of Weisberg (2005). (Center) Prediction of the age-18 value by the age-9 value, with the fitted regression line. (Right) The added-variable plot for the age-2 measurement indicates that it contains no additional information.

L4.5.3.3 Special Case III: Linear Discriminant Analysis

We have dealt with two special cases so far – a predictor list that mixes groups with measurements, and prediction by repeated measurements of the same quantity. There is a third commonly encountered special case of this same family of computed least-squares fits: the application to a whole series of binary hypotheses, one for each specimen of a data set. Suppose, indeed, that we have two groups of specimens (call them A and B) and a set of one or more quantitative measurements – a *vector* $X = (X_1, X_2, \ldots)$ of observations on every specimen. A new specimen shows up, and we want to *classify it* – to assign it to one or the other of the groups, A or B.

We have enough machinery in place to proceed rationally in this context. To say the unknown specimen belongs to group A is one hypothesis, that it belongs to group B, the other hypothesis. Each of these hypotheses will have a likelihood given the data. We want to assign every new specimen to the group for which its membership is the hypothesis with the larger likelihood. Then it can be shown (Mardia et al., 1979, Chapter 11) that the log-likelihood for the hypothesis of belonging to group A (vis-à-vis group B) is strictly proportional to the predicted value from the regression of group number on the measurements that are supposed to characterize the two groups, as long as the within-group covariance matrices are the same. In other words,

assuming identical within-group covariance matrices, the optimal (maximum-likelihood) solution to the *classification problem*, the assignment of a new specimen to a group, is the assignment based on the sign of the regression residual from the *(apparently nonsensical)* regression of a group dummy variable, -1 or $+1$, on the vector of available measurements for this purpose.

This derivation is originally Ronald Fisher's, from 1936 or so, using a data set of three measurements of a sample of iris flowers that you can find at many Web sites. The regression is nonsensical because, by assumption, it isn't the values of the X-variables that determine the group – it is the identity of the group that accounts for the values of the X's.

To exemplify this approach along with the preceding topic, added-variable plots, I offer a recomputation of a classic 19th-century data analysis, Hermon Bumpus's (1898) measurements of sparrows found on the ground after a severe snowstorm in Providence, Rhode Island. There were 57 adult male birds, of which 35 survived after being brought into the laboratory to warm up. Bumpus's publication is a prototype of more than a century's worth of successor studies that attempt to study evolution "in the wild" by examining the relative risk of mortality by specific environmental events (cf. Endler 1986; Weiner 1995). My reanalysis of the 1898 data proceeds in three graphic displays, Figures 4.38 through 4.40. The data are from a generous posting to the web by Peter Lowther of the Field Museum, Chicago, at http://fm1.fieldmuseum.org/aa/staff_page.cgi?staff=lowther&id=432.

Figure 4.38, the only one of these figures having an iconography that Bumpus himself might have appreciated, shows the cumulative distributions of all nine of the raw measurements he made. For each variable, the distribution for the survivors is

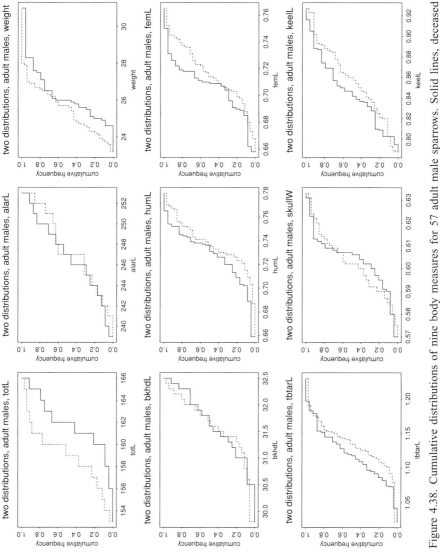

Figure 4.38. Cumulative distributions of nine body measures for 57 adult male sparrows. Solid lines, deceased birds; dotted lines, survivors. (Data from Bumpus, 1898 via Peter Lowther's website.)

shown in dashed line, and for the deceased birds in solid line. Apparently the greatest such shift is for the first measure (upper left panel), total length of the bird from beak to vent.

Selecting that variable as our first predictor for the "regression" that will eventually supply our likelihood function, we examine the added-variable plots as specified in Section L4.5.3.2 to extract the variable that offers the strongest *additional* information predictive of mortality. From the plots, Figure 4.39, we see that the variable we want is the one that now finds itself in the second row and second column, namely, femoral length. (Another round of these plots, not shown, indicates no good reason to add any predictor after these two.)

Following the formulas of the text, we convert the regression of survival on the two selected measures, total length and femoral length, into a plot of log-odds for the hypothesis of belonging to the group that survived. The resulting plot, Figure 4.40, shows a satisfactory discrimination. The birds that survived differ from the birds that did not in a combination of lower total length and greater femoral length. That is, while the birds that died tended to be longer, the survivors tended to have larger femurs both before and after adjusting for this total length. One might speculate that survivors proved better at hanging on longer to their branches (bigger legs and also less torque around the perch). The surprise here, in other words, is that the coefficients of the optimal discriminator are opposite in sign. It is a proportion, not a summary magnitude, that best predicted the survival or nonsurvival of these birds.

L4.5.4 The Daunting Epistemology of Multiple Regression

What is required if regressions are to sustain strong numerical inferences? The preceding three examples all exploit the best possible setting of multiple regression, when the model is "known" to be true and powerful (so that the null model is ridiculous in context) and only its parameters need estimating. Many caveats apply to interpretations of the "direct effects" d computed here from the correlated-predictors figure using the language of orthogonal (counterfactual) predictions – "other things being equal" – corresponding to the setup in the uncorrelated-predictors figure. In a methodological discussion of 2005, my longtime colleague Paul Sampson and I reviewed all the ways in which this whole methodology often manages, in all but the least clumsy hands, to destroy the possibilities of abduction and consilience both at once (Bookstein and Sampson, 2005). Other experts likewise reject the method for routine application to the biological sciences (cf. Anderson, 2008), but it may be more constructive to list the additional criteria that must be checked if inferences from these models are to have any force in numerical inference – if they are to prove both abductive and consilient.

Corresponding to the literature of passive smoking (environmental tobacco smoke) that will be our exemplary context (next section), we state these requirements in a relatively specific language, that of the epidemiology of disease risk. (For other domains of application, just translate some of these technical references into the new jargon. For instance, the points apply with equal force, Clive Bowman assures me, in the context of drug regulation.) In this context, coefficients in equations will

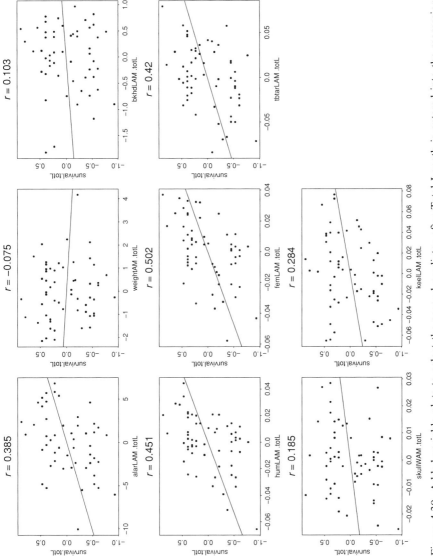

Figure 4.39. Added-variable plots to select the second predictor after Total Length is entered into the regression (included in the prediction). See text.

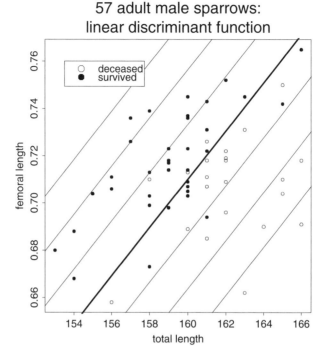

Figure 4.40. Discrimination by Total Length and Femoral Length. The oblique lines are isolines of predicted log-odds of survival. See text.

calibrate the *risks* of various deleterious outcomes with respect to "exposures" to their potential causes, along with "covariates" such as lifestyle factors. In this language, declarations that environmental conditions increase the risk of human diseases are to be taken seriously only if the following criteria are met:

- Biological plausibility. The exposures must involve substances known to cause the disease outcome in laboratory studies.
- Dose–response relationships. In most studies, the groups of highest exposure or dose should have the highest disease rates.
- Broad evidence. Typically this means convergent findings from multiple independent studies of somewhat varying methodology.
- The effects are still there after adjustments for measurement error (for all exposures, but especially for exposures that entail undesirable outcomes).
- The effects are still there after adjustments for potential confounds, such as differences between exposed and unexposed in other known factors of the disease under study. This will involve auxiliary comparisons of the differences between diseased and nondiseased on the same factors, but these are not the same comparisons.
- Statistical significance. This is the least important of the six.

In effect, the entire remaining argument of this section can be put in one terse decree: come as close as possible to a single perfect inference like Snow's. That enormous ratio of death rates is a rare occurence, unless you are Richard Doll and Bradford Hill (1956) discovering the causal role of smoking in lung cancer, so the rest of this section concentrates on the second part of Snow's example, the elimination of alternative explanations, which is, of course, the other premise required for an abduction.

L4.5.4.1 Biological Plausibility: Measuring the Right Thing
In studies of known teratogens, such as lead or alcohol, biological plausibility is in one sense obvious given familiarity with the corresponding high-dose syndromes. But it is not enough that the measurement, for example, by concentration in blood or serum, be appropriate; it has to appear appropriately in the detailed quantitative argument. The issue here is not the biological plausibility of the hypothesized mechanism but the biometrical plausibility of the choice of measurement given the mechanism. The failure of biological plausibility that we see most often, in fact, appears perfectly innocently early in manuscripts, when, for instance, a sample is divided into "a low-exposure group" and "a high-exposure group," averaging within each separately (or carrying out an analysis of variance, or similar multivariate computation).

There is no reason to believe that any average over a "high-exposure group" is meaningful, even where there is a clear dividing line between low and high, which, unless the dose-response curve takes on the unusual form of a sigmoid curve, there usually isn't. Regardless of wording, a high-exposed group is not homogeneous, even in Snow's case – different people drink different amounts of tap water. Also, in terms of our basic abductive reasoning structure, many factors in addition to dose could differ on average between the low-dose and high-dose groups, and any of those could account for some or all of the mean difference observed (see Section L4.5.4.5). There is thus no real surprise possible at the p-value of any single significance test of such a group difference, even on extensively "adjusted" outcomes (after all, no guarantee is possible that some important covariate has not been overlooked). Regardless of p-values, **almost no nonexperimental group mean differences can be surprising** without considerable further argument. Surprise would be entailed only in the actual *magnitude* of that mean difference, like Snow's 6:1 example, and even then one would need to argue that other plausible causes of differences in this outcome were not in fact operating: a scientific argument, not a statistical one. (Note, also, that Snow's measure of exposure, even though a binary variable, was measured at the level of the individual case of cholera, not at the level of the neighborhood.)

All of this assumes some actual quantitative estimate of the dose, not just a yes–no report. Other measures of "dose" sometimes turn out actually to be other *responses*, such as maternal toxicity (intoxication, in fetal alcohol studies). There are enormous individual differences in the competing processes reported here and even more enormous individual differences in the conversion to an actual quantity that might be biologically plausible (at the placenta, for prenatal effects; past the blood–brain barrier, for postnatal ones; in the bronchi, for lung cancer). In general the more reductionist the data, the better.

There also needs to be a biological plausibility to the outcome. If we are talking about neurotoxins, the best measurements for a study of prenatal effects would be of the relative size of a brain part or a brain function directly – volume of the hippocampus, or shape of the corpus callosum, or myelination, or neural speed of signal propagation. At the other extreme one might be referring to an enormously multifactorial descriptor such as a high-school achievement score; a biographical measure like that is unlikely ever to be informative about any specific mechanism(s) of developmental damage. In between are the sensible compromises, such as neonatal brain size, which is multifactorial but for which the strongest predictors (birth size and gestational age) are known in advance, or specific psychometric profiles tapping particular brain functions, regions, or networks (memory test batteries, attention tests). Even in low-dose studies, which typically exploit samples of normal and nearly normal individuals, the outcome measure needs to be "tuned" to the expected mechanism of environmental insult in this way. For studies of diseases that manifest heterogeneously, the heterogeneity interferes with any statistical assessment of causes; if the inferences can be sharpened by a subclassification (e.g., for lung cancer, the histological type), so much the better. An analogous concern in clinical drug studies centers on the concept of a "surrogate endpoint" in studies that would take too long to carry out if the investigators had to wait for a disease process to run its full course, for example, for substantial numbers of participants to actually die. Blood cholesterol is a surrogate endpoint for mortality, for instance, in some studies of the class of drugs called statins. See Horton (2003) and the response by McKillop (2003).

L4.5.4.2 Dose–Response Relationships

The solution to the problem of measuring doses or outcomes as categories is simple: don't do this, but work harder to generate original data in quantitative form. Often it is straightforward to ask all of the usual questions about biological effects at the level of explicitly quantitative scores. But we are not here recommending any automatic approach here via correlations or other linear models. Those approaches reduce to averages under a logic that is just as flawed as that of grouping: to wit, averages over an entire sample, regardless of heterogeneity.

Recall Edgeworth's identity from Section L4.1, for the special case of a bivariate regression of y on x that goes through the origin – no dose, no response. The computed slope $\Sigma xy / \Sigma x^2$ is the same number as $\Sigma\left((x^2)(y/x)\right) / \Sigma x^2$, the weighted average of all the case ratios y/x with weights x^2, the square of the dose (the predictor). There are two inferential issues here. First, it does not necessarily make any sense to weight the cases at dose 10 one hundred times as much as the cases at dose 1; unless the true model is precisely linear, they are not that much more informative. Second, it is not necessarily plausible a priori that there be linearity of a biological effect over an entire order of magnitude. So strenuous an assumption requires an explanatory mechanism, and that issue, too, must be justified beforehand.

Because the linearity of dose–response relationships is thereby in question for most potentially useful studies, the science requires an explicit examination of dose–response relationships that treats dose as a continuous variable and rescales it to correspond to realized effect on a likewise quantitative outcome. By "explicit" I mean graphical, not algebraic: displays of the raw scatterplots of all the main outcome measures or composite indexes on all of the main dose measures. The curves fit to these scatterplots can have thresholds, plateaux, or a variety of other features. I am not recommending that one interpret these individual wiggles unless they correspond to specific compartments or other plausible mechanisms of their own (although substantial failures of monotonicity should of course be pointed out); I simply note that it is not necessary to make *any* assumptions about the form of an empirical bivariate relationship. The resulting diagrams can be very challenging for interpretation. Response curves by dose, for instance, often go haywire exactly at zero dose, as people whose lives involve complete avoidance of a substance or a condition often prove to be unusual in other ways as well. Dose–response modeling along these lines can be done for more than one predictor at the same time, though things become graphically opaque if there are more than two predictors.

In the dose–response domain, a numerical inference will be justifiable only to the extent that dose and response are scaled sensibly. A finding that arises when a continuous dose measure is used in categorical fashion, for instance, raises concerns. It might be the case that the finding arises *only* after categorization, in which case serious questions arise as to how the dividing value was set; and either way, the implied grouping of the sample into the "low-dose" versus the "high-dose" subsamples is troublesome, as I already noted under the heading of "biological plausibility." One should be particularly skeptical of speculative claims regarding the existence of one or another "threshold" for an effect of interest. Any such claim needs to be confirmed ("consiliated," were that actually a correctly spelled participle) by reference to other types of curve-fitting, other dose measures, other outcomes, and some plausibility as to a mechanism that is buffered up to some level but then gets triggered. One powerful form of reassurance is to note that one's dose–response curves look like the relevant bits of somebody else's that was published earlier, using a similar dose measure or similar outcome measure. Both of the passive smoking articles discussed in Section E4.6 exploit this version of consilience. The article to be reviewed in E4.6.2 is consistent with linear *extrapolation* of the carcinogenesis risk toward the zero level of dose, whereas for the article of E4.6.1 the excess risk of heart attacks is consistent with the *measured risk* of platelet aggregation according to experiments in animals and in test tubes.

L4.5.4.3 Broad Evidence

When Snow tabulated his death rates over multiple neighborhoods, one water company dominated the cholera death list in most of them. Such consistency makes it more difficult to argue that his finding owes to other causes even had the water supply not been randomized (because any such other cause would have to apply with reasonable consistency over all 32 neighborhoods). Finding the same pattern in each of several independent subsamples (for instance, a geographical disaggregation) supplies a sort of internal replication (though, of course, the same confounds could still account for the pattern in each of the subsamples). Where data sets are too homogeneous or too small for this technique to be fruitful, one can imitate Snow by measuring dose and outcome more than once. Dose can be measured by integrated dose (lifetime or gestational total), peak dose, dose at each of a range of temporal windows, and so on. Findings that are consistent in sign and approximate magnitude across this sort of variation thereby become much more persuasive than findings based on only one choice, especially if there is a profile along the coefficients that confirms (is consilient with) some a-priori property of these variables as measured wholly independently of the causal nexus under study (for instance, some physical property). Of course, once those multiple dose measures have been accumulated, one is obliged to admit it: no "cherry-picking"! The benefits of redundancy of outcome measure are weaker, as approximate equality of coefficients can arise from common confounding by some unmeasured factor, whereas their *in*equality would lead to well-founded doubts about the specificity of one's preferred hypothesis.

L4.5.4.4 Adjustment for Bias

By this I mean bias in the individual measurements. In the studies reported in Section E4.6, the smoking status of the "nonsmoking" spouse could have been the subject of a deception. As the spouse's own smoking would affect lung cancer rates a lot more than her spouse's, the limits of this bias needed to be calibrated. An analogy from another field may be useful here: the correlation between social class and measurement error on the SAT, through the mechanism of commercial test preparation courses, renders the prediction of college grades from SAT scores nearly meaningless.

There are equally clear analogies in studies of human teratogens. Anything that affects a propensity for the data collector to be deceived, or to be honestly mistaken, needs to be considered, and, if possible, its worst-case magnitude estimated. A common closely related problem is bias of sampling frame according to individual characteristics. This problem often arises in human studies in connection with samples gathered clinically in disadvantaged populations, in which not only follow-up but even the propensity to appear in the screening sample in the first place can strongly covary with the same condition (for example, alcoholism) that launches the medical process under study. On the other hand, belonging to a particularly *advantaged* population can also lead to biases (Feinstein, 1988), for instance, if one aspect of that advantage is a medicalized life style (in today's United States, one with adequate insurance), increasing the chances that any serious condition will be detected earlier than it otherwise might have been.

A numerical inference is justified only to the extent that two other hypotheses can be ruled out: that the putatively causal agent, or a proxy for it, is affecting the reported findings via measurement *error* or sample *nonrepresentativeness* in addition to, or in the worst case rather than, by some causal process entailing the measures per se.

L4.5.4.5 Adjustment for Confounds and Covariates, and Significance Testing
By this point we have arrived back at somewhat familiar territory. Measured outcomes, whether or not in keeping with the preceding four strictures, may well be susceptible to causal processes beyond the diseases that are the outcomes of interest, and the study design, whether experimental or not, may well have left these competing processes unbalanced over the dose measures of primary interest. We need corrections of all the reported "effects" of dose on outcome by reference to all of these other potential causes. In fact, because no randomization can be perfect, this logic applies equally to experimental and to nonexperimental studies.

The first thing an author needs to do is to list the potential confounding factors. The confound need not have been explicitly measured in each study to which it might apply; it is enough to be able to put an upper limit on its magnitude. For example, dietary carotene affects rates of lung cancer, and most studies of passive smoking do not measure dietary carotene. But there are studies that measure it in smokers and nonsmokers, and other studies that measure its effect on lung cancer; combining these studies, it became possible to place an upper limit on the possible confound of the computed relative risk of spousal smoking (about 0.01 of the 0.30 excess risk, according to Hackshaw et al., 1997: see the discussion at Section E4.6.2).

If there is no main effect, do *not* be tempted to produce one by adjustment. For passive smoking, the articles to be reviewed find detectable effects at realistic dose levels (namely, real life with a smoker). If there is no significant dose-response correlation, *a manuscript should never be able to persuade you there is a finding there just by "adjusting."* Instead, the authors will need to work a great deal harder – they will need to explicitly *control* for the other factors, by doing analyses within explicit subsets of the data. The apparent alternative, multiple regression, is **never** persuasive. We will demand a verification of the assumptions underlying the multiple regression analysis, and the study will not have done so; indeed, its investigators probably have never seen anyone ever do so. One of the reasons that multiple regression is the single most abused technique in the entire statistical toolkit is that its assumptions are never stated, let alone verified, in the articles that employ it. One advantage of the added-variable method, Section L4.5.3.2, is that it forces you to attend to at least one of those assumptions for every variable you claim to be important.

It is worthwhile to review these assumptions:

- Each predictor affects the outcome linearly and additively, and with a coefficient independent of the values of all the other predictors.
- The prediction error around the composite linear predictor is independent of all the predictors and has the same variance for every case in the data set.
- For any sort of generalization to a new sample, such as the causal imputation for a corresponding intervention study, each predictor variable must also be arguably a locus of true intervention or experimental control.

The situation is even worse for "stepwise" regressions, in which the list of predictors is free to vary. (The modern literature of stepwise regression consists almost entirely of monitory notes and severe warnings – see, for instance, Anderson, 2008: for a sophisticated exposition of the alternatives, see at Section 5.3.) One sometimes sees studies in which the effect of the putative dose is made visible only after adjustment for six covariates, or for dozens. Any claim based on regressions like these is absurd, in view of the literally hundreds of verifications of assumptions that have not been carried out. If there was a main effect there before the adjustment, please demonstrate it via scatterplots; if it was not there before the adjustment, please start over with a different study design.

There is considerable interest among statisticians in the predictable ways in which the three bulleted assumptions above usually fail and how the resulting multiple regressions or covariate adjustments can predictably mislead. For instance, a prediction that is actually a pure interaction – an effect only for underweight females, for instance – might be represented in a multiple regression as a combination of a significant effect for weight and a significant effect for sex, with no detection of the interaction that is actually the true signal. When true predictors are correlated, the multiple regression necessarily overestimates the difference of their separate coefficients, leading to problems in interpreting the signs of predictors (problems that were responsible for inventing the stepwise method in the first place).

This is important enough that it bears repeating: *no multiple regression is ever persuasive by itself.* No abductive force is generated by any multiple regression that finds an "effect" not already present in simpler data presentations, such as controlled subgroup scatterplots, or that goes unconfirmed by a wide range of consiliences across substantially different measurements or samples resulting in closely similar path coefficients. The defensible applications of multiple regression in environmental risk studies are instead in connection with *challenges* to a claimed effect, namely, the rival hypotheses of covariate effects instead. As a rhetoric for parrying that challenge, nevertheless, the multiple regression must be technically persuasive in light of its own assumptions. The larger covariate effects must correspond to roughly linear added-variable plots and cannot show substantial clustering by values of other predictors in or out of the model. One must state these assumptions, and check them to the best of one's ability.

It is obviously *far better* to choose as outcome variables candidates that are not too susceptible to effects of other causes than the teratogen in question; this makes the subsetting by those other causes much more promising and much easier to understand. For instance, when outcomes are formal composite scores, such as full-scale IQ, it is often the case that covariates affect the "general factor" underlying the individual scores. In that case, it is often helpful to produce a residual from the summary score for each original measure before analysis proceeds. When outcome measures are all measures of length or size, it is similarly useful to convert to the language of shape variables (ratios of size variables) prior to further statistical analysis; when diseases are multifactorial, choose an appropriate subsetting (before the analysis begins, of course). In either of these special cases, confounding remains possible, but its magnitude is likely reduced, rendering a study's main hypothesis more robust against noise and other explanations.

We now turn to two studies that, though using adjustments for confounding variables, appear to substantially meet these requisites, and so permit strong numerical inferences. While both studies report their statistics in the form of multiple regressions, the inferences follow from the discarding of all predictors except the one of interest (spousal smoking). Reduction of the multiple-predictor context to prediction by a single measurement is in general far more conducive to abduction (focused surprise) as well as consilience (across multiple studies, across the divide between animal experiments and observations on humans).

E4.6 On Passive Smoking

I review a pair of papers, consecutive in their original publication venue, that wrestle multiple regression around into a form that can be used to drive an abductive numerical inference. Each is a consilience, the first with an experimental literature and the second with an earlier epidemiological literature, and each one handles the abductive aspects arithmetically, by showing that consilience cannot be achieved for those other explanations – their statistical signal amplitude is just not large enough. Both papers emerge from all the arithmetic qualified to speak with remarkable authority about the inferences they have drawn, regarding a preventable risk affecting the public health. Taken together they are as good a pair of examples of the use of multiple regression to drive strong numerical arguments as I have ever seen.

Each of the two exploits in addition a novel form of consilience that was surely not available to Peirce and of which perhaps Wilson was unaware when he wrote *Consilience*. This is the approach to scientific quantification that is usually called *meta-analysis*, the explicit averaging or other arithmetic manipulation across studies of reported quantities that are supposed to be measuring "the same thing." In keeping with our underlying attitude of fundamental skepticism, there is a mechanism for querying that specific assumption (a χ^2 [chi-squared] statistic, to be mentioned briefly in connection with the Law exegesis), and, when it shows the presence of a signal, further analyses that assort those reported effect sizes according to characteristics of the study design at hand (nature of the sample, details of measurement, etc.). The consilience aspect here is inseparable from the explicitly social process by which articles out of different laboratories become clustered together via indexes and reviews. (It is therefore a formal response to the surprisingly risky practices of relying on fallible judgments and separate significance tests in summarizing those same lists of articles. See, in general, Cooper et al., 2009.) A somewhat different social mechanism for the same result, the collection of multiple numerical adumbrations of the same process, will concern us in Section E4.7 when we discuss alternatives to the Milgram approach of mega-experiments.

While meta-analysis is ordinarily encountered in the social sciences and the medical sciences, it is not limited to them for any logical reason. Edward Tufte, a political scientist turned graphic designer, published a beautiful book of scientific graphics 30 years ago (Tufte, 1983; there have been four more from him since) that included a collection of examples he considered superb. I have excerpted one of these in

Figure 4.41 because it represents a less frequently encountered style of meta-analysis that should not be overlooked. The chart shows the findings of over a hundred separate studies of one single physical property, the thermal conductivity of copper. As you see, different experimental preparations share an upper envelope that is the "recommended" curve. The meta-analysis is not of the *average* of the measurements at any given temperature, but of the *maximum* such measurement across the competent experiments. This would seem to be of considerable potential usefulness in such domains as "evidence-based medicine," where particular techniques can be expected to be shown effective only for patients with particular genotypes and for studies that rely on competent technicians for their laboratory measurements. I look forward to seeing this potentially compelling chart design in a biomedical science soon.

The examples in this section deal with health effects of passive smoking, and in the interest of full disclosure I should explain how they got to be mentioned in a broad-spectrum essay like this one. Seventeen years ago I was sitting in a faculty office in Ann Arbor, Michigan, minding my own business as an academic multivariate statistician teaching mostly precursors of the material in Chapter 7, when a phone call came in from an epidemiological colleague, Kenneth Warner, originally an officer working under the U.S. Surgeon General, and then a faculty member of the University of Michigan epidemiology department. (Today he is Dean Emeritus of their School of Public Health.) I was shortly going to be telephoned, Warner told me, by a Florida lawyer, Stanley Rosenblatt, who specialized in medical torts; he was a serious fellow; I should accept the call. I did so, and ended up applying my knowledge of Partial Least Squares (Section 6.4) in an unexpected context, a $5 billion civil suit by nonsmoking flight attendants of the 1970s against the major tobacco companies. And, as an academic mathematical statistician, I was actually called to testify in open court as an expert about the meta-analysis of environmental tobacco smoke – the quality of the government's reasoning in EPA (1992). From an AP press report that day in 1997:

> "I quickly realized that it [the EPA report] was one of the best examples I had ever seen of a careful, scientifically correct use of numbers in a public health question," said Bookstein, a researcher at the University of Michigan. Bookstein uses the EPA study in his classes. He tried to put his testimony in layman's terms but the jargon was weighty, including relative risks, confounders, and covariates....

The lawsuit was settled for $349 million in October of that year (Josefson, 1997), but probably not as a result of anything I said.

This episode whetted my interest in generalizing the forms of statistical reasoning I had been developing and teaching. Also, it stimulated my interest in the specific rhetoric of the adversarial system, which lies far outside the academy and where arguments are couched in an entirely different metric (billions of dollars) and an entirely different rhetoric (legal responsibility for singular events) than in peer-reviewed science. There is another mention of a forensic application, this time in criminal trials, in Chapter 7, or see Bookstein and Kowell (2010). Meanwhile, please keep in mind

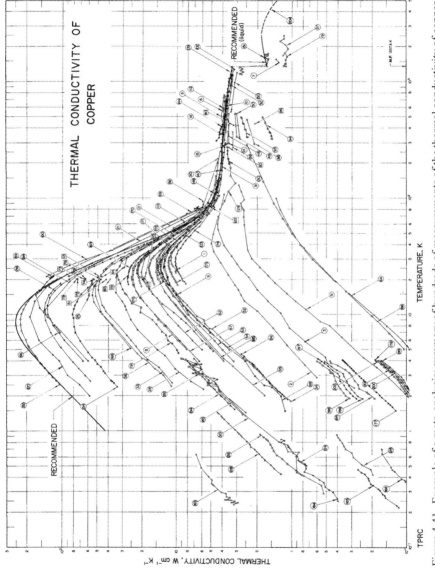

Figure 4.41. Example of a meta-analysis: summary of hundreds of measurements of the thermal conductivity of copper. (From Ho et al., 1974, p. 244, via Tufte, 1983.)

that the logic of abduction and consilience being foregrounded in this essay may be one of the more effective bridges between scientific and legal reasoning.

E4.6.1 Passsive Smoking and Heart Attacks

Law et al. (1997) have two objectives: to estimate the risk of ischaemic heart disease that is induced by environmental tobacco smoke (ETS), and "to explain why the excess risk is almost half that of smoking 20 cigarettes per day when the exposure is only about 1% that of smoking." (The word "ischaemic" means "lacking in blood supply," so that ischaemic heart disease is the condition in which some of the heart muscle is dead or dying from lack of oxygen.) The ultimate intention is to demonstrate that there is "no satisfactory alternative interpretation of the evidence reviewed here than that environmental exposure to tobacco smoke causes an increase in risk of ischaemic heart disease of the order of 25%," which is about the same as the effect of ETS on lung cancer but which accounts for a great many more deaths, inasmuch as there is much more heart disease around than lung cancer. "The effect of ETS is not trivial, as is often thought. It is a serious environmental hazard, and one that is easily avoided." So the authors' purpose is clearly persuasion, not mere reporting.

To carry this rhetorical point the paper is divided into no fewer than *five* separate studies, each quantitative but of a different methodology. A first study is of the genre known as meta-analysis: a summary quantification of 19 of the 20 previously published findings on the effect of ETS on ischaemic heart disease, concluding that the excess risk due to living with a smoker is about 30%. A second study, likewise a meta-analysis, reviews the dose–response studies of the effects of *direct* smoking on heart disease, and finds the risk of smoking one cigarette per day to be similar (about 39%, versus 78% for smoking a pack a day) and similarly disproportionate to the actual dose of any toxin. Some of the findings of these studies are excerpted in Figure 4.42.

Two separate analyses supply mutually consistent estimates of the component of this excess risk that might be attributable to concomitant causes instead of to the ETS per se – one study of the effect of stopping smoking on ischaemic risks, and another summarizing the effects on heart disease risks of the dietary differences known to differentiate smokers from nonsmokers; either of these amounts to about 6%. These substudies incorporate an admirable innovation adapting the method of multiple regression, Section L4.5.2, to this context of meta-analysis. It is presented most clearly in Table 2, pages 975–976, and the accompanying text. I reproduce this table as Figure 4.43. Recall from the lecture that we produce our estimated "direct effects" from raw regression slopes (like those in Figure 4.42) by subtracting products of the indirect effects on *other* variables by the direct effects of *them* on the ultimate outcome variable of interest.

There is no mandate that all these quantities be estimated using the same data resources. And so Law et al. construct these estimated adjustments from a meta-analysis of their own, which cleverly combines estimates from the literature of *both* of the numerical factors that multiply to give that adjustment: the product of the

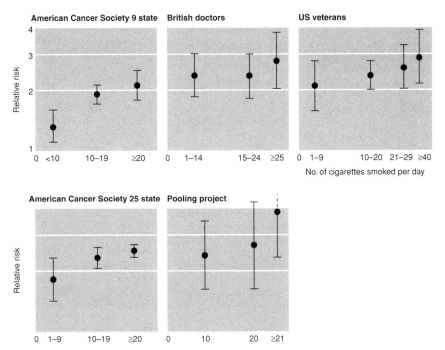

Figure 4.42. Relative risks of myocardial infarction by spouse's smoking, five studies. (After Law et al., 1997.) Republished with permission of BMJ Publishing Group, from BMJ 1997; 315, *The accumulated evidence on lung cancer and environmental tobacco smoke*, A.K. Hackshaw, M.R. Law, N.J. Wald; permission conveyed through Copyright Clearance Center, Inc.

difference in dietary consumption, smokers versus nonsmokers (and so presumably their spouses), columns 4 and 5 of the table, times the effect of such a difference in consumption on the risk of ischaemic heart disease, column 2 of the table. As these products are not themselves corrected for their own indirect paths via the direct effect of passive smoking, they may be overestimates. But there is no harm to the authors' argument in overcorrecting in this manner, as it results in even more confidence in the final quantity estimated, the (remaining) direct effect of passive smoking per se. Note also that the entire process of adjustment for all those fruits and vegetables, which are here computed separately and summed, is also confirmed as a whole (page 976, under "indirect estimate"), by a consideration of the process by which the excess cardiac risk reverses, namely, permanent cessation of smoking. Table 3, yet another meta-analysis, summarizes a range of such studies to come up with a remaining risk of 6%, which is attributed to behaviors prior to today, at a time when "dietary change was not widely advocated on health grounds". "Indirect" and "direct" adjustments are numerically the same: this is very reassuring (consilient) indeed.

The first four component studies serve mainly to set up the Peircean abduction – the "surprising fact" observed – that the excess risk associated with ETS goes far beyond what is proportional to the actual extent of smoke exposure. But, using Peirce's template, if the excess ischaemic risk were actually attributable to changes in platelet

Table 2 Estimates of increased risk of ischaemic heart disease in smokers relative to non-smokers, attributable to lower consumption of fruit and vegetables

Marker of consumption of fruit and vegetables	Relative risk of ischaemic heart disease for decrease in consumption of 1 SD	Difference in consumption (smokers minus non-smokers)			Estimate of relative risk of ischaemic heart disease	
		No of studies	Difference (proportion of 1 SD)			
			Median*	Largest	Median	Largest†
All fruit	1.16 (1.02 to 1.31)	5	−0.22	−0.33	1.03	1.09
All vegetables	1.23 (1.08 to 1.40)	6	−0.12	−0.25	1.03	1.09
Carotenes	1.06 (1.03 to 1.11)	9	−0.20	−0.34	1.01	1.04
Vitamin C	1.05 (1.00 to 1.09)	13	−0.24	−0.49	1.01	1.04
Vitamin E	1.05 (1.02 to 1.10)	6	−0.12	−0.27	1.01	1.03

* Median differences correspond to consumption lower by 10–15% in smokers than non-smokers.
† Based on upper confidence limit of relative risk estimate and largest difference between smokers and non-smokers.

Figure 4.43. The central table in Law and colleagues' (1997) reduction of their multiple regression to a simple regression, by estimating the maximum possible value of the correction for the indirect effect via diet. (After Law, 1997, Table 2.) Republished with permission of BMJ Publishing Group, from *BMJ 1997*; 315, *The accumulated evidence on lung cancer and environmental tobacco smoke*, A.K. Hackshaw, M.R. Law, N.J. Wald; permission conveyed through Copyright Clearance Center, Inc.

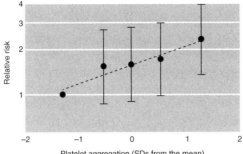

Figure 4.44. Resolution of the abduction: mapping the apparent jump of the low-dose myocardial risk to the evident jump in platelet aggregation at the same low doses. (After Law et al., 1997.) Republished with permission of BMJ Publishing Group, from *BMJ 1997*; 315, *The accumulated evidence on lung cancer and environmental tobacco smoke*, A.K. Hackshaw, M.R. Law, N.J. Wald; permission conveyed through Copyright Clearance Center, Inc.

aggregation, the ratio would no longer be surprising. In their fifth study, likewise a literature review, it is found that an increase of 1 standard deviation in platelet aggregation increases the risk of an ischaemic event by 33% and also that one effect of any single exposure to ETS upon platelet aggregation is to increase it by more than 1 standard deviation for at least a period of time. (See Figure 4.44.) Thus if the mechanism for the excess risk were platelet aggregation, the ratio of half (39% vs. 78%) would be expected as a matter of course. As the authors themselves summarize the situation, "so large an effect from a relatively small exposure, though unlikely on first impression, is supported by a great deal of evidence," including animal studies, studies of pipe and cigar smokers, and the like. In this way all three clauses of Peirce's syllogism are represented, and the abduction is closed consiliently.

A very interesting auxiliary component of this paper arises right at the end, where the authors find it necessary to justify the omission of one published study from the series that they convey to the meta-analysis. They need to vitiate the "surprise" of a study that found no effect at all of ETS exposure on the risk of heart attacks. For this purpose they combine every form of rhetoric that is at their disposal in this journal context. They point out that the study they propose to overlook was published by "consultants to the tobacco industry," who thereby have "a vested interest," and then raise three specific numerical criticisms: disagreement between the omitted studies and a reanalysis of the same data "commissioned by the American Cancer Society (the owners of the data)," inconsistency with any confounding from the dietary differences, and inconsistency with all of the other data reviewed in the course of the present paper. Having thus demolished the dissenting study, the authors can proceed (in the very next sentence) to the assertion that there is "no satisfactory alternative" to their claim of a 25% excess risk. In this way, the specific logic of a Peircean abduction based on quantification becomes conflated with general principles of propaganda, reminding us, if reminder were needed, that numerical inference is a species of rhetoric, of persuasive speech.

E4.6.2 Passive Smoking and Lung Cancer

Turn now to the next paper in the same issue of *BMJ*, Hackshaw et al. (1997), on the effect of passive smoking on the risk of lung cancer. This is actually the more classic topic within the passive smoking research domain, being the main subject of the U.S. Environmental Protection Agency's (EPA) report of 1992 that originally implicated environmental tobacco smoke as a carcinogen. That report enumerated other general domains of risk, such as childhood asthma, but emphasized cancer because at the time proof of carcinogenesis was the easiest road to regulation in the United States. (As I write, the EPA has just received congressional authority to regulate greenhouse gases as well, such as carbon dioxide; it certainly did not have that power in 1992.)

Like the situation for the paper by Law et al., the paper by Hackshaw et al. begins with a meta-analysis of the existing literature pertinent to the central consilience to follow. This comes in the form of the attractive graphical presentation at the left in Figure 4.45: a plot of the estimated relative lung cancer risk (to the nonsmoking wife) of living with a smoking husband as a function of the year through which the literature was tabulated. This image of an "inverted Christmas tree" powerfully communicates all three of the following propositions: there is a central tendency across replicated studies, perhaps indicating a true "underlying" risk level; it shows no secular trend (to the time scale of the underlying papers); and the precision with which this value is estimated is improving over time. (The χ^2 statistic I mentioned earlier flags one of the papers in the meta-analysis as being inconsistent with the other 36: a study showing an "improbable" protective effect of cigarette smoking, in which the true effect "was probably obscured by another cause of lung cancer, indoor cooking using open coal fires with little ventilation.")

The consilience will be between the risk implied by the Christmas tree, amounting to an extra 24%, with the risk implied by a *linear* extrapolation down toward zero of the fairly well-established dose–response dependence of lung cancer on smoking for active smokers, as shown at the right in Figure 4.45 for one study deemed by these authors to be of particularly high quality. In a fine demonstration of the "interocular traumatic test" (Edwards et al., 1963; see Epigraphs), the regression line fitting these points obviously would pass through the point (0,1) of no additional risk at log dose level zero. We need not bother with the arithmetic of actually fitting a regression line. (This is another example of the rhetorical "power of zero" that we have already seen in connection with the infectious origin of ulcers, Section E4.4.)

Hackshaw et al. go on with some technical adjustments of the underlying estimate of 24%. As already noted in connection with the article by Law et al., this arithmetic constitutes the production of an estimated "direct effect" d by explicit subtraction of a series of estimated indirect paths using regression slopes or group differences collected knowledgeably from the literature (via small meta-analyses of their own). One of these adjusts for the (known) frequency of lying about being a smoker in these studies; that adjustment would reduce the relative risk to about 18%. A further adjustment for the known additional effects on lung cancer of the dietary differences between smokers and non-smokers yields an upper limit for additional adjustment

E4.6 On Passive Smoking

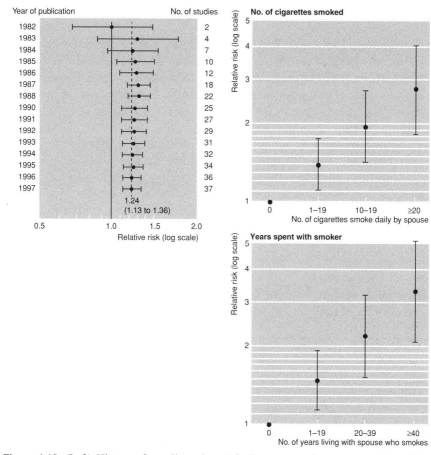

Figure 4.45. (Left) History of a policy-relevant finding: range of summary relative risk of female lung cancer by spouse's smoking, 37 studies, 1982 through 1997, by year. (Right) Contrast with Figure 4.42: two dose–response curves, lung cancer risk by spousal smoking, that show the appropriate (potentially consilient) smooth rises from 1 (no excess risk). (After Hackshaw et al., 1997.) All bars are 95% confidence intervals. Republished with permission of BMJ Publishing Group, from *BMJ 1997*; 315, *Environmental tobacco smoke exposure and ischaemic heart disease: an evaluation of the evidence*, M.R. Law, J. K. Morris, and N.J. Wald; permission conveyed through Copyright Clearance Center, Inc.

of about 2% (one standard deviation of a dietary shift alters the risk by about 20%, from one group of studies, and families with smokers differ from families with nonsmokers by about 0.12 of a standard deviation, from another group of studies). A final adjustment has the effect of cancelling out both of these, by noting that the spouses of nonsmokers nevertheless have exposure to smoke elsewhere than at home. Judging by measurements of urinary cotinine (a metabolic endproduct of nicotine, as well as an anagram), they are exposed at the rate of roughly one-third of the wives of smokers, and so the true effect of the passive smoking, relative to a baseline of zero exposure, is 1.5 times 18%, or (after the dietary adjustment and after all computations have been carried out to one more significant digit) 26%, just about where we started.

The principal abduction of this article by Hackshaw and colleagues now follows on pages 985–986. If the effect of smoke on lung cancer risks is linear (and it looks as if it is, from the right side of Figure 4.45), then the excess risk we expect in the nonsmoking wife of a smoker ought to be computable by simple arithmetic. You would simply multiply the *known* excess risk of lung cancer in active smokers (about 1900% – the rate of lung cancer in smokers is roughly 20-fold that in nonsmokers) – by 1%, a factor corresponding to the urinary cotinine concentration in exposed nonsmokers relative to that of smokers. (This number itself is the summary of values from seven different studies – another small meta-analysis.) There results an "expected" relative risk of 19%, compared to the "observed" 26%. "The indirect (19%) and direct (26%) estimates of excess risk are similar," the authors conclude, in an explicit demonstration of Kuhn's argument about the treatment of "reasonable agreement" in scientific publications (recall Section 2.3). In summary, "all the available evidence confirms that exposure to environmental tobacco smoke causes lung cancer," and "the estimated excess risk of 26% corresponds to several hundred deaths per year in Great Britain." The article's own abstract makes the point in a sharper rhetoric that highlights the underlying role of the abduction in all this:

> The epidemiological and biochemical evidence on exposure to environmental tobacco smoke, with the supporting evidence of tobacco-specific carcinogens in the blood and urine of non-smokers exposed to environmental tobacco smoke, provides compelling confirmation that breathing other people's tobacco smoke is a cause of lung cancer. (Hackshow et al., 1997, p. 980)

The crux here is the word "compelling," which conveys the force of the abduction as the authors hope it will be felt by a disinterested reader (in this case, anybody except a tobacco company lawyer).

Note also the use of linearity as a specialized explanatory trope. In the analysis of the heart attack data of Law et al. (Section E4.6.1), the whole point was the departure from linearity of the risk at low doses of smoke. Rather, the authors had to find a quantity (platelet aggregation) that *was* linear both in dose and in induced risk, and channel the argument through that mechanism. This is not to claim that processes in the real world need to be linear, but that statistical computations involving explanations via linear regression must be. Recall the argument in Section L4.1 that a regression coefficient is, at root, just a type of weighted average. For that average to make sense, the cases over which the average is taken must be homogeneous. This means that the general method exploited here cannot be expected to apply to systems in which the true explanation is characterized by bifurcations, or catastrophes, or other intrinsically nonlinear phenomena (see Tyson et al., 2003) unless, as in the analysis by Law et al., the crucial nonlinear step is identified, cut out of the process, and reduced to linearity by explanations based on auxiliary data sets.

L4.6 Plausible Rival Hypotheses

The foundation of the method of abduction begins with surprise, and surprise begins with the incompatibility of your data with the competing hypotheses. For this launch to be effective there needs to be some discipline for systematically surveying those competing hypotheses. Statisticians are taught several of the "standard modes" by which statistical pattern analyses can fail in practice (heterogeneous data, heterogeneous noise, secular trends, instrument calibration failures, and so on), as just reviewed in Section L4.5.3, but these are relatively abstract properties of the numerical data resources. There is also a suite of less abstract properties of the alternate explanations, properties that have been most usefully listed, perhaps, in a handbook chapter by Donald Campbell and Julian Stanley that is fifty years old. References for this subsection include that chapter (Campbell and Stanley, 1963/1966), Campbell and Ross (1968), and Cook and Campbell (1979).

Campbell, his colleagues, and I are all dealing with the issue that is called, over in their discipline, *validity versus invalidity* of an inference from a study of any sort. *Internal invalidity* is a matter of the extent to which other explanations than one's preferred one could explain the same data; *external invalidity*, the extent to which a finding might be so specific to a particular inferential setting as to be uninterpretable outside that domain. Their approach fits quite well under the rubric of abduction that underlies this book, as one's own hypothesis must explain the data better than anybody else's can, and so a systematic search through such forms of invalidity is a helpful adjunct toward strengthening one's own abductions if they are valid at all.

The particular type of study that Campbell and Stanley are interested in involves series of measurements over time. At some moment something happens – some intervention is carried out – following which the series of measurements continues to accumulate. *Arithmetically*, there are a variety of patterns of the resulting time series (or group difference, etc.), of which my Figure 4.46 provides an interesting summary of the range of these numerical possibilities. Most represent forms of *internal* validity or invalidity, as they refer to arithmetic internal to the data record itself. The more subtle and therefore more interesting forms of invalidity are the *external* ones, having to do with the possibility of quite different interpretations. The topic of the Campbell–Ross (1968) discussion, for instance, is a drop in highway death rates in the state of Connecticut in 1956 following the imposition of a change in the legal handling of speeding tickets. The first step is naturally to place the change in its actual historical context, the time series in the solid line in Figure 4.47. We notice that the rise from 1954 to 1955 was actually greater than the decline from 1955 to 1956. This interferes with an abduction in two ways: first, to suggest that if the concern was for magnitude of change from year to year, then the explanandum should be the rise, not the drop; second, to suggest that the passage of the law might have been a consequence of the previous year's rise and thus just a phenomenon of "regression to the mean" as reviewed in Section L4.3. A survey of experience in adjacent states, likewise in Figure 4.47, finds another state with the same profile – did Rhode Island, too, pass a new speeding law?

264   4 The Undergraduate Course

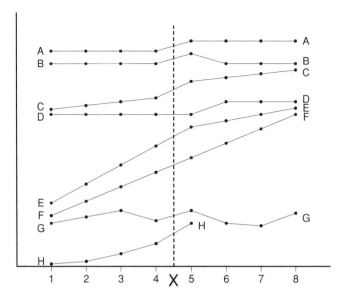

Figure 4.46. Eight contexts for the same year-on-year change, differing widely in their support for an inferred effect. (After Campbell and Stanley, 1963.) Campbell and Stanley comment on this range of patterns as follows: "The legitimacy of inferring an effect varies widely, being strongest in A and B and totally unjustified in F, G, and H."

It is also quite sensible, in the course of any abduction, to scan for other effects that might appear to be stronger than the effect about which one was originally asking. Campbell and Ross's Figures 5 and 6 (my Figure 4.48) show two such effects, each one more obvious than the effect on highway deaths. Note that all four have the same

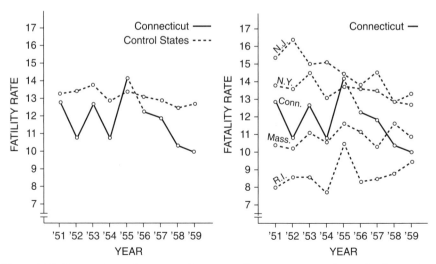

Figure 4.47. Two versions of a comparative context for a year-on-year change (in Connecticut highway deaths). (After Campbell and Ross, 1968.)

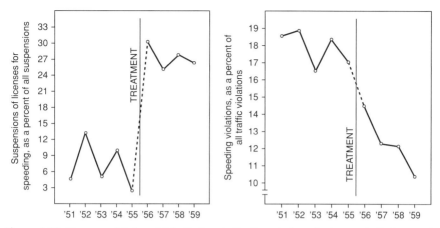

Figure 4.48. Two associated social indicators indicating obvious (but unwanted) changes in associated social systems. (After Campbell and Ross, 1968.)

order statistic, with a p-value of about 0.01. By comparison, the p-value of the time series interruption in the original units, highway deaths, is only about 0.10, strongly suggesting that we are at risk of misrepresenting the process under discussion when the discussion degenerates into a focus on NHSST in this way.

Campbell and colleagues go on with a list of many different generic plausible rival hypotheses that can arise whenever one is examining an uncontrolled social "experiment" in this sense, what they call a *quasi-experimental design.* These threats to the validity of your preferred interpretation, operative over a wide range of contexts across the social and biological sciences, include

- *History*, other things that happened at the same time as the intervention
- *Maturation*, ordinary processes cued to the passage of time
- *Testing*, the simple effect of being measured twice
- *Instrumentation*, meaning change in a machine, a bureaucratic form, a technology, a definition, or a questionnaire within the measurement series
- *Regression toward the mean*, when the timing of an intervention is suggested by the earlier value itself ("children of tall parents," Section L4.3)
- Differential *selection*, whereby the comparison groups are sorted for a criterion correlated with the measured outcome
- *Mortality*, here meaning loss of respondents that is differential by treatment rather than totally at random

Other disciplines offer a similar range of preformulated rival hypotheses (sources of nonsurprise, interceptions of abduction) in their own vocabularies. A typical guide for reviewers of medical and epidemiological research, for instance (Elwood, 1998) lists 15 of these (his Table 9.1, page 219), of which three are rival hypotheses in the sense here (conditions interfering with the *major* premise of the abduction) – "Are the results likely to be affected by observation bias?/ confounding?/ chance variation?" – while five others are aimed at weakening the *minor* premise of the abduction, the strength with which the data support *your* hypothesis – "Is there a correct time relationship? Is the relationship strong? Is there a dose–response relationship? Are the results consistent within the study? Is there any specificity within the study?" A final four deal specifically with consilience: Are the results consistent with other evidence, particularly from study designs at least as powerful? Does the total evidence suggest any specificity? Are the results biologically plausible? For a major effect: is the account consistent with the distribution (over time and geography) of exposure and outcome? Note the omission of the kind of consilience that is specifically arithmetical (numerical) from the rhetoric considered by Campbell and colleagues, corresponding to its relative absence from their area of application, which is public policy studies.

E4.7 Milgram's Obedience Experiment

The main resource for this example is Milgram, 1974.

The setting of this research is a drama, with an unwitting actor on whom a script is imposed without his prior awareness or consent, and with the researcher Stanley Milgram as dramaturg. The script varies; the study is of the actor's response. Here's Milgram's own plot summary:

> Two people come to a psychology laboratory to take part in a study of memory and learning. One of them is designated as a "teacher" and the other a "learner." The experimenter explains that the study is concerned with the effects of punishment on learning. The learner is conducted into a room, seated in a chair, his arms strapped to prevent excessive movement, and an electrode attached to his wrist. He is told that he is to learn a list of word pairs; whenever he makes an error, he will receive electric shocks of increasing intensity.
>
> The real focus of the experiment is the teacher. After watching the learner being strapped into place, he is taken into the main experiment room and seated before an impressive shock generator. Its main feature is a horizontal line of thirty switches, ranging from 15 volts to 450 volts, in 15-volt increments. There are also verbal designations which range from **slight shock** to **danger – severe shock**. The teacher is told that he is to administer the learning test to the man in the other room. When the learner responds correctly, the teacher moves on to the next item; when the other man gives an incorrect answer, the teacher is to give him an electric shock. He is to start at the lowest shock level (15 volts) and to increase the level each time the man makes an error.
>
> The "teacher" is a genuinely naïve subject who has come to the laboratory to participate in an experiment. The learner, or victim, is an actor who actually receives no shock at all. The point of the experiment is to see how far a person will proceed in a concrete and measurable situation in which he is ordered to inflict increasing pain on a protesting victim. At what point will the subject refuse to obey the experimenter?
>
> Conflict arises when the man receiving the shock begins to indicate that he is experiencing discomfort. At 75 volts, the "learner" grunts. At 120 volts he complains verbally; at 150 he demands to be released from the experiment. His protests continue as the shocks escalate, growing increasingly vehement and emotional. At 285 volts his response

can only be described as an agonized scream.... Each time the subject hesitates to administer shock, the experimenter orders him to continue. To extricate himself from the situation, the subject must make a clear break with authority. The aim of this investigation was to find when and how people would defy authority in the face of a clear moral imperative. (Milgram, 1974, pp. 3–4)

E4.7.1 Calibration as Consilience

The nature of those "orders to continue" is itself scripted. The experimenter uses the following "prods" whenever the subject indicates a wish not to proceed or seeks advice on whether to proceed:

Prod 1. Please continue, *or* Please go on.
Prod 2. The experiment requires that you continue.
Prod 3. It is absolutely essential that you continue.
Prod 4. You have no other choice, you *must* go on.

If Prod 4 is unsuccessful, the session is ended; otherwise it continues until the subject has pressed the switch at 450 volts three times in a row.

The drama of this setting is, as Milgram himself acknowledges, difficult to convey in print. There is a film of this experiment, and it is gripping (as is the film of the Stanford Prison Experiment): you are watching some innocent adult citizens of New Haven as they suffer without understanding. The fake shock generator, labeled *Shock generator, type ZLB, Dyson Instrument Company, Waltham, Mass., output 15 volts–450 volts*, is now in the Smithsonian Institution, in the company of Julia Child's kitchen. This study of Milgram's is, in my judgment, the best, most informative, most startling, and most compelling psychological experiment of all time. And it takes the form of an abductive consilience, precisely along the lines this book is recommending.

In the *first* condition, there is no vocal complaint from the victim. He is in another room where he cannot be seen, nor can his voice be heard, only his answers. At 300 volts he pounds the wall; after 315 volts, he is not heard from again. Twenty-six of 40 subjects obey the experimenter's orders to the end. Although this is the finding most people remember if you say "the Milgram experiment," the infamous 65% obedience rate, it is not the main point of the study. There are 18 more variations to come.

In the *second* condition, subjects could hear the victim. As Figure 4.49 shows, 8 subjects became disobedient when these protests began, and another 7 around the time they ceased, leaving 25 who still went all the way to the end of the voltage scale. (This condition is the one summarized by Milgram in the text extract that opens this section.)

In the *third* condition, the subject was in the same room as the victim, both visible and audible. Ten of 40 disobeyed at the first substantial protest, and 16 went to the end of the scale. In the *fourth* condition, the subject must force the victim's hand onto the plate. Now 16 subjects become disobedient at the level of first protest, whereas 12 go to the end of the scale.

Maximum Shocks Administered in Experiments 1, 2, 3, and 4

Shock level	Verbal designation and voltage level	Experiment 1 Remote (n = 40)	Experiment 2 Voice-Feedback (n = 40)	Experiment 3 Proximity (n = 40)	Experiment 4 Touch-Proximity (n = 40)
	Slight Shock				
1	15				
2	30				
3	45				
4	60				
	Moderate Shock				
5	75				
6	90				
7	105			1	
8	120				
	Strong Shock				
9	135		1		1
10	150		5	10	16
11	165		1		
12	180		1	2	3
	Very Strong Shock				
13	195				
14	210				1
15	225			1	1
16	240				
	Intense Shock				
17	255				1
18	270			1	
19	285		1		1
20	300	5*	1	5	1
	Extreme Intensity Shock				
21	315	4	3	3	2
22	330	2			
23	345	1	1		1
24	360	1	1		
	Danger Severe Shock				
25	375	1		1	
26	390				
27	405				
28	420				
	XXX				
29	435				
30	450	28	25	16	12
	Mean maximum shock level	27.0	24.53	20.50	17.88
	Percentage obedient subjects	65.0%	62.5%	40.0%	30.0%

* Indicates that in Experiment 1, five subjects administered a maximum shock of 300 volts.

Figure 4.49. Milgram's first four experiments: table of outcomes (level above which subjects became defiant, if they did). (After Milgram, 1974, Table 2.)

I will not review all of the other conditions enumerated in Milgram (1974), but here are a few more. In the *seventh* condition, the experimenter leaves the room; 9 of 40 subjects go to the end of the scale. In the *ninth* condition, the learner emphasizes that he has a heart condition; 16 of 40 go to the end. In the *eleventh* condition, the subject chooses a shock level; now the subjects average only 82 volts, and only one goes to the end of the scale. In condition *fifteen*, there are two experimenters, who disagree; no subject goes past voltage level 165. In condition *eighteen*, the subject merely checks correctness of the "answer," while somebody else "administers" the shocks; in this condition, 37 out of 40 subjects watched as the stooge went to the end of the shock generator scale.

Choice of a quantity for reporting. It is helpful to examine the table in Figure 4.49. This is one of five that together convey the complete data record of the entire 636-subject project. It is an interesting exercise in "propaganda" (to use the word Milgram applied to one of his scripts) to examine the reduction of information applied to these charts prior to their overall summary. As you see, what is tabulated is a count, the level of putative shock at which the subjects who disobeyed decided to disobey. As Milgram notes, the rows of tables like these with the largest total counts are the "150-volt" level, at which the "learner" first seriously objects, and then the 300-volt level, at which "learners" scream in agony, or the adjacent 315-volt level, at which they cease to emit any signals at all. If the subject is to be forced to make a completely unanticipated moral decision, these are the levels at which the moral force has most obviously intensified. Milgram chose, however, not to report at this level of detail in his summaries or his discussion, but to simplify the data greatly by a numerical summary of the data in each column. Two such summaries, "mean maximum shock level" and "percentage obedient subjects," appear at the bottom of the table. For instance, in experiment 1, the setup explained a few pages earlier, the average shock level achieved was 405 volts, and 26 out of 40 subjects went all the way to the end of the machine's scale. In contrast, in experiment 4, where the subject had to press the learner's hand onto the shock plate, the average peak shock inflicted was 268 volts, and only 30% (12 subjects) went to the end of the scale.

Milgram chose the second summary statistic, the percent who were obedient to the end, as the quantity highlighted in all his scientific communications. Why? In a preliminary survey, people stated that surely hardly anybody would go up the voltage scale beyond 150, the point of the learner's first objection, and surely no sane person would go right to the end of the scale. The mean expected maximum shock level in these pilot surveys was 120 volts (versus that actual 270) and the mean expected "percentage of obedient subjects" was about 0.1%, one in a thousand, corresponding, perhaps, to informants' estimates of the percentage of actual sociopaths in New Haven that year. Either way, the results were far more discrepant from these expectations than Snow's Lambeth-to-Southwark ratios. But the mean maximum shock level is an explicit function of an accidental aspect of study design, the printed upper limit of the scale on the mock shock generator, in a way that the percentage of obedient subjects is not. No defiant subject chooses to defy past the 375-volt level, and so the report of percentage obedience seems much more *generalizable* than that of mean maximum shock; it is likelier to support a powerful abduction referring to the underlying question of interest (obedience, not voltage).

Force of the abduction. Milgram's original finding – that in experiment 1 a full 65% of his subjects remained obedient right up to the end of the session – was shocking (no pun), even to him – but that was not the main thrust of his study. His intention was to demonstrate *control* of this percentage by manipulation of the script being followed by the "experimenter" and the "learner." Figure 4.50 summarizes the whole study, all 19 experiments, by sorting along this single dimension. Confirming the discussion of graphical rhetorical force in Chapter 3, it illustrates that power in a particularly poignant way. (This is ironic inasmuch as the figure is not in fact Milgram's; it is from Miller [1986], a book written after Milgram's premature death.)

The aspect of consilience that is conveyed visually in this chart is conveyed verbally in Milgram's own discussion. To his way of thinking, there is a latent dimension of the "authority of the experimenter" that can be manipulated by Milgram (the true experimenter). The consilience here is not with other data but with the reader's subjective understanding of what it must have been like to be a subject in such an experiment. At one end, the script could be designed to impose no moral cost for an act of disobedience (e.g., experiment 12, "learner demands to be shocked," or experiment 17, where three "subjects" have to agree on the shock and the two stooges appear disobedient). At the other end of the scale, in experiment 18, "a peer administers shocks," all a subject had to do was stand by while another ostensible subject ran the shock generator; in that condition, nearly all subjects remained obedient. Experiment 1, the most widely publicized, appears to be one of six (out of 19) that generated roughly the same two-thirds rate of obedience. Consilience here, in other words, is via a *subjective mental assessment* of the "force of the situation," which is, after all, a metaphor, not a quantifiable concept. But one might reverse the logic, and treat the Milgram percentage as the appropriate quantification of this "force."

Ethical aspects. One indication of the importance of Milgram's study is its continuing effect on ethical reasoning in the arena of psychological research. Almost as soon as the original (one-condition) study was published, a San Francisco psychologist, Diane Baumrind, attacked it from an ethical point of view (Baumrind, 1964), claiming that Milgram risked severe psychological damage to his subjects by inducing irresolvable moral stresses and self-loathing in the course of the scripted sessions. (Today we might call this "unwanted knowledge," specifically, unwanted self-insight: demonstration of the evil that ordinary people like ourselves are capable of.)

In the course of his 1974 book, Milgram attempted a defense of his own procedures. In an Appendix, "Problems of Ethics in Research," he justified his standards as follows:

> All subjects received a follow-up questionnaire regarding their participation in the research.... 84% of the subjects stated they were glad to have been in the experiment, 15% indicated neutral feelings, and 1.3% indicated negative feelings.... Four-fifths of the subjects felt that more experiments of this sort should be carried out, and 74% indicated that they had learned something of personal importance as a result of being in the study.... In my judgment, at no point were subjects exposed to danger and at no

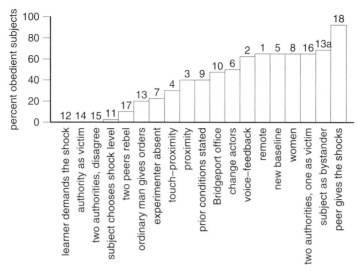

Figure 4.50. Chart of defiance rates, all 19 Milgram experiments, emphasizing the abduction (complete control of the rate of obedience by the script). (After Miller, 1986, Figure 3.3.)

point did they run the risk of injurious effects resulting from participation. If it had been otherwise, the experiment would have been terminated at once.

The central moral justification for allowing a procedure of the sort used in my experiment is that it is judged acceptable by those who have taken part in it. Moreover, it was the salience of this fact throughout that constituted the chief moral warrant for the continuation of the experiment. (Milgram, 1974, pp. 195, 199; italics in original)

Zimbardo (2007) reviews the Stanford Prison Experiment from this point of view, helpfully pointing out that he himself had completely missed the clues about ongoing damage to his subjects until his research assistant (eventually to become his wife) called the situation to his attention.

Milgram's after-the-fact logic – "let *me* decide about the risks here" – was thoroughly rejected in the course of the Belmont Report of 1979, the current guide to research involving humans, which emphasizes "informed consent, beneficence, and justice" and led shortly to codification of the rules that all American researchers today are legally mandated to use when running any study involving observations or measurements of humans. The investigator is not qualified to assess issues like these on behalf of the subject, for he is biased by the brilliance of the design and the power of the enlightenment potentially to be achieved. The risks must be assessed prior to the beginning of the study, by a committee of disinterested evaluators who are not all scientists nor even all academics. Subjects (not only experimental subjects exposed to some intervention, but also anybody else who is a source of data via medical records, questionnaire, etc.) must give advance permission, in writing, for their exposure to any foreseeable risks they may be running. Their after-the-fact judgment of whether the study was "worthwhile," which was Milgram's main defense in the 1974 book, plays no role in this a-priori evaluation. In other words, not only was Milgram's published reasoning discredited, but the rebuttal of his argument has been institutionalized in a massive bureaucracy. It is partly due to the sheer power of Milgram's finding, and the possibility of unwanted knowledge it conveys, that the Belmont Report's rules have universally been extended to the behavioral sciences as well as the biomedical sciences. The upshot, then, is that Milgram's seminal work brought about its own prohibition on replication.

The irony here is particularly painful inasmuch as the impact of the work is inseparable from the profundity of the issues that it taps, about the meaning of moral behavior and the authority of the man or woman in the white coat.[14] Contemporary research into moral reasoning consists primarily of studies of verbal (Colby and Kohlberg, 1987) or neuronal (Greene, 2005) responses to moral dilemmas like the story of Heinz and the druggist or the Trolley Problem. (For an interesting overview of one unexpected social context for this work, see Rosen, 2007.) No sufficiently robust substitute has been found for Milgram's visible and tangible realization of a real moral choice in the laboratory, and some (including myself) would argue that the protocols that protect subjects from undesirable self-insight have become overextended. The argument would be a lot easier if it were the tradition in social psychology to produce conclusions as powerful as Milgram's, and so the general abductive weakness of this field is partly to blame for the restriction of its own research possibilities.

E4.7.2 Critique of Pseudoquantitative Psychology

The force of Milgram's experiment is easier to appreciate when it is contrasted with more typical studies from the same discipline, experimental social psychology. Consider, for instance, a finding of more typical magnitude from the same discipline and indeed from the same domain of application, military psychology. Figure 4.51 is a summary of the empirical meta-analytic side of an otherwise brilliant methodological critique, Greenwald et al. (1986). Most social-psychology study designs are fatuous in that they systematically do *not* attempt to determine analogues of "experimental force" (which is no less dead a metaphor today than it was in Karl Pearson's time; see Section 7.8) for producing the conditions of greatest interest to them (violent behavior, etc.). The actual topic of this discussion is the so-called "sleeper effect," the tendency of a second bit of propaganda variously to enhance or attenuate the effect of a first dose of propaganda after some time lag. The original context of this research was a major study (Hovland et al., 1949) of the effect of propaganda on members of the U.S. military during World War II (and thus neatly aligned with Milgram's particular policy concern, the behavior of ordinary citizens in collaborating with Nazi exterminators during the Holocaust). Anthony Pratkanis, whose work is being

---

[14] Zimbardo himself, whose Stanford experiment produced the same behavior in students *who had studied Milgram's paper*, noted that his own experimental outcome perfectly predicted the behavior of U.S. captors at the Iraqi prison Abu Ghraib in 2004, a connection denied by the overseers of that prison system (Zimbardo, 2007).

summary of "sleeper effect" studies

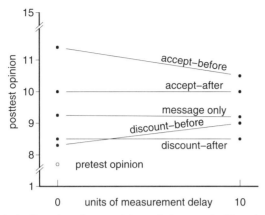

Figure 4.51. Typical effect sizes for a social psychology study. Note that the change scores range from −1.0 to +0.5 on a 15-point scale. (After Greenwald et al., 1986.)

criticized here, was one of the authors of the critical article from which this figure is extracted.

The *superficial* message is the difference in slopes among those various line segments, which subdivide the experiments being summarized by the type of propaganda sequence involved. The *real* issue here is indicated only indirectly by those paired hash marks \\ on the vertical axis: the effects reported here ranged from +0.5 units to −1.0 units *on a scale ranging from 1 to 15*. If the results are meaningful at all, it can be solely in terms of statistical significance. Recall that Milgram's results span virtually his entire scale, from 0% to 93% obedience. In contrast, in Fig. 4.51 neither consilience nor abduction is numerically possible – no other domain predicts the magnitudes of these differences, and hardly any theory is incompatible with them (in view of their very low amplitude).

It would be straightforward to extend this critique to a more general overview of styles of numerical inference that are common in today's social sciences. Such an argument, which if taken to any length would divert us from the main theme of this book, would emphasize the ways that numerical reasoning in these domains, because of its nearly universal reliance on significance testing, typically circumvents both the abduction and the consilience steps of the methodology I am foregrounding. The following examples were chosen from a much broader universe of possible choices.

Multiple outcomes with no focus of measurement. Richard Nisbett and Dov Cohen (1996) report the results of multiple related experiments dealing with differences of stereotyped behaviors between American students from different geographical regions. In another book, Nisbett (2004) similarly accumulates multiple experimental findings related to a similarly generalized account of differences between Western and Asian cognitive and cultural stances. In both cases the narrative consists of a collection of reports of statistically significant differences, of magnitudes on the order of those in Figure 4.51, for a wide range of studies none of which can be regarded as randomized over the principal explanatory factor (namely, culture of origin). Hence the narratives of books like these, while intriguing and suggestive, and a rich source of anecdotes for use in conversations about politics, do not admit of either abduction or consilience. The accumulation of multiple studies does not actually permit any greater level of surprise than the first one or two alone, and the agreement among them is solely qualitative, not consilient with any assessments of these cultural differences by other channels. There results a somewhat totalizing claim – in a great variety of domains, from mother–child interactions to professional conferences, it might be possible to find traces of these cultural differences – but it is almost impossible to forecast in advance which effects would be larger and which smaller, which more reliable and which less, which subject to introspection or intervention (as by behavior modification regimens) and which not. The results, then, are not in any format that can be treated as the target of a quantitative challenge, and, *for that reason alone*, are not effective as a source of productive new hypotheses or cross-disciplinary speculations.

We can usefully contrast this sort of research accumulation with the far sharper sequence of studies recently reported by DeHaene (2009), on the neurobiological origins of a recently evolved but nevertheless exclusively human behavior, namely, reading. From within the interdisciplinary domain of cognitive neuroscience, De-Haene narrates a surprisingly intricate sequence of strong inferences that results in a succession of consiliences between animal studies and human studies and, within human studies, between studies of normal people and studies of dyslexics, studies of alphabetic languages and studies of pictographic languages, and studies of perception against studies of neural circuitry. Abduction is managed by the design of disconfirming experiments, not significance tests, and consilience is across species, human competences, or levels of analysis, depending on the nature of the pattern claim. In my view the result superbly exemplifies how one builds a methodology by filtering ordinary curiosity through the possibilities of an integrated science that permits quantifications from below even when some of the data, such as word recognition, is molar (holistic) from above. In this extended series of studies, consilience and abduction are everywhere, and the resulting inferences (e.g., about the comparative costs in time for reading different kinds of nonwords) count as the sort of strong numerical inferences on which this book looks favorably.

Multiple theories that never confront one another. An otherwise fine recent book about evolutionary psychology, Agustín Fuentes's *Evolution of Human Behavior* (2009), tabulates some 38 different theories of the evolutionary origins of human behavior patterns, but never refers to any studies in which some of them were rejected in favor of others. Abduction is thus made inaccessible: every theory explains *something* – every study listed is, in effect, just another novel out of Balzac's *Comédie Humaine*, another description on the hearing of which one can only shrug and mutter, "That could happen." To have 38 theories of human behavioral evolution is no better than to have none at all. In the absence of any methodology of consilience

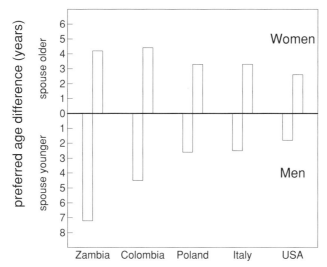

Figure 4.52. Typical evolutionary-psychological analysis of ostensible human "universals." No explanation is proffered of the inconstancy of these contrasts. (After Buss, 1993, Figure 6.)

based in neurological or other exogenous domains, the mass of studies, even though individually quantitative, amount to a mere narrative summary of human possibilities. No meta-analyses are possible, nor any generalizations.

A similar critique can be mounted against cross-cultural studies of human behavioral traits or reported preferences, such as are reported in, for instance, Buss et al. (1990) or Buss and Schmidt (1993). There is no real possibility of either abduction or consilience regarding the cultural differences being displayed here. It is from textbooks that we learn the accepted methodology of a field, and so it is appropriate to notice how, for instance, Buss (2009), quoting Figure 6 of Buss and Schmidt (1993), asserts that in several different cultures men prefer to marry younger women, and women, older men (Figure 4.52). But the age differences are not equal and opposite within cultures, nor are they the same across different cultures. The textbook offers little discussion of these variations, so any associated inferences cannot be numerical ones: they replace the numerical signal by just a couple of bits, the signs of the sex-specific average differences. Likewise counts of cultures that share particular traits, or crosstabulations of one custom against another (e.g., uncle/niece incest taboo [yes/no] vs. matrilineality [yes/no]), after the fashion of the old Human Relations Area Files on which I was trained at William James Hall of Harvard University, do not conduce to numerical inference. Absent possibilities of either abduction or consilience, we cannot use them to delve deeper into the processes they purport to describe. In settings like these, bar charts or contingency tables do not actually support numerical inference, but only pretend to do so by virtue of the p-values against the null hypothesis that accompany them. They have not thereby become quantitative scientific arguments.

It is useful to compare Buss's approach to presentations involving numbers with that of Frank Livingstone in Section L4.4. Livingstone was likewise concerned with evolutionary origins of observed data. He found a pattern of variation in the frequency of the sickle cell trait over the map of Africa, and found that the endemicity of malaria accounted for some but not all of the details of this pattern. He thereupon turned to additional hypotheses that *could* explain the deviations of the data from the malaria-based prediction, and confirmed two based in historical forms of explanation (migration, patterns of agriculture). This accords with an old proverb, "The exception proves the rule," in which the English word "prove" is used in the sense of "test" (the sense of its German cognate *probieren*). In this sense, apparent exceptions provide a fertile ground for variously strengthening or assailing abductions; they are not always mere "noise."

Buss, in contrast, is a much lazier rhetorician. The text accompanying this bar chart "explains" nothing of these details – not the averages (for either sex) shown in the figure, not the discrepancy between the male preference and the female preference (which, if based on actual marriages, should sum to zero), not the variation across cultures. This sort of presentation cannot be considered a serious effort at numerical inference, and the juxtaposition of many others like it is no help. The rhetoric here is, instead, designed to reassure the true believer in the underlying theory, not to convince a skeptic. Evolutionary biology is a fundamentally quantitative theory, but Buss, like most of his colleagues in evolutionary psychology, makes no attempt to use any of the tools of numerical inference whose application to actual evolutionary biology was so hard won by Sewall Wright's generation and those who came just after (for a recent critique along these lines see Bolhuis et al., 2011). When Buss's displays *are* accompanied by the machinery of inference, as in the figure preceding the one from his 2009 textbook that I have excerpted (his Figure 4.4, page 116), the only quantification of the inference is a row of p-values across the bottom of the chart.

From the point of view of this book, this is not a rational argument at all. The appropriate method would instead treat the contents of the chart seriously as numerical information, justifying the particular value of the average of those bars in the graph and then proceeding to explain the parts of the information displayed that are *not* explained by the underlying theory: the lack of symmetry between male and female and the obvious variation across cultures. These are the residuals from the author's main theory, and, just as Livingstone explained the residuals from the malaria explanation by recourse to information about migrations and about agriculture, Buss needs to find some consilience in the complete pattern of this chart, not just the signs of the averages of the bars.

Multiple outcome variables. In their notorious book of 1994, *The Bell Curve*, Richard Herrnstein and Charles Murray attempted a systematic loglinear analysis of a huge data set, the National Longitudinal Study of Youth, that had been assembled for other purposes over some 30 years. The bulk of their book dealt with a suite of parallel analyses over a sample of American "whites" (Euro-Americans) that mostly took the form of multiple logistic regressions[15] of various outcomes on measured parental IQ and measured parental socioeconomic status – just the sort of analysis deconstructed in Section L4.5.2. Herrnstein and Murray found that in most of these regressions, the salience of parental IQ far exceeded that for socioeconomic status, and took that as evidence that American society was a meritocracy in a particularly destructive and hypocritical way, claiming equality of opportunity but making it impossible by not investing enough resources in the lives of those intellectually underequipped from birth. The policy implication, Chapter 22, followed on a long American tradition of applied social science (compare, for instance, Lynd, 1939): the prescription that true equality of outcome requires the honest discussion of human differences.

Herrnstein died shortly after his book was published, but Murray went on to comment, in various venues, that he was astonished at the firestorm it sparked. Their analyses were relentlessly reasonable, he insisted, and yet nobody was willing to take them seriously as a basis for policy. That was partly because of outrage over the book's politically inconvenient and scientifically sloppy coverage of racial differences, but more because the book's main argument was not remotely as powerful as its authors believed it to be. (It can't have helped that in the Afterword to the 1995 edition, page 569, Murray boasted that the findings here "would revolutionize sociology." Of course they did no such thing.) The accumulation of somewhat similar findings over multiple correlated outcomes (such as labor participation, marriage, and "middle-class values") is not of itself a consilience, and there is no quantitative theory anywhere that would account for the actual numerical values of the regression coefficients at which these authors' models arrived, either the central tendency or the many exceptions. The analysis thus permitted neither abductive nor consilient reasoning, but instead illustrated one of the main points of Section L4.3.4: statistical significance *per se* is not actually a form of understanding, nor any form of honorable persuasion.

Experiments with an insufficiently structured hypothesis. In some classic studies, there is a clear numerical outcome, but no consilience. Robert Rosenthal and Lenore Jacobson, for instance, ran a beautiful experiment recounted in their *Pygmalion in the Classroom* (1968). This was a true randomized experiment in which a set of elementary school students, of various levels, were given an intelligence test, the results of which were intentionally misreported to teachers in a randomized pattern unrelated to actual child performance as scores for "expected intellectual growth." On follow-up measures a year later, the children arbitrarily labeled "ready to bloom" averaged better year-on-year change than their classmates not so labeled. This is now called the *observer-expectancy effect*. It is memorable, and ironic, but there is no parallel theory that predicts the extent of this effect as a function of starting

---

[15] Logistic regressions are regressions for which the dependent variable is a binary variable, one taking only two values, say, 0 and 1. They are normally fitted as *logistic* models, predicting the *logit* of the outcome, the quantity $\log p/(1-p)$, where p is the probability of getting a 1 from a certain combination of predictors. For an accessible introduction, see Andersen and Skovgaard (2010).

age, starting IQ level, and teacher sophistication, all variables that the authors found to affect their statistics. So no numerical inferences are possible from these data, properly speaking; their function is purely hortatory, to provoke the guilty feelings of pedagogues.

Or "reason and freedom," or the management of social systems? Continuing in this wise, one finds that hardly any research in the psychosocial disciplines aligns with the principles of abduction and consilience on which this book is centered – examples like Milgram's are very, very rare. Among the exceptions are some old and new interdisciplines that emphasize the consilience per se between behavioral observations and measurements made by physical machines – agreement of numerical patterns across suites of measurements made within the methodologies of the two contributing sciences separately. Neuroeconomics, for example, is aimed at consilience between functional brain imaging and economic behavior in competitive contexts involving money. See, for instance, Glimcher et al. (2008).[16] From the point of view of numerical inference, the emergence of such fields as neuroeconomics and behavioral economics is to be commended in light of the old struggle (so old that it has a German name, the *Methodenstreit*), about whether economics is an observational science at all – are its propositions empirical, or normative? See, for instance, Machlup (1955). Another large emergent interdiscipline is social network analysis, founded in a sense by Milgram himself in his studies of the "six degrees" phenomenon (Watts, 2004), where researchers try to unearth identities of scaling between networks of social contacts and the analogous models that arise from studies of physical phenomena. See, in general, Barabási (2003).

But a better explanation of this dearth of examples might come from an analysis of the logicocognitive structure of those fields themselves. In 1984, at the outset of his deliciously informal and idiosyncratic *Notes on Social Measurement*, the sociologist Otis Dudley Duncan published a telling list of 13 potential "numéraires," actual quantifications around which methodologies of social measurement have coalesced. The list begins with voting and continues with enumerating, money, social rank, appraising competence or performance (including college grading), graduating rewards and punishments, probability, random sampling, psychophysical scaling, index numbers, utility (?), correlation matrices, and measures of social networks.

Some interesting concepts *fail* to appear on this list – race, ethnicity, happiness, opinions, and even preferences (as you might have suspected from that question mark after the word "utility" above, a question mark that was Duncan's before it was mine). Here Duncan is tacitly invoking some mathematical social science literature arguing that some of the fundamental modes of observation of psychosocial phenomena simply do not conduce to the pattern languages designed for prototypes from the natural sciences. The economist Kenneth Arrow, for instance, received a Nobel Prize for his

---

[16] This field escapes the curse of unwanted knowledge (see earlier under Milgram) inasmuch as the money used in the study, while real, is considered a benefit, not a cost, of participating. There are equivalent studies in animals that use various zoological analogues of money (see, e.g., Hauser, 2006), studies that come under the related heading of evolutionary economics.

"general possibility theorem" (see Arrow, 1951), a generalization of the Condorcet paradox, that no "social welfare function" (algorithm for combining samples of individual rankings into a collective ranking) is possible that meets five requisites, each one reasonable separately (for instance, that individuals can order their preferences in any arrangement they like, that if one individual raises his relative preference for some outcome the collective ranking cannot respond by *lowering* the rank of that outcome, or that there is no "dictator" whose personal preferences automatically determine the "collective" decision). The psychologist Clyde Coombs, working in light of Arrow's theorem, dug deep into the structure of reported preferences and showed how under most feasible forms of reporting these states of consciousness, the conventional mathematical models, which involve points on an intersubjective continuum, could not be unambiguously derived from verbally mediated data (Coombs, 1956). Coombs's observations bear critical implications for those psychological and social sciences that rely on correlational analysis of verbal or behavioral reports of mental states. The list of such domains includes, for instance, psephology (statistics of elections) and the evolutionary psychology of perceived attractiveness. In independent research efforts that founded behavioral economics, the economist Maurice Allais and the team of psychologists Daniel Kahneman and Amos Tversky showed that in making judgments under uncertainty most experimental subjects fail to follow the rules of *Homo economicus* (Allais and Hagen, 1979; Kahneman, 2011). These studies probably encompass the reasons why Duncan questioned "utility" as one of the 13 mensurands.

All these have been logical critiques, in one sense or another, but we might also choose to proceed sociologically. Duncan, who was, after all, a sociologist, goes on to emphasize that these measurements have a shared purpose quite different from the purposes of abduction and consilience. Their goal, instead, is the *management of ongoing social systems*. This version of the social sciences shares Milgram's concern with "bringing the large into the laboratory," but centers not on the niceties of that laboratory but on the applications in studies to manage allocations of scarce and variable resources (public buildings, roads, bureaucrats) or to ameliorate the problems of a society in the large; most often, beginning from the very earliest 19th-century roots, studies of various forms of human misery, particularly in cities. This approach de-emphasizes any role for "surprise" and eschews numerical consilience in favor of an entirely verbal alternative. The focus instead is on the self-incriminating role of knowledge of social structures: to know their flaws is to glimpse a view of how to remedy them.[17]

One *locus classicus* of this approach, which goes a long way toward explaining the relative dearth of positive examples in this section of my book, is the celebrated

---

[17] But there are also well-known horror stories when social management schemes are applied based on ostensibly biological or psychobiological measurements. See, *inter alia*, Bowker and Star's (1999) chapter on the South African schemes of ethnic classification. One could also argue that the study of the "sleeper effect," Figure 4.51, was part of an applied investigation of the motivation of soldiers (Hovland et al., 1949), and thus would ultimately be used to the detriment of the original subjects themselves.

Appendix, "On Intellectual Craftsmanship," in Mills (1959). Here Mills emphasizes the centrality of ordinary 19th-century analysis of crosstabulations of binary variables to the understanding of the principal structuring features in modern American society, to wit, class, status, and power. The argument to which this is an appendix explicitly states that the purpose of social science is to turn ordinary human troubles into social problems: from these applications of reason, we gain freedom. The divergence from the themes of Chapters 2 and 3 here could not be more stark.

Beyond these few mathematical or sociological remarks, the present book does not deal with this massively alternative approach to the psychosocial sciences. There may be some other version of numerical inference that works in these contexts, or there may not. I only conclude that the works of applied social science that I *do* note favorably appear to be those, like Gaddis's book of 2002, that have no melioristic flavor, or those, like Milgram's, where the attempt at amelioration actually failed.

E4.8 Death of the Dinosaurs

My last retelling in this "undergraduate" chapter is of a paper astonishingly clear in its rhetoric of numerical inference in spite of having to maintain a chain of nearly a dozen of them in a row: the great article by Alvarez et al. (1980) about the cause of the Cretaceous–Tertiary extinction. The paper, which deserves to be more widely known by historians and sociologists of science, is a nearly perfect example of the rhetoric of strong inference as in Platt (1964). The topic is paleontology, but the senior author, Luis Alvarez, was a Nobel laureate in physics (in 1968, for unrelated work on elementary particles), and so the paper shows the protean character of the respect for numbers that Kuhn declared to characterize the physicist particularly.

E4.8.1 Both Kinds of Abductions

Our authors begin with a terse summary of the explanatory task: the great extinction 65 million years ago, perhaps half of all genera then living, at the geological boundary between the Cretaceous and the Tertiary periods (hereinafter "the C–T boundary"). "Many hypotheses have been proposed to explain [these] extinctions, and two recent meetings on the topic produced no sign of a consensus" – a hint of a Kuhnian "anomaly," even though each of the six hypotheses merits only a short phrase on its own. They note the difficulty of explaining an event like this using paleontological "and therefore indirect" evidence, or even by using what little indirect physical evidence there is (stable isotope ratios, for instance), but note that these patterns are "not particularly striking and, taken by themselves, would not suggest a dramatic crisis." This surprisingly brief introduction, less than half of page 1095, closes with the set-up of a classic Peircean abduction:

> In this article we present direct physical evidence for an unusual event at exactly the time of the extinctions in the planktonic realm. None of the current hypotheses adequately accounts for this evidence, but we have developed a hypothesis that appears

E4.8 Death of the Dinosaurs 281

to offer a satisfactory explanation for nearly all the available paleontological and physical evidence.

At this point the authors introduce the data that will drive a striking series of strong inferences to follow. The study began, they note, with the "realization" that platinum group metals, including iridium (Ir), are much less abundant in the earth's crust than in meteorites. It had been suggested, indeed, that measures of iridium might supply a time scale for the duration of crustal exposure to this accretive process, and so it was reasonable to assess iridium content in one particularly interesting and very narrow exposure of the C–T boundary near Gubbio, in the Umbrian Apennines.

I let the authors speak (page 1097):

> Twenty-eight elements were selected for study.... Twenty-seven of the 28 show very similar patterns of abundance variation, but iridium shows a grossly different behavior; it increases by a factor of about 30 in coincidence with the C-T boundary, whereas none of the other elements as much as doubles.

This was a surprise indeed. More sustained measurements resulted in the Ir profile reproduced here as Figure 4.53. Note several inventive aspects of the visualization here, particularly the highly nonlinear scaling of the vertical axis (the depth in the section).

The authors now launch on a brilliant succession of strong inferences based on extraordinarily careful measurements, some retrieved from a wide range of published literature and others their own (sometimes the readouts of instruments developed in the course of this very investigation). I extract four successive abductions from pages 1097–1104.

1. "To test whether the iridium anomaly is a local Italian feature," they assessed a different exposed C–T boundary layer near Copenhagen, one that was deposited under seawater. A more sophisticated series of measurements of 48 elements was produced, and the finding was of a 160-fold increase in the concentration of iridium at the boundary. The authors take care to argue that this exceeds the probable iridium concentration of a seawater column itself by at least 20-fold. Furthermore, "recent unpublished work" at a third site, in New Zealand, finds a 20-fold concentration of iridium, again, right at the C–T boundary.
2. "To test whether the anomalous iridium at the C–T boundary in the Gubbio section is of extraterrestrial origin," the authors check whether *any* terrestrial source could increase iridium content by that factor without increasing the concentration of all the other trace crustal elements by a similar factor; they conclude that it could not. The same is argued for the Danish site.
3. "We next consider whether the Ir anomaly is due to an abnormal influx of extraterrestrial material at the time of the extinctions, or whether it was formed by the normal, slow accumulation of meteoritic material" followed by some concentration process. An argument is pursued that the iridium anomaly is exactly coincident with the extinctions (we shall see this argument again in

Fig. 5. Iridium abundances per unit weight of 2*N* HNO₃ acid–insoluble residues from Italian limestones near the Tertiary - Cretaceous boundary. Error bars on abundances are the standard deviations in counting radioactivity. Error bars on stratigraphic position indicate the stratigraphic thickness of the sample. The dashed line above the boundary is an "eyeball fit" exponential with a half-height of 4.6 cm. The dashed line below the boundary is a best fit exponential (two points) with a half-height of 0.43 cm. The filled circle and error bar are the mean and standard deviation of Ir abundances in four large samples of boundary clay from different locations.

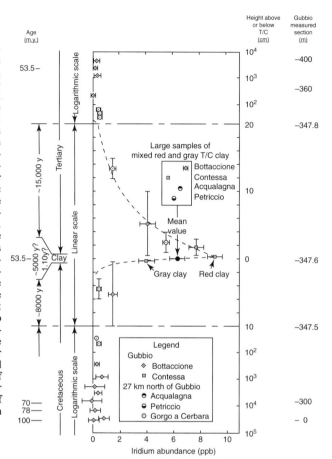

Figure 4.53. The fundamental signal driving the abduction in Alvarez et al. (1980). This started the abduction; the discovery of Chicxulub closed the syllogism. (From Alvarez et al., 1980.) Republished with permission of American Association for the Advancement of Science, from *Science* 6 June 1980: *Extraterrestrial Cause for the Cretaceous-Tertiary Extinction*, Luis W. Alvarez, Walter Alvarez, Frank Asaro, Helen V. Michel; permission conveyed through Copyright Clearance Center, Inc.

12 years, in connection with the Chicxulub crater report of 1992), and that the extinctions were a worldwide extraordinary event and thus should have a worldwide cause, such as the Ir anomaly now known from three widely separated sites. "In summary," they conclude, "the anomalous iridium concentration at the C–T boundary is best interpreted as indicating an abnormal influx of extraterrestrial material."

4. The authors then turn to the popular hypothesis that the C–T extinctions were the consequence of a nearby supernova. They compute the probability of such a supernova as approximately one in a billion, *and yet do not reject it on that ground*. Instead, they rejected "positively" (note the implied downgrading of merely probabilistic reasoning) by considering two other independent pieces

of evidence, each the endpoint of a long metrological chain of its own: the relative concentration of ^{244}Pu, which was assessed at less than 10% of where a recent supernova should have set it, and the ratio of two iridium isotopes, which was found to *match* its value for solar system (terrestrial crustal) materials. (This latter finding required the development of an entirely novel assessment technology.) "Therefore," they argued, "the anomalous Ir is very likely of solar system origin."

So what extraterrestrial source within the solar system "could supply the observed iridium and also cause the extinction"? After considering and rejecting a number of possibilities, "finally we found that an extension of the meteorite impact hypothesis provided a scenario that met most or all of the biological and physical evidence."

That was the abduction; now for the consilience, pages 1105–1106. The authors accumulate a vast number of previously unrelated quantitative facts from diverse other expert sources. Four citations support their estimate that an asteroid with a diameter of 10 km should hit Earth about every 100 million years. This is consistent with the views of others who are expert in the size distribution of large terrestrial impact craters. (The discussion near footnote 63 is particularly charming as regards the corresponding expertise. "Rather than present our lengthy justification for the estimates based on the cratering data, we will simply report the evaluation of Grieve, who wrote, 'I can find nothing in your data that is at odds with your premise.' ... This section of our article has thus been greatly condensed now that we have heard from experienced students of the two data bases involved.") The explosion of Krakatoa, 125 years ago, yielded at least qualitative observations about the effects of large quantities of atmospheric dust and the consequent disturbances.

Alvarez et al. then calculate the size of the asteroid in four different ways: from the net iridium it must have delivered (a long metrological chain, evaluated as "$M = sA/f$" where s is the surface density of iridium at Gubbio, A is the surface area of Earth, and f is the concentration of iridium in carbonaceous chondrites, the type of meteorite this is expected to be) ending up in an estimate of its diameter as 6.6 km; from the diameter that would give us five recurrences in half a billion years (about 10 km – this is because the paleontologists consider there to have been five mass extinction events in that time period, of which the C–T is the most recent); from the thickness of the nonconforming clay layer itself (another long metrological chain, this one using "the clay fraction in the boundary layer, the density of the asteroid, the mass of crustal material thrown up per unit mass of asteroid, [and] the fraction of excavated material delivered to the stratosphere"), resulting in an estimated diameter of 7.5 km; and from the need to darken Earth by something like a thousand times as much dust as was released by Krakatoa. The biological effects of any such impact would be qualitatively equivalent to what is observed at the C–T boundary. Note the difference of this rhetoric from Snow's, which was similarly concerned with inference about a unique historical event. After showing that nobody else's theory can account for the iridium anomaly – it remains "surprising" – they change rhetorics to reassure that the fact would follow "as a matter of course" from their asteroid explanation. Namely, if there had been such an impact, the impacting object must have had a size, and the three different estimates of that size show reasonable agreement (from 6.6 to 10 km diameter, again illustrating Kuhn's point about the meaning of the word "reasonable").

E4.8.2 The 1992 Confirmation

The abduction in this example is particularly strong, as it was aimed at the demolition of those competing hypotheses. The consiliences are relatively weaker, as they are attempting to apply essentially uniformitarian data bases to an inference about a unique event. Still, a confirmation would be very welcome: "we would like to find the crater," the 1980 authors concluded. Yet it isn't any of the three known craters of that size – they are of the wrong age – and there is about a two-thirds chance that the object fell into the ocean (which would render it nearly impossible to locate even had it not been subducted). But 12 years later, after the death of Luis Alvarez, the assignment was completed – the associated crater was indeed located (Kerr, 1992). The closing argument was actually the demonstration that some strange glass globules found in Haiti, which must have been impact debris, could be attributed to a huge crater newly discovered (by petroleum geologists) off the coast of Yucatán. Peirce's "surprising fact" reduces here to a simple coincidence of numerical values: a measured age of 65.06 ± 0.18 million years for the glass, versus 64.98 ± 0.06 million years for the crater. This is another numerical inference, and a most welcome one. The significance test is obvious – a subsidiary null hypothesis, that the dates are equal, is obviously true: the ages of the impact, the impact debris, and the mass extinction are all the same. Note that the role here of statistical significance testing is the *correct* role: to *confirm*, not to infirm, a hypothesis (Section L4.3.4).

"It's almost beyond imagining," said one converted critic, "that one of the largest known impact craters and the impact deposits in the Caribbean could have the same radioisotopic age and not be part of the same event." "It looks to me like this is the smoking gun," said the surviving Alvarez author. "This should let us stop arguing about whether there was an impact and start working on the details of the impact." The consonance with Perrin's and Snow's closing arguments is striking: successful quantitative rhetoric consists in closing an argument, whether about a disease, an atomic configuration, or a mass extinction. Notice that the aspect of the search that might have appeared to have the greatest similarity to an actual Popperian confirmation – the 1980 paper estimated the diameter of the crater at about 180 km, and it was found to have a diameter of about 200 km – was not, in the event, the pivotal parameter of the affair; it was rather the coincidence of dates, which matched much more closely, thus affording substantially more "surprise" to be reduced. In the twenty years since this inference from Chicxulub, little of this consensus has changed (Schulte et al., 2010).

Interim Concluding Remark

According to the Table of Contents, this spot marks the end of the "undergraduate course." As diverse as our examples may have seemed during a first read-through, they have a great deal in common. They are in this book because they share an appeal to the two core engines, abduction and consilience, driving effective numerical inferences. They are specifically in Part II because they share one further fundamental structural criterion: a concern with "dependence" of some specific measurement (or rate, or risk) as the principal thrust of the investigation.

This tacit motif – the concentration of one's attention on some specific numerical assessment – runs throughout most the examples of Part II. To Snow, it was the decision to use risk ratios of cholera in preference to either neighborhood counts or etiological details of individual cases; to Millikan and to Perrin, obsession with a single physical constant estimated and re-estimated by complex algebraic transformations of very lengthy series of actual instrumented observations; to Watson and Crick, a very short list of parameters of the helix (radius, pitch), tested using a long list of interatomic binding distances from a classic textbook; to Law and to Hackshaw, graded relative risks of one disease as a linear function of one biographically accessible cause, the estimated individual history of exposure to environmental tobacco smoke; to Milgram, the rate of defiant behavior (a novel social measurement; compare Duncan [1984], or Mills [1959]) script by script. Yes, this certainly is one principal narrative style of modern quantitative science: find a number that can serve as a crux for abductive or consilient rhetoric, report what accounts for it, publish that report, and defend it against its detractors.

Sometimes that monadic theme is obvious given the other tropes in the literature of the subject area. Inasmuch as the laws of physics involve constants, for example, it is understandable that experiments are set up to produce their values and show that they are unvarying. Or, if a dread disease (cholera, or lung cancer, or sickle cell anemia) is the focus of our attention, we know that the analysis of specific risk or protective factors is warranted on moral grounds.

More often, we have seen decisions to *re*focus the reader's (or the community's) attention on an alternative quantity, perhaps one that is more salient to the core conceptions of the theory, perhaps one that is more conducive to powerful numerical inferences. Millikan and Watson and Crick had little choice in this matter, but Snow carefully argued for the primacy of house-by-house death rates over ecological summaries; the ulcer researchers, for the primacy of the rate of recurrence; Hackshaw and Law, for the importance of *both* the slope and the intercept of the relative risk line; Milgram, for the single index that is the fraction of defiance; Alvarez and colleagues, for the size of the meteorite, which was the only clue the iridium afforded as to how the search should be performed; Perrin, explicitly, for the rate of rarefaction of his "atmospheres" of gamboge balls in water, likewise the only way of spanning so many orders of magnitude of physical scale in the course of his successful attempt to persuade.

At the same time, we have seen a great many homogeneities of style, beginning with the announcement that an abductive consilience has indeed been concluded, a problem has been solved, a discipline has moved on. The combination of abduction and consilience conduces to very effective reports, reports that close issues, over a huge range of disciplinary contexts, from the very broad (warming, or seafloor spreading) to the very narrow (finding the *Scorpion*, or a prophylactic against ulcer recurrence, or the specific geometry of the DNA molecule). There is a heterogeneity, I admit, in the polarity of the initial abduction. Some of my examples begin with an unexpected discrepancy, others with an unexpected agreement. (We saw one of each in Section E4.6, the two different styles of inference about passive smoking risks for the two different diseases.) But in all these examples the power of the summary scientific argument, the explicit or implicit evocation of the feeling of what Perrin tellingly called "the decisive proof," is inseparable alike from the initial surprise, from the reduction to that surprise afforded by the abduced hypothesis, and from the support that hypothesis finds from agreement with new experiments or old, replications or appeals to other levels of analysis.

What our examples have all shared, in other words, is the role of numerical evidence in providing a sense of rhetorical closure. It is the numerical inference, far more than any associated verbal ratiocination, that closes issues – that permits us to turn from what Kuhn called "crisis science" back to normal science and its routine unskeptical accumulation of confirmatory results. Numerical inference, I argued following a point of Kuhn's in Part I, is peculiarly suited to closing debates of this sort, and what most of the examples in Part II share is this quality of *closure* by overwhelming force of numbers (the pun in this phrase is intentional), from cholera deaths through earthquake maps, from rates of ulcers through rates of disobedience, from unexpected iridium through unexpected symmetry of nucleotide counts. They have also shared the *absence* of any role for null-hypothesis statistical significance testing, that sorry atavism that pervades so many of our undergraduate curricula.

Yes, it is the numbers that carry the examples in this part. In a sense, that is what makes this syllabus accessible to undergraduates.

But what if a problem does not conduce to a primary focus of measurement on a single scale in this way? What if the concern is with properties of systems for which we do not yet agree on crucial quantifications, but for which the concern is instead with health, or stability, or shape, or network redundancy? *There are other grand themes in statistical science*, of which the one that drives Part III of this book, the examination of complex organized systems in terms of the multidimensional *patterns* interlinking their measurements, has no use for the asymmetry between "response variables" and "predictors" or "explanatory factors," nor likewise for the idea of "choosing" from a list of variables. (The same contrast has appeared in the literature of statistical pedagogy as the antinomy between studies of dependence and studies of *interdependence*.)

From the next chapter on, this book will not attempt simplistically to reduce these more complex problems to single variables. Instead, still wielding the two principles, abduction and consilience, as the main intellectual tools, we will turn to a completely separate ground for statistics, no longer in terms of the least sum of squares of some quantity or other but instead by exploiting the information content of entire patterns over a wide range of geometrical representations and metaphors. Except for abduction and consilience, then, Part III starts over, with a whole new sense of what constitutes "order" or "disorder" for systems that are *not* conducive to modeling by Gaussians.

The work is mathematically deeper, but the problems are just as important, and the role of quantification in their solutions, and the role of the rhetoric of those quantifications in the reporting of those solutions for public understanding of science and for the public weal. To those more complex rhetorics, matched to the more complex system queries, we now turn.

Part III

Numerical Inference for General Systems

Make everything as simple as possible, but not simpler.
— Albert Einstein (attrib.)

Introduction to Part III

Part III is more technical than parts I and II, but the change of rhetoric is unavoidable. As in earlier books of the tradition within which I'm working (Jeffreys's *Theory of Probability* of 1939, Jaynes's *Probability Theory* of 2003), the argument is that certain procedures based on arithmetic lead to valid reasons for believing in some scientific explanations – that some arithmetically based inferences about true states of the world should be thought of as reasonable, as rational. When the scientific assertions are complex, as they will be in most of the examples in this part of the book, so are the arguments that defend the rationality of the arithmetic applied there. In particular, the mathematical language of the patterns that arguments like these invoke – patterns like singular vectors, random walks, or distributions in shape space – is not part of everyday spatial metaphors, everyday conversation and gesture, or everyday persuasion, the way it was in Chapter 4.

Just as scientists need reasons for asserting that new evidence is consilient with a particular explanation, so they need reasons for asserting that a finding is indeed surprising on other explanations. When data are numerical, so, usually, are these reasons – extremeness or other unique characteristics of particular statistical descriptions or geometrical patterns. Reasoning about the extremeness of the pattern descriptors is typically less familiar and more technical than the reasoning about cause and effect that attaches the data to the abduced explanation. The mathematical text that follows concentrates, then, on the protocols that support the claims of surprise, not the experimental approaches (typically falling under headings such as "reductionism") that support the (equally important) claims of consilience. The difficulty of these mathematical foundations is highly variable; hence so, too, is the text I have provided. As before, there will also be the occasional digression into the stylistics of these mathematical arguments per se.

Still, because parts of this material will be hard going for a substantial fraction of readers, it may be helpful if this "graduate course" begins with a

Summary. When searching for explanations in more complex systems than those considered in Chapter 4, the dominant roles of abduction (Chapter 3) and consilience (Chapter 2) persist. But now the forms of inference they drive demand a new logical maneuver, the algorithmic or human *selection* of one possibility or emphasis, or a few, from a list, a map, or any of a variety of other manifolds and diagrams. The associated statistical methods apply information theory to select among alternative arithmetical models of the same data. Sometimes the algebra of these methods incorporates one of the classical approaches collectively called "multivariate statistics"; more often it does not.

Chapter 5 begins with three diverse examples that open an overview of the general area from a logical and epistemological perspective. I then introduce the idea of an *organized system*, which helps us convert our main themes, abduction and consilience, into styles of reasoning that suit this context. A general framework, information theory, permits one to assess numerical hypotheses in this domain with respect to some extended models of randomness that are likewise introduced here. Chapter 6 introduces the geometry and algebra of the singular-value decomposition (SVD), the principal pattern engine for our work, and then the three main descriptive statistical techniques radiating from it: principal components analysis of data matrices, principal coordinates analysis of pairwise distances (differences), and Partial Least Squares analysis of associations among adequately structured measurements from multiple domains. Chapter 7 reviews geometric morphometrics, the methodology with which I am most closely identified, from this point of view, and a final section sketches a variety of other prototypical applications as well.

5

Abduction and Consilience in More Complicated Systems

5.1 Analysis of Patterns in the Presence of Complexity

Continuing in the style of argument explored across the examples in Chapter 4, we ask: Can abduction and consilience bring us to clear and authoritative explanations even in analyses of more complex systems? Yes, sometimes they can. As in Section E4.3, we detect them via empirical *constraints*: when multiparameter data or summaries that could otherwise arise anywhere in their joint measurement space are restricted to a very small fraction of that manifold of possibilities – clusters, curves, etc. separated by systematic voids. Let me put off all definitions until a bit later and just present three quite different examples of what I am referring to.

5.1.1 The Hertzsprung–Russell Diagram

Astrophysics is unusual among the sciences: the oldest natural science yet by far the most limited in its range of observational channels. Harwit (1981) points out that any astronomical data resource, however extensive, is at root an organized list of basic facts, instrument-derived "elementary astronomical observations," that have only seven descriptors each: the type of "carrier" (physical entity) to which the instrument is responding; the wavelength or energy of that carrier; the angular resolution, spectral resolution, and temporal resolution of the instrument; the date and time of the observation; and the direction in which the instrument was pointing (along with directionality or polarization, in some cases). Regarding that first facet, type of carrier, in turn, there seem to be only five possibilities: electromagnetic radiation, cosmic-ray particles, solid bodies (meteors and meteorites such as the one that left the iridium for Alvarez to measure in Section E4.8), neutrinos, and, although hitherto only in theory, gravitational waves. The first of these five domains, electromagnetic radiation, is the most frequently exploited, owing to its huge range of detectable wavelengths, but over most of that range its signal can be observed only by raising instruments above the atmosphere.

This would seem a remarkably depauperate domain for either abduction or consilience, and yet the explanations afforded by astrophysics are among the most elegant in all of science. Much of this is because of the authority the equations of physics have to explain the range of astrophysical phenomena, sometimes after they are encountered (supernovas), sometimes before (black holes), and much of *that* is because of the powerful relationships of those seven parameters with one another. For instance, color combines with brightness to underlie the accepted theory of stellar evolution, the theory that also accounts for the existence of carbon in the universe and thus ultimately for the existence of life forms like you. The first of these quantitative constraints to be recognized was reified in one fundamental scientific tableau, the **Hertzsprung–Russell diagram**, which is itself, like the universe as a whole, mostly empty space.

The Hertzsprung–Russell diagram[1] (so named in 1933 after Ejnar Hertzsprung and Henry Norris Russell, who began its study independently before 1910) is not a "star map," but a scatterplot of stars according to two conceptually simple but remarkably fundamental quantities: "absolute magnitude" (or "luminosity") and effective color (or surface temperature). Measurement of the former of these occupied the community of astronomers for some time. It is relatively easy to tell how bright a star appears to be (by counting photons or by other photographic means), but to know what its output of energy *actually* is, one needs to know in addition how far away it is, and that proved quite difficult to approximate until the advent of digital imaging quite recently. Initial measurements (beginning as early as Bessel's of 1835) were by parallax – the apparent motion of a nearby star relative to stars more distant as Earth moves around a diameter of its orbit over a six-month interval. Later measures were of relative brightnesses within galactic clusters, whose members are all at roughly the same distance from Earth, and, even later, calibrations by *Cepheid variables*, stars whose apparent brightness fluctuates with a period that depends on their actual size and thus permits an indirect measure of distance by examining the frequency of this periodicity. Then for any cluster that includes at least one Cepheid, the distances of *all* the stars can be estimated as the distance of the Cepheid, and their apparent brightnesses are all corrected at once into the absolute magnitudes needed for the plotting.

Figure 5.1 here satisfies the Interoptical Trauma Test. Those startlingly broad swatches of empty space in the graphic beg for an explanation, along with that handsomely sinuous curve up the middle. The explanation of that curve, called the Main Sequence, followed by the end of the 1930s. There is indeed a relation among mass, luminosity, and radius for stars of uniform chemical composition: stars generally do not leave the Main Sequence until they begin running out of hydrogen to "burn" (to convert into helium). Thus a wholly static scientific visualization (stars mostly do not change their brightness or their distance from us on any human scale) is explained by a theory of stellar evolution on a time scale of billions of years. This

---

[1] These notes are based mainly on the interesting account of the history of observational astronomy by Harwit (1981).

5.1 Analysis of Patterns in the Presence of Complexity 293

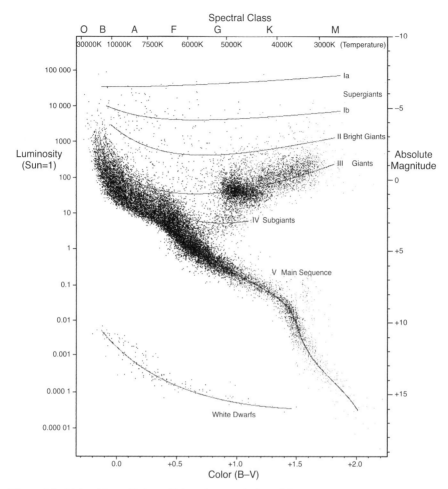

Figure 5.1. Richard Powell's beautiful modern rendering of the Hertzsprung–Russell diagram. The original is in color as http://en.wikipedia.org/wiki/File:HRDiagram.png. Horizontal: effective surface temperature (color). Vertical: luminosity as a multiple of the luminosity $L_\odot$ of the sun.

is no less "miraculous," in Wigner's metaphor, than the equally outrageous claim that the diameter of the proton in an atomic nucleus, the length of the ruler on your desk, and the distance to that quasar in your telescope can all be measured in multiples of a meter, over a range of scales of some 10^{39}.

5.1.2 The Ramachandran Plot

As hinted in the sentence that closed the preceding section, we move now from the almost unimaginably large stars to the almost unimaginably small: the molecules that underlie most living processes. The next example lies at the foundation of the modern molecular study of proteins.

Though they haven't been around as long as the stars have, proteins have been studied for nearly as long as astrophysics, and the field in which those studies are centered, *proteomics*, has drastically expanded recently as regards data resources and instrumentation alike. From the wide range of analytic tools and tactics on which the field relies (cf. Tramontano, 2005) I highlight the *Ramachandran plot*, after Ramachandran et al. (1963), which summarizes the local geometry of protein structure in a remarkably efficient way. To understand its construction, we need only a bit of geometry: the definition of two angles in terms of the local structure of the protein molecule. A protein is assembled by folding a linear polypeptide made up from a series of amino acid *residues*, one of twenty different choices, attached to a carbon atom (Cα) that is linked on one side to a nitrogen atom (N) and on the other side to another carbon atom (C) lacking a residue but linked in turn to the N of the next unit along the polypeptide. The residue sequence is one of three different ways of recording information about the protein molecule's structure that turn out, in practice, to overlap strikingly in their information content. A second method is by locating all of its atoms with respect to one another in three-dimensional space – this is difficult to observe instrumentally and even more difficult to model – whereas an intermediate level, describing the relation of each structural unit to its predecessor and its successor by two *dihedral angles*, is remarkably effective at bridging the gap.

If you haven't seen a dihedral angle before, think of it this way. Draw a jagged line connecting four points in the order 1–2–3–4 in the shape of an irregular Z on a piece of paper. Fold the paper along the edge between points 2 and 3, and open the fold to some angle between 0° and 180°. The angle between the two halves of the piece of paper is called the *dihedral angle* of the planes 1–2–3 and 2–3–4, the dihedral angle around the 2–3 edge. As Figure 5.2 indicates, it is the same as the angle by which point 4 has to rotate out of the plane 1–2–3, which is also how far point 1 has to rotate away from the plane 2–3–4. A polypeptide has two of these angles at every residue (Figure 5.2), the angle ψ (psi) around the C–Cα bond at the residue and the angle ϕ (phi) around the Cα–N bond at the residue. As applied to protein molecules, this representation was invented (discovered?) by Ramachandran and colleagues in 1963, hence its name. It is in a quite interesting coordinate system, the pair of angles that specify position on a *torus* (a doughnut shape, a bagel). That means that the top edge of the plotting region in Figure 5.3 represents the same data as the bottom edge, point for point (e.g., there is only one region labeled G in the figure, not two). Likewise, the left edge has precisely the same information as the right edge. This toroidal geometry drives some fancy statistical modeling (Mardia et al., 2007) that does not involve the actual geometry of a torus or the broader family of surfaces (the cyclides; Maxwell, 1868) in which it is embedded. The concern in proteomics is with the torus as a scheme of labelling, not a shape. See also Mardia (2013). For a version of *this* figure in glorious color, please see Ch. Fufezan's version at p3d.fufezan.net.

Formally, the logic of inference compatible with the Ramachandran plot (Figure 5.3), is identical to that conveyed by the Hertzsprung–Russell diagram (Figure 5.1), even though the former deals with structures that are about 10^{20} times smaller. In either context we are struck by the dominance of the "empty space" in

5.1 Analysis of Patterns in the Presence of Complexity   295

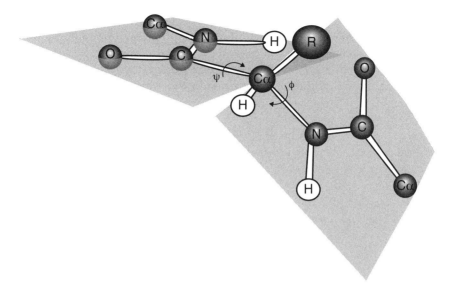

Figure 5.2. The Ramachandran angles Φ and Ψ along the backbone of a polypeptide of amino acids. (From http://www.cmbi.kun.nl/mcsis/richardn/pictures/ex_fig3.jpg.)

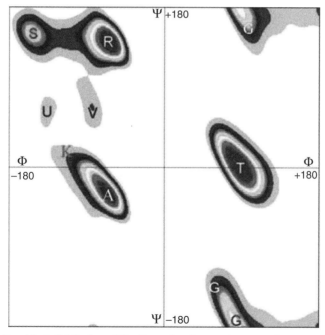

Figure 5.3. Empirical distribution of the Ramachandran angles over a large number of actual proteins. Note the huge preponderance of empty space in this plot. Key to labeled peaks (shape states): S, β-sheets; R, poly Pro II; U, V, bridging regions; K, 3_10-helices; A, α-helices; T, right-handed helices; G, glycine. (From Hovmöller et al., 2002.)

the plot, the pile-up of empirical data on a small fraction of the available parameter domain. Regarding the Ramachandran plot, our surprise at this concentration initiates an abduction that concludes with a constructive theory of protein form: the relative dominance of α-helix and β-sheet features, corresponding to the modes at upper left and center left, over most of the polypeptide once it has folded itself stably. The corresponding "evolution" is over nanoseconds to minutes, rather than the billions of years it takes the stars, but the logic of the numerical inference is nevertheless the same: the only way to account for the remarkable intensity of the clustering in plots like these is a physical explanation in terms of physical equations and physical laws. (In other contexts the explanation will be historical instead, like the crags in the Friedrich painting on the cover.) There's a lot more to say about these angles, which vary as a function of both the amino acid residue in question and the large-scale structure of the folded protein molecule as a whole. See Hovmöller et al. (2002).

5.1.3 The Face of Fetal Alcohol Syndrome

The preceding two examples arose in *exact sciences*, where the instruments giving us their measurements have a precision steadily increasing over time and where explanations are usually in terms of equations and physical laws. It would be good to have an example from a discipline of a different cognitive structure. Here is one from my own work in teratology (the study of birth defects). There will be more about this field and its statistical issues in Section 6.4 and even more in Chapter 7.

Jones et al. (1973) reported on a series of infants born at a Seattle hospital who had the same unusual faces (Figure 5.4). All showed poor Apgar scores (neurological maturity at the moment of birth) along with failure to thrive later, all were small, and all had alcoholic mothers. This would be quite a surprise unless the maternal alcoholism had something to do with the other defects of these infants. The authors therefore suggested a possible *syndrome*, which they named fetal alcohol syndrome.[2]

In a history helpfully recounted by Streissguth (1997), one of the authors of the original Jones paper, it soon became clear that this congeries of deficits had been reported a few times before in fugitive literatures – in fact, that it applies to at least a few children per every thousand born (and up to a few per hundred, in populations at particularly high risk; May et al., 2009). As such fetal damage is wholly avoidable, research began immediately in several disciplines (neuroscience, behavioral teratology, epidemiology) to pursue the consiliences of this work. By now there have been thousands of scientific articles about the conditions under which the risks of this form of damage are highest (in short, when the fetus is exposed to episodes of binge drinking rather than the same amount of alcohol over the course of steady

---

[2] But this was late in the 20th century, and there couldn't be any effect of maternal alcoholism or it would have been noticed well before then, correct? The surprise value of the observations themselves was heightened by the apparent contradiction of the abduction with every reasonable tenet of birth defect surveillance up until then. Compare the reception of the abduction regarding the infectious origin of ulcers, Section E4.4. In fact, for decades obstetricians had recommended that their patients *drink* alcohol to relax themselves during the more stressful times in a pregnancy.

5.1 Analysis of Patterns in the Presence of Complexity 297

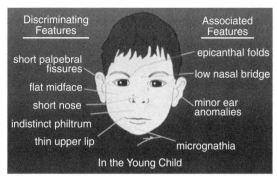

Figure 5.4. The three main features of "the face of FAS" are generally taken to be the first, fourth, and fifth in the left-hand column: eyelid length, philtrum sculpture, and upper lip thickness. (From Streissguth, 1997.)

drinking) and about the mechanisms that are responsible for the damage (see Guerri et al., 2006: the basic problem is connected with migration of neuroglia – alcohol kills some embryonic neurons and confuses others regarding the direction in which they are supposed to be going). Typical presentations of the consequences of this exposure for the children labeled with what is now called a "fetal alcohol spectrum disorder" (compare "autism spectrum disorder") are those in Figures 5.5 through 5.7. Figure 5.5, from a 1996 report that interviewed the guardians or caregivers of more than 400 people with one of these diagnoses, shows that people with this condition (to be precise, one of these two conditions; see later) average a systematic deficit in IQ and also a deficit in diverse psychometric scores even below what is predicted by those deficits in IQ. Figure 5.6, from the data summarized in Streissguth et al. (1998), shows the remarkably high rate of variously troublesome character traits in these same

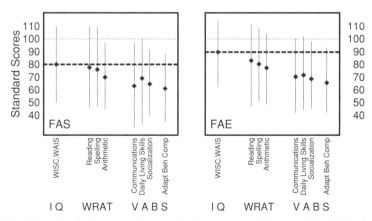

Figure 5.5. IQ may be the *least* sensitive standard assessment of the cognitive deficits in FASD. Deficits in other standard composite measures of applied intelligence and social intelligence, such as arithmetic achievement or adaptive behavior, average deeper than deficits in IQ. Sample sizes: FAS, 178; FAE, 295. (From Streissguth et al., 1996.)

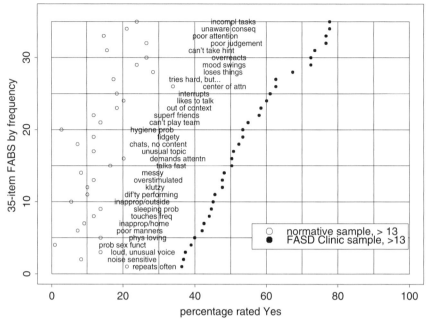

Figure 5.6. Characterological data from the study of Streissguth et al. (1998). 472 FASD versus 186 "normals," of whom, however, 13 actually had mothers with "alcohol problems." The odds ratios of a reported defect, FASD versus normative sample, average 3:1. "> 13": older than 13 years of age.

patients, by comparison to a sample of much more normal people, and Figure 5.7, from the same source as Figure 5.5, documents the appalling rate of serious real-life problems in these patients – note, for instance, the lifetime incidence of nearly 60% for a history of imprisonment. (What other birth defect leads to imprisonment for more than half its victims?) People with fetal alcohol damage are subject to the death penalty in America, and at least one has already been executed. For a recent report of average assessments for children with these diagnoses, see Astley et al. (2009a, b).

Yet it turns out that the original observation, the connection of those facial signs with the brain damage that accounts for all of these other nasty effects, was a lucky guess. The facial signs are sufficient to infer the neonatal damage, but they are not necessary. Children with this particular combination of facial features, making up the subset labeled as "FAS" (fetal alcohol syndrome) in Figures 5.5 and 5.7, are virtually certain to have brain damage, yes; but even within the class of the exposed children who do not have all three main aspects of "the face" of Figure 5.4, those with the typical brain damage, making up the "FAE" subset (fetal alcohol effects), nevertheless have all the expected deficits of psychological functioning and real-life outcomes. In Figure 5.6, the FA groups are pooled (as they averaged no difference on the scale totaling these items) and the diagnosed patients averaged 20 of these

5.1 Analysis of Patterns in the Presence of Complexity 299

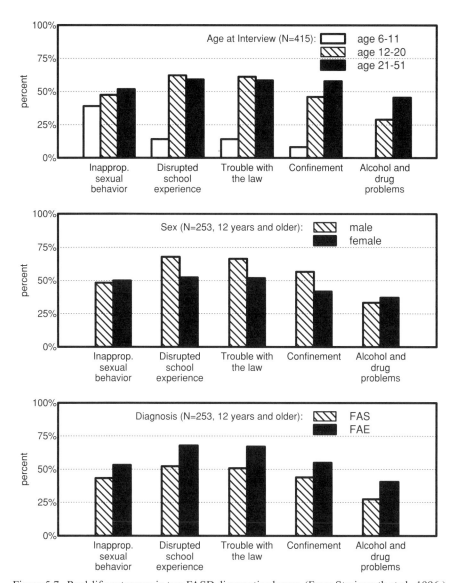

Figure 5.7. Real-life outcomes in two FASD diagnostic classes. (From Streissguth et al., 1996.)

35 characteristics, versus 6.6, only a third as many, for the unselected Seattle townsfolk. The fetal alcohol face (Figure 5.4), in other words, is very effective at predicting the behavioral deficits and other peculiarities, but in no way can be considered to be their cause. Even more ironically, those diagnosed patients *without* the face turn out to have *worse* lives than those *with* the facial signs, in spite of suffering comparable extents of prenatal brain damage (Figure 5.7).

This example shows how our joint logic of abduction and consilient inference can operate far from the domain of exact equations that characterized the preceding two examples. What measurements could be less "exact" than the ratings in Figure 5.6 or the "secondary disabilities" tabulated in the bars of Figure 5.7? And yet the same principle of inference by reducing surprise applies, and the explanation that is suggested thereby is one that has led to thousands of successful research studies and dozens of tenurings of faculty over the 40 years since the original discovery.

5.1.4 Preliminary Comments Suggested by the Preceding Examples

Although the examples in Chapter 4 were mainly "univariate," one important measurement or one pair of measurements at a time, do not be tempted to construe this chapter and the next as "multivariate," at least not in the usual textbook sense ("vector-valued variables," see at Section 5.2.2). We are after bigger game than that highly conventionalized approach. What we now bring to the discussion is the additional conceptual step of *selection*.

We will encounter many ways to choose the "largest" signal, the signal that drives the numerical inference. Among the alternatives will be:

- Largest regressors, most serious risk factors, etc. (that is, selecting from some unsorted list of predictors)
- Focus or peak on a map or a spectrum
- The first two or three axes of an N-dimensional ellipsoid (Section 6.3)
- A place to break/intervene on a network, a string, a series
- Gap-separated peaks on a scatterplot (the Ramachandran plot)
- A curve in a scatterplot (the Hertzsprung–Russell visualization of the Main Sequence)
- "The best-fitting model" (topic of Section 5.3)
- The best m_0 hypotheses out of an original set of m (Benjamini and Hochberg, 1995)
- Ranges of totals over checklists (psychiatric diagnoses, the "Big Five" personality factors)

These foci of inspection can be highlighted one or many at a time, and when there is more than one they can be selected all at once or instead serially: first one, then a second adding the most insight in some sense to the explanation based on the first, a third then optimally improving on the second, and so on.

The goal of this chapter and the next is to survey the passage from arithmetic to explanation for data sets like these, data sets too complex to be summarized by single quantities or pairs of quantities specified in advance. In these domains, data analyses typically emphasize the selection of one possible focus of numerical description from among many (perhaps a great many) possibilities. We will try to discriminate the sound from the unsound across strategies for inferences like these. *Not all numerical inferences from data accord with this declaration of purpose.* Many applications of statistics are not concerned with explanations, and many other modern methodologies, such as finite element analysis of engineered structures, most of bioinformatics, and both "big data" and structural equations models as they are exploited in business and finance, appear to have no formal inferential basis linking data to explanations. Those are methodologies that share an emphasis on forecasting, control, or metaphors of visualization instead.

It is easy to find good examples of numerical science that are unconcerned with numerical inference. In front of me as I draft these pages is the July 2, 2010, issue of *Science*, including a nice article (May, 2010) by the mathematical biologist Robert M. May on a new estimate of the total number of species of arthropods. There is no explanation associated with this number (although there is an explanation of the decision to change the rules for estimating this number). See also Zimmer (2011).

In other domains, such as paleoanthropology, it is unwise to attempt refining one's arguments in any crucible as intense as the abductions and consiliences put forward with admiration here. Nothing in those data resources can stand up to these requirements, owing to the extreme scarcity and nonrepresentativeness of human fossils. The paleoanthropologist relies on X-rays, for instance, rather as the art historian does – as a stimulus for speculative thought only mildly constrained by quantitative expectations. If a scientific fact, according to Epigraph 14, is "a socially imposed constraint on speculative thought," then fields like this behave as if they have hardly any such constraints. At the same time, nobody is proposing shutting down whole departments or wings of museums. On the whole, it may be more reasonable to consider the future of fields like evolutionary psychology or paleoanthropology as if they were branches of the humanities that happen to rely on expensive machines (like 3D cameras or the solid medical imaging equivalent) or expensive social surveys. These are fields that remain at the focus of intense public interest even though they rarely proffer quantitative explanations.

Among the other branches of quantitative science for which neither abduction nor consilience seems relevant are some that pursue "how many" or "how much" questions rather than "why" questions – questions for which forecasts rather than reasons are required. Typical disciplines on this list might include public health, the administrative sciences, and bioinformatics. This orientation to results instead of explanations may obtain even when the same data would be conducive to explanations instead (the "under what circumstances?" approach to applied science; see Greenwald et al., 1986).

5.2 Abduction and Consilience in General Systems

When the titles of Part III as a whole, and this section in particular, refer to "general systems," what do they mean?

"General Systems Theory," the economist Kenneth E. Boulding wrote in 1956, "is a name which has come into use to describe a level of theoretical model-building which lies somewhere between the highly generalized constructions of pure mathematics and the specific theories of the specialized disciplines" (page 197). Ten years later I explored the toolbox for this style of scientific work when I took Boulding's undergraduate course in general systems at the University of Michigan. In a slightly more evocative language, it is the methodology that deals with interdisciplinary or multidisciplinary "problems arising whenever parts are made into a balanced whole" (Bode et al., 1949, p. 555): the tools that allow the trained mind to "effectively move back and forth between the perfectly describable Platonic world of theory and the fuzzy world of practice" (Richardson, 2004, p. 127). In other words, it is an epistemology of the way(s) in which mathematics attaches to the world beyond the world of physics. It is not "a general theory of practically everything" – that would be vacuous – but, in an analogue of the Mendeleev periodic table (Section 2.5), it serves as a device for pointing out similarities and analogies across all the systems sciences, from biology on up – the focus of interest of the majority of the scientists in most interdisciplinary teams.

Boulding (1956) continues with two brilliantly eclectic selections of such similarities and analogies; any statistical methodology for these must necessarily overlap with our toolkit here. A first set detects commonalities across fields in terms of their methodologies of the time. One of these is the notion of a *population* undergoing processes of birth, aging, and death. A second is the interaction of an entity (say, an organism) with its *environment* in order to maintain *homeostasis.* Indeed, Boulding defines "behavior" as the individual's restoration of that homeostasis after some environmentally induced perturbation.[3] A third universal approach is *growth*, especially in the formula "growth and form" that the biologists had already imbibed from D'Arcy Thompson (1917/1961). A fourth is *information*, in the sense of the physics or semiotics of communication. But a second, independent list, much more interesting from our 21st-century point of view, entails the classification of theoretical systems and constructs in terms of a hierarchy of *complexity.* This ladder begins at the lowest level with *frameworks* (static structures) and *clockworks* (of simple deterministic dynamics; compare Popper, 1965), and extends upward by the addition of information exchange, self-maintenance of structure (open systems), and so on through levels for which, Boulding admits, there are no adequate theoretical frameworks at all. Some of these ideas derive from Weaver (1948), who characterized the general systems sciences as the sciences of "organized complexity," in contrast to the sciences of "disorganized complexity" that are adequately described by "statistics of variation around average behavior" (his characterization of thermodynamics and by extension every other discipline that identifies variation with Gaussian models of noise). I return to this second list in Chapter 8, after Part III has surveyed a sampling of the necessary algebraic structures.

The essay by Bode et al. suggests that statistics should comprise a full 10% of the training time in the college curriculum of the man *[sic]* who intends to work as a "scientific generalist" in this way, and yet the description of the associated specific statistical tools is so vague it may as well have been omitted:

> In statistics we propose four semester courses, which can probably be found at a few institutions where the training in statistics has been well organized and developed: one or two semester courses in elementary mathematical statistics, one semester in design

---

[3] This definition shows signs of age, omitting, for example, everything that currently comes under the heading of cognitive sciences.

and analysis of experiment with practical work, and one or two semesters in advanced mathematical statistics, including multivariate methods and the use of order statistics. (Bode et al., 1949, p. 557)

In a sense, this whole Part III constitutes a 21st-century response to that shadowy 20th-century characterization: a survey of the statistical methods most broadly appropriate for holistic or multidisciplinary scientific work today. The statistics of organized complexity complements and supplements the Gaussian models that sometimes suit the fully disorganized complexities of high entropy, the Boltzmann world of diffusion and white noise. Note, too, that Boulding, Bode and their friends were writing before the dawn of scientific computing, without tools for analysis of images and networks. The world of computational statistics has changed utterly in the intervening half-century, and with it the world of methodology on which Deutsch et al. (1971, 1986) centered.

Then how do we build a method for numerical inference in this more complex, 21st-century context? By persevering in the use of the two most central concepts of Parts I and II, abduction and consilience, I submit, but by giving them broader, more flexible, generalized meanings.

5.2.1 Abduction and Consilience in a More General Context

To this point, abduction has meant

quantitative surprise, followed by quantitative resolution

and consilience has meant

getting the same number more than once.

Both need to be modified for these more complex system settings.

"The surprising fact C is observed," Peirce begins (Epigraph 9). In a complex system, what could make a pattern observation surprising? In quotidian life, it is a confluence of events that appear to be meaningfully related without any obvious causal connection (Diaconis and Mosteller, 1989). In the world of numerical phenomena that is the concern of this book, it is instead the emergence of a pattern whose "amplitude," in some sense yet to be specified, rises well above what is routine in the investigator's discipline or line of work.[4] Part of the job of the statistician is to compute the prior odds whereby simple coincidences of a combinatoric style can be reduced to commonplaces and thus obviated as grounds for further reflection or investment of intellectual resources. Readers probably know the "birthday problem" – how many people need to be at a party before there is a 50:50 chance that two have the same birthday? (23.) We have analogous ways of reducing the

---

[4] We have already (Section L4.3.4) noted Paul Meehl's discussion of the "crud factor" – the well-known fact that in psychology "everything is correlated with everything else." If there is nothing surprising in a correlation of 0.25 or so, neither is there any possibility of an abduction that accounts for it in some other way (for instance, as the product of two path coefficients each of value 0.5).

surprise in the observation of series of points within $\pm x°$ of straight lines (Kendall, 1989) or closely neighboring pairs in point processes, messages buried within long strings of characters (e.g., the notorious "Bible Code," Witztum et al. [1994]; McKay et al. [1999]), and other processes equally conducive to modeling by pure randomness. Chapter 6 extends these models to today's most frequently encountered and at least sometimes plausible representations of multidimensional numerical phenomena. There are analogous computations, too complicated for inclusion in this book, for dealing with clusters on maps of varying density (e.g., epidemics that may be over cities, suburbs, or exurbs) or on other kinds of algebraic or geometric structures.

Our interest is never in statistical significance testing. The objections to the consideration of only one hypothesis at a time (and that one a null, a hypothesis unconnected to any explanation) that were reviewed in Section L4.3.4 are even more important in these more complex contexts, where the deviation of the data from a model of meaninglessness can appear in any of indefinitely many aspects of a pattern. We require that surprise be *genuine*, like the obvious densities and voids in the distributions of the stars or the proteins (Figures 5.1 and 5.3) or the really odd fact that all seven babies in the 1973 Seattle study were light in birthweight, not quite right in the head, odd-looking, and born of alcoholic women. Two might have been a coincidence, but seven was simply too many to be ignored.

Likewise consilience is subtly redirected in contexts like these. In systems with an indefinitely wide variety of potential quantifiers, one must take care in deciding exactly what has been agreed *on*. In a complex [high-dimensional] situation there are an indefinitely wide range of possible forms of "agreement"; the contract must be carefully drafted in advance of any computation. For Figure 5.1, the consilience is with exact astrophysical theory; for Figure 5.3, it is with empirical verifications over what are now thousands of "solved" proteins (proteins for which the position of every atom has been mapped to 2 Å or so of precision in 3D, by crystallographic imaging). For Figure 5.4, it is a more semantic network, the construction first of what is called an *animal model* for fetal alcohol syndrome (an experimental treatment that, when applied to a mouse fetus, reliably produces facial deformities homologous to those of the afflicted human child), but subsequently, and more importantly, the confirmation (Section 7.7) that the pattern of brain damage is highly predictive of the pattern of behavioral deficit. For that reason, images of the FAS brain can sometimes serve as part of a "mitigation argument" – an argument for diminished culpability – in American death penalty trials (Bookstein, 2006; Bookstein and Kowell, 2010).

Because the number of possible parametric summaries grows so much faster than sample size, there is no "law of large numbers" for pattern analysis, and sheer mass of primary data does not correlate much with the yield of explanations. Consider Figure 5.8, for instance, which shows what might be described as "the largest fact in the universe" – it *is* the universe, in the sense of showing the peak frequency of the microwave background radiation left over from the Big Bang. (The white band across the middle is the Milky Way, our galaxy of residence; it is opaque to the instruments recording this quantity in every other direction.)

5.2 Abduction and Consilience in General Systems   305

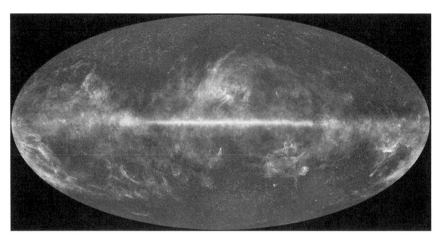

Figure 5.8. The largest possible image extent: a sky map of the microwave background radiation of the entire universe, courtesy of the European Space Agency. (For a color version, see http://www.esa.int/images/PLANCK_FSM_03_Black.jpg)

Once the missing geometrical coordinate (distance from Earth) is restored, one may appreciate how startlingly nonuniform is the distribution of matter around us at this same cosmic scale – the "complex web of clusters, filaments, and voids" that completely surrounds us here in the Milky Way (Courtois et al., 2013, p. 1). See especially Courtois's Figures 8 and 9, which show, among other features, the Great Wall of Galaxies discovered by Margaret Geller and John Huchra (1989).

Such maps are not subject to any obvious abductions in this explicit, extended form. One needs a quantitative model *first* (say, a model about the scale and amplitude of those fluctuations) before any intelligent question can be asked. A similar frustration arises if one tries to ask questions about the general metabolism of the animal cell, the subject of a magnificent chart from Roche Applied Science that can be viewed and searched at http://www.expasy.ch/cgi-bin/show_thumbnails.pl. This reader finds the chart as a whole to be quite incomprehensible in terms of its row-column coordinates – it is not even legible without the use of a digital "magnifying glass." As a reference work, or some version of what used to be called a library's "card catalog," it seems to lack the hooks to the retrieval tools that would help specific queries about how our metabolism reacts to novel situations or diseases or how it came to be what it is. The entire classic literature of graphical design, from Brinton (1914) through Bertin (2010), also lacks tools for coping with this kind of complexity. A recent masterpiece of modern design, Rendgen and Wiedemann (2012), explores a variety of contemporary uses of display space for such retrievals, but it is still the case that one must understand the appropriate pattern language in advance to exploit the proferred tools effectively. See also Lima (2011).

Advanced bioinformatics (the "ChIP-chip" readouts), atomic physics (bubble chamber and other image records of cosmic particles), functional magnetic resonance imaging (millions of image voxels observed every few seconds for half an hour over multiple subjects), ... – the sheer volume of data resources available to the analyst is growing faster than exponentially, only exacerbating all our problems of pattern insight. Even the methods of this book, the best I know, are applied only sporadically in the fields of greatest need.

5.2.2 More Advanced Schemes for Data

Nevertheless we need to locate, or generate for ourselves, pattern languages rich enough to permit comprehensible talk about the extent of surprise afforded by a particular empirical manifestation of some unexpected structure.[5] There will be a quite general approach to information criteria for judging among multiple hypotheses at Section 5.3. The rest of this subchapter deals with two classes of more elementary methods closer to the level of ordinary or augmented visual perception where the appreciation of pattern ordinarily begins. One of these models, discussed in this subsection, is the tabular metaphor, data as matrix. The other, in the next subsection, centers on more sophisticated models for what might count as a "completely random" (i.e., explanatorily worthless) process.

What is a matrix? How does it represent data? The core metaphor of *multivariate statistics*, one of the great technology transfers of 20th-century applied mathematics, applies a particularly efficient algebraic notation, the *matrix*, to denote a wide variety of manipulations of one common data type. Data matrices have no real epistemology (not outside of quantum mechanics, anyway). They serve us mostly as bookkeeping devices, a simplification of notation that arose around 1900 in the literature of least squares – they make it easier to write down the algorithms corresponding to certain numerical maneuvers. The application to least squares computations converged with some applied mathematics that had been evolving for some time to deal with the centrality of these same forms in "linear algebra," the mathematics of vector spaces (a generalization of the representations of ordinary three-dimensional mechanics).

This may be the reason there seems to be no published history of multivariate statistical analysis – its ideas do not typically arise from those of applied technoscience, which *does* have a history, but instead from a relatively closed collection of 19th-century mathematical maneuvers designed for applications in physics and engineering long before the sciences existed that are most often seen to use conventional linear models today (e.g., social psychology, epidemiology, econometrics). When I

---

[5] The main attempt to formulate a general language of pattern in probabilistic/statistical terms is Ulf Grenander's *General Pattern Theory* of 1994 (for a simplification of this 900-page treatise, turn instead to Grenander, 1996). The word *pattern* has so many different meanings (Moyers et al., 1979) that it is difficult to claim any primacy for the scientific construal, let alone the statistical one. Grenander's book joins other books with the word "pattern" in their titles that represent overlapping approaches to the same concept: for instance, Christopher Alexander, *A Pattern Language* (1997), about how we view houses, neighborhoods, and cities; Peter Stevens, *Patterns in Nature* (1974), about biological textures; or Philip Ball, *Nature's Patterns* (2009; in three parts, "Shapes," "Flow," and "Branches"). Books like these tend to be stunningly beautiful; they serve to prod the imagination via the sheer visual pleasure they impart.

5.2 Abduction and Consilience in General Systems

was a sociology graduate student, around 1970, the core seminars in methodology all tried to fit the round pegs of real social systems data and research styles into the square holes of matrix formalisms. (For instance, we were explicitly instructed to pay no attention to Mills [1959], who asserted that anything worth saying about modern social systems can be communicated using ordinary crosstabulations.) It was a kind of brainwashing, really, with no attempt at a justification nor any legitimation of the urge even to *ask* for a justification.

Mathematically, a matrix is a simple rectangular array of numbers, a table with some count of rows and some count of columns where neither the rows nor the columns have any further structure. If the counts of rows and of columns are the same, they might be specified to be the same list, in which case the *diagonal* of the matrix, the cells with row number equal to column number, becomes a further potentially useful entity. Algorithms that deal with the matrix structure rest mainly on rules for $+$ and for $\times$. There is a kind of data to which this format is most appropriate: the so-called "flat file," cases by variables, where every measurement is labelled by precisely two tags, which case (or specimen, or subject) it is and to which question or probe this number is the answer. But such a format is hardly ever suited to the questions that arise in the sciences of complex systems, where, usually, there is a structure all their own to the questions, a separate structure to the "cases" (if they are discernible at all as separate entities), and far more information associated with concatenations of these and other labels than can be represented as a single numerical entity. In the morphometrics work to be reported in Chapter 7, for instance, the original datum for a particular person consists of an entire solid image, perhaps 8 million rectangular bricks, each bearing a uniform shade of gray. The bricks do not correspond from person to person in any obvious way – the construction of a canonical far-from-obvious correspondence is, in fact, the main work of morphometrics. After that modification, this may be the only branch of applied statistics where a matrix *does* represent data in any epistemologically satisfactory fashion.

Figure 5.9 explores several of the ways that these remarkably useful rectangular arrays of numbers can lead to guides to inference in the form of diagrams. At upper left is an example of a matrix as *operator from vectors to vectors*, in two versions: above, the matrix $\begin{bmatrix} 0 & 1 \\ -1 & 0 \end{bmatrix}$ for a 90° clockwise rotation as it operates on a two-vector; below, in what amounts to just another mathematical pun similar to those we already examined in Chapter 4, it operates instead on a picture.

At upper right a matrix is shown as a *gridded surface*, here, the deviation of the midplane of the mouse brain in Bowden et al. (2011) from planarity. Grid points not inside the midsagittal outline of the canonical mouse brain are left blank. In this setting, the matrix is just a way of keeping track of two subscripts.

At center left is one example of the matrix as *covariance structure*. The eight arrows explicitly visualize the regressions of Henning Vilmann's eight rat neurocranial landmark points on Centroid Size by segments connecting the predicted location of each point at an extrapolation 50% past minimum size to the predicted location

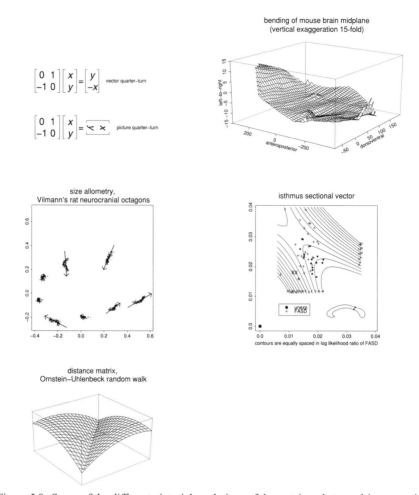

Figure 5.9. Seven of the different pictorial renderings of the matrices that can drive numerical inferences.

of the same point at an extrapolation 50% past maximum size. Plotted under these arrows is a complementary diagram, matrix as *data structure*, namely, the 16 shape coordinates of those same points, ordered as 8 Cartesian pairs. The mean of this structure was used to center the rendering of the arrows, which represent the biological phenomenon called *size allometry* or *allometric shape*. For the anatomical context of these locations, see Figure 6.25; for a different statistical treatment of those same underlying locations, see at Figure 7.5 and the corresponding text.

At center right is the sketch of a matrix as a *quadratic form*, which is to say, an operator that turns any vector of appropriate length into an ordinary real number. Plotted here is the likelihood ratio in favor of a hypothesis of deformation consistent with fetal alcohol spectrum disorder (FASD), versus the competing hypothesis of anatomical normality, for the 45 adult male subjects of the Seattle fetal alcohol brain study (Section 7.7.4). Quantification is reduced to one single shape coordinate pair, the transect of the callosal midcurve shown in the inset (compare Figure 7.25). The surprise is that this quadratic form is hyperbolic, not elliptical: it goes to large positive values in two directions and large negative values in two others. Solid circles, unexposed brains; plus symbols, brains for subjects previously diagnosed with an FASD. XX is the location of the corresponding transect for a convicted murderer who was known to have been exposed to high levels of alcohol before birth. The rhetoric of the analysis here, which supported his classification as FASD at odds of over 800 to 1, was discussed in Bookstein and Kowell (2010). The algebra of these elegant figures will be discussed in more detail in Section 6.5.

Finally, in the lower left panel of Figure 5.9, we see an example of the square matrix that is a *pattern of squared interspecimen distances*. The example here is the modification of the geometry of a high-dimensional random walk (Bookstein, 2013b) by an additional term that regresses forms weakly backward toward a fitness maximum.

The five panels of Figure 5.9 thus illustrate seven wholly different contexts for this protean mathematical object. There are really manifold different contexts for their application in numerical science, not just one (the "data matrix") or two (the "covariance matrix" also). The information content and rhetorical applications that make one of these senses useful are not the same as those that make another one useful, and none of them can claim to represent data from complex systems in any fashion that is automatically useful. What makes *any* of them useful is, instead, the extent to which consilient patterns of numerical inference can be represented with their aid. Thus the usefulness of the formalism is a function of the examples built with its aid in this and the next two chapters.

For instance, there are a variety of preexisting canonical forms for *selecting* interesting descriptive features that simplify the matrix into one or another report that can, under some circumstances, become the focus of an abduction. There are two general classes of these descriptors, differing in their semantics. *Eigenanalyses* supply a series of *directions* in the vector space of the columns of the matrix; *profiles* assort the columns themselves by derived properties such as salience for prediction. We encounter both of these types in later examples.

In spite of the splendid range of the examples in Figure 5.9 (among innumerable other instances I might have chosen – one produces a *lot* of these over a 35-year career!), in my view the underlying machinery for encoding data structure is unacceptably impoverished in most applications. Even in the familiar special case where the rows constitute a "random sample," for instance, we are afforded no notational possibilities for the *other* set of labels, the columns. Today's natural and social sciences that rely mainly on matrices to encode data sets thus end up in desperate need of escapes like the network graphs to be considered below. If the rows and columns of a square matrix come in a numerical order, for instance, one can encode that order with the aid of an additional matrix looking like

$$\begin{bmatrix} 0 & 1 & 0 & 0 \\ 0 & 0 & 1 & 0 \\ 0 & 0 & 0 & 1 \\ 0 & 0 & 0 & 0 \end{bmatrix}$$

that indicates the successor row or column for each original row or column, or a matrix like

$$\begin{bmatrix} -1 & 1 & 0 & 0 \\ 0 & -1 & 1 & 0 \\ 0 & 0 & -1 & 1 \\ 0 & 0 & 0 & 0 \end{bmatrix}$$

that subtracts each row from its successor. One can add further subscripts to a matrix, replacing the two-dimensional system (row by column) by a three-way analysis (row by column by time), or a four-way analysis, or something higher. Or one can fill the cells of a matrix with little geometric objects that are more complicated than simple numbers – in magnetic resonance imaging, they are little 3-vectors; in diffusion tensor imaging, they are little 3×3 matrices of their own, leading to various seductive visualizations but not, to date, any generally useful pattern language. Interestingly, the square matrix (last row of Figure 5.9) that represents some of the adjacency information in networks, graphs, and dyadic dissimilarities can be transformed back to a hypothetical row-by-column form by imputing a pattern for the missing index term, the columns. We will deal with data resources of this format in Section 6.5.

Still in this conventionalized row-by-column setting, one important application of matrices arises in connection with the information methods of Section 5.3. Many scientific hypotheses involve the values of one or more decimal *parameters* that we would like to estimate from data, and under fairly general conditions of random sampling the uncertainty associated with any such estimate of a list of parameters is represented pretty well by the negative of the matrix of second partial derivatives of the net log likelihood of the data near the estimate. Specifically, the inverse of this matrix is proportional to the expected covariance of the parameter estimates, as a vector, in the imaginary world where they can be computed over and over from endlessly new samples of the selfsame process. This *information matrix* (the name is Ronald Fisher's) is not a model for scientific measurement, but it is a useful way of communicating the implications of one's data for one's parameters. If matrices are much more important in the statistics of "flat" (cases-by-variables) data than in most other applications of numerical inference to the natural sciences, it is because in the natural sciences, where data arise most often from machines, typically we know far more about our cases and our questions than just the row number or the column number, respectively.

Though this book strongly counsels against any reliance on significance tests, it is nevertheless remarkable to note how widely diversified they have become in the hands of typical qualified users. The kinds of pattern entities that multivariate significance tests deal with are a rich underlayment for the much more important questions that deal with patterns among those descriptions. Any graduate text of multivariate analysis, such as Mardia et al. (1979), Krzanowski (1988), or Morrison (2004), will offer, among other tools:

- Tests for a preassigned vector, such as zero, to stand in for the computed profile of something (e.g., prediction coefficients) over a set of columns (variables)
- Tests for inconstancy of these coefficients as something else is varied (that is to say, *interaction terms*)
- Tests for the pattern of covariances among the columns of such a matrix (on the assumption that it is scientifically appropriate to describe the variables on the list by something as straitjacketed as a covariance) – tests for covariances all zero, zero between lists of variables, proportional to some standard matrix, or otherwise conducive to some simple verbal summary. As noted, these tests go much better in application to estimated parameters, which *must* be describable by a covariance, than to raw data.
- Tests for a variety of differences between groups defined in advance of the analysis, such as males versus females, or people who prefer vanilla versus people who like chocolate, or people versus chimpanzees, or rats (or patients, or bacteria) from treatment group 1 versus from treatment group 2

Nevertheless, in general the techniques that represent data sets as ordinary matrices support neither abduction nor consilience. Their information content is just too reduced for that; they are instead prophylactic. In bureaucratic contexts, they can supply superficially rational justifications of protocols for the investment of further resources under conditions of scarcity, for reimbursement of medical procedures in America, for assignment of scarce space in peer-reviewed scientific journals, and so on. As statisticians, we emphasize the mathematics of $+$ and $\times$ as they apply to the square matrices far more than is justified by anything in the world of data to the matrices of which we are pretending to align this formalism. For instance, Tyson (2003) is a wonderful survey of the models that really describe how biological systems are working over time within the space of a cell, and the corresponding models don't look anything like matrices; Coombs (1956) is a wonderful survey of the true epistemology of measurement in psychology, and the corresponding models don't look anything like flat matrices; the Grenander (1994) approach to pattern grammars looks nothing like matrices; the Grenander–Miller (2007) approach to analysis of medical image deformations doesn't look anything like matrices; and so on. *None of the interesting problems in the sciences of organized systems appear to be facilitated by denoting empirical data in terms of flat row-by-column matrices.* Matrix notation is simply inadequate to convey the complexity of real scenarios in organized natural or social systems.

In a deeper mismatch, the operations of abduction and consilience are not based on + and ×, the operations that drive most of matrix algebra, at all. Decision rules involve, after all, a discontinuity – the most severe category of nonlinearity – right at the core of the process. There is a mathematical model for these, the *catastrophe*, which isn't notated effectively in terms of matrices. In a sense, to abduce is to search for the places where the matrix formalism most obviously and spectacularly fails: events to call to the attention of subsequent arithmetic. Where do we look for these? If an image: a focus, a hot spot, a funny shape, an alignment or other strange syzygy. If there are two images, we might model the relation between them as a diffeomorphism and search for extrema of derivatives – a crease (see Section 7.7.3). If there is a network, we look for its disconnections, its multiple connected components, or the easiest places at which to intervene (links to cut) to *make* it disconnected.

The principal challenge of good systems statistics is different from the "matrix mechanics" that characterizes routine significance-test-based inquiry: it is to move beyond that matrix default formalism as soon as possible. Variables can be ramified beyond the caricature of rows and columns in many different ways: as maps, for instance, or networks, images, spectra, or branching diagrams. In these ramifications usually lie the potential abductions and consiliences that we hope will arise in the applications of numerical reasoning to complex systems. The corresponding data may arise from machines, from people, or from people overseeing and annotating the outputs of machines; what matters is the logic by which numbers are claimed to lead to insights and explanations. The tools of this approach arise from the assumptions (a strongly ordered set of columns in the data, a small number of underlying pure substances undergoing mixture), not from the axioms of the 19th-century matrix machinery we are borrowing; the application is of a wholly different logic than the algebra of + and ×. The sense of surprise driving abductions in this domain depends a good deal on *counting* structures: one where there should be many (peaks in a spectrum, for instance), or many where there should be one (ellipses in a density plot). A least-squares display that takes the form of two crossed lines, for instance, instead of the prototypical ellipse (for an example, see Bowman and Jones, 2010), strongly hints at an explanation in the form of two distinct processes correlated to some unspecified (mild, in this example) extent. But a contingent multiplicity like this is quite difficult to denote in the matrix-algebraic setting; so is discontinuity/catastrophe; so is coincidence.

5.2.3 Models for Randomness Complement Our Models for Order

Though the matrix models for data and data-driven questions are thus somewhat oblique to our explanatory purpose, a different alternative is doing splendidly these days: an approach by explicit quasiphysical modeling of randomness. The distinction is between constrained relationships and historical contingencies. The phenomena of random walks and random fields make the distinction painfully clear.

Among my favorite Romantic paintings is Caspar David Friedrich's *Der Wanderer über dem Nebelmeer*, the painting on the cover of this book. The scene here is partially real – the visitor gazes out from a specific vantage point on the Kaiserkrone in the Elbsandsteingebirge in Bohemia, not too far from Vienna. To indulge in a pardonable anachronism, the painting serves this chapter very well as metaphor of the dilemmas arising when the quantitative scientist tries to apply notions of abduction and consilience to scenes as realistically complex as this one. In the scene, there are crags, and there are clouds. The clouds, we understand, are stochastic on a very small scale. The crags look individuated, with names and with cairns at their tops, but in fact they are also stochastic themselves – historically contingent, albeit at an enormously larger scale (namely, the orogeny of the Alps). What they have in common is the origin in a random process, but the methods we use to characterize that randomness vary wildly over the scales at which these phenomena might be encountered. What is random at one scale (the pattern of peaks in the Alps) is wholly deterministic at another (the paths to the top, according to the contour map), all the while the fog swirls around. As Popper (1965) says, all clocks are clouds, all clouds are clocks. Popper meant this as a discussion of physical reality, but it applies in discussions of abduceable data as well. The duality between apparently individuated entities (mountains, species, anatomical components) and persistently patternless noise characterizes most scientific studies at any level of observation from atoms on up.

Complementing this fine painterly metaphor of Friedrich's is a set of formal tools from probability theory that can be imagined as a sermon on all the different ways of generating meaninglessness. Meaninglessness of pattern is, after all, itself a pattern, one for which the mathematical languages available to us have been particularly well-developed over the 20th century. (The characterization of mathematics as the "science of pattern" or "the formal study of patterns" is a commonplace of the semipopular literature: see Thurston [1994] or Devlin [1994].) If mathematics is the search for order (both old forms and new), then one construal of statistics is as the formal study of *disorder*, and the two approaches cannot be disentangled. The role of probability theory in our domain of "numbers and reasons" is as the formal mathematics of the latter class of models, the models of pure disorder. To this mathematics corresponds, furthermore, an actual physics, the statistical thermodynamics of entropy. We see presently, for instance, that the disorder of gases as a given temperature (the Maxwell–Boltzmann distribution, Section E4.3) can be represented by theorem as having the highest entropy of any distribution of a given variance. See also Lindsay, 1941. Such models *must* be in the margins of our awareness, or even nearer, whenever we are to begin an abduction – to claim that a finding is "surprising" on any rational grounds. Later, we need operations like + and × to navigate the "matter of course" reconstruction of the finding and then the consiliences of its recurrence in other investigations.

5.2.3.1 The Bell Curve, Revisited

Section L4.2 explained one origin of the bell curve in a physical randomizing process, the bouncing of a ball down the pins of Galton's quincunx (Figures 4.13–4.15), which is equivalent to arbitrarily many repeated samples from the distribution of the count of heads in flips of an adequately large handful of coins. That was a distribution in one spatial dimension. I went on to mention Maxwell's argument that this *must* be the distribution of the velocities of elementary interacting entities such as gas molecules given only two seemingly nonmathematical axioms, axioms about symmetry rather than formulas: the distribution of velocities must be the same in every direction, and the velocity in one direction must be independent of the distribution of velocity in every perpendicular direction. These appear to be completely unrelated approaches to the same mathematical object, so that their agreement on the formula seems to be another illustration of Wigner's "unreasonable effectiveness" of mathematics in the physical sciences. But that is not the case: we can characterize the Gaussian distribution as unique using arguments making no reference to spaces, rooms, or symmetry. Jeffreys (1961, p. 101) puts the point pointedly:

> The normal law of error cannot be theoretically proved. Its justification is that in representing many types of observations it is apparently not far wrong, and is much more convenient to handle than others that might or do represent them better.

The reason it is to be seen as "convenient" is the enormous range of kinds of argument from which it appears to emerge as a reasonable formalism. Here are several, mostly from Jaynes (2003), Chapter 7.

Gauss's characterization. In Section L4.3 we argued that *if* variation obeys the Gaussian law *then* the average is the maximum-likelihood estimate. The main reason the distribution is called "Gaussian" is that Gauss provided the *other* inference: if we want the average to be the likeliest true value, *then* the probability distribution must be the bell-curve one. His argument goes roughly as follows. Let $g(u)$ be the logarithm of the symmetric probability density we are looking for, and suppose our sample consists of n values $x_1, x_2, \ldots x_n$ all equal to the same value u and one value that is $-nu$ (in order to average 0). Then for 0 to be the maximum of likelihood for the mean of all $n + 1$, the derivative of the log likelihood $\sum g'(\mu - x_i)$ must be 0 at $\mu = 0$ – we must have $g'(-nu) = -ng'(u)$. So the derivative of the log of the probability density must be a linear function of u through 0. The exponent must thus be something positive times $-u^2$.

The reproducing characterization. Suppose we are interested in *white noise* (the fluctuation of a physical quantity, such as the voltage v in a circuit) that is a function of many other things, and suppose that the empirical distribution of this fluctuation takes the same form in every experimental measurement except for an amplitude. In other words, suppose that the form of white noise is the same up to a scale factor regardless of the mechanism by which it is produced, and that adding another source to the generator changes the scale without otherwise changing the form of the distribution. What distribution shape can satisfy this condition as we change σ^2? If we write this probability distribution as $p(v|\sigma^2)$, it is not elementary but nevertheless straightforward (see Jaynes, 2003) to show that p must satisfy the *diffusion equation* $\frac{\partial p}{\partial(\sigma^2)} = \frac{1}{2}\frac{\partial^2 p}{\partial v^2}$. But the solution of this equation is in fact the Gaussian law $\frac{1}{\sqrt{2\pi}\sigma}\exp\left(-\frac{v^2}{2\sigma^2}\right)$.

5.2 Abduction and Consilience in General Systems

The chi-square. A surprisingly useful aspect of the bell curve is via the *square* of its value, or the sum of squares of a number of these. In a development first suspected by Karl Pearson (he of the correlation coefficient) around 1900, it can be shown that the information content of a "sufficiently large" data set that summarizes samples of independent, identically distributed "cases" for the question of whether a simpler model fits better than a more ramified one, one with additional parameters, is approximately distributed as one of these sums of squares, the number of squares in the sum being the same as the number of independent parameters being estimated. More specifically, for many applications to studies aggregating data from large samples of identically distributed cases, -2 times the logarithm (base e) of the *likelihood ratio* (recall Section L4.3), the probability of the data on the first model vs. the probability of the data on the best-fitting version of a second, is distributed as a sum of as many independent squared Gaussian variables as there were additional parameters to be estimated in the more carefully specified model. This formulation includes the chi-square for tests on contingency tables (crosstabs) and also all the extensions mentioned under Section 5.2.2 in connection with the general multivariate Normal model. If the underlying quantities being squared really *did* come from k different bell curves of mean 0 and variance 1, then the sum of their squares is called a *chi-square distribution on k degrees of freedom*. The distribution has a mean of k and a variance of $2k$, and proves a very handy guide to what is meant by "unsurprising" in a variety of practical contexts. See, for instance, the discussion of T. W. Anderson's test for sphericity at Section 5.2.3.4.

Maximum entropy. We would like our models of noise to be models of ignorance: they should not be stating anything that has information content. We see in Section 5.3 that there is one single formula for information content that applies across the universe: the formula $\int_{-\infty}^{\infty} -p(x) \log p(x)\,dx$ for a continuous distribution, or $-\sum p_i \log p_i$ for a discrete distribution. As it happens, the Gaussian distribution has the greatest entropy (the least information) of *any* distribution of given mean and variance: $\frac{1}{\sqrt{2\pi}\sigma} \exp\left(-\frac{(v-a)^2}{2\sigma^2}\right)$ minimizes $\int_{-\infty}^{\infty} -p(x) \log p(x)\,dx$ over the class of all distributions $p(x)$ of mean $\mu = \int x p(x)\,dx$ and variance $\sigma^2 = \int (x-\mu)^2 p(x)\,dx$.[6] "For this reason," Jaynes goes on,

> inferences using a Gaussian distribution could be improved on only by one who had additional information about the actual errors in his specific data beyond its first two moments and beyond what is known from the data.... [The reason] for using the Gaussian sampling distribution is not because the error frequencies are – or are assumed to be – Gaussian, but because those frequencies are unknown.... Normality [is] not an assumption of physical fact at all; it [is] a description of our state of knowledge. (Jaynes, 2003, p. 210.)

---

[6] *Proof*, after Rao (1973, p. 163). We use Gibbs's Inequality, to be introduced a few sections further on, that for any two probability densities p and q we must have $-\int p \log p\,dx \leq -\int p \log q\,dx$. Choose q as the Gaussian distribution with $\log q = \alpha + \gamma (x-\mu)^2$ for $\alpha = -\log(\sqrt{2\pi})\sigma$, $\gamma = -\sigma^2/2$, and we find $-\int p \log p\,dx \leq -\int p(\alpha + \gamma(x-\mu)^2)dx = -(\alpha + \gamma \sigma^2)$. But this is just the entropy of the Gaussian on the given mean and variance. This is another typical mathematical proof style, in which the answer "comes out of nowhere" but is easily verified once we have it.

This comment applies to "95% of the analyses," namely, whenever the data are presumed to arise as independent samples identically distributed.

Here the ubiquity of the Gaussian distribution appears ineluctable. To change the answer, we are going to have to change the question: to ask about types of data that are different in design from the accumulation of independent observations, identically distributed, that all underlie the justifications of the Gaussian. That is to say: we wish to ask about statistics for organized systems, where separate components of the data set are entwined with one another, not independent and certainly not identically distributed. We wish to cease sampling over interchangeable cases and instead begin sampling over domains of an integrated entity (an organism, a social system, a mind). Empirically, this change of focus is easy to accomplish: there are a great many other fundamental set-ups in the statistics of general systems. Here, in less detail, are some of the alternative probability models that arise to model "pure chance" in these other different settings.

5.2.3.2 Random Walk

One simple way to intercept the assumptions that make the Gaussian distribution so disorganized is to contravene the independence of successive observations. If we are thinking in terms of coin flips, this means to keep track of the *running sum* of heads minus tails: from the flips $X_1, X_2, \ldots X_k, \ldots$, where each X is either $+1$ or -1, we construct and attend instead to the totals $S_1 = X_1$, $S_2 = S_1 + X_2, \ldots$, $S_k = S_{k-1} + X_k = X_1 + X_2 + \cdots + X_k, \ldots$. There has already been an introduction to this concept, both in one dimension and in many dimensions, back in Chapter 2. There we discussed the arcsine law, which describes the location of the extreme value of the walk and also the fraction of time it spends above the x-axis, and we sketched what happens when we apply principal components (Section 6.3), with their assumption of identically distributed cases, to data that actually obey the entirely different law here. But one question has been overlooked: *given* a model for those increments X_i, say, a coin flip, or a bell curve, *does* the empirical time series show a range consistent with it? Interesting explanations apply whenever it does *not*, and those explanations arise in two versions: when the series shows too *much* range (in which case we suspect the walk of having a "drift," a preferred direction of change), and when the series shows too *little* change (in which case we suspect the underlying process is actually one of canalization, or stabilizing selection, or the *Ornstein–Uhlenbeck* equations[7] that bias the increment to have the opposite sign from the current state).

If the original data are distributed as a bell curve with mean 0 and some variance σ^2 (and indeed even if they aren't, as long as k is large enough and σ^2 exists – this is

---

[7] The Ornstein–Uhlenbeck process can be imagined a modification of the simple heads-minus-tails random walk (Chapter 2) that keeps its variance finite over time. In one congenial version, instead of flipping one coin over and over (the interpretation of the quincunx), we imagine selecting a coin from a large sackful of black coins and white coins, such that whenever we pick a white coin we replace it with a black coin, and vice versa. Lande (1976) introduced this model into the study of evolutionary processes.

5.2 Abduction and Consilience in General Systems

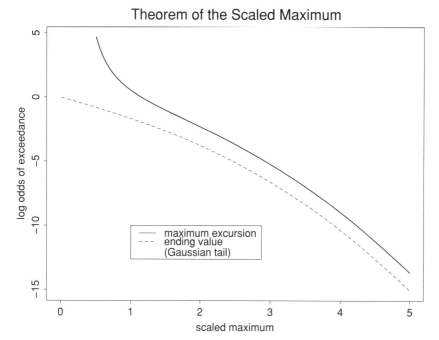

Figure 5.10. The Theorem of the Scaled Maximum calibrates the probability of *ever* exceeding a multiple of the expected standard deviation as a function of that multiple, as the length of the random walk increases without limit. Horizontal: scaled maximum of the walk, $\max_{1 \leq k \leq n} |S_k|/\sigma \sqrt{n}$ where σ is the variance of the single increment. Vertical: log-odds of exceeding that multiple somewhere, in the limit of an arbitrarily long walk ($n \to \infty$). The Gaussian tail curve, drawn in as the dashed line for comparison, starts at log odds zero (a 50:50 chance) for the average and drops off in familiar fashion. The excursion tail (solid curve) begins at likelier than not (e.g., odds of about 6:1 for an excursion of at least $0.75\sigma \sqrt{n}$ somewhere) and falls to 50:50, even odds, at about $1.15\sigma \sqrt{n}$. For large excursions its limiting value approaches precisely four times the Gaussian tail.

one version of the Central Limit Theorem), the distribution of any particular partial sum S_k is going to be a bell curve with mean zero and variance $k\sigma^2$. But that is not our question. Rather, the issue is that of the total range of the process over its full length – the expected distribution of the quantity against which we might under some circumstances mount quite a powerful abduction indeed. The distribution is shown in Figure 5.10, and its formula comes from the

Theorem of the Scaled Maximum. For any $x > 0$, define $F(x)$ as the fairly complicated expression

$$F(x) = 1 - \frac{4}{\pi} \sum_{k=0}^{\infty} \frac{(-1)^k}{2k+1} \exp\left(-\frac{\pi^2}{8}(2k+1)^2 x\right).$$

Then for any random walk for which the elementary step X has a finite variance σ^2 and a symmetric distribution (i.e. $p(X = x) = p(X = -x)$, all x), we have

$$\lim_{n \to \infty} \text{prob}\left(\max_{1 \leq k \leq n} |S_k| \leq \sigma x \sqrt{n} \right) = 1 - F(x^{-2}).$$

The infinite series for F actually arises from an elementary argument that breaks the random walk the first time it reaches the level x, and reflects it; then breaks it again the first time it reaches $-x$ on the way down, and reflects it there; and so on. There is also a fairly elegant step at which the square wave that is multiplying the Gaussian distribution involved in this argument gets expanded as a Fourier series, and another where one looks to the published tables for the value of the definite integral $\int_{-\infty}^{\infty} e^{-pt^2} \cos qt \, dt$. Bookstein (1987) explains the history of the theorem and provides the details for which we do not have the space here.

For readers who have not seen an argument like this one before, or who are accustomed to using statistical tables only in conventional form, a short review of the logic here might be helpful. We begin with a class of scenarios (in this case, random walks, cumulative sums of independent draws from some distribution symmetric around zero). We are searching for the abduction-launching surprise that ends up replacing this scenario, which is presumably of little interest, with another that involves some sort of wiser explanation than sheer random sampling (e.g., a trend, a bias in the increments, a restoring force that keeps the empirical range restricted). From the description of the kind of quantity on which a narrative is likely to seize we turn to the mathematicians to generate a reasonable model for that quantity under the assumed nonexplanatory distribution. We then compare our observed pattern magnitude to the distribution given in the chart, and proceed accordingly with the rhetoric. You are probably used to this logic as applied to z-scores (observations that are interesting when they are a few standard deviations out along a postulated bell curve, the way Hansen used them for his argument in Appendix 3.A), and you may further have seen analogous tables or charts that govern descriptors like F-ratios, t-ratios, nonparametric statistics, and the like. The maximum excursion distribution here is the first of a series in this chapter that you have probably *not* seen, because they arise in a context of systems more complex than the simplistic linear models that lead to z's, t's, and F's.

5.2.3.3 Over Networks: The Precision Matrix

Continuing with the promise of this section, let us now consider a quite different kind of system, where the underlying matrix is a square construction of relationships between pairs of entities. The entities could be names of molecules, and the relationships the coefficients of the chemical-kinetic equations that produce them as a function of the concentration of other molecules in their pathways; or the entities could be measured variables and the relationships the "special factors" that account for strong entrainments of the values of one against the other independent of larger-scale controls of the situation. In the former situation, if there is a special relationship between the concentrations of molecule X and molecule Y, and we are

5.2 Abduction and Consilience in General Systems

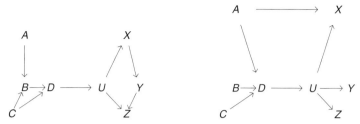

Figure 5.11. Two different network structures. (Left) The network separates at the link $D \to U$. (Right) A different network does not separate there – there is an additional process linking measurements A and X. $A, B, \ldots Y, Z$: measured variables. $\to$: causal connections or processes. See, in general, Pearl (2009). These were directed links (arrows); there are analogous symbolic representations with undirected links, such as networks of friendships.

concerned (perhaps for reasons of health) to control the concentration of Y, it would be reasonable to proceed via interventions that affect the concentration of X. In the latter situation, if we are interested in the hierarchy of system dimensions, it would be reasonable to search for processes local to the X, Y pair of descriptors in addition to any other processes that affect the system as a whole. See Figure 5.11.

But to proceed with either of these approaches to consilience, we need first to be surprised by the value of a quantity representing the strength of that pairwise association. Hence we need a formula for extracting them, a whole set at a time, so that we can decide if one or a few are substantially more worthy of attention than the others.

When the links between nodes of the network graph are covariances, there is an unusually simple approach that will often work right out of the box. Take the usual covariance matrix Σ. Find its *inverse* Σ^{-1}, the matrix that, when multiplied by Σ (on either side), yields the identity. (This is often called the *precision matrix* or the *concentration matrix* corresponding to the original covariance structure Σ.) Divide each element of Σ by the square root of the product of the diagonal elements in its row and in its column. The resulting quantities, multiplied by -1, are the correlations of each pair of the original variables after regressing out *all of the other variables in the study*. This approach is sensible when the ultimate explanation of the network's behavior is intended as a superposition of "general factors," applying to large subsets of the measurements if not all of them at once, atop "special factors," which represent processes local to small sublists. In the approach via concentration matrices, "local" is taken to mean "dyadic."

I exemplify this approach using a set of six variables that were first discussed by Sewall Wright in 1922 as an example of another method entirely. The correlation matrix R below is for six length measurements in a data set of 276 leghorn chickens. The most accessible canonical source these days is volume 1 of his late-life masterpiece *Evolution and the Genetics of Populations* (1968). Wright's analysis modifies the principal components approach of Section 6.3 to highlight pairs of original measures that show unusually high correlations, in this case measures on the same structure (head–head, wing–wing, leg–leg), resulting in a stack of four separate explanations, one "general" and three "special." But the construction of these modifications was uncomfortably subjective, and the method has never proved very popular.

In this analysis of Wright's, the correlation matrix

$$R = \begin{pmatrix} 1.000 & 0.584 & 0.615 & 0.601 & 0.570 & 0.600 \\ 0.584 & 1.000 & 0.576 & 0.530 & 0.526 & 0.555 \\ 0.615 & 0.576 & 1.000 & 0.940 & 0.875 & 0.878 \\ 0.601 & 0.530 & 0.940 & 1.000 & 0.877 & 0.886 \\ 0.570 & 0.526 & 0.875 & 0.877 & 1.000 & 0.924 \\ 0.600 & 0.555 & 0.878 & 0.886 & 0.924 & 1.000 \end{pmatrix}$$

is approximated as $g \otimes g + s_1 \otimes s_1 + s_2 \otimes s_2 + s_3 \otimes s_3$ where $g = (0.636, 0.583, 0.958, 0.947, 0.914, 0.932)$, $s_1 = (0.468, 0.468, 0, 0, 0, 0)$, $s_2 = (0, 0, 0.182, 0.182, 0, 0)$, and $s_3 = (0, 0, 0, 0, 0.269, 0.269)$ and, for sound but technical reasons, the diagonal entries of 1.000 are not part of the approximation. (The notation "$\otimes$" means "outer product," the matrix whose entries are pairwise products of the two vectors either side of the $\otimes$.)

As by happenstance the highlighted subsets of the original Wright example are all pairs of variables,[8] we can apply the method of concentration matrices instead. The concentration matrix derived from Wright's original correlation matrix is

$$R^{-1} = \begin{pmatrix} 1.880 & -0.629 & -0.298 & -0.299 & 0.105 & -0.350 \\ -0.629 & 1.746 & -0.779 & 0.462 & 0.065 & -0.377 \\ -0.298 & -0.779 & 10.172 & -7.018 & -1.531 & -0.687 \\ -0.299 & 0.462 & -7.018 & 10.237 & -1.035 & -2.029 \\ 0.105 & 0.065 & -1.531 & -1.035 & 7.902 & -5.139 \\ -0.350 & -0.377 & -0.687 & -2.029 & -5.139 & 8.568 \end{pmatrix},$$

so that the matrix with the negative of the completely partialled correlations off the diagonal is

$$\begin{pmatrix} 1.000 & -0.347 & -0.068 & -0.068 & 0.027 & -0.087 \\ -0.347 & 1.000 & -0.185 & 0.109 & 0.017 & -0.097 \\ -0.068 & -0.185 & 1.000 & -0.688 & -0.171 & -0.074 \\ -0.068 & 0.109 & -0.688 & 1.000 & -0.115 & -0.217 \\ 0.027 & 0.017 & -0.171 & -0.115 & 1.000 & -0.625 \\ -0.087 & -0.097 & -0.074 & -0.217 & -0.625 & 1.000 \end{pmatrix}$$

in which the largest entries are evidently in the [1,2] and [2,1] cells, the [3,4] and [4,3] cells, and the [5,6] and [6,5] cells. In other words, the three largest negative entries in the normalized concentration matrix are in the same positions as the three special factors S_1, S_2, S_3 in Wright's version of the analysis, and now we see that two of them, for the limbs, are roughly equal at a value double that of the third (for the

---

[8] In the earliest modern discussion of this method, Bookstein et al. (1985), this restriction is lifted.

head). The concentration of these residuals in the larger two pairs of these numbers would be surprising enough to start an abduction rolling were the analysis not nearly a century old already. Consilience would probably require the sort of detailed knowledge of morphogenetic control mechanisms that arose only toward the end of the 20th century.

In terms of the logical model sketched at the end of the previous section, the discussion of the maximum excursion statistic, the quantity driving the abduction here is likely to be the value of that largest partial correlation from the precision matrix, perhaps tested on its own by conversion to a Fisher's z with p's (over)corrected for the number of correlations off the diagonal, or perhaps simply compared to the other elements in the matrix. We would immediately discover that we had two of these candidate features, not just one, and that they are deeply homologous in a biological sense (forelimb and hindlimb, wing and leg). The rhetoric, then, would interpret them as twinned from the beginning, a possibility that was not anticipated at the time the formulas were created.

Mathematical note 5.1: a constructive proof. The assertion with which this section began – that the normalized inverse Σ^{-1} of a covariance matrix has the elementwise interpretation we've been exploring – is not proved in most of the standard textbooks. The original observation appears to be due to Guttman (1938), whose demonstration is unbearably clumsy. Cox and Wermuth (1996) offer a breezy derivation based on the identity that the covariance matrix relating the vector of variables X_i to the vector of variables $\Sigma^{-1} X_i$ is $\Sigma^{-1} \Sigma = I$, the identity matrix, so that the variables $\Sigma^{-1} X$ must be proportional to the residuals of each of the original variables X after regressions against all the others. But there is pedagogic merit in a demonstration treating this construction as the simple matrix manipulation it actually is, independent of any empirical meanings. Here's the derivation I hand out to my students.

It relies on three identities. (Below, $'$ is the matrix transpose operator.)

1. Let A and C be symmetric matrices, let C be invertible, and let B be any matrix conformable with A and C such that $A - BC^{-1}B'$ is invertible. Then the inverse of $M = \begin{pmatrix} A & B \\ B' & C \end{pmatrix}$ is $\begin{pmatrix} D & -DBC^{-1} \\ -C^{-1}B'D & C^{-1} + C^{-1}B'DBC^{-1} \end{pmatrix}$ for $D = (A - BC^{-1}B')^{-1}$. (Proof: their product is $\begin{pmatrix} I & 0 \\ 0 & I \end{pmatrix}$ where the partition is the same as in the matrix M.)

2. The inverse of the 2×2 symmetric matrix $\begin{pmatrix} a & b \\ b & c \end{pmatrix}$ is $(ac - b^2)^{-1} \begin{pmatrix} c & -b \\ -b & a \end{pmatrix}$. (Proof: a special case of (1.) [This is a joke: of course, just multiply it out.])

3. Define $\Sigma = \begin{pmatrix} \Sigma_{XX} & \Sigma_{XY} \\ \Sigma_{YX} & \Sigma_{YY} \end{pmatrix}$ to be the covariance matrix of the concatenation of two blocks of centered (mean-zero) variables X and Y on the same cases. (Thus each dimension of Σ is $\#X + \#Y$, the sum of the count of X-variables and the count of Y-variables.) Then the covariance matrix of the residuals of the X's on all the Y's at once, from the usual multiple regression, is $\Sigma_{XX.Y} = \Sigma_{XX} - \Sigma_{XY} \Sigma_{YY}^{-1} \Sigma_{YX}$. (Proof: the regression coefficients of the X-block on the Y's, taken together, are $B_{X|Y} = \Sigma_{XY} \Sigma_{YY}^{-1}$, and the covariance of $X - B_{X|Y} Y$ telescopes in a familiar way.)

The result we seek follows quickly from these three familiar propositions. We can certainly renumber our original list of variables so that the off-diagonal element $\Sigma^{-1}{}_{ij}$ we are inspecting is $\Sigma^{-1}{}_{12}$. Apply proposition (1) above to the partition for which the X's are the first two variables of the original list and the Y's are the rest. Then, by comparing the matrix in (3) to the definition of D in (1), we see that the upper left 2×2 block of the inverse of Σ matches the definition of D^{-1} when D is the matrix $\Sigma_{XX.Y}$, the covariances of the X's after regression out of all the Y's. Writing this matrix (which is of course symmetric) as $\begin{pmatrix} \Sigma^{-1}{}_{11} & \Sigma^{-1}{}_{12} \\ \Sigma^{-1}{}_{12} & \Sigma^{-1}{}_{22} \end{pmatrix}$ the matrix $\Sigma_{XX.Y}$ we seek is its inverse. By proposition (2), this is $E^{-1}\begin{pmatrix} \Sigma^{-1}{}_{22} & -\Sigma^{-1}{}_{12} \\ -\Sigma^{-1}{}_{12} & \Sigma^{-1}{}_{11} \end{pmatrix}$ where E is $\Sigma^{-1}{}_{11}\Sigma^{-1}{}_{22} - (\Sigma^{-1}{}_{12})^2$. The covariance of the residual of X_1 on all the original variables numbered above 2 with the residual of X_2 on the same list of predictors is the $(1, 2)$ element of this matrix, which is $-\Sigma^{-1}{}_{12}/E$. To get their correlation, we divide this covariance by the product of the standard deviations of the two residuals in question, that is to say, the square root of the product of the two diagonal elements involved. In the course of doing this, all the factors of E cancel each other, and hence the partial correlation of X_1 with X_2, adjusting for all the other observed variables, is $-\Sigma^{-1}{}_{12}/\sqrt{\Sigma^{-1}{}_{11}\Sigma^{-1}{}_{22}}$.

But this is what we set out to prove. And we now know where the minus sign comes from: it arises from the 2×2 matrix inverse formula, property (2).

For a recent instance of this theorem's playing a role in a professional dispute, see Mitteroecker and Bookstein (2009b).

5.2.3.4 Over Ellipsoids: T. W. Anderson's Distribution

We see in Section 6.3 how useful it is to locate the directions in n dimensions that correspond to the diameters of an ellipse in two dimensions. The ellipsoids whose diameters we are searching for are usually specified by information matrices that describe combinations of estimated quantities of particularly high or low precision, or by covariance matrices describing combinations having particularly high or low variance. But to say anything about these directions, they have to exist – the corresponding features cannot be "the same in all directions," or it wouldn't make sense for us to claim that the data are selecting directions for us to inspect, abduce from, etc. As a result, we need a probability distribution for meaninglessness in this setting, of sets of two or more vectors all at $90°$ to one another.

The formula $np \log(a/g)$ I'm calling to your attention here was first published by T. W. Anderson in a difficult article of 1963 that included formulas for answering a good many questions of this general flavor. It is one version of an approach to the analysis of the so-called *scree plot* (the run-out of eigenvalues corresponding to the nonmeaningful principal components: named thus because of its resemblance to the rubble at the base of a mountain slope breaking up as a result of weathering). Anderson's derivation is rigorous in terms of chi-squares, but for our purposes we can resort to the usual logic of a likelihood-ratio test as sketched a few pages above. Remember that for most such tests, on a null model of no true signal for i.i.d. samples, the logarithm of the likelihood ratio, multiplied by -2, is distributed as a chi-square variable (a sum of squares of independent random variables all distributed according to the same bell curve).

Our question was,

> What is the evidence in this data set of a set of k perpendicular directions that are well-defined in the sense of bearing underlying linear combinations with numerically different expected variances and zero covariances?

If the eigenvalues, those variances, are really different, then the directions exist, rather like the diameters of any ellipse that is not a circle. It is straightforward to to show (cf. Mardia et al., 1979, p. 107) that if the eigenvalues were all equal the likeliest estimate of that equal value is their average – call it a (for "arithmetic mean"). The log likelihood ratio for the hypothesis of different eigenvalues is the difference between the likelihood on equal eigenvalues and that for any set of eigenvalues (i.e., the best ones for that specific data set). On Anderson's formulas, on a true model of identical eigenvalues the difference of log-likelihoods is the startlingly simple expression $-\frac{np}{2} \log \frac{a}{g}$, which must be a negative number since the arithmetic mean cannot be less than the geometric mean for any set of positive numbers (see later).

Then the usual large-scale approximation applies: on the simpler hypothesis, -2 times the log likelihood ratio for the simple hypothesis (all the same eigenvalue) with respect to the more complex hypothesis (different eigenvalues) is distributed as a chi-square with degrees of freedom equal to the number of parameters we are estimating. How many of those parameters are there? Count them: if there are p principal components under consideration, then they can rotate in $p(p-1)/2$ different ways (the number of dimensions of the space of rotation matrices in p Euclidean dimensions),[9] and there are a further $p-1$ ways that the eigenvalues can be different around the common value we were assigning them in the simpler model. So the total number of degrees of freedom for the chi-square corresponding to our log-likelihood ratio is $p(p-1)/2 + p - 1 = (p-1)(p+2)/2$. For the special case of $p = 2$ consecutive eigenvalues, this reduces to just 2. (Notice, counterintuitively, that this formula does not involve the number of variables in the data set with whose principal components we are concerned.)

Now proceed numerically. For a chi-square on two degrees of freedom, sum of the square of two variables each distributed Normally with a standard deviation of 1, the expected value is 2. We can set a very low bar indeed against abduction:

> no eigenvector is entitled to interpretation whose value differs from its successor by a ratio less than that for which $2n \log(a/g)$ equals that expected value.

In other words: if the hypothesis of no signal is actually a *better* explanation of the pattern in the data than anything with scientific content, go with the null, and describe it as "more likely than not" to be actually the case. These thresholds are straightforward

---

[9] The first eigenvector can be rotated to set any $(p-1)$ direction cosines independently. The second must be perpendicular, and so has $(p-2)$ degrees of freedom remaining, and so on. The total available for rotation is thus $(p-1) + (p-2) + \cdots + 2 + 1 = p(p-1)/2$.

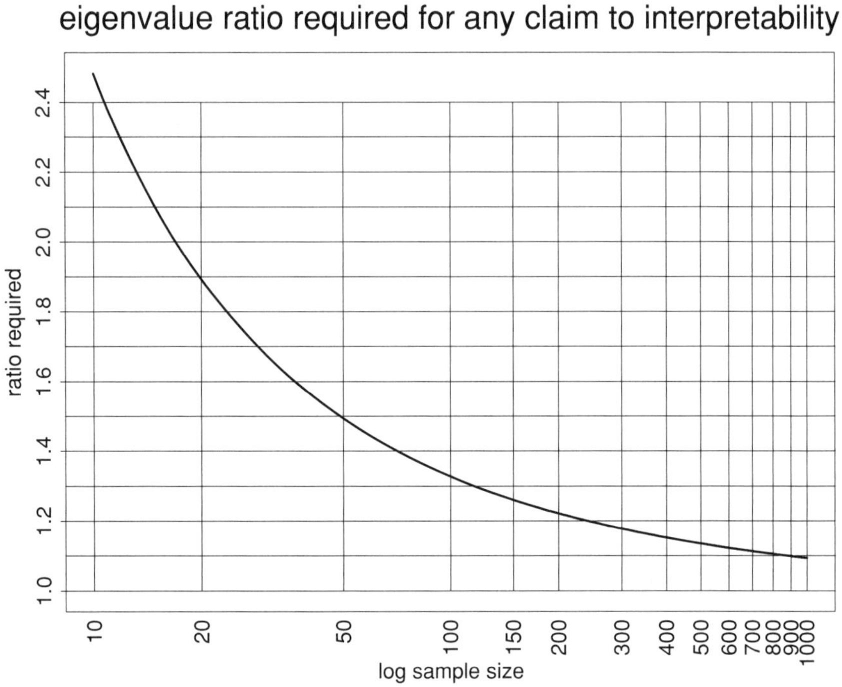

eigenvalue ratio required for any claim to interpretability

Figure 5.12. Minimum ratio of successive eigenvalues for their meaningfulness separately to be more likely than not.

to compute by simple iterations, resulting in the chart of Figure 5.12. As sample size grows without limit, the ratio of two consecutive eigenvalues corresponding to $2n \log(a/g) = 2$ is approximately $1 + \sqrt{\frac{8}{n}}$. Note that this is *not* a significance test, for which the ratio required would be nearly twice as far from unity.[10] It is only a criterion for meaningfulness in any discussion associated with the empirical patterns the principal components analysis is supposed to be supporting. We will see examples of this in the course of Chapters 6 and 7.

This is a fairly sophisticated flavor of question. We are interested in the extent to which individual directions of particularly high or particularly low variance justify the effort to search for a scientific explanation (e.g., for biological data, in terms of growth, or in terms of natural selection; for environmental data, in terms of some global circulation model; etc.). We wish to explore only those dimensions that are surprisingly distinct from the residual noise in precisely the sense of the theorem.

Mathematical note 5.2: Why does the arithmetic mean exceed the geometric mean? As one more aperçu into how mathematicians think, here is a pleasantly elementary proof of the "fact" (the theorem) that was taken for granted in the preceding discussion. The proof, by the great teacher George Pólya (already introduced in Chapter 3),

---

[10] This version is demonstrated by Mardia et al. (1979), Example 8.4.2.

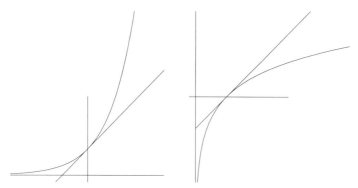

Figure 5.13. Two versions of the same identity. (Left) The curve, $y = e^x$, is never lower than the line $y = x + 1$. This is the version of the inequality used for the arithmetic-geometric mean theorem. (Right) The curve, $y = \log x$, is never higher than the line $y = x - 1$. This is the version of the inequality used for the Gibbs Theorem, Section 5.3.1. In both panels, the perpendicular lines are the coordinate axes. The two panels are related by a reflection across the line $y = x$.

starts with an "obvious" fact appearing at first glance to have nothing to do with the question at hand.

It is "obvious," the mathematicians say, that

$$e^x \geq x + 1,$$

with equality only for $x = 0$. (Why is this "obvious"? Because $e^x = x + 1$ for $x = 0$, the slope of e^x equals the slope of $1 + x$ at $x = 0$ [both have slope 1], and the function e^x has second derivative positive everywhere [namely, e^x] and so must be bending away from its tangent line everywhere. See Figure 5.13, left.)

Suppose we have a collection of numbers $x_1, \ldots, x_k$, where k is any sample size. Write a for the arithmetic mean of the x's, $a = \sum_{i=1}^{k} x_i / k$, and apply the inequality on e^x to each of the terms $\frac{x_i}{a} - 1$. (This idea, the crux of the proof, "comes out of nowhere" in the usual forward narration; it is defended by the way it makes a crucial term cancel below.) For each i, we get $e^{\frac{x_i}{a} - 1} \geq \frac{x_i}{a}$. Multiplying all these inequalities together, we get

$$\prod_{i=1}^{k} e^{\frac{x_i}{a} - 1} \geq \prod_{i=1}^{k} \frac{x_i}{a},$$

where the notation $\prod$ (the Greek capital letter pi, in a large font) is an equivalent of $\sum$ (a big capital sigma) that refers to multiplication instead of addition.

But

$$\prod_{i=1}^{k} e^{\frac{x_i}{a} - 1} = e^{\sum_{i=1}^{k} (\frac{x_i}{a} - 1)} = e^{\left(\sum_{i=1}^{k} \frac{x_i}{a}\right) - k} = e^{\frac{\sum_{i=1}^{k} x_i}{a} - k} = e^{k-k} = e^0 = 1,$$

since we defined a as $\sum_{i=1}^{k} x_i/k$. So

$$1 \geq \prod_{i=1}^{k} \frac{x_i}{a} = \frac{\prod_{i=1}^{k} x_i}{a^k} = \left(\frac{g}{a}\right)^k,$$

where g is the geometric mean $\sqrt[k]{\prod_{i=1}^{k} x_i}$. If the kth power of $\frac{g}{a}$ is less than 1, the ratio must itself be less than one. That is,

$$\frac{\sqrt[k]{\prod_{i=1}^{k} x_i}}{\sum_{i=1}^{k} x_i/k} \leq 1,$$

or

$$\sqrt[k]{\prod_{i=1}^{k} x_i} \leq \sum_{i=1}^{k} x_i/k,$$

which is what we were trying to prove. Following through the steps above, it is clear that the equality obtains only if all the terms $\frac{x_i}{a}$ equal 1, that is, only if all the x's are equal; otherwise the inequality is, as they say, "strict."

5.2.3.5 Over Grids: The Intrinsic Random Field

The schema of principal components just reviewed is usually applied to samples generated from data matrices, but it need not be. It can apply just as well to any other approach that uses a variance–covariance structure to systematize the pattern of information content of an empirical system.

Here the surprise will come from deviations from the null model just sketched at Section 5.2.3.4 for a most cleverly constructed set of system descriptors indeed, a set of locations of points in a plane that correspond on biological grounds (that is, as *homologous points*, see Chapter 7), across all the cases of a sample.

We have a grid transformation, perhaps one of those in Figure 5.14. Probably we are *surprised* to see visually striking features here – the crease from upper left to lower right in the center of the 3 × 3 panel of Figure 5.14, the pair of what might appear to be growth centers along the middle of the grid to its right. Under ordinary circumstances observations like these would launch an abduction. But as it happens, none of them are real. We are victims of our own retinas, seeing signals where they certainly do not exist. Why "certainly"? Because the ten grids here have been explicitly generated by a process conveying no information about features at *any* scale. See Bookstein (2007), and, for the equations, Mardia et al. (2006).

In a system organized with such complexity (multilocal, multiscale), the approach to abductions is the same as it has been hitherto in this chapter: to build a symmetric basis for assessing departures from a hypothesis of meaninglessness in the variation. Here, in accordance with the schemata of systematics and other image-based classification/feature extraction sciences, the nature of the abduction being pursued is that of a feature of *any* geometry detected at *any* geometric scale. The background model, then (equivalent of the spherical distribution of eigenvalues for T. W. Anderson, or

5.2 Abduction and Consilience in General Systems 327

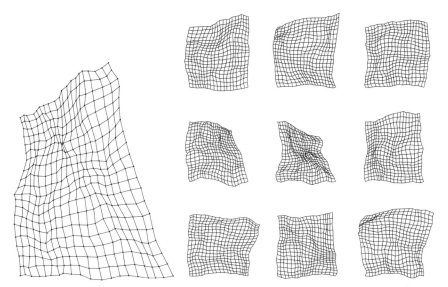

Figure 5.14. Intrinsic random deformations having no information at any position at any scale. (Left) The example examined quite closely in the next figure. (Right) Nine more examples like the one at left.

the interchangeability of successive increments in the random-walk models), must be a distribution on these grids that assigns the same improbability to any linearized feature (displacements of the grid nodes) at *any* scale. It may be somewhat surprising that such a distribution exists, but it does. (Figure 5.15 shows, for instance, that the distribution of shapes of little quadrilaterals, in the grid at left in Figure 5.14, is indeed precisely the same at three separate geometric scales, corresponding to squares of sides 1, 2, and 4.) Hence the grids in this figure and the next do indeed exemplify the nothing-doing distribution whose inapplicability is necessary if any abduction is to begin.

Supposing still that this is our model, then to examine whether an abduction can begin (a finding of some feature at some scale), one refers an empirical data

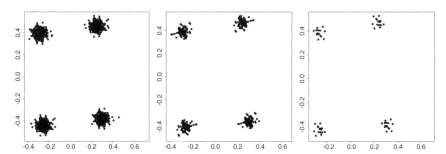

Figure 5.15. Distribution of shapes of quadrilaterals at different scales from the left exemplar in Figure 5.14. The pattern of shape variation is self-similar.

set to the T. W. Anderson procedure for checking sphericity against *this* model of randomness, using a basis set for the location data that is constructed according to the rules in Chapter 7. When that is done, and checked, the data set in Figure 5.16 shows, quite surprisingly, a feature at a quite small scale: the one shown in Figure 5.17. Section 7.7.3 explains how gratifying this finding actually was, as the data set actually consisted of two groups, doctors and their schizophrenic patients, so that the discovery of this particular feature converted the group difference from a mere statistically significant feature (the finding in the original publication) to the most powerful summary available for what is surprising about the data set as a whole, which is a much more compelling characterization. Put another way: it is one thing to confirm a group difference designed into a data set, but it is far more persuasive to discover such a difference using techniques that know nothing about any grouping system a priori. It is mildly astonishing that the unpromising-looking data in Figure 5.16 conduced to a summary this coherent, powerful, surprising, and suggestive.

5.2.3.6 Two Approaches to Images

Section 5.2.3.5 dealt with grids. There is a lot more to images, especially medical images, than just the locations of points; there is also the image contents pixel by pixel. Two powerful methods of constructing null distributions deserve mention here even though their mathematics is too complicated to exposit in this chapter.

If a random process is situated in two or more dimensions, the local values direction by perpendicular direction may be uncorrelated, in which case we get just the Maxwell–Boltzmann distribution of velocity mentioned in Sections E4.3 and L4.3, or we may instead be examining a quite different kind of process for which the two or three dimensions of the image are interlinked. Two versions of this spectrum have proved enormously fertile in the sciences this book is covering. In both versions, the image is presumed to be a dynamical system (even if it is only observed once), within which individual pixel values are always changing under control of some overall energy function expressing the cost of local information. In one approach, the *two-dimensional Ising model*, there are only two colors for pixels, white and black, and the energetic cost of a scene is the sum over every pair of adjacent pixels of either a cost for identity (white–white, black–black) or a cost for discrepancy (white–black). These models, which are intended for application to metallic alloy states and to ferromagnetism, prove to have a truly remarkable mathematical behavior whereby a quantity analogous to temperature controls the chances of their uniting into coherent domains of very large scale versus instead congealing into multiple smaller regions of contrasting uniform color. The parameter, in other words, specifies a *phase transition* between long-range order and long-range disorder. See, for instance, Baxter (2007).

In the other approach, most elegantly developed by the late Canadian probabilist Keith Worsley, each pixel in an image is modeled using a summary gray-scale value, which is more than one bit of information. Again it is the relation between each gray value and its neighboring gray values, aggregated over all the pixels of the image, that determines an overall energy and thereby a probability that this time goes

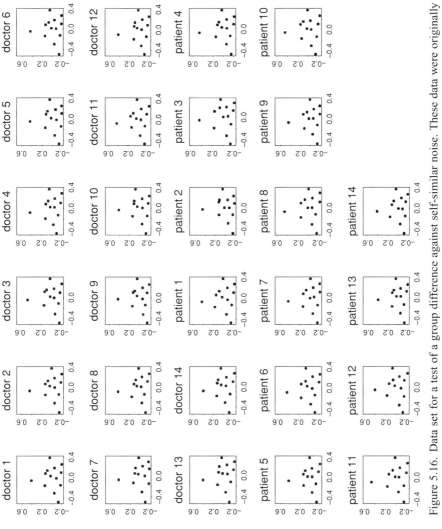

Figure 5.16. Data set for a test of a group difference against self-similar noise. These data were originally published in DeQuardo et al. (1996).

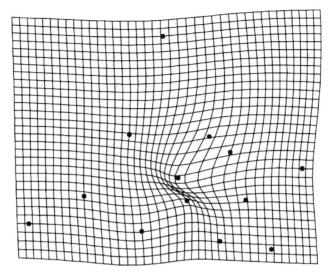

Figure 5.17. The first principal component of the data from the preceding figure against the model of self-similar noise. Compare the actual group difference (Figure 7.19).

as a Boltzmann factor $e^{-kE/T}$. The Worsley distributions are particularly useful for deciding whether a medical image corresponding to some sort of intervention, such as an image of cerebral blood flow during some assigned cognitive task, is or is not showing one or more regions of particularly intense response to that intervention: in other words, deciding whether the image is or is not a useful source of information about the system that is ostensibly doing the responding. For the information to be scientifically useful, it must first be surprising on a model of no-information. See Taylor et al. (2009).

5.2.4 A Concluding Comment

Evidently there are a wide variety of approaches to the nature and scope of quantitative reasoning in complex systems. It is not that a reduction to significance tests is needed, Gaussian assumptions, tail-probabilities, and the like: far from it. Rather, the statistician, wearing the hat of a probability theorist much of the time, listens carefully to the scientist and then collaboratively constructs the machine that will permit the launching of an abduction if and only if the data are indeed surprising on a background of whatever was known before a study was begun. If your training centered on flat data structures (rectangular matrices) and significance tests, it is because you were taught to seek surprise in only that one highly stereotyped and unpersuasive way. (How surprising could a claim of statistical significance possibly be when in some disciplines nearly every published paper reports such an event?!) After sketching all the other ways that a Normal (Gaussian) distribution might be of use to you, I delved into a variety of other possibilities (studies of eigenvalues, networks, deformation grids) that are quite a bit more useful as prototypes in the search for surprise in complex systems. In the next subsection we review the method complementary to this approach, the search for consilience: for the hypothesis that best supports the data you've got.

5.3 Information and Information-Based Statistics

The principal references for this section are Burnham and Anderson (2002) and D. R. Anderson (2008). The text is adapted from Bookstein (2009c), where earlier versions of Figures 5.18 and 5.19 first appeared.

5.3.1 Information as a Mathematical Quantity

Recall the discussion in Chapter 3 about that stunningly prescient and deservedly famous *Science* article of 1890 by T. C. Chamberlin, as modified for the second half of the 20th century by John Platt (1964). As scientists we are advised to proceed always in the context of multiple hypotheses, some of which (ideally, one of which) would be adequately supported by the data while all the others would be rejected *by the same data*. In the presence of complex data sets describing ever more complex systems, how, exactly, do we manage that selection?

The answer, as usual in matters of fundamental statistical innovation, is indirect. From the eruption of statistical thermodynamics at the turn of the 20th century, we draw from Josiah Willard Gibbs (American mathematical physicist, 1839–1903) the theorem usually called "Gibbs's inequality," which states that if $P = \{p_1, \ldots, p_n\}$ and $Q = \{q_1, \ldots, q_n\}$ are any two probability distributions (which we write in the discrete notation suited to sums instead of the continuous notation suited to integrals), then what is now called the *relative entropy* of P with respect to Q, $\sum_1^n p_i \log(p_i/q_i)$, is a positive quantity unless P and Q are the same distribution, in which case the formula yields a value of 0.[11] That this is called a "relative entropy" strongly suggests a relationship to that concept from mathematical physics, which had been developed at just about the same time Gibbs was flourishing by his Austrian counterpart Ludwig Boltzmann.

Just after World War II, Claude Shannon (1916–2001) rediscovered the formalism of expressions like these in the context of *information theory*. Shannon worked for the Bell (Telephone) Laboratories, and so understandably was most concerned with data compression and transmission, but his ideas have ramified into a great many neighboring fields, such as cryptography, natural-language processing, neurobiology, quantum computing, and the application that concerns us in this book, numerical inference. See Gleick (2011).

---

[11] The proof is a surprisingly easy rearrangement following from the fact that for natural logarithms (logs to the base e) we always have $\log x \leq x - 1$ (which is itself just a rearrangement of the identity $e^x \geq x + 1$ that launched our discussion of the arithmetic-geometric mean inequality; see Figure 5.13). Assume two lists of probabilities p_i, q_i each totaling 1. From the inequality, $-\sum_i p_i \log \frac{q_i}{p_i} \geq -\sum_i p_i\left(\frac{q_i}{p_i} - 1\right) = -\sum_i q_i + \sum_i p_i = 0$, as we needed to show.

Shannon's basic contribution was published as Shannon (1948) but was completed mostly during World War II. For our purposes here, it is sufficient to explain the fundamental discovery as if it were a verbal metaphor (although it is not – it actually has a physics, see, for instance, Feynman et al. [1996], Brillouin [1962], or Vedral [2010]). The following explanation is précised from the version in Jaynes (2003, pp. 347–350).

Information is a decrease in uncertainty, just as the British physicist Maxwell suspected when he invented "Maxwell's demon" in 1867. Suppose that we wish a formula – an exact mathematical expression – for the "amount of information" that is borne in a particular *probability distribution* $P = \{p_1, \ldots, p_n\}$. By "amount of information" we mean a quantity that states, in some sense, "how much more we know" after we have sampled from that distribution than before. In the most famous example of this formula, we select from just two choices, heads or tails of a fair coin, and that amount of information turns out to have an even more familiar name, the single *bit* that now physically corresponds to the individually addressable domains of doped silicon that your latest laptop computer has about a hundred billion of.

The requirements on this formula for the information $H(p_1, \ldots, p_n)$ turn out to be quite simple, and there are only three of them. First, H must be a continuous function of all the p's. Second, if all the p's are all equal to the same fraction $1/n$, H must be an increasing function of n. Third, the "composition law": the information in a single observation must come out the same no matter how we organize the task of actually selecting one of them. It is this third assumption that has the philosophical bite, the one that actually does all the work: essentially, it is the claim that information has to obey a quasiphysical conservation law. Because the logarithm is the only reasonable function obeying $H(mn) = H(m) + H(n)$, the theorem that H *must* be proportional to Shannon's formula $-\sum p_i \log p_i$ follows very quickly. This log-additivity of the logarithm is the same principle that had been used for slide rules (Figure 4.25) for about 400 years; it plays many other roles in applied mathematics as well (see, e.g., Section 6.5).

The top half of Figure 5.18 illustrates the essence of the Shannon approach, that third requisite. Here, the information in one toss of a fair die (the information in a choice among the numbers 1, 2, 3, 4, 5, 6 each with probability 1/6) must be some function $H(6)$ of the number 6, but it must also be the sum of the information $H(3)$ in a choice among three possibilities (faces 1 or 2, or 3 or 4, or 5 or 6) and the information $H(2)$ in the further selection of one alternative from each of the pairs just listed. The only function that satisfies this rule for general m, n in place of the foregoing 3 and 2 is proportional to the function $\log n$ or, for more general probability distributions $(p_1, p_2, \ldots, p_i, \ldots, p_n)$ among n outcomes, the function $H(p_1, p_2, \ldots, p_n) = -\sum p_i \log p_i$. If logs are taken to base 2, the value of the formula is in units of bits, which have the familiar physical meaning.

Soon after Shannon's original publication, the statistician Solomon Kullback (1907–1994) and the engineer Richard Leibler (1914–2003) realized that Shannon's

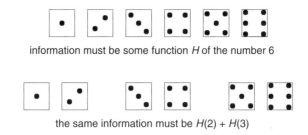

information must be some function H of the number 6

the same information must be H(2) + H(3)

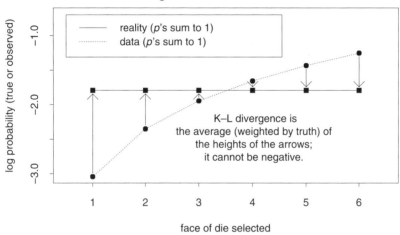

Figure 5.18. Two views of information theory. (Top) Shannon's third requirement for a measure of information, illustrated by a two-step approach to the information in one toss of a fair die. (Bottom) The Kullback–Leibler divergence, illustrated using two probability distributions on dice.

formula, as applied to the data compression problem, was in fact a version of the Gibbs formula expressing a quantity that an engineer would have a reason to be interested in. If Q is an *assumed* distribution underlying some data (in our application, it will be a statistical model), and if P is the *true* distribution (i.e., "reality"), then the formula $KL(P|Q) = \sum_i p_i \log(p_i/q_i)$ is proportional to the number of additional bits you would need for the data compression if you were using the wrong distribution Q instead of the correct distribution P. It is, in other words, a *divergence* that measures the discrepancy of your model from the truth in a particularly practical way. It is thus called the *Kullback–Leibler [K–L] divergence*, or, a little inaccurately, *K–L distance* (inaccurate in that the formula doesn't give a true distance measure – the divergence of Q with respect to P isn't quite the same as the divergence of P with respect to Q that was just defined). Gibbs's Inequality assures us that the quantity itself must be positive. It has also been called an *inefficiency* when the wrong model Q is used to represent reality P, and also the information "lost" when Q is used to represent P. Any of these is an acceptable metaphor.

The formula is illustrated in the lower half of Figure 5.18. Here the horizontal line represents the probabilities of outcomes for the toss of a fair die – a probability of 1/6 for each face. The dotted line corresponds to the toss of a remarkably unbalanced die for which the probability of i spots is proportional to i (hence probabilities of $1/21, 2/21, \ldots, 6/21$). Gibbs's Inequality states that the weighted sum of the directed lengths of the arrows in the figure, each arrow the logarithm of the ratio between the probabilities of the possible outcomes on the two hypotheses, must be positive when the weighting is by the probabilities of the distribution in the numerator of the probability ratios. The Kullback–Leibler divergence, which quantifies the information gained when the distribution of the data is used to represent reality, is this weighted sum when the weights are taken as the true probability distribution (i.e., "reality"). The scientist responsible for a numerical inference therefore wants this distance to be as small as possible – wants the model to be as close as possible to the (unknown) reality that generated the data. Akaike's theorem (Figure 5.19) allows us to approximate this distance as it serves to differentiate among alternative models.

5.3.2 From Information to Numerical Inference

It is time to explicitly link these abstractions to the context of this chapter, the pursuit of abduction and consilience in numerical data from complex systems. D. R. Anderson (2008, p. 52) describes the architecture of the bridge:

> I will assume the investigator has a carefully considered science question and has proposed R hypotheses, the "multiple working hypotheses," all of which are deemed plausible. A mathematical model (probability distribution) has been derived to represent each of the R science hypotheses. Estimates of model parameters and their variance-covariance matrix have been made under either a least-squares (LS) or a maximum-likelihood (ML) framework. In either case, other relevant statistics have also been computed (adjusted R^2, residual analyses, goodness-of-fit tests, etc.). Then, under the LS framework, one has the residual sum of squares (RSS), while under a likelihood framework, one has the value of the log-likelihood function at its maximum point. This value (either RSS or max log L) is our starting point and allows answers to some of the relevant questions of interest to the investigator, such as: • given the data, which science hypothesis has the most empirical support, and by how much? • what is the ranking of the R hypotheses, given the data? • what is the probability of, say, hypothesis 4, given the data and the other hypotheses? • what is the (relative) likelihood of, say, hypothesis 2 versus hypothesis 5? • how can rigorous inference be made from all the hypotheses (and their models) in the candidate set?

"This is multimodel inference," Anderson concludes. But it is also *strong inference*, in the sense of Platt, likewise encompassing Chamberlin's "method of multiple hypotheses," both as adapted to this especially powerful context of numerical reasoning.

It is quite remarkable that one theorem, whose first Web-accessible appearance seems to be Akaike (1974), supplies the foundation for *all* of these considerations. Let us take the K-L formula $\sum_i p_i \log(p_i/q_i)$, where the p's are reality and the q's are one of our variably corrupt approximations, and simply expand the logarithm: we have $KL(P|Q) = \sum_i p_i \log(p_i/q_i) = \sum_i p_i \log p_i - \sum_i p_i \log q_i$. We don't know what the first term might be (because we don't know true reality, P), but if we are going to be comparing across multiple models it will always be the same number, so we can ignore it.

Our task is to compute the second term, $-\sum_i p_i \log q_i$. As we still don't know the p's, this appears still to be impossible, but Akaike found a formal identity that applies whenever the models Q are near enough to the truth P: an identity for the *expected value* of this expression *over data sets like this one* (i.e., sampled from the same "true" distribution). It looks like we are using the probabilities q_i at the fitted (least-squares, or maximum-likelihood) model. **But these are the wrong probabilities**, because the very act of fitting a model – any model – overestimates its accuracy with respect to ultimate reality. Akaike proved that a correct *estimate* of $-\sum_i p_i \log q_i$ – what you would get for multiple samples "like this one" from the same true distribution – is equal to the *observed* $-\sum p_i \log q_i$ – what you got at the best fit – *minus 1 for every parameter you estimated*. It was a familiar fact (to statisticians, anyway) that the value of the expression $-\sum_i p_i \log q_i$ for the more familiar kind of models we use, models like regressions, could be written equivalently as either $\log L$, the logarithm of the actual maximum value of the computed likelihood, or, for linear Gaussian models, its actual evaluator $-n \log \hat{\sigma}$. In this way we arrive at a formula about "distance from reality" purely in terms of the data and the model. "Reality," knowable or not, will turn out to have been hidden in the relationships among multiple models.

Figure 5.19, using a "true simulation" of a relatively weak linear model (one having a correlation only about 0.3), sketches the underlying geometry of Akaike's theorem. The expected value of the K–L divergence of the true model from the data is not the likelihood of the fitted model at its maximum but instead the likelihood along the heavy line in the last panel, which is the line for the contour where the χ^2 is equal to its expected value for the likelihood surface, corresponding to the number of parameters (here, three: intercept, slope, standard deviation) being estimated. The K–L divergence formula is just this adjusted maximum log likelihood, and so we arrive at the celebrated *Akaike Information Criterion*, or AIC (after Hirotugu Akaike, 1927–2009, whose theorem it was):

$$\text{AIC} = \text{expected K–L information} = \log \text{maximum likelihood} - K,$$

where K is the number of parameters being estimated.[12] This is perhaps the single most fundamental advance in all of statistical science of the last half-century: the remarkably straightforward link between statistical principles and formal information theory. The relationship between Boltzmann's entropy and K–L information, the dominant approaches in information theory, and maximum likelihood, the dominant logical toolbox in statistics, supplies a reassurance that is welcome to practitioners of either bounding discipline. For statisticians, it permits the reformulation of the quest: it is to model not the data but the information *in* the data, and this is how it can be

---

[12] Because of the relation to the χ^2 I just mentioned, Akaike advised multiplying all of these formulas by -2. Without that correction, the AIC is in units of information (in effect, bits, times 0.673).

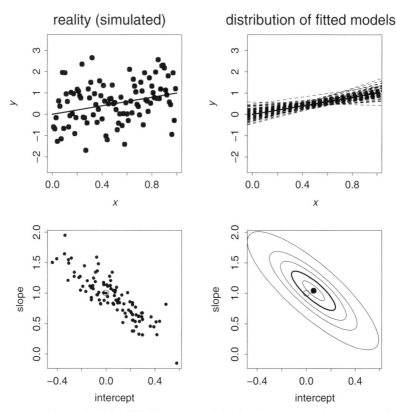

Figure 5.19. Schematic of the AIC. The open circle in the lower two panels represents the true parameters of the line in the upper left panel. The likelihood surface is only an approximate rendering. See text.

done, as long as you have more than one plausible model of what that information should look like.

The panel at upper left in Figure 5.19 shows a data set simulated from a "true" model $y = x$, the solid line in the figure, together with additional noise in y of considerable variance. (That solid line is there for didactic purposes only. In most general complex systems, such a distribution is unknowable. Instead we would have only a longer or shorter list of plausible quantitative models, each of which we shall formalize as a parameterized probability distribution of its own.) The fitted model to any such sample is, of course, the ordinary regression. At upper right are shown 100 regression lines corresponding to the sampling distribution of samples of 99 like the one at upper left. Each such model has three parameters (intercept a, slope b, and standard deviation σ for the fitted equation $y = a + bx + \epsilon$ where ϵ is normal with mean 0 and variance σ^2). At lower left are scattered two of these three parameters, the intercept and the slope, for each of the 100 replicated samples, together with the "true" parameters (coordinates of the open circle), which under normal circumstances are inaccessible to human reason.

This completes the set-up for exploiting the AIC (lower right panel). From the act of fitting all those models to the same data set we actually generate a *likelihood surface*, diagrammed here via its elliptical contours in the lower-right panel. The filled circle is at the "contour" of size zero that is exactly at the maximum of mathematical likelihood of the estimate here, and the other ellipses are all copies of one particular elliptical shape that comes from the algebra of the maximum-likelihood computations near that maximum. What Akaike proved is that the *expected* value of this likelihood over repeated samples from the true distribution (which, remember, "really" has its parameters at the open circle) is the likelihood along the heavy line, corresponding to a χ^2 statistic with one degree of freedom for every parameter being estimated. *This value is the expected information distance between the model in question and "reality" as represented by the sample data here.* The values of this AIC, after some adjustments, thus make possible a formally rational calculus for the assessment of multiple models as they might explain the same data, and also the averaging of those models for even better prediction or understanding. By comparison, null-hypothesis statistical significance testing (Section L4.3.4) is able to offer neither of those desirable formalisms.

One aspect of the tie between this decision protocol and our concern for explanation in complex systems can proceed via the discussion spanning Sections 5.2.3.4 and 6.3 about indistinguishable eigenvalues. We seek, precisely, a methodology for deciding when it is that the data at hand do not permit us to tell the difference between signal and noise. In D. R. Anderson's phrasing, for any sufficiently complex data set, large effects can be picked up even with "fairly poor analytic approaches, e.g., stepwise regression," moderate-sized effects require better methods, and beyond them the "interactions and slight nonlinearities." But ultimately we must enter the domain of "huge numbers of even smaller effects or perhaps important effects that stem from rare events." Within this hierarchy of tapering, Anderson advises, one must be quite disciplined in deciding when to stop analyzing the present sample and instead reserve these hints for "larger sample sizes, better study designs, and better models based on better hypotheses." Akaike's method is perfectly suited to this reality of tapering effects, in that some hypotheses cannot be distinguished based on the particular finite data set at hand, and so persist as complementary competitors past the end of any study.

The transcription of the Akaike calculus into the context of abduction and consilience begins when one's list of working hypotheses is transformed into a list of statistical models all fitting the same data to some greater or lesser extent. (It is assumed that this fit is pretty good, or else that one of the models is "true enough." Applications to models none of which are anywhere near the data, like studies of correlations that are considerably closer to 0.0 than 1.0, are not relevant to the kind of numerical inference with which this book is concerned.) After one small technical change in the formula, the replacement of the term $-K$ by $-K(n/(n-K-1))$, where n is the sample size, the method is to assign each model (each hypothesis) a log-likelihood that is proportional to the extent to which its adjusted maximum log-likelihood (that is, its value of the AIC after correction) falls short of the maximum over all the hypotheses. In this way, the adjusted likelihood of the optimal parameter set *under* its hypothesis becomes the likelihood *of* its hypothesis within the set R of all that are being considered at the time.

In a helpful notation, write $\Delta_i = \text{AICc}_{max} - \text{AICc}_i$ for the divergence of the ith hypothesis from the maximum. (We use the version of the formula without the additional factor of -2.) One of these Δ_i is 0. That is the likeliest of our hypotheses, *but it is not the one we select:* at least, not yet. We also investigate all the other Δ's, concentrating on those less than about 7 (i.e., likelihood ratios of less than $e^7 \sim 1000$), to investigate the relationship of that minimum value to the values for most of the other hypotheses. All of the Chamberlin (1890) desiderata can be expressed in this same consistent notation. The likelihood of any single hypothesis is proportional to $e^{-\Delta_i}$; the probability of the same hypothesis within the set of all R is $e^{-\Delta_i} / \sum_j e^{-\Delta_j}$; the evidence ratio of hypothesis i with respect to hypothesis j is $e^{(\Delta_j - \Delta_i)}$. The exponentials of adjusted log-likelihoods are indeed a very plausible way of keeping track of all of your hypotheses in the range of about three orders of magnitude of plausibilities among the hypotheses themselves by examining the models for adjusted AICs varying over a range of about 7.

5.3.3 An Example

As an example of the AIC technique I turn to the first published data set that demonstrated the use of the thin-plate spline as a relaxation engine for morphometric data from curves. We will learn a good deal more about this technique in Chapter 7. For now, let us simply take for granted that the data set in Figure 5.20 has been competently assembled and preprocessed as described in Bookstein (1997a). Traced here are the outlines of the midline corpus callosum in a sample of 25 brain images from the study reported in DeQuardo et al. (1999). The actual data points were produced by my longtime mathematical and computational collaborator William D. K. Green. Twelve of the images came from psychiatrists and the other 13 from their patients, so it would be comforting if we could tell the groups apart by something other than coat color.[13] The points are *semilandmarks* (SL's), locations that have been allowed to slide along the curve. This form is highly regulated in the course of neurogenesis (Bookstein et al., 2002), so it meets the requirements for a system of "organized complexity" at which the techniques of Part III are all aimed. In the center of the figure, all 25 outlines are set down in the least-squares *Procrustes registration* described in detail in Section 7.3.

The standard model for analysis of data of this sort is a model for group mean shape difference as shown at the far right in Figure 5.20, along with the numbering system for the semilandmarks. From the figure it is apparent that the groups may differ in their average shape, particularly near semilandmarks 5 and 15.

---

[13] In America, psychiatrists, like other medical doctors and like biological scientists, are conventionally imagined to be wearing white coats, and are depicted so in movies and advertisements.

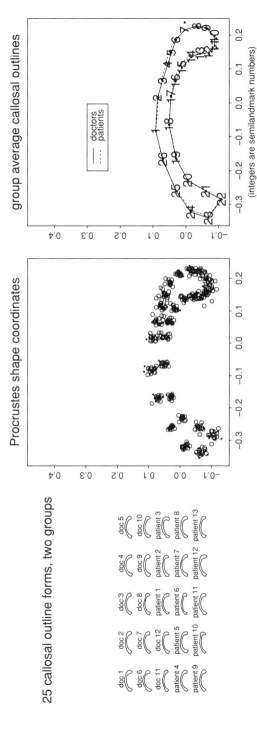

Figure 5.20. The 25 outlines of the data analyzed in DeQuardo et al. (1999), a different study from the one that supplied the data for Figure 5.17. (Left) As separate polygons. (Center) As a Procrustes semilandmark superposition (see Section 7.6) after each set of points is centered, rotated, scaled, and slid along the average outline to a position of best fit. (Right) Magnitude of group mean differences by position around the outlines at left: the two group average shapes, with the semilandmarks numbered.

339

A reasonable initial approach to the information in data like these is via perusal of the pointwise t-ratio for group differences in the coordinate perpendicular to the average lie of the curve. For that computation, the highest signal strength is at SL 15, where the specific t-ratio is 3.20, conventionally significant at 0.002 on 23 df. But the contrast conveyed by this variable is a *global* one – the Procrustes fit that produced the coordinates in Figure 5.20 used the entirety of these outlines, including the parts a long way away and also the parts that were mainly noise. A more nuanced, local approach might prove more powerful statistically and also more fruitful scientifically.

The description of group difference in the direction normal to the average curve involved a total of **three** parameters: one for the group mean difference, one for the residual variance around that difference, and one for the subscript (the index) of the landmark to which we are attending. To do better, we should attempt to fit a more flexible model, and afterwards examine whether the improvement in quality of fit meets the AIC criterion of increasing the likelihood beyond the cost of the additional parameter(s) involved. An adequate approach for this data, suggested in Bookstein (1997), replaces the global Procrustes fit driving the shape coordinates with a "varilocal" version. At every semilandmark, we construct the set of circles centered there through every other semilandmark, and sort these circles by the total number of semilandmarks in their interior. If we further require the circles of interest to include both of the neighbors of the focal landmark on its arc (so that we can continue to compute perpendiculars to the chord, the normal displacement driving our arithmetic), there are a total of 603 of these circles eligible to be examined. The best-fitting analysis among these 603 is the analysis shown at left in Figure 5.21, corresponding to a t-ratio of 4.86 for group mean difference in the shift of landmark 15 in the direction normal to the average outline. This is for a neighborhood of nine points around SL 15, and, as you see, the separation between the doctors and the patients is uncannily good – better, even, than the analysis in the original publication. But we have added two additional parameters (a center and a radius) – is this improved model actually telling us anything?

As the Procrustes analysis sets scale somewhat arbitrarily, we need to restandardize, here by setting that coordinate normal to our mean outlines to pooled variance unity. When the two different versions of the shift at SL 15 are standardized in this way, -2AICc for the nonfocal version, Figure 5.20, evaluates to -2.973,[14] and -2AICc for the focal version, Figure 5.21, using a neighborhood of only eight neighbors in addition to SL 15, evaluates to -4.844. Recall that the zero of these values is without meaning; only their difference matters. The difference of corrected AIC's indicates that the focal model should be thought of as roughly $e^{(4.844-2.973)/2} \sim 2.5$ times likelier than the simpler model. Comparing the rightmost two panels in Figure 5.21, we see how very much better the separation of groups is in the focal analysis. The separation is clean enough, in fact, to justify a causal exploration of the obvious

---

[14] Anderson supplies a most helpful shortcut: when models are fitted by least squares, as these two were, then we have $-2\text{AICc} = N \log \hat{\sigma}^2 + 2K \frac{N}{N-K-1}$, where $\hat{\sigma}^2$ is the residual variance from the least squares fit. Here the K's differ by 2.

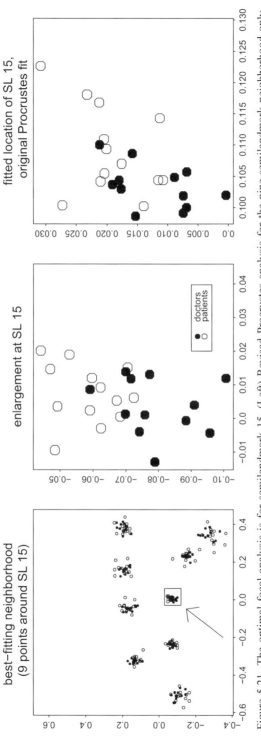

Figure 5.21. The optimal focal analysis is for semilandmark 15. (Left) Revised Procrustes analysis for the nine-semilandmark neighborhood only. (Center) Enlargement of the square at the left, the distribution of semilandmark 15 only. (Right) Enlargement of the scatter for the 15th SL in the center panel of Figure 5.20. The focused interpretation at center has a higher AIC – a higher likelihood even after the adjustment for its two additional parameters.

342 5 Abduction and Consilience in More Complicated Systems

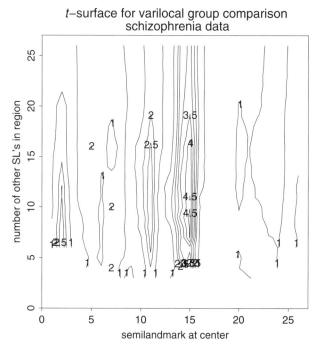

Figure 5.22. The likelihood surface in the neighborhood of the maximum group separation signal strength is a sharp ridge, meaning that it is the centering of the focal region, not its radius, that is the crucial empirical issue. In other words, the AIC corresponding to the model in Figure 5.21 does not really involve two additional parameters, but only a bit more than one, and the adjusted difference Δ is closer to 5 than to 2.

anatomical suggestion. This justifies the exercise of adding that pair of additional parameters – we may indeed have learned something useful about the system that manages this form. (As it happens, the shift at SL 15 is not a cause of the disease, but a consequence of the sort of treatment that was typical late in the previous century.) Figure 5.22 shows the entire likelihood surface for all these circle models. The maximum is on a very narrow ridge in the direction of uncertainty of radius. The choice of SL 15 is definitive, but the circle could include anywhere from 7 to 12 landmarks without substantial loss of evidentiary value. Hence the correction of K by two in the AIC formula is an overcorrection, and the improvement is even better than stated.

We might exploit the ordering of this little data set in two distinct ways. So far our approach has been one-dimensional, by position along the curve: this gave us the local tangent direction to the curve and thus the normal direction in which we measured displacement. Another ordering is two-dimensional, involving the space "inside" the curve, as in Figure 5.23. Such an analysis is consistent with the requisite to be introduced in the next chapter that variables describing complex systems should best be equipped with an a-priori ordering of their own. It will prove a desideratum of most of our multivariate analyses in complex systems. The crude assumption of

5.3 Information and Information-Based Statistics

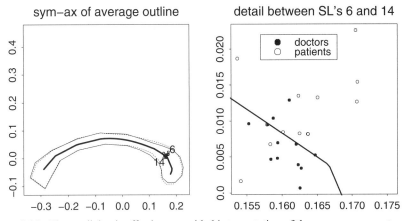

Figure 5.23. The medial axis affords a two-sided interpretation of the same group contrast.

multivariate Gaussian distributions will apply mostly to the "disorganized complexity" of Boltzmann (thermodynamic) systems, not the organized complexity of more differentiated systems.

As there are two organizing systems for the point data here, the one-dimensional version around and the two-dimensional version embedded, then we should examine signals aligned with the second possibility as well. To this end we use a technique from the late 20th century that has intrigued image analysts ever since its original publication: the *medial axis*, *symax*, or *symmetric axis* invented by the mathematical psychologist Harry Blum (1973) as a plausible model for the way the primate retina processes outline forms. The symmetric axis is a polyline (a connected chain of straight segments) running precisely up the middle of any closed outline in two dimensions. Using the algorithm of Bookstein (1979), I produced this structure for the averaged arch form (Figure 5.23, left panel) and then located the middle of each specimen with respect to this template. Signal strength for shift was operationalized as the relative difference between group averaged positions of the corresponding transect in the shared coordinate system. The maximum of that signal, over 13 labeled transects along this axis, was for the midpoint of the segment from SL 14 to SL 6. From its scatter around the mean symax, Figure 5.23 (right), we extract a signal strength corresponding to a t-ratio of 2.58, which is barely more than half the signal amplitude in Figure 5.21 and is inferior to the t-ratio of 3.20 pertaining to SL 15 in the original unfocused Procrustes fit. From local averaging along the symmetric axis (thus involving one more parameter, for the interval covered by this averaging) the best one can do for such a two-sided shift signal gives a t-ratio of 3.49 with the same number of parameters as our 4.68 for the optimal Procrustes neighborhood approach. Thus, in spite of the temptation (Figure 5.20, right) to see the phenomenon as a displacement of the interior, in terms of its information content it is more clearly interpreted as an erosion of one surface, the underside, only. The upper arch opposite is too variable to afford any additional signal detection here.

Systems described by suites of variables organized a priori, like these shape coordinates, usually require preprocessing to get rid of nuisance variation. For data that come as spectra, the nuisance variation consists of machine calibrations like temperature, initial concentrations, and so forth. For shape coordinates, there are four standard nuisance variables built into the Procrustes method (orientation, scale, and two for position), and when the data come from unlabeled outlines, like these callosal outlines, there is also a "nuisance functional," the labeling that matches up a discrete subset of points across the many specimens (see Section 7.6). But, as an automobile's bumper sticker would put it, "one man's nuisance variation is another man's signal." For the purpose of signal detection in organized systems, such as these callosa, no presumption of this sort should go unexamined. In this example, we checked for the possibility of a local registration on the median line to substitute for the more conventional least-squares possibility. Relaxation of the characterization of the registered coordinate system proved unhelpful in this case, but it was worth exploring nonetheless.

5.4 A Concluding Comment

This ends the prologue to Part III. We began with three classic examples of organized complex systems (stars, proteins, and a neurological birth defect), briefly moved retrograde to characterize such systems in general verbal terms, and then explored the range of focal descriptions for typical phenomena within this domain. We reviewed some of the standard graphics for complex systems – charts with empty space (Figures 5.1, 5.3), extensive lists of related measures (Figures 5.4–5.7), and maps or networks with information everywhere (Figures 5.8, 5.9, 5.11, 5.16). We saw that there can be rules for making arithmetic rational as applied to focal numerical inferences from many of these types of charts (e.g., Figures 5.10, 5.12), including an expanded justification of the Gaussian distribution in terms of these additional arithmetical criteria, and we have reviewed a quite general class of decision procedures for deciding if an enrichment of a description is worth the cost of the additional parameters that were required to enrich it (the Akaike Information Criterion, Figure 5.19). The chapter closed with an example that combined a distributed data resource and the AIC to arrive at an inference about the focal nature of a neuropsychiatric group difference while simultaneously deprecating a different focal description of the same phenomenon. Throughout, we took care to separate the symbolic machinations of the applied mathematician from the much less playful interrogations of Nature by the scientist.

While free play with ranges of descriptors along these lines can be both exhilarating and satisfying, it lacks the computational predictability and stability of a teachable toolkit. The next chapter turns to the most fundamental computational engine that applies to domains like these, sometimes as model for signals and sometimes as model for the background against which signals are to be perceived: the singular-value decomposition along with its associated geometrical diagram styles. The singular-value decomposition applies to a variety of summary structures, not only to matrices. Imposing no Gaussian assumptions of any sort, it enables a transition from the flat and overly symmetric structure of matrices to the far richer domain of scientific ordinations (examinations of multidimensional similarity and dissimilarity) and the immensely varied choices of rhetoric, numerical encoding, and numerical inference that they afford.

> # 6

The Singular-Value Decomposition: A Family of Pattern Engines for Organized Systems

To find numerical structures that are both surprising and explanatorily useful in complex organized systems, we need a general-purpose numerical pattern engine. This chapter, building on the foundation in Chapter 5, introduces a range of variations on one theme, the *singular-value decomposition* (SVD). Section 6.1 is about an elegant geometric diagram, the doubly ruled *hyperbolic paraboloid*. Section 6.2 describes the algebra of the singular-value decomposition, whose diagram this is. Sections 6.3, 6.4, and 6.5 explicate three important application contexts of this tool – principal components analysis, Partial Least Squares, and principal coordinates analysis – with variants and worked examples. Chapter 7 is occupied mainly with demonstrations of the craft by which all this may be interwoven in one context, morphometrics, that very nicely matches the theme of organized systems on which Part III concentrates. What makes morphometrics so suitable an example is its systematic attention to the parameterization of the variables that go into the SVD pattern engine, a parameterization that the SVD is specifically designed to inspect and indeed to try and simplify.

6.1 The Hyperbolic Paraboloid

How can geometry help us apprehend numerical explanations – help us turn arithmetic into understanding – in more complex systems?

Most readers who have ever taken a statistics course have seen straight lines used as if they were explanations. And if you've taken more than one statistics course, probably you also remember learning how circles are compared to ellipses or spheres to ellipsoids as part of the logic of conventional data analysis (cf. Section 5.2.3), and likewise the generalization of lines to planes or hyperplanes (Section L4.5) as part of the way one is usually taught linear models for applications across all the sciences. Spheres define equidistance, ellipsoids characterize variance as a function of direction, and planes give you something to measure distance *from* when one approaches numerical reasoning in these popular but stereotyped ways. Ellipsoids and planes are geometrical objects with potentially useful descriptors, but they do not exhaust the useful geometries for statistics of systems – in fact, they are somewhat

6.1 The Hyperbolic Paraboloid

oblique to the most useful applications, applications to organized complexity, right from the outset. There is a geometrical reason for this: spheres and planes have too much symmetry. Specifically, planes map onto themselves by sliding in two directions, and spheres rotate onto themselves via a three-parameter group of rotations. We need substrates for our descriptions that do not offer so much degeneracy.

For our purpose here, which is the conversion of numbers into explanations, neither ellipsoids nor planes prove as useful as a less familiar geometric structure, the *hyperbolic paraboloid*, that maps onto our prior notions of structured explanation a great deal more directly. Because they are doubly ruled (see later), hyperbolic paraboloids accommodate information about variables a great deal more explicitly than planes or ellipsoids can. This labeling feature allows hyperbolic paraboloids to serve effectively as underlay for carefully crafted visualizations in three different analytic contexts that, taken together, suffice for much of the territory of the organized systems sciences. The present subchapter introduces the geometry of this useful explanatory surface, and the next shows the algebra by which data is mapped onto it.

Ellipse, parabola, hyperbola – what do they have in common? Geometrically, they are all *conic sections*, cuts of a cone by a plane. Algebraically, it can be shown (this was one of the early triumphs of analytic geometry) that they all have equations of the general form $ax^2 + bxy + cy^2 + dx + ey + f = 0$, most of which can, by change of origin of coordinates and shear, reflection, or rotation of the Cartesian axes, be brought into one of the three forms $x^2 + y^2 - f = 0$, $x^2 - y^2 + f = 0$, or $x^2 = y$. These specify ellipses (including circles as a special case), hyperbolas, and parabolas, in that order. The ellipse and hyperbola together constitute the "general case," two quadratic terms; the parabola is less general by one degree of freedom.

There is more than one way of getting to the hyperbolic paraboloid that will prove so useful to us here.

1. *The surface $z = xy$.* Consider the simple mathematical function that is multiplication – $z = xy$ – as a surface (Figure 6.1). It has an interesting geometry. Through every point $(x_0, y_0, x_0 y_0)$ pass two distinct straight lines $(x_0, t, x_0 t)$ and $(t, y_0, t y_0)$ that lie entirely in the surface, and these lines must span the tangent plane to the surface at $(x_0, y_0, x_0 y_0)$. Then the tangent plane to the surface at the origin is the xy-coordinate plane itself. The other two coordinate planes cut the surface along the Cartesian x- and y-axes, while every other plane on the z-axis cuts the surface in a parabola. As we rotate that vertical plane around the z-axis, from the x-axis around to the y-axis the curvature of the parabola along which the surface is cut rises from 0 (the straight line) up to a maximum for the diagonal, where the surface cut has the equation $z = r^2/2$, and back down to 0 for the plane through the y-axis. Continuing the rotation leads us to a *downwardmost* gape of the parabola, $z = -r^2/2$, after which curvature attenuates back to zero along the x-axis again. If you rotate this entire $z = xy$ surface by 90° around the z-axis, you get back a reflection of the original surface in the xy-plane. (We will see another instance of this same rotational antisymmetry in Section 7.5: the surface for the "purely inhomogeneous" thin-plate spline.)

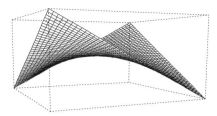 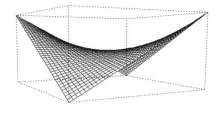

Figure 6.1. Two views of the same surface $z = xy$, the surface that we use to visualize each dimension of a singular-value decomposition. The loci that look like straight lines drawn on the surface are indeed exactly straight lines in space; the surface is woven from two families of these.

2. The *sorting of the conic sections* into the three classes (ellipse, hyperbola, parabola) by their equations in some affine system can be carried out in 3D just as in 2D. Instead of just two general cases, we get three: $x^2 + y^2 + z^2 = f > 0$, $x^2 + y^2 - z^2 = f > 0, x^2 - y^2 - z^2 = f > 0$, up to shear and rotation – these are the ellipsoid, the hyperboloid of one sheet and the hyperboloid of two sheets (the latter two surfaces are shown in Figure 6.2). Up to a scaling of axes, either of the hyperboloids can be generated by rotating an ordinary plane hyperbola around one of the bisectors of its pair of asymptotes. Following these three general cases are the two kinds of *paraboloids* $x^2 + y^2 = z$ (the elliptic paraboloid) and $x^2 - y^2 = z$ (the hyperbolic paraboloid), and then some even more special cases, for a grand total of 17. Because $x^2 - y^2 = (x + y)(x - y)$, the surface we are interested in can be rewritten as $z = uv$ for new axes $u = x + y$, $v = x - y$. Thus it is the same (except for a 45° rotation) as the surface that was already explored in (1).

> The hyperboloid of one sheet (Figure 6.2, left center) is the general case of a curved surface that is *doubly ruled* (i.e., that contains the entirety of two straight lines passing through each point of the surface). The hyperbolic paraboloid shares this property. Doubly ruled surfaces are easy to construct, either out of children's construction string or out of real building materials (if you do a Google Image search for either "hyperboloid" or "hyperbolic paraboloid," back will come some pictures of building roofs). The hyperboloid of one sheet can also be "constructed" virtually, by spinning any line around an axis to which it is skew. In this situation the other ruling is generated by the reflection of the line in the plane through the axis and the point of closest approach of the lines.

3. On the right in Figure 6.2 is a helpful construction that arrives at the hyperbolic paraboloid by a different approach. Begin with an elliptic paraboloid, as in the bottom right frame. Beginning from the lowest point, it rises upward according to two principal curvatures that can be thought of as parabolas pointing upward in the xz-plane and the yz-plane. If we flatten one of these parabolas to a straight line, we get the parabolic cylinder (center right panel). Finally, if we arrange the parabolas so that they point in opposite senses – upward along the x-axis, say, and downward along the y-axis – then we get the hyperbolic paraboloid, upper right. Because of this alternation of sign of its principal curvatures,

6.2 The Singular-Value Decomposition 349

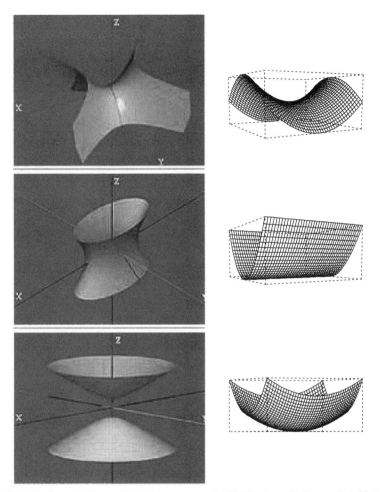

Figure 6.2. The hyperbolic paraboloid in context. (Left) The hyperbolic paraboloid (top) is distinct from both the hyperboloid of one sheet (middle) and the hyperboloid of two sheets (bottom). Figures from the Wikipedia commons. (Right) The same hyperbolic paraboloid (top), now drawn in the coordinates of the principal curvatures instead of the rulings that were used for Figure 6.1, extends the progression from elliptic paraboloid (bottom) through parabolic cylinder (middle).

the hyperbolic paraboloid, unlike the elliptic paraboloid and the hyperboloids, *cannot* have fourfold rotational symmetry. A 90° turn turns it upside down, as already noted. If you move in close to this surface, it approaches its own tangent plane, where the rulings turn it into a graphical model for analysis of variance.

6.2 The Singular-Value Decomposition

Why are we bothering with the description of the hyperbolic paraboloid? *Because **any** data matrix can be decomposed as a stack of hyperbolic paraboloids.* We will see this in three different contexts, but they all have the same algebra, which is the algebra of the *singular-value decomposition*, the SVD. It is the single most useful tool in all of multivariate data analysis, and its mathematics is not too difficult.

Suppose that instead of choosing one column of a flat data matrix as the "dependent variable," and some or all of the others as "independent variables," we wish to fit the data matrix *as a whole*, the entire rectangular pattern, still by least squares. Is there any algorithmic way to do this? Yes, of course, or modern data analysis would not exist.

6.2.1 The SVD Theorem and Its Role in the Description of Organized Complexity

We need an (intuitive) notion of *simplicity* in order to launch a discussion of what it is about real data sets that makes them so often infuriatingly or charmingly complicated. In a context of matrix algebra, the usual way of working through multivariate analysis, this can begin with the simple quantity that is the **rank** of a matrix.

Here the simplest nontrivial structure is a matrix of rank **one**. A rank-1 matrix is a matrix $S = (S_{ij})$ that takes the form $S_{ij} = a_i b_j$ for some vectors a and b for all i and j. By throwing in a normalizing constant, we can write this as $S = duv^t$, where u and v are unit vectors (which will not have the same count of elements, unless S was a square matrix) and d is some positive number. Geometrically this means that if you change the surface drawing version of the matrix so that the rows are spaced as the values of u and the columns as the values of v (instead of integer by integer) then the picture is always the same surface, the hyperbolic paraboloid that we have just been talking about.

A rank-two matrix is a matrix that isn't rank one but that is the sum of two that *are* rank one; a rank-three matrix isn't rank two but is the sum of a rank-two matrix and a rank-one matrix; and so on.

Obviously any $m \times n$ matrix is the sum of the mn matrices that have all entries but one equal to zero, so any $m \times n$ matrix has rank no more than $m \times n$. But that estimate is way too high. In fact, the rank of a $m \times n$ matrix is no more than m or n, whichever is less. Let's phrase this useful mathematical fact in a way that seems quite distant from any practical scientific investigation. (But it isn't.)

Let S be a general $m \times n$ matrix arising in the course of some statistical study. (Maybe it is a data matrix, or maybe a covariance matrix, or a zero-diagonal matrix of dissimilarities.) In what ways, if any, can we write S as the sum of two matrices $S = S_1 + S_2$ with $\text{rank}(S_1) = 1$ and $\text{rank}(S_2)$ <$\text{rank}(S)$?

In other words, can we separate S into two parts, one of which has minimum complexity, without increasing the total complexity? You can imagine that if we can do this once, we can do it again, maybe r times, and then stop, or maybe even stop earlier if we peel off these partial descriptions S_i in some optimal order. If that is what you are thinking, you are right.

The answer to this question follows from the **singular-value decomposition**. This is a theorem, originally suspected around 1889,[1] that states that any rectangular matrix S can be decomposed in essentially just one way into a triple product

$$S = UDV^t,$$

where U is orthonormal $m \times m$, D is diagonal $m \times n$, and V is orthonormal $n \times n$. (Orthonormal means that the matrix times its own transpose is the identity; geometrically, this is the kind of matrix that produces rotations. And a nonsquare matrix is diagonal if all of its entries d_{ij} for $i \neq j$ are 0; we write d_i for the remaining diagonal entries d_{ii}. Another way of thinking of the "ortho" part of the word "orthonormal" is that the columns of U have dot products of 0 with one another as lists, that is, they are perpendicular as vectors out of the origin, and likewise the columns of V.) We write the d's in descending order of magnitude, with d_1 the largest. In this context, the columns of U and V are called the *left* and *right singular vectors*, and the nonzero entries of D are the *singular values*. The answer to our original question is that the matrices S_1 we asked about constitute a finite set, the $r = \min(m, n)$ matrices $d_i U_{\cdot i} V_{\cdot i}^t$, and for each the corresponding matrix S_2 is just $\sum_{j \neq i} d_j U_{\cdot j} V_{\cdot j}^t$. (In this notation, $U_{\cdot j}$ stands for the elements of U with second subscript j, in other words, for the jth column of U. We will also write this as U_j whenever the implied positioning of that subscript is clear, as in the rest of this section.)

It is these r distinct structures that we can "peel off" from the starting empirical matrix S without our description extending what may already be a complicated situation. For data analysis we start with the pattern $d_1 U_1 V_1^t$, then $d_1 U_1 V_1^t + d_2 U_2 V_2^t$, and so on.

What does any of this have to do with hyperbolic paraboloids? Each of those terms $d_i U_i V_i^t$ is a sample of the function $z = xy$ over some finite set of U's and V's. All these terms, then, have exactly the same picture, which is the picture of your basic hyperbolic paraboloid, Figure 6.3 (compare Fig. 6.1), within which the lines corresponding to the entries of the vector U_i within the x-rulings have been highlighted, and likewise the lines corresponding to the entries of the vector V_i along the y-rulings. For this example, the matrix we are drawing is

$$\begin{pmatrix} 48 & 64 & 72 & 88 & 100 \\ 480 & 640 & 720 & 880 & 1000 \\ 912 & 1216 & 1368 & 1672 & 1900 \\ 1056 & 1408 & 1584 & 1936 & 2200 \\ 1248 & 1664 & 1872 & 2288 & 2600 \end{pmatrix}$$

which is the rank-1 product of $u = (2, 20, 38, 44, 52)$ and $v = (24, 32, 36, 44, 50)$. This engraving on the hyperbolic paraboloid is our commonest model of a **simple**

---

[1] As per Reyment and Jöreskog (1993, p. 62), who go on to cite Eckart and Young (1936) for the "full extension to rectangular matrices."

352   6 The Singular-Value Decomposition

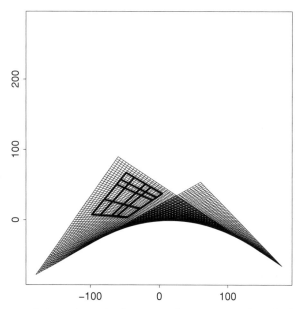

Figure 6.3. Every rank-1 matrix can be drawn exactly as a rectangular selection of the rulings of a hyperbolic paraboloid. This is our geometric model of a simple explanation as the tabular arrangement wanders here and there over the common grid. See text.

explanation. (For a slightly different version, involving addition instead of multiplication, see Section 6.4.3.6.)

These lines can be numbered or labeled corresponding to the a-priori content of the matrix row or column (case or variable) in question. There is one of these structures for each singular vector pair that is deemed worthy of interpretation. Successive selections $U_1, U_2, \ldots$ make an angle of $90°$ as vectors – recall that is one interpretation of the characterization of the matrix U as "orthogonal" – and likewise successive V's.

Some identities relating S and the elements of its SVD can be useful:

$$U^t S = DV^t,$$
$$SV = UD,$$
$$U^t SV = D,$$
$$\Sigma d_i^2 = \Sigma_{ij} s_{ij}^2,$$

where the last identity means that the sum of the squared singular values is the sum of the squared entries of the original matrix – thus the SVD is, in a sense, a *decomposition* of this sum of squares into a rank-1 piece for each subscript i that "explains" (reconstructs) a total of d_i^2 of that total sum of squares.

And the following, only slightly deeper property is often surprisingly useful:

Least-squares property. If the d's are in descending order of magnitude, then for each rank j between 1 and $r - 1$, the approximation

$$M = \Sigma_{k=1}^{j} U_{\cdot k} d_k V_{\cdot k}^t$$

minimizes

$$\Sigma_{i,j}(S_{ij} - M_{ij})^2$$

over the class of matrices M of rank not greater than j. This means that the SVD is functioning as a sort of explanation of the matrix S as a whole. If S is describing an organized system, then the SVD of S is a promising approach to the search for structure, subsystems, or simplifications, and, because it is wholly parametric, it articulates with the principles of abduction (surprise) and consilience (multilevel measurement) that we have been emphasizing all along.

It's worth convincing you of this for the first level in particular, the rank-1 representation $S \sim d_1 U_{\cdot 1} V_{\cdot 1}^t$. Suppose we have the minimizing U for this least-squares problem, and seek the minimizing V. The minimization of $\Sigma_{i,j}(S_{ij} - U_{\cdot 1} d_1 V_{\cdot 1}^t)^2$ can clearly be carried out column by column, and for each column j the minimizer $d_1 V_{j1}$ is evidently the ordinary regression coefficient (without constant term) for the regression of $S_{\cdot j}$ on the vector $U_{\cdot 1}$. We talked about how to get this coefficient – it is the weighted average of the ratios $S_{\cdot j}/U_{\cdot 1}$ – in Section L4.1. But that formula is just the same as the corresponding row of the equation $SV = UD$ above – the entries of the SVD *are* the regression coefficients of the least-squares interpretation.

Please note the radical difference between this version of least squares and the more common version relating to an expression like $y = a_1 x_1 + a_2 x_2 + \cdots$ claimed to predict one variable, the "dependent variable," from a list of one or more others that are imagined to be "independent," Section L4.5. The prediction version is minimizing a sum of squared "errors" (discrepancies) in the value of one variable, y. By contrast, the SVD is minimizing the sum of squared discrepancies in the values of every entry in the entire data matrix. Whereas the regression equation is one formula that leaves errors of prediction, the SVD is a whole stack of components that eventually is guaranteed to reconstruct the data matrix perfectly, without any remaining error at all. As several wise people have already commented over the course of the 20th century, in the world of complex systems there are no such creatures as "independent" and "dependent" variables, anyway – that formalism makes no scientific sense as a rhetoric of explanation in these domains. It can be of use only as a calculus of prediction, and even in that context has severe, often disastrous philosophical limitations and associated practical pitfalls and pratfalls. See Pearl (2009).

If we are faced with curves as data, typically we will sample them at a series of appropriately spaced points, turn the samples into a data matrix, and submit that matrix to this analysis. There is a standard graphic for the SVD, the *biplot* (Gower and Hand, 1995; Gower et al., 2011), that displays pairs of both sets of loadings, the U's and the V's, on the same diagram; or we can draw them separately for clarity or for annotations. If we want to study variation around an average, we average the columns of the matrix and subtract off these averages before doing the fitting. We can do the same for the rows, or for both rows and columns, as we will see; but we are not required to do so.

> SVDs can earn you a lot of money. The famous *Netflix Prize* of 2009 was an award of a cool $1 million from the Netflix Corporation to the first team of scientists who could demonstrate a 10% improvement over the previous algorithm for forecasting how much customers like movies. The prize was won jointly by a large team of computational statisticians – Robert Bell, Martin Chabbert, Michael Jahrer, Yehuda Koren, Martin Piotte, Andreas Töscher, and Chris Volinsky – and the winning technology was a clever modification of the SVD, averaged over a variety of different models for human preference orderings. See http://en.wikipedia.org/wiki/Netflix_Prize and http://www.netflixprize.com.

In Section 6.3 we apply the SVD to mean-centered data matrices. In this context it is usually called *principal components analysis (PCA)*. In Section 6.4 the SVD is applied to general rectangular correlation and covariance matrices (along with some generalizations); in these contexts it is usually called *Partial Least Squares (PLS)*. And in Section 6.5 we apply the SVD to dyadic distance (dissimilarity) matrices; here it is usually called *principal coordinate analysis*. At root, these are all the same technique. The SVD is indeed the main workhorse of statistics for systems analysis, the pattern engine you need to have if you can't have more than one.

6.2.2 Example: Viremia in Monkeys Exposed to SIV

What follows here is an SVD of actual original data, without any mean-centering or scaling. It shows the power of an otherwise conventional multivariate analysis taking advantage of the fact that the variables came in with an a-priori ordering. (Recall that the salience of such orderings is one of the essential characteristics of complex systems.)

The analysis in Figures 6.4 and 6.5 uses some data previously published in Letvin et al. (2006). (I am the twelfth *al* of the *et al*'s who wrote this article.) The original study, part of the National Institutes of Health's (NIH's) research program on AIDS, involved 30 NIH experimental macaques of which 24 were vaccinated variously and then all were inoculated with a calibrated dose of SIV (simian immunodeficiency virus). Diverse aspects of their blood were measured, of which I have selected only the viremia data (blood plasma concentrations of the viral RNA). Note the log scale for viremia along the vertical axis – the raw data ranged from 125 to 24,200,200, hence this transformation. I have restricted the subsample here to just the 23 monkeys with complete data from the experimental groups all of whose animals survived at least 288 days post-inoculation. In particular, this reanalysis omits the control [unvaccinated] animals, as they did not systematically survive this long. (And so these figures cannot serve in any way as a substitute for the published analysis.)

The experimental design measured the monkeys at least 22 times each, but the measures were not evenly spaced in time. They came more frequently at the outset of the experiment than afterwards, at a spacing that was roughly exponential with

6.2 The Singular-Value Decomposition 355

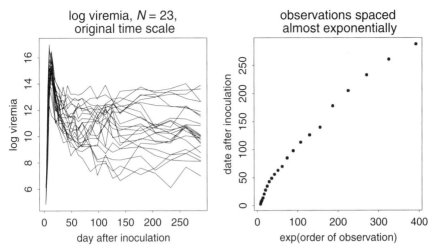

Figure 6.4. Part of a data set about viremia in experimental monkeys. (Left) Time curves of viremia (to a log scale) for 23 monkeys observed over 22 ages post-inoculation. (Right) The observation dates are approximately loglinearly spaced in time. (Data from Letvin et al., 2006.)

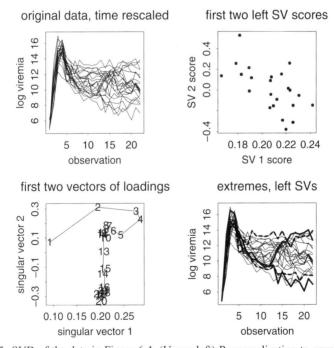

Figure 6.5. SVD of the data in Figure 6.4. (Upper left) Renormalization to equal spacing in postinoculation dates. (Upper right) First two left singular vector scores (scatterplot of the first two columns of U). (Lower left) First two right singular vector scores (scatterplot of the first two columns of V), connected in order of measurement dates. (Lower right) Enhancement of the upper left panel indicating the time courses of the animals with the highest and lowest scores on left singular vector 1 (dashed lines) and left singular vector 2 (solid lines).

sequence number (Figure 6.4, right panel). The horizontal axis on this chart is actually $e^{1.9087+.1846*\text{data\_wave}}$, where the exponent is the regression prediction for log of measurement date on measurement wave number.

Figure 6.5 shows the SVD-based analysis of log viremia against measurement number for the 22 measurements after time is normalized according to this rescaling. The analysis is thus of a 23-by-22 rectangular matrix, and its relation to the original data is like that of a log-log plot. The viremia scores have *not* been mean-centered, as it is their absolute level and its time course that matter most for the science (and the monkeys). Viremia was measured in counts per milliliter (cubic centimeter) of blood, and as the "per milliliter" unit is arbitrary the corresponding log transformation has no natural zero. As the SVD is formally multiplicative in the mean profile (SV 1), the effect of setting the unit of viremia needs to be checked. In the present situation, changing this unit, say to cubic millimeters, makes no difference for the pattern analysis to follow.

In Figure 6.5, the upper left panel simply restates the raw data from Figure 6.4 with the times of observation of viremia now evenly spaced. The singular values of this data matrix are 251.91, 16.55, 9.63, and 9.02. As already noted, the first value is meaninglessly large, as it is a function of the location of the zero for the log function. The third and fourth of these values are nearly enough equal that we should stop interpreting the science after the second. The lower left panel shows the remarkably unexpected pattern[2] displayed by the first two columns of the V matrix. Clearly the system of viremia management in these monkeys is divided into two phases: an initial shock, measurement days 3, 7, 10, and 14 (waves 1, 2, 3, and 4), after inoculation, and then all the remaining dates, 21 through 288, during which the loadings return to a constant on the first SV but drop fairly steadily on the second. The plot at upper right shows the case values corresponding to these two patterns. The correlation between these scores is nearly -0.6 – as these variables were not mean-centered, they are not principal components. There was no reason to expect their correlation to be zero, and it isn't.

The most interesting panel of this composite is the panel at the lower left, the ordination of loadings on the first two right singular vectors V. At lower right I have reproduced the plot from upper left with four specific time-series highlighted: the highest and lowest on SV1, marked with the bold dashed lines (– – –), and the highest and lowest on SV2, marked with the bold solid lines. Component 1 is evidently tracking the mean viremia levels: lowest on the first measurement date, highest at dates 3 and 4, and then of relatively constant average levels from the seventh measurement appointment on to the end. The score on this SV is a gently weighted average of all the viremia measures, highest for the monkey whose history follows the upper bold dashed line of the panel at lower right, and lowest for the monkey of lowest average level, the lower heavy dashed line. The second SV, the one on which the original log viremia measurements drop steadily from the seventh

---

[2] Over 40 years of applied multivariate analysis I have never before generated a scatterplot taking the form of a question mark. One might refer to this data analysis as self-skeptical.

measurement onward, is evidently picking up the contrast between early and late level, the contrast between the two trajectories drawn in continuous heavy line in the same lower right panel.

Thus this SVD has pulled out the two most salient aspects of the empirical data distribution automatically. The sensitivity of this particular pattern engine matches well the way in which a competent expert in the field would examine these data prior to explicit consideration of particular hypotheses pertinent to the experimental groups in the actual design. The SVD thereby supersedes the a-priori orthogonalization of the same 22-wave measurement, which would not be so attuned to the strong systematic aspects of the empirical time course here (the exponential acceleration of that early rise, followed by a partial fall-off and then a variety of individual time courses that combine a mean level with a mean slope that is predicted by that mean level to some extent). This SVD is likewise much more promising for any subsequently more detailed study of time courses like these using more sophisticated system understandings or additional information (here, that might be the details of the four different vaccinations, or the data on immune function that were also part of the *Science* publication). For the full context of this experiment, see Letvin et al. (2006). The sensitivity of the SVD to emergent details of the scientific setting, as shown here, is typical of its applications all across the sciences of organized complexity.

6.3 Principal Components Analysis

A good reference for this section in respect of applications is the 1993 book by Reyment and Jöreskog (a second edition, but with a new title). For statistical methodology, the reader would do well to consult the magisterial reference manual by Mardia et al. (1979), Chapter 8.

6.3.1 Geometry of the Variation of Axes of an Ellipsoid

The application we were examining in Figure 6.4 was a cases-by-variables example; we continue in this context. Figure 6.5 suggested one standard method for thinking about the contents of the U array. Here we investigate the corresponding conventions for V.

We are in a vector space different from the one in which the data arose. It is instead the so-called *dual space* to that vector space, the space of linear combinations of the originally measured variables. One models elements in this space of directions not by a sphere but by an ellipsoid E for which distance from the origin in the direction α is the sum of squares of the corresponding linear combination – the variance in the other domain, the quantity $\alpha^t S^t S \alpha$ (where S here is the *centered* data matrix, after every column is shifted to mean zero). The singular vectors V are the principal axes of this ellipsoid E, and the singular values are now the squared lengths of those axes. $S^t S$ is N times the *covariance matrix* of the original data, and the SVD can be applied to this matrix just as easily as to S itself. This would make it appear as if there were two sets of d's (the singular values), but those for $S^t S$ are just the squares of those for

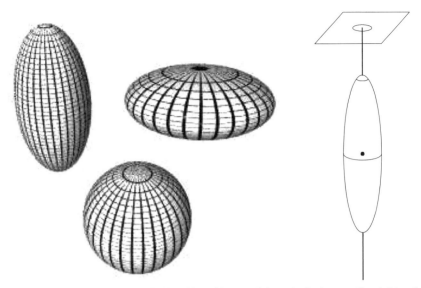

Figure 6.6. (Left) An assortment of ellipsoids, with one of the principal axes of each hinted at by a Mercator chart. (Right) The standard error of any such axis is usefully represented by a distribution of piercing points described by the *remaining* axes as scaled by Anderson's formula (see text). Equivalently, this error distribution can also be represented by a distribution in some plane tangent to an enlarged copy of the ellipsoid.

S itself (because if $S = UDV^t$ then $S^tS = VDU^tUDV^t = VD^2V^t$, which is itself in SVD form). See Figure 6.6 (left). This is called **principal components analysis**. Now the diagonal elements of D are called *eigenvalues* and the columns of V are the *principal components*.

Principal components are perpendicular and of unit length as vectors in their space – that is just what it means for the matrix V to be "orthonormal." But, also, the corresponding *scores*, the variables SV_j where S is the original measured data matrix after centering – are **uncorrelated as variables**. (*Why?* For $j \neq k$, we have $(SV_j)^tSV_k = V_j^tS^tSV_k = V_j^tVD^2V^tV_k = D_{jk}^2 = 0$ because the matrix D is diagonal; but this quantity is just the numerator of the usual formula for the correlation.) This is, in fact, an alternative way to *define* the principal components of a data set S: as the variables that are both perpendicular as formulas and uncorrelated as scores.

We are using algebra to produce some quantities that were not directly measured. They must therefore have some numerical uncertainty that arises from their dependence on accidents of sampling (among other features). What kind of plus-or-minus can we put around these vectors? As the panel on the right side of Figure 6.6 suggests, this can take the form of an ellipse of its own (or, in higher dimensions, an ellipsoid) in the tangent plane or hyperplane to E at the piercing point of the axis we're asking about. We could also draw it directly on the surface of E; the figure illustrates both of these possibilities.

What are the size and shape of these derived ellipses, exactly? Again the difficult article by T. W. Anderson (1963) comes to our aid with the crucial formula. Write

$\lambda_1, \ldots, \lambda_p$ for the eigenvalues of the covariance matrix and $V_1, \ldots, V_p$ for the corresponding eigenvectors (principal axes). Then (Figure 6.6) the variability of the jth axis is entirely in the direction of the space spanned by all the *other axes*. Asymptotically, the variation in each direction V_k of this tangent plane, $j \neq k$, is Gaussian with variance $\frac{1}{N-1} \frac{\lambda_j \lambda_k}{(\lambda_j - \lambda_k)^2}$ independent of all the other axes.[3] That last expression reduces to $\frac{1}{N-1}$ times the reciprocal of $(\frac{\lambda_j}{\lambda_k} - 1)(1 - \frac{\lambda_k}{\lambda_j})$. When $\lambda_1 \gg \lambda_2, \ldots, \ldots$ (the case of a cigar), these coefficients are roughly equal to $\frac{\lambda_j}{\lambda_1}$, the ratio of fall of the eigenvalues. (The symbol "$\gg$" is read "much greater than.") When $\lambda_1 \gg \lambda_2 \gg \lambda_3 \ldots$ (like a flattened cigar), the $k = 2$ term in this expression, in turn, dominates all the later terms, meaning that the uncertainty of PC1 is mainly in the direction of PC2. By comparison, the variance of PC1 itself, in this formula, begins with a coefficient that is approximately $\frac{\lambda_2}{\lambda_1}$, and so is reduced from that of the other PC's by at least this ratio. In either case, whenever the eigenvalue associated with PC1 is hugely dominant, the corresponding eigenvector is hugely more stable than the others.

6.3.2 Under What Circumstances Can PCA Make Sense? Pattern Spaces for Variables

As Kuhn said in Epigraph 7 of this book,

> The route from theory or law to measurement can almost never be travelled backwards. Numbers gathered without some knowledge of the regularity to be expected almost never speak for themselves. Almost certainly they remain just numbers.

In our context of "numbers and reasons," this proposition applies with some bite to the specific technique, principal components analysis, that is the concern of this subsection. Principal components analysis is a tool for "parsimonious summarization," in a felicitous phrase from Mardia et al. (1979). In a context of explanation, such a purpose seems pointless should it come after data collection, that is, as a component solely of the analysis. Instead it needs to come *before* the analysis, in the course of designing a measurement scheme intended to support a strong inference about one theory in comparison to others. If the final presentation is of linear combinations of variables, the machinery for interpreting those linear combinations needs to be part of the *science*, not just the statistics.

In this way, the technique of principal components analysis prods us to contemplate a domain we have not yet touched on much: *what* to measure originally. Where do variables come from, and what does it mean to take a sample of them for the purpose of constructing a new composite that is explanatorily more helpful than any of the original measures separately? What kinds of scientific questions allow us to postpone this issue of their "summarization" until this late in the scientific process? Principal components analyses, like factor analyses, are capable of making scientific sense to the extent that the original variables *can* be construed as susceptible to a process of linear

---

[3] In other words, the same principal components that summarize the data, in this approach, summarize the sampling variation of the principal-component summary itself: a very convenient recursion.

combination of this flavor. The best scenario is the *structured list of variables* arising in some domain of possible questions or probes that has a mathematical structure all its own. Such an understanding needs to be in place long before the onset of data collection for the instant study – the variables need to have been structured before *any* data have been collected – and so schemes like these constrain the language of scientific summaries and explanations at least as much as do any patterns empirically encountered. What is important is not that the new axes are a "rotation" of the old in the geometry of the SVD. The epistemological issue is rather that of the linear combination of variables per se.

In what ways might our measurements arise as a list on which weighted averages make sense? They might be –

1. Samples from a formal population indexing the range of possible measures, such as spelling words sampled from a dictionary or arithmetic problems generated by a pseudorandom number generator.
2. Physically analogous measurements differing only by one or more scalar parameters: image energy as a function of wavelength in a spectrum, or gray value in a medical image as a function of x-, y-, z-coordinates once these have been standardized, or physical properties at a grid of points below Earth's surface or in the air above. One can take integrals over intervals of spectra like these, or over areas of images, or the equivalent in 3D.
3. In morphometrics (see Chapter 7), the *shape coordinates* that arise as displacements of standard points of a template under the operation of the mathematical maps standing in for anatomical correspondence when we are talking about the locations of points instead of parts.
3a. In the morphometrics of curves, in particular, the two *simultaneous* orderings applicable to the same data, arc length and spatial position, as demonstrated in the example closing Section 5.3 (the midline corpus callosum outline curve in schizophrenia).
4. Responses to probes that have a discrete parameter of their own: the first, second, ... repetitions of a test procedure, or the response to probes of physiological stress set to a series of rising biochemical parameters.
5. In time-series analysis, observations at a discrete series of *times* (1, 2, 3, ... seconds) or a discrete series of *lags* $t - 1$, $t - 2$, etc. preceding the present moment t.
6. Responses to probes that submit to just about any general parametric scheme that we can handle by "Domino PLS," see at Section 6.4.3.3: for instance, consumer ratings of a variety of beers, which can be scaled by the fractional contents of a shared list of possible ingredients and flavorings as well as by properties of those consumers, such as mouth flora.

In all these contexts, principal components analysis goes well only if the variables incorporated are all in the same units. (The nonenforcement of this requisite was one of the main reasons that the multivariate morphometrics of the 1970s was replaced by the geometric morphometrics of the 1990s, where all shape variables are now shape coordinates that all have the same unit, namely, Procrustes distance.) When variables do not come in natural units or dimensions, it is tempting to force them into that mold by reducing each one to a "z-score," distance from the mean in units of standard deviations. But this is rarely conducive to consilient explanations, for which units need translating from context to context just as much as findings do.

Variables with units can often have a natural zero, and when a measure is amenable to such an interpretation it is often helpful to carry out the SVD or the equivalent PCA on the measured values with respect to this natural origin. (This is particularly true when sums of squares have a physical meaning themselves, as when the squared amplitude of a waveform stands for its energy.) When the data matrix submitted to an SVD is uncentered in this way, the first right singular vector extracted will generally correspond to the average or the mean profile that would otherwise be subtracted off, and the corresponding left singular vector conveys the multiple of this profile that best characterizes each case in turn. Second and subsequent SVs are constrained to be orthogonal to the mean in the sums-of-squares geometry the SVD is using. When data have first been mean-centered, there is no such constraint on the first singular vector, which could well correspond to variation in the direction of higher or lower general average (i.e., a singular vector nearly proportional to $(1, 1, \ldots, 1)$), or instead might have a direction that is orthogonal to the mean even in the reduced geometry.

In all these contexts, the role of a principal component is that of a *pattern* linking values over the whole set of variables. This corresponds to the analytic context of an organized system, in which there is always coordination between different aspects of its integration and for which other variables, that is, responses to other probes, could have been measured, but weren't. The coefficients of these vectors, then, specify patterns of meaning in their own right – profiles that rise to a peak and then fall, or deformations that are most pronounced in some region of the form, or physiological responses that peak at some combination of volume and concentration. In other words, the coefficients of the PCs themselves serve as system measurements, eventually to be subjected to explanations in their own domain – another locus of consilience.

One interesting special case is the study of the principal components of particularly *small* variance. Patterns like these are expressing near-identities among the variables, linear combinations that, by seeming to be clamped in value near zero, express near-identities of linear combinations of the measurements they represent the values or the derivatives of. Such a search for invariants in fact characterized the first appearance of this method in print, work of Karl Pearson's around 1900. In this form the method is still used, as for instance in constructing a midline (axis of mirror symmetry) from a set of paired and unpaired points taken on a face or a photo of a face, or constructing a local tangent plane from a sampling of surface points in order to make more precise the computation of differential properties of the surface as it bends away from that plane.

Another interesting special case corresponds to the exemplary ellipsoid at upper left in Figure 6.6: a form with one axis much longer than all the others. In applications, this might arise when all measurements are strongly affected by the same exogenous factor. It could be, for instance, a large collection of size measurements that are all affected by a child's growth, or a large collection of cognitive capability assessments that express (albeit to varying extents) the same factor of overall intelligence or quickness. When an analysis of this sort is carried out in each of a list of covariance structures arising in separate samples, one often sees it asked whether the resulting first principal axes are "all the same" in their direction. This is not quite a well-formed question, however, as they can vary in length (singular value) as well as direction in the space within which they are computed. A complementary method, Flury's approach of *common principal components* (Flury, 1988), asks whether a list of separate covariance structures can be represented as a collection of ellipsoids on the same axes but varying in the lengths (specific variances) of those axes, meaning, again, the singular values. Such models are weaker, in general, than the factor models that supplement them by additional a-priori system information, such as the identities of the factors that are known to be susceptible to independent empirical interventions, and weaker, too, than the approach by principal coordinates of covariance (Section 6.5.3) that treats changes in length and changes in direction jointly as commensurate adjustments of the covariance structure as a whole.

6.3.3 Example

Chapter 7 introduces the concept of *shape coordinates*, a successful contemporary formalism for bringing quantitative anatomy under the sway of strong numerical inference. We will see how the methodology there represents varying biological shapes as a specialized set of Cartesian coordinates in two or three dimensions. Such a suite of variables meets the requirements of the itemized list in the preceding section – indeed, it is the specific concern of item 3 there. Taking this suitability for granted until the next chapter, we exemplify principal components analysis by an application to a representation of this type in a data set of $N = 29$ midsagittal skull forms. ("Sagittal," in craniometrics, is the dimension from left to right; the "midsagittal" or "midsagittal plane" is the plane up the middle of the solid head, the plane with respect to which we are approximately symmetric.) These data were originally published in Bookstein, Gunz et al. (2003), to which please refer for the scientific context of the work. Principal components of shape coordinates are usually called *relative warps*, after their introduction in Bookstein (1991); that will be our language here, too.

Principal components analysis is carried out as an SVD of the mean-centered shape coordinates. (The correlation option makes no sense because these variables are already in the same unit, Procrustes distance, as will be explained in due course, and a conversion to correlations would destroy this commensurability.) For morphometrics, the biplot is typical as far as the columns of U (the scores) are concerned but takes advantage of a good deal of diagrammatic artistry in respect of displaying the contents of V.

A sequence of analyses is displayed all at once in Figure 6.7. At upper left is the initial analysis for the starting data set of $N = 29$. The first few singular values of the analysis are $0.236, 0.0349, 0.0159, 0.0129, \ldots$, and so it is appropriate to look

6.3 Principal Components Analysis 363

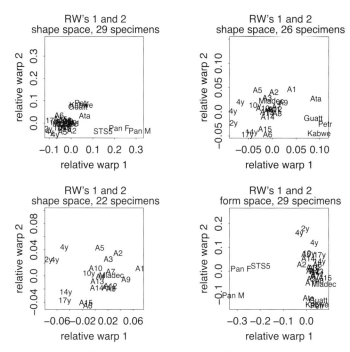

Figure 6.7. Four different PCAs of some skull forms. (Upper left, upper right, lower left) A sequence of three PCAs of shape, plotted by the first two columns of U (the first two *scores*) on $N = 29$, 26, or 22 forms, respectively. (Lower right) Analysis of a slightly different list of measurements, which augments the preceding list by one more variable, for Centroid Size (see Section 7.2). The ordination interior to the chimpanzee subsample differs considerably, illustrating the dependence of principal component analyses on the precise list of variables involved. A, Adult *Homo sapiens*, numbered; n y, child aged n years.

at the first two dimensions (singular vectors) only.[4] The structure of this scatter is informative, but not in a way that recommends the interpretation of the principal component analysis (the SVD) per se. We see three forms quite a distance from the others. From the labels, two of these are chimpanzees (of the genus *Pan*: one female, one male) and the third is the famous australopithecine specimen Sts 5 discovered at Sterkfontein in 1947, the specimen affectionately known as "Mrs. Ples" ("Ples" for "Plesianthropus," the initial scientific nomenclature, and "Mrs." because it was thought at the time to be a female [though as of this writing the issue of its gender is not yet settled]). At the left of the plot, in the direction opposite Pan M out of the grand mean at (0, 0), is a mass of forms labeled either "Hs" followed by a number – these are all adult *Homo sapiens* specimens – or, for the six juvenile humans, by their age in years (y). Off to one side of the line between these two groups is a clear cluster of three forms all given informal names – these three are all Neanderthals, as is a fourth form, Atapuerca (a particularly archaic Neanderthal), plotting nearby.

---

[4] The ratio $0.0159/0.0129$ falls short of the threshold for $N = 29$ shown in Figure 5.12.

The information this plot conveys about the analysis by principal components is thus explicitly self-refuting. What it says is that the method of principal components should not be applied to these data. Any findings in respect of those linear combinations are necessarily a function of the relative proportions of *H. sapiens*, *H. neanderthalensis*, and the other species in the sample, and these proportions are not biological phenomena. A numerical inference via PCA can make sense only if applied to a group that might be a legitimate sample of something.

So we toss out Mrs. Ples and the chimps and repeat the data centering and the SVD, resulting in the figure in the upper right panel. The first principal component is aligned with the difference between the average Neanderthal and the average *H. sapiens* – this cannot be news, and suggests, once again, that the analysis continues to be without merit. We also delete the four Neanderthals, then, leaving only the 22 *sapiens* (21 modern forms, plus one archaic specimen, "Mladec," from the Mladeč cave in what is now the Czech Republic). In the resulting scatter (lower left panel) of the first two columns of U, we see, finally, a continuum from the children on the left to the adults on the right, suggesting, at last, a biological meaning for component 1. We see no structure in RW2, but that is all right because the singular values of this 22-skull data set, 0.0286, 0.0144, 0.0110,..., suggest (by calibration against Figure 5.12) that no dimensions after the first are interpretable anyway.

We will return to this 22-specimen SVD in a moment, but first we revert to the original data set of $N = 29$ to examine what happens if we shift from "shape space" to "form space," in the language of Section 7.4, by augmenting the data set by the additional variable representing geometric size (by way of its logarithm). The result, Figure 6.7 lower right, is quite different from that for shape space, upper left, with which we began. The singular values are 0.0954, 0.0804, 0.0259,..., implying that these two dimensions bear all the interpretative weight we can get from this SVD. Now it appears that the male chimp does *not* simply represent an extrapolation of the difference of the female chimp from the humans, but, rather, he differs from her in a direction not too far from the direction along which the human adults differ from the three infants (the 2 y specimen and the two 4y's). The subscatter of all the humans at lower right looks somewhat like a shear of the corresponding configuration at lower left, except that the archaic Mladeč specimen has shifted position in the direction of the Neanderthals. It is still the case that the chimpanzees, and then the Neanderthals, need to be ostracized. There is considerably more explanation of this specific example in Chapter 4 of Weber and Bookstein (2011).

Back in the lower left panel of Figure 6.7, it remains to display the other part of the biplot, the information conveyed in the matrix V. As this was the unidimensional analysis, we are concerned only with the first column of V. Figure 6.8 shows the layout of the actual measurements used for the shape coordinates here – the Cartesian locations of the twenty points named there, located on the midplane of the skull as hinted by the outline dots between them – and Figure 6.9 combines the two different standard plots for the columns of V. As this is a pattern vector, it can be visualized in either polarity, toward the right in Figure 6.7 or toward the left (i.e., infants to adults or adults to infants). In the left column of Figure 6.9 are the representations

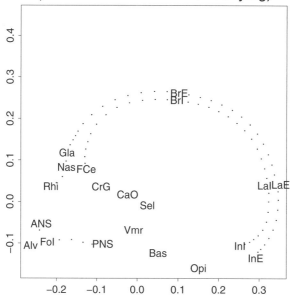

Figure 6.8. Guide for interpreting the V side of the biplot for the $N = 22$ example. The principal components are of the Procrustes coordinates of the twenty landmarks indicated here on a typical form. Dotted lines sketch the curving of this form between the landmarks. For a color version indicating also the conventional division of the bones of the midsagittal skull, see Figure 1 of Bookstein, Gunz et al. (2003). Landmark names and operational definitions: *Nas*, nasion, highest point on the nasal bones in the midsagittal; *Gla*, glabella, most anterior point of the frontal in the midsagittal; *BrE*, *BrI*, external and internal bregma, outermost and innermost intersection points of the coronal and sagittal sutures; *LaE*, *LaI*, external and internal lambda, outermost and innermost intersections of sagittal and lambdoidal sutures; *InE*, *InI*, external and internal inion, most prominent projections of the occipital bone in the midsagittal; *OpI*, opisthion, midsagittal point on the posterior margin of the foramen magnum; *Bas*, basion, midsagittal point on the anterior margin of the foramen magnum; *Sel*, sella turcica, top of dorsum sellae; *CaO*, canalis opticus intersection, intersection point of a chord connecting the two canalis opticus landmarks with the midsagittal plane; *CrG*, crista galli, point at the posterior base of the crista galli; *FCe*, foramen caecum, anterior margin of foramen caecum in the midsagittal plane; *Vmr*, vomer, sphenobasilar suture in the midsagittal plane; *PNS*, posterior nasal spine, most posterior point of the spina nasalis; *FoI*, fossa incisiva, midsagittal point on the posterior margin of the fossa incisiva; *Alv*, alveolare, inferior tip of the bony septum between the two maxillary central incisors; *ANS*, anterior nasal spine, top of the spina nasalis anterior. *Rhi*, rhinion, lowest point of the internasal suture in the midsagittal plane.

by actual vector elements (loadings of the singular vector), each drawn at the mean location of the corresponding landmark in the appropriate (x- or y-) direction. At right are the equivalent representations by thin-plate spline of the (mathematical) page on which these configurations are drawn. See, again, Chapter 7. These look

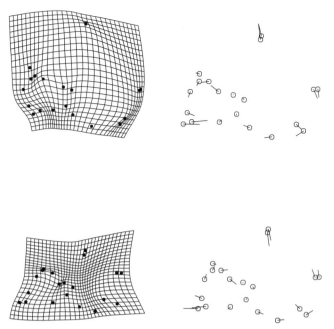

Figure 6.9. Standard geometric representations of column 1 of the matrix V derived from the space of shape coordinates here for the 22-specimen, 20-landmark representation. (Upper row) the negative direction (infantilization). (Lower row) the positive direction (hypermorphosis). (Left) Interpretation as a single feature of an organized system, namely, the relative enlargement or diminution of the area occupied by brain in comparison to the area involved in masticatory functions. (Right) Coefficients of the individual shape coordinates, interpreted as simultaneous shifts of all points by the same multiple of the little vectors shown.

exactly as the physical anthropologist would expect them to look. In one direction this dominant principal component points to the integrated process of "infantilization" – transformation of the average form toward the typical infant form – and in the other direction toward the opposite integrated process, that of (hyper)maturation (what the evolutionary biologist might call "hypermorphosis"). We confirm this in the plot of scores, Figure 6.7 (lower left), when we notice that at the left, with low scores on the first relative warp, are the youngest forms, aged 2 and 4 years, whereas all the adult forms are right of center, toward the positive end of this same coherent descriptor. There is no surprise here, no abduction, no new discovery. Rather, the grids of Figure 6.9 are consistent with everything we know about the pattern of growth in the human head, not just qualitatively but, to a surprising extent, quantitatively.

6.4 Partial Least Squares and Related Methods for Associations Among Multiple Measurement Domains

Terminological note. The version of Partial Least Squares discussed in this section is called "PLS-svd" or "Bookstein-style PLS" in the literature to distinguish it from

"PLS-regression," a differently iterative approach pertaining to the prediction of just one outcome variable at a time. Confusingly, each method was originally published by somebody named Wold: one version of PLS-svd by Herman Wold, the Swedish econometrician, and the first version of PLS-regression by Herman's son Svante, a chemometrician. Both versions have their origins in Sewall Wright's method of path analysis (the narrative here), in Karl Jöreskog's approach to maximum-likelihood analysis of covariance structures, and in the refinements of econometric statistics around 1950 that are now usually called two-stage least squares and three-stage least squares. The name "Partial Least Squares" is somewhat clumsy, but not as clumsy as Wold's original suggestion, Nonlinear Iterative PArtial Least Squares, for which the acronym is NI-PALS. I was introduced to path analysis as a graduate student in sociology around 1972, met Herman Wold when he attempted to explain NIPALS at a Harvard seminar around 1973 without any mention of the SVD, and then became active in transforming it from a method of theory confirmation to a method of numerical inference in work through the 1980s and 1990s. The method of Section 6.4.2 was first published in Streissguth et al. (1989), but our understanding of it is still evolving (see, e.g., Mitteroecker and Bookstein, 2008). See also Bookstein et al. (1996).

6.4.1 Path Analysis Revisited

Recall the explication of multiple regression in Section L4.5.2. We examined a set of four regression coefficients (proxied by differences of conditional means) and applied some algebra that made some graphical analyses of "paths" come out right. The algebra that produced the equations we solved turned out to be the same as the so-called *normal equations* that specify a least-squares fit of a plane to three-dimensional point data, and so what resulted turned a least-squares prediction of one selected variable into an explanatory network accommodating a selection of regressions among them in pairs. So the regression that was ostensibly an "explanation" (in the usual jargon) of one column of the raw data matrix is also an explanation (without the quotes) of the whole pattern of correlations.

But other explanatory patterns are possible, and for purposes of analysis of organized systems most of these alternatives are more helpful than those of Section L4.5.2. The simple arithmetic whereby graphs of correlations or covariances can be rearranged along these lines is the nucleus of a remarkably powerful approach to explanatory statistics for applications to general systems – probably the most powerful general approach we've got.

Our story begins with Sewall Wright (1889–1988), a remarkably long-lived American geneticist and biometrician, who, by 1920, had invented an algebraic system for converting regressions into explanations in a variety of relatively simple systems, of which that at the right in Figure 6.10 is typical. A good authoritative source is Chapter 13 of Wright (1968), volume 1 (this diagram is a variant of the one in his Figure 13.2 there). In language that nicely aligns with our theme of strong inference, Wright introduces his method as follows: it is

> ...an attempt to present a method of measuring the direct influence along each separate path [in a system] and thus of finding the degree to which variation of a given effect

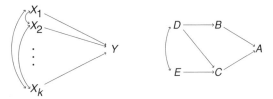

Figure 6.10. From normal equations to general path models. (Left) For the normal equations connected with the prediction of Y by the X's, as in Section L4.5.2 (compare Figure 4.33). (Right) A more nuanced example; for one of the corresponding equations, see text.

is determined by each particular cause. The method itself depends on the combination of knowledge of degrees of correlation among the variables in a system with such knowledge as may be possessed of the causal relations. In cases in which the causal relations are uncertain, the method can be used to find the logical consequences of any particular hypotheses in regard to them. (Wright, 1968, p. 325)

For reasons that need not detain us here, Wright is working with correlations, interpreted as regressions of variables all scaled to unit variance. In a world in which all such regressions are presumed linear, we can convert any scheme of variables connected by a set of lines with arrows in a causally sensible way (i.e., no loops, no backward arrows) into an explanation by the following rule:

The correlation between any two variables in a properly constructed diagram of relations in equal to the sum of contributions [products of path coefficients] pertaining to the paths that one can trace from one to the other in the diagram without going back after going forward along an arrow and without passing through any variable twice in the same path. (Wright, 1960, p. 302)

For instance, at right in Figure 6.10, the equation pertaining to the correlation of A with C is $r_{AC} = p_{AC} + p_{AB}p_{BD}p_{CD} + p_{AB}p_{BD}r_{DE}p_{CE}$. Here the r's are ordinary correlations (Section L4.2) and the p's are "path coefficients," numbers written atop individual arrows in the figure to represent the postulated effect of changing just the variable at the left end of the arrow. Applied to the scene at left in the same figure, the rule gives us back the normal equations we already had: if $r_{x_i y} = \beta_{x_i y} + \sum_{i \neq j} r_{x_i x_j} \beta_{x_j y}$, all i, then $\beta = R^{-1} r_{Xy}$ where $\beta = (\beta_{x_1 y}, \ldots, \beta_{x_k y})$, $r_{Xy} = (r_{x_1 y}, r_{x_2 y}, \ldots, r_{x_k y})$ and $R = (r_{ij})$, a matrix with 1's down the diagonal.

Explaining measurements by factors. Access to this more general algebra of path coefficients leads directly to a variant of principal component analysis, *factor analysis*, that modifies the diagonal of a covariance matrix for the purpose of producing path coefficients that now quantify explanatory relevance. Using Wright's rules, the explanatory model by a single factor, Figure 6.11 (left), would hope to account for each correlation r_{ij} as the product $a_i a_j$ of the path coefficients by which the two variables in question regress on the unknown factor. In short, we are modeling $r_{ij} = a_i a_j, i, j = 1, \ldots, 6, i \neq j$; this is 15 equations in six unknowns. We can nevertheless fit these equations by least squares. Algorithmically, we search for a vector

6.4 Partial Least Squares and Related Methods 369

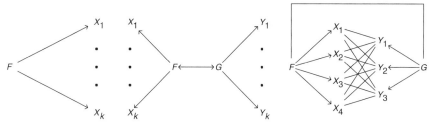

Figure 6.11. From path models to factor models. (Left) The pure single-factor model. (Center) The related model in which one pair of factors explains the relation between two separate lists of variables (or perhaps their residuals from a single-factor model like that at the left). (Right) The equivalent explanation, diagrammed and fitted as the SVD of the cross-correlation matrix of the X's by the Y's. "Two-block Partial Least Squares" is another name for the computation and interpretation of this SVD; see text.

$(a_1, a_2, a_3, a_4, a_5, a_6)$ that is an eigenvalue not of

$$R = \begin{pmatrix} 1.000 & 0.584 & 0.615 & 0.601 & 0.570 & 0.600 \\ 0.584 & 1.000 & 0.576 & 0.530 & 0.526 & 0.555 \\ 0.615 & 0.576 & 1.000 & 0.940 & 0.875 & 0.878 \\ 0.601 & 0.530 & 0.940 & 1.000 & 0.877 & 0.886 \\ 0.570 & 0.526 & 0.875 & 0.877 & 1.000 & 0.924 \\ 0.600 & 0.555 & 0.878 & 0.886 & 0.924 & 1.000 \end{pmatrix}$$

but of

$$R' = \begin{pmatrix} a_1^2 & 0.584 & 0.615 & 0.601 & 0.570 & 0.600 \\ 0.584 & a_2^2 & 0.576 & 0.530 & 0.526 & 0.555 \\ 0.615 & 0.576 & a_3^2 & 0.940 & 0.875 & 0.878 \\ 0.601 & 0.530 & 0.940 & a_4^2 & 0.877 & 0.886 \\ 0.570 & 0.526 & 0.875 & 0.877 & a_5^2 & 0.924 \\ 0.600 & 0.555 & 0.878 & 0.886 & 0.924 & a_6^2 \end{pmatrix},$$

a modification substituting exact reconstructions a_1^2 through a_6^2 for the diagonal elements, the ones in which we are not interested. In other words, we are applying the least-squares property of the SVD to *the offdiagonal cells only* in the original matrix R.

There results a first factor with coefficients (0.665, 0.615, 0.953, 0.942, 0.923, 0.941) in place of the first principal component (0.743, 0.698, 0.948, 0.940, 0.929, 0.941) of R. Either can be compared to the "general factor" (0.636, 0.583, 0.958, 0.947, 0.914, 0.932) we extracted from this same matrix via the precision matrix route, Section 5.2.3.3, an explanation that incorporated three "special factors" (head, wing, leg) in addition to the general factor here. We already had those special factors; now we have the background against which they are, indeed, "special."

What is the value of the explanatory factor? We are explaining six observed measurements, case by case, by a seventh that we have not observed. Using the usual likelihood-based machinery (Section 5.3) for estimation of anything, we maximize the likelihood of observing the six observed scores we have for regressions on a predictor with the regression coefficients specified. The solution proves to be

$$\hat{F} = \frac{\sum_1^6 a_i X_i}{\sum_1^6 a_i^2}$$

which, as you might recognize from Section L4.1, takes a form similar to a weighted average, except that the weights are now the diagonal terms we ignored in the reconstructed matrix. The purpose of the denominator is to guarantee that this score has unit variance.

6.4.2 Two-Block Partial Least Squares

To make all of this relevant to abduction and consilience in analyses of complex systems, it is good to start with a generalization of the idea of the *arrows* in Figure 6.11 – a generalization not of nodes but of measurements. Eventually we will be arguing for consilience in the factor models via the consistency with which the factor loadings relate to prior knowledge of path coefficients observed in other ways.[5]

Suppose we have two measurement domains, one of m X-variables and the other of n Y-variables, that we believe are tied together by a single causal path after the fashion of Figure 6.11 (center). The X's and Y's separately are intended to meet the requisites for a linearly combinable, profile-able set, Section 6.3.2, and furthermore are presumed *not* to meet that requirement if combined, that is, they are measuring two *different* aspects of some underlying system. The claim about the causal path is meant in the sense of "accounting for correlations" we have been exploring for some pages now. What we have postulated are a vector of path coefficients p_i of the X's on a factor F and a vector of path coefficients q_j of the Y's on a factor G such that the coefficient $r_{x_i y_j}$ observed between the ith X-variable and the jth Y-variable is the product of coefficients corresponding to three of the paths in the diagram: the equation is $r_{x_i y_j} = r_{FG} p_i q_j$.

We have mn equations (one for each X for each Y) but only $m + n + 1$ parameters to fit (the m p's, the n q's, and one path r_{FG} from F to G), so we cannot fit all the equations exactly. Naturally, then, we turn to the standard first cut for problems of this sort, the method of least squares. In this context, we are asking for a least-squares approximation to this rectangular matrix by an outer product of two vectors – a matrix of rank one.

---

[5] This is a domain in which factor analysis per se, the generalization of schemes like that in Figure 6.11 (left), is particularly weak as a scientific rhetoric. In my experience, and always excepting the work of Louis Guttman (see Section 7.8) and his Ann Arbor epigone James Lingoes, factor analysis rarely leads to strong inferences (abductions) of the sort this book is concerned with, and is also only rarely conducive to arguments from multilevel consilience. For instance, the most powerful signal in all the factor analysis literature, Spearman's factor G of general intelligence, is still completely swathed in controversy as regards the possibility of strong inference based either in consilience with its causes – neurohistological or molecular characteristics of the underlying brain, or perhaps social class? – or its effects, including real-life consequences of being smart. And this variable has been under study for an entire century!

But we know how to do this already! It is the same singular-value decomposition already introduced in Section 6.2, and results in the same hyperbolic-paraboloid surface for each pair of singular vectors, on which the coefficients of those vectors serve to select a finite handful of straight lines.

When singular vectors are derived from correlation or covariance matrices they can serve as coefficients of linear combinations of variables [always assuming that it makes sense to consider linear combinations of the variables, as at Section 6.3.2, in the first place]. In this setting their interpretation is radically enriched. Let S, $m \times n$, now be a covariance matrix relating two sets of centered (but not necessarily z-scored) variables, a block of m X's and a block of n Y's on a sample of N specimens:

$$S_{ij} = N^{-1} \Sigma_k X_{ki} Y_{kj}, \quad S = X^t Y / N.$$

As before write $S = UDV^t$. In addition, construct the matrices

$$LV_X = XU, \quad LV_Y = YV$$

(see Figure 6.12) that rotate the original centered measurements into the coordinate system of the singular vectors. Here "LV" stands for "latent variable," an ancient usage from the social sciences. These have the same formula as the numerator of the factor estimate F for the single-factor model that we have already seen in connection with Figure 6.11.

Then:

- Each column of U is proportional to the covariances of the block of X's with the corresponding column of the matrix LV_Y, and conversely. (By construction, $(XU)^t Y = U^t X^t Y = U^t UDY^t = DY^t$, likewise $(YV)^t X = D^t U^t$.) For reason of this proportionality, the entries of singular vectors computed from correlation matrices are often called *saliences*.
- The covariance between LV_{Xi} and LV_{Yi} is d_i. (Check: $N^{-1}(XU)^t_{\cdot i}(YV)_{\cdot i} = (U^t SV)_{ii} = (U^t UDV^t V)_{ii} = d_i$.) In particular, the covariance between LV_{X1} and LV_{Y1} is d_1, the maximum possible for any pair of linear combinations, one on each block, with coefficient vectors of unit length.
- An additional graphical structure becomes available beyond the U and V vectors: scatterplots of the latent variable scores. You can plot U's against U's and V's against V's or U's against V's. Any of these can show trends, outliers, or clusters in the same way that they do for principal components scores.
- Subtle scientific signals often can be revealed directly when facets of the data design that are not coded in the matrices X or Y (for instance, subgrouping variables not explicit in the measures, or an ordering implicit but not explicit in the names of the variables) become obvious in the scatterplots of the LV scores or the biplots of the singular vectors U and V.

In this way, the machinery of the singular-value decomposition allows us to accommodate this whole line of development within the explanatory framework of this book. The purpose of this style of Partial Least Squares analysis is to simplify paired nodes from a larger system into a stack of one or more associations between

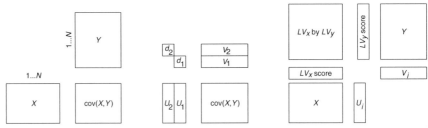

Figure 6.12. The basic structure of a two-block Partial Least Squares (PLS) analysis. (Left) From two blocks of data to their covariance matrix. (Center) From the covariance matrix to its first one or two singular vector pairs. (Right) From the singular vectors to paired latent variable scores and their scatterplots. Each right singular vector V_i is proportional to the covariances of the corresponding latent variable score LV_X with the variables of the Y-block, and, conversely, each left singular vector U_i is proportional to the covariances of the corresponding latent variable score LV_Y with the variables of the X-block. If the singular vectors U and V are unit vectors, the covariance of the ith paired LV_X and LV_Y scores is d_i (center panel).

underlying dimensions that are assumed to account for the *measurements*, not just the *correlations*. We went to all that trouble of measuring redundantly just *because* we expected to get more information out of these composite "latent variables" than out of any single empirical measurement no matter how carefully designed and calibrated.

The method seems to work best when at least one of the blocks of data involved comes from a machine, where it was indexed by a real, objective parameter (wavelength, or pixel subscripts, or the like), and when the associated quantitative theory is rich enough to make use of patterns among the coefficients of the singular vectors that emerge (we will see an example of this shortly) in addition to patterns that emerge among the estimated LV scores for F and G. In other words, the lower the rank of the meaningful part of the SVD (the fewer the conceptual aspects required to understand it), the higher the scientific usefulness of the corresponding analysis: a possibility that is not present when all we are fitting are least-squares regressions to individual outcome scores.

An example. From the Pregnancy and Health Study run by Ann Streissguth, a large longitudinal study of prenatal alcohol exposure and offspring development, I extract a small fragment involving questionnaire data from 1974–75 about intake of alcohol by pregnant women versus IQ scores of the child at age 21. These are *not* the people we were talking about in Section 5.1.3, people with actual diagnoses of prenatal brain damage. These are people without any diagnosis at all, even now, 40 years after it was first ascertained that maternal drinking had put some of them at a small but computable risk of brain damage.

The measurement sets meet the logical requirements for an analysis by linear composites at Section 6.3.2. The nine IQ subscale scores are all aspects of the same underlying item domain, which, as mentioned in note 5, has been the most carefully scrutinized domain in all of psychometrics since the initial construction of IQ for the purpose of assessing "feeblemindedness" more than a century ago. The ten alcohol scores probe different ways (quantity, frequency, variability) of assessing the same alcohol challenge to the fetus. They come in two groups of five in respect of the time during pregnancy to which they pertain (early in pregnancy, or instead during midpregnancy), and for each of those two periods assess Drinks per Occasion (DOCC), Maximum Drinks per Occasion (MAX), Monthly Occasions (MOCC), Absolute intake of Alcohol per week [in fluid ounces] (AA), and Volume Variability (VV). The sample for this demonstration is the most highly exposed 250 offspring of the study as actually executed. The IQ scores were measured by trained psychometrists. Alcohol exposure was coded from women's responses to a brief questionnaire whose centrality to the original study design was carefully concealed. (This was in 1974, when hardly anyone beyond Streissguth and her co-authors knew that prenatal alcohol exposure could be bad for the fetus.) For design details, see Streissguth, Barr et al. (1989).

The correlation matrix driving the PLS analysis, 10 alcohol variables by nine IQ variables, is

$$R = \begin{pmatrix} -0.097 & -0.110 & -0.117 & -0.134 & -0.155 & -0.193 & -0.207 & -0.224 & -0.226 \\ -0.086 & -0.098 & -0.103 & -0.119 & -0.137 & -0.171 & -0.184 & -0.198 & -0.201 \\ -0.077 & -0.088 & -0.093 & -0.107 & -0.124 & -0.154 & -0.166 & -0.179 & -0.181 \\ -0.074 & -0.084 & -0.089 & -0.103 & -0.119 & -0.148 & -0.159 & -0.172 & -0.174 \\ -0.059 & -0.067 & -0.071 & -0.081 & -0.094 & -0.117 & -0.126 & -0.136 & -0.137 \\ -0.049 & -0.055 & -0.058 & -0.067 & -0.078 & -0.097 & -0.104 & -0.112 & -0.113 \\ -0.037 & -0.042 & -0.044 & -0.051 & -0.059 & -0.073 & -0.078 & -0.085 & -0.086 \\ -0.029 & -0.033 & -0.035 & -0.040 & -0.046 & -0.057 & -0.062 & -0.066 & -0.067 \\ -0.005 & -0.006 & -0.006 & -0.007 & -0.008 & -0.010 & -0.011 & -0.012 & -0.012 \\ -0.004 & -0.004 & -0.004 & -0.005 & -0.006 & -0.007 & -0.008 & -0.009 & -0.009 \end{pmatrix}.$$

The order of the alcohol variables (the rows): AAP, AAD, MOCCP, MOCCD, DOCCP, DOCCD, VVP, VVD, MAXP, MAXD. The order of the IQ variables (columns): Arithmetic, Digit Span, Information, Similarities, Block Design, Picture Completion, Object Assembly, Picture Assembly, Digit Symbol. Figure 6.13 displays the SVD of this matrix and Figure 6.14 the singular vectors and the corresponding LV scores. The singular values are 1.085, 0.325, 0.158, 0.130, . . . showing good separation between first and second and also between second and third, but not thereafter. Dominance of the first indicates that the fit of the rank-one matrix to the correlation matrix R is good, with 88% of its total sum of squares accounted for by the representation at lower left in Figure 6.13.

Notice how strong are the patterns in the top row of Figure 6.14, no matter the weakness of the actual LV–LV correlation (about 0.28). Consilience in this example comes from the information in the saliences, the scatters in the top row. (Following the usual biplot convention, the axes here are each scaled as the square root of the corresponding singular value.) The alcohol vectors clearly emerge in two sets: those that pertain to maximum (binge-related) drinking (DOCC, MAX, and VV) the more salient, and those that relate to steady drinking (AA and particularly MOCC) less

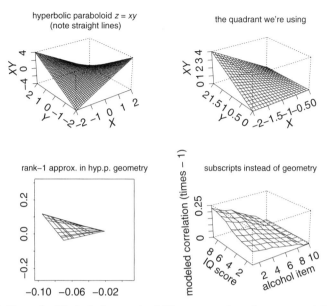

Figure 6.13. Example: 10 alcohol scores by 9 IQ measures. I, surface representation.

so. Also, the during-midpregnancy scores (the ones with abbreviations ending in D) are less informative than those pertaining to early pregnancy (ending in P). Over on the IQ side, we see two scores in particular, Arithmetic and Block Design, that are most sensitive to these ostensible effects of prenatal alcohol. The dominance of these algorithmic subscales of IQ in assessment of the effects of alcohol dose is consistent with profiles of deficit for adolescents and adults who have actually been diagnosed with fetal-alcohol–related brain damage, and the greater salience (relevance) of the peak-dose–related measures is consistent with animal experiments, with the known buffering action of the placenta, and with knowledge at the neurocellular level about how alcohol does the damage it does. Thus both sets of singular vectors are adequately consilient, as far as such claims *can* go in weaker sciences like psychology.

We return to this data set at Section 7.8, adding a better explanatory variable than alcohol exposure and making much more sense of the scatterplots in Figure 6.14. Only at that point will an abduction become possible: the correlations we see are (surprisingly) due to the combination of high alcohol dose with low social class, not with either ostensible factor separately.

By this applied-mathematical recycling, the hyperbolic paraboloid we used as a model for a raw data matrix has been recommissioned as a pattern engine for covariances or correlations of redundantly measured variables. Again there is a close connection with formal arguments from consilience, namely, pattern matching against a-priori subgroupings or clusters. What emerges is a profound alternative multivariate praxis for systems analysis, one that promises to incorporate much of the ancillary knowledge directly, as we see in the next few subsections.

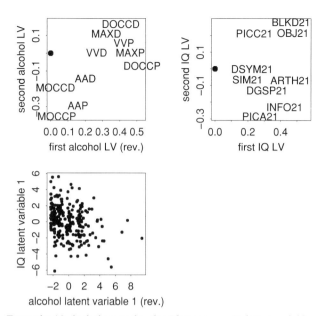

Figure 6.14. Example: 10 alcohol scores by nine IQ measures. II, latent variables and saliences. rev., scale has been reversed to ease the interpretation.

6.4.3 Extensions and Modifications

The technique just demonstrated can ramify in a great many directions. This subsection introduces three.

6.4.3.1 Re-Expressing the SVD Algebra

I have not yet explained how to compute an SVD from first principles.[6] For use by the first two of these extensions it will be helpful if we actually provide such an algorithm, at least for the first singular vector pair, the columns U_1 of U and V_1 of V. The algorithm is iterative, and I introduce it here in the form that Herman Wold originally introduced it in his papers of the 1960's (cf. Wold, 1982).

The approach is as in Figure 6.15, where I write "cov(X,Y)" for the matrix of covariances of the X's by the Y's and "cov(Y,X)" for the transpose, the covariances of the Y's by the X's. In ordinary notation, these are the matrices Σ_{XY} and Σ_{YX}, respectively. Begin with any guess at the coefficients of the first left singular vector – the first column of the matrix U. This guess may as well be taken as the appropriate multiple $(1, 1, \ldots, 1)/\sqrt{k}$ of a unit vector with all elements the same, corresponding to a latent variable score that is an unweighted average of the k variables in the X-block. Denote this vector as $U_1^{(1)}$, the first iteration of our estimate.

To get the first iteration $V_1^{(1)}$ of the corresponding column V_1 of V, multiply $U_1^{(1)}$ by the matrix Σ_{YX} and then normalize to unit length: $V_1^{(1)} = v/|v|$ where $v = \Sigma_{YX} U_1^{(1)}$.

---

[6] Of course, in practice you simply ask your favorite computer statistical package to do it for you.

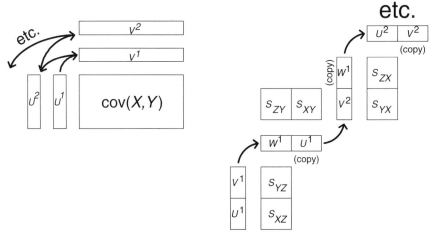

Figure 6.15. PLS by iterations of matrix multiplication. (Left) For two blocks of variables (obsolete algorithm for the SVD). The blocks are X and Y, and the singular vectors (coefficients of the latent variables) are denoted U and V. The algorithm follows the bouncing circular arc outward. The graphic of U^1, for instance, bordering the matrix Σ_{XY} means the operation $\Sigma_{XY}^t U^1$, the sum of products of U^1 by the dimension of Σ_{XY} with which it is compatible. (Right) For the saturated three-block model, Bookstein et al. (2003). The blocks are now X, Y, and Z and the quasi-singular vectors are denoted U, V, W. Again the algorithm follows the bouncing circular arcs drawn in heavy line. Each step corresponds to a temporary two-block model in which one of the original blocks, first Z, then Y, and finally X, is faced with a temporary superblock consisting of the other two blocks. To diagram this as a closed loop, one would draw it over two cycles on a Möbius strip.

Then compute every subsequent pair $U_1^{(i)}, V_1^{(i)}$ of estimates from the preceding pair by setting $U_1^{(i)} = u/|u|$ with $u = \Sigma_{XY} V_1^{(i-1)}$, the normalized estimate of U_1 given the current V_1, followed by a repetition of what was already described, the generation of $V_1^{(i)}$ from $U_1^{(i)}$ by premultiplication by Σ_{YX} followed by normalization again, and so on.

Figure 6.15, at the left, diagrams exactly this sequence of computations, omitting only the normalizations. It can be proved that this algorithm converges to the first pair of singular vectors unless your starting guess was extremely unfortunate. A variant of this algorithm, in which the blocks X and Y are the same, is one ancient method, the *power method*, for computing principal components by matrix multiplication alone.

6.4.3.2 More than Two Blocks

Whereas the singular-value decomposition and the theorem on which it is based are limited to ordinary matrices, the algorithm just reviewed – a carefully structured sequence of matrix multiplications and normalizations – is not. Suppose, for instance, that we have *three* blocks of measurements X, Y, Z on the same sample of cases. Generalizing the prediction task in the central panel of Figure 6.11, we can imagine that there are three linear composites of the data, one per block, and each is tasked with predicting each of the other two. There corresponds the sequence of matrix multiplications shown at right in Figure 6.15, a sequence that, again, converges quickly in all but the most improbable data sets. (The figure omits instructions regarding normalization of the vectors U, V, W in the event that the counts of variables in the three blocks are discrepant. For these details, see Bookstein et al. [2003], Appendix.)

One can easily convert a variety of path diagrams on explicitly measured variables, such as that at right in Figure 6.10, into analyses of latent variables by iterated loops of matrix multiplications and normalizations after this fashion. For instance, simple causal chains (measurement of the same block of variables at a sequence of ages) can easily be fitted to this formalism. For an example, see Streissguth et al. (1993).

6.4.3.3 "Domino PLS" for Reference Properties of Subjects or Measurements
Another, quite different example of this same type of iterated looping matrix multiplication is due to the Norwegian chemometrician Harald Martens (2005). Suppose we have an ordinary data matrix of scalar assessments, for instance, ratings of beers by citizens of Norway, and for each variable *and also for each subject* there is an otherwise unrelated information resource taking the form of an additional tabulation: for the beers, perhaps, the spectrum of the wheat from which they are brewed, or the pattern of mineral content of the water; for the raters, perhaps, a demographic profile (age, sex, income) or maybe mouth flora. There results an L-shaped diagram and an algorithm that *predicts* the entries of the quasi-singular vectors describing one cross-block prediction by the properties of the corresponding variables as encoded in the ancillary data sets. Second and higher dimensions of these analyses can be particularly interesting, as they uncover patterns whereby, for instance, people with unusual mouth flora might rate beers in a predictably unusual way depending on their ingredients in a nonstandard fashion.

Martens refers to this tableau as a "domino" diagram, not after the Latin word for the Deity but for the game that matches 2×1 rectangles by the count of their dots in rectilinear paths over a gaming table.

6.4.3.4 Using a Matrix S of Information in Place of the Data
Partial Least Squares applies a singular-value decomposition to a matrix of associations between *variables*. Earlier, we reviewed how the same SVD could apply to simplify a matrix of raw data instead. In an elegant compromise between these two methods, Oliver Delrieu and Clive Bowman (2006) apply the data-matrix decomposition except that data values have been replaced by their information content relative to some outcome. In their approach, an extensive raw data matrix of single nucleotide polymorphisms (SNPs) – essentially, alleles over a large range of separate genes considered all together – is replaced, *datum by datum*, by the information content *of* the allele in question for some consequent outcome, such as the prevalence of the allele in a group with a chronic disease vis-à-vis its prevalence in those without the disease, or the success or failure of a certain treatment (e.g., with a drug under development) in a defined clinical population. The replacement entry, computed after aggregation

over the entire sample, may be the logarithm of the relative frequency with which that allele is found in the cases versus the controls, or perhaps (Bowman et al., 2006) a more sophisticated function of the available data, such as a log-likelihood measure on the two models (for cases, versus for controls). The resulting matrices are then submitted to SVD analysis just as if these measures of information had been the variables originally measured.

6.4.3.5 For Tables of Counts: Correspondence Analysis
Correspondence analysis (from the French *Analyse des correspondances*, Benzécri, 1973) is a variant of the SVD that relates closely to chi-square analysis of a contingency table (an ordinary crosstabulation of counts by row versus column category). Suppose we have a matrix S of counts, meaning that entry s_{ij} is the number of individuals encountered who are described both by category i of the row variable and by category j of the column variable. Write r_i for the row totals $r_i = \Sigma_j s_{ij}$, c_i for the column totals $c_j = \Sigma_i s_{ij}$, and N for the table total count, $N = \Sigma_i r_i = \Sigma_j c_j$. Then, if we write e_{ij} for the *expected value* $r_i c_j / N$ of the ijth cell (expected, that is, on the assumption of independence of the row classification and the column classification), we can notate the standard chi-square statistic, test statistic for (one version of) this independence hypothesis as $\chi^2 = \Sigma_{ij} \frac{(s_{ij} - e_{ij})^2}{e_{ij}}$.

This formula is N times a somewhat more interpretable formula $\Sigma_{ij} \frac{((s_{ij} - e_{ij})/N)^2}{e_{ij}/N}$ where the numerator is the square of a *normalized residual* for each table entry, its deviation from the expected count as a fraction of the table total count, and the denominator is the expected cell fraction of the table total, namely, the product $r_i c_j / N^2$ of the row fraction r_i/N and the column fraction c_j/N. Correspondence analysis is the SVD of the matrix whose entries are the square-roots of the summands in this normalized chi-square formula, the quantities $\frac{(s_{ij} - e_{ij})/N}{\sqrt{e_{ij}/N}}$. The analysis is thus a decomposition of this "normalized chi," cell by cell, into a stack of individual tables each of which accounts for its singular value's worth of that normalized chi-square by a rank-1 pattern pertinent to those residuals – a combination of coordinated + and – patterns pairing the "non-independent" part of specific rows with specific columns.

Example. Consider the matrix of counts

$$\begin{pmatrix} 1 & 2 & 3 \\ 4 & 5 & 6 \\ 7 & 8 & 9 \end{pmatrix}.$$

The pattern of expected values, as a fraction of the table total, is

$$\begin{pmatrix} 8 & 10 & 12 \\ 20 & 25 & 30 \\ 32 & 40 & 48 \end{pmatrix} / 225,$$

and so the residuals $s_{ij} - e_{ij}$, before normalization, prove to be

$$\begin{pmatrix} -1 & 0 & 1 \\ 0 & 0 & 0 \\ 1 & 0 & -1 \end{pmatrix} /75.$$

The normalized residual matrix, according to the formula above, works out to

$$\begin{pmatrix} -0.070711 & 0 & 0.035355 \\ 0 & 0 & 0 \\ 0.057735 & 0 & -0.028868 \end{pmatrix}$$

which is a matrix of rank 1 proportional to the tensor product $(\sqrt{3}, 0, -\sqrt{2}) \otimes (2, 0, -1)$. The cells of count 1 and 9 on the main diagonal are disproportionately low, and those of counts 3 and 7 are disproportionately high, and all five of the other cells are exactly where they should be (on the independence hypothesis).

6.4.3.6 For Tables of Outcomes: Additive Conjoint Measurement (ACM)
The ACM approach replaces the multiplication operator of the chi-square formalism by an addition, corresponding to the different explanatory purpose it is designed to serve. The explanation here is adapted from Krantz et al. (1971), volume I, chapter 6. Consider the toy example in Figure 6.16. In keeping with the real example it was intended to accompany, from Coquerelle et al., 2011, I will refer to "teeth" and "jaws" as the objects of study here. Teeth will be represented by "stages" (A, B, C, or D), and each jaw by a single summary size measure, which is a first principal component in the actual publication and so will be called "PC1" here too.

The toy data are shown at upper left in Figure 6.16. We have ten cases on which we have observed morphogenetic stages for each of two teeth. Each stage can take on one of four values A, B, C, D. As the figure shows, the stages are somewhat correlated across the teeth, but the correlation is imperfect. The observed value of jaw PC1 rises with stage of each tooth, but not linearly. All this is typical of examples over growing complex systems, except that here the jaw PC1 values are all integers, for ease of mental arithmetic.

The upper right panel of the figure shows the crucial construction. For each tooth and for each stage, we average the values of jaw PC1 over the subsample having that stage for that tooth. These averages are variously over two or three of the toy cases, and we have plotted the resulting *calibration* in heavy dashed line for each tooth separately. As averages, these lines are each least-squares in the estimate of the height of each little column of points separately: that is one way to think about the phrase "least squares" in "Partial Least Squares." As for the "partial," either of the two curves in this panel pertains to only *part* of the data. The heavy dashed lines in the figure stand for a gently nonlinear rescaling of the original data as if each tooth, by itself, was predicting jaw PC1. These curves are monotone, corresponding to the growth process they are attempting to help us understand.

380   6 The Singular-Value Decomposition

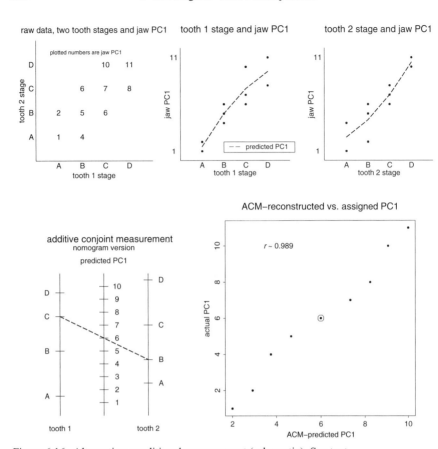

Figure 6.16. Alternating conditional measurement (schematic). See text.

The "additive" part of "additive conjoint measurement" arises in connection with the maneuver that is the subject of the lower left panel. Here, taking advantage of the fact that there are only two teeth in the toy example, we show the addition graphically, in the form of a *nomogram*. The heavy dashed line connects the point for the scaled prediction from tooth 1 with the point for the scaled prediction for tooth 2; their average, which is where this heavy line crosses the middle vertical axis, is the prediction of jaw PC1 case by case. This prediction, too, takes the explicit form of a Partial Least Squares computation, as it is in fact the average of the predictions according to each predictor separately; that is another way of explaining what Partial Least Squares is actually doing – averaging univariate predictions.

We assess the adequacy of this "model" by the scatterplot in the lower right panel, which compares our reconstructed jaw PC1, the intersections of the heavy line with the middle axis in the nomogram, to the original jaw data typed at upper left. The fit is good, with a correlation of about 0.989. (The point in the middle corresponds to two of the original cases, and thus is emphasized by the additional circle around the dot.)

6.5 Another Tableau: Dissimilarities as Primary Data

Recall that the singular-value decomposition (SVD) of Section 6.2 applies to any rectangular matrix. We have already mentioned one application to a square symmetric matrix: the principal components analysis, Section 6.3.

There is another important version of this special application to square symmetric matrices: the method of *principal coordinates analysis* (PCO). In this case, the rows and columns of the matrix S stand not for variables from the original data set but for specimens, and the entries of S are not covariances (or correlations) between variables but distances or dissimilarities between those specimens in pairs. Where a covariance matrix has variances down the diagonal, a dissimilarity matrix has 0's down the diagonal (because no matter how we define dissimilarity, the dissimilarity of a specimen from itself ought to be zero). The entries can be Euclidean distances in some vector space (the commonest example in the literature), or covariance distances between covariance matrices corresponding to rows of S that stand for whole subsamples, which will be the subject of the worked example here. An example that is rapidly becoming common is Procrustes distances in the appropriate shape manifold (see Chapter 7) – we have already shown three examples of this analysis, in passing, in Figures 6.10 through 6.12.

6.5.1 From Distances to Coordinates: The Underlying Theorem

Remember the fundamental *least-squares property* of the SVD: if the rectangular matrix S is decomposed as $S = UDV^t$, with the entries d_i of D in decreasing order, then for each rank j between 1 and $r-1$, the approximation $M = \Sigma_{k=1}^{j} U_k d_k V_k^t$ minimizes $\Sigma_{i,j}(S_{ij} - M_{ij})^2$ over the class of matrices M of rank not greater than j. In other words, $d_1 U_1 V_1^t$ is the best-fitting rank-1 matrix to S, $d_1 U_1 V_1^t + d_2 U_2 V_2^t$, the best-fitting one of rank 2, and so on.

If the matrix S is symmetric, the same on either side of the diagonal, then its SVD must be symmetric, too. The orthonormal matrices U and V must be the same, so that using the representation $S = UDU^t$ we are decomposing it as a sum of terms $U_i d_i U_i^t$ of outer products of a vector with itself.

Now let me pretend to change the subject for a minute. Suppose you have a set of k ordinary decimal numbers $x_1, \ldots, x_k$ that total zero. Consider the matrix of their pairwise products:

$$S = \begin{pmatrix} x_1 x_1 & x_1 x_2 & \ldots & x_1 x_k \\ x_2 x_1 & x_2 x_2 & \ldots & x_2 x_k \\ & & \ldots & \\ x_k x_1 & x_k x_2 & \ldots & x_k x_k \end{pmatrix}$$

where the diagonal is just the squares of the original x's. This matrix certainly has rank 1: it matches the criterion in the definition in Section 6.2.1 if we take the same list of x's for both vectors, the a's and the b's.

Let us fiddle with this matrix a bit in ways that will make sense shortly. Pull out the diagonal, the vector $(x_1^2, x_2^2, \ldots, x_k^2)$. Multiply the original matrix S by -2. To every row of $-2S$ add the original diagonal of S, the vector we just pulled out, and also add it to every column. You end up with a matrix of which the ijth entry is just $(x_i - x_j)^2$, the squared difference of the corresponding pair of entries in the vector of x's.

It is more interesting to run this trick in reverse: not from coordinates to squared distances but from squared distances back down to coordinates. Suppose you have a set of points anywhere in a plane or in space. You don't have their coordinates, but you have the squared distances $(x_i - x_j)^2$ among them in pairs, and you know that they lie on a straight line. How can you most easily put them back on that line in the right places? Start with the matrix D of their squared interpoint Euclidean distances. D has zeroes down the diagonal, but otherwise its entries are all positive, so every row has a positive average. Construct a vector whose entries are these averages. Subtract this vector from every row of D, and also from every column of D, and then, to every entry of the matrix that results, add back in the grand average of D (which is the same as the average of those row averages). Finally, multiply the result by $-1/2$.

What you get turns out to be the matrix S we started with – a rank-one matrix whose only nontrivial singular vector is the set of point coordinates along a line with which we began. See Figure 6.17.

Here is how to see this most easily. Write $\overline{x^2}$ for the average of the squares of the x's, and remember that the average of the x's as a whole vector was set to zero by assumption. Then the average of the entire matrix D, all of the values $(x_i - x_j)^2$, is just $2\overline{x^2}$. The typical entry of the matrix we started with was $(x_i - x_j)^2$. The sum of the entries in row i is $\Sigma_j(x_i - x_j)^2 = kx_i^2 + \Sigma_j x_j^2 - 2\Sigma_j x_i x_j = kx_i^2 + k\overline{x^2} - 2x_i \Sigma_j x_j = kx_i^2 + k\overline{x^2}$, because $\Sigma_j x_j = 0$. The row average is this divided by k, which is $x_i^2 + \overline{x^2}$. Similarly, the column average is $x_j^2 + \overline{x^2}$. Subtracting these both from the original value, $(x_i - x_j)^2$, gives us $-2x_i x_j - 2\overline{x^2}$, and adding back the average of all of D, which is $2\overline{x^2}$, to every entry leaves us with $-2x_i x_j$, so that multiplying by $-1/2$ gets us back to the matrix $S = (x_i x_j)$, the rank-one matrix where we started.

In a matrix notation that is no help to the intuition but that is at least typographically concise, if D is the matrix of squared distances we started with, then the rank-one matrix at which we arrive (for data that originally arose as coordinates of points on a single line) is $-HDH/2$, where H is the *centering matrix* with diagonals all equal to $(1 - 1/k)$ and all other entries $-1/k$.

The method of principal coordinates is the SVD of the matrix $-HDH/2$ where D is any reasonable distance measure. The basic theorems are John Gower's, from a great *Biometrika* paper of 1966. Remember that the SVD, in this application to symmetric matrices, is guaranteed to fit it by a stack of components $U_i d_i U_i^t$ where U_i is the ith column of the orthonormal matrix $U = V$ of the SVD theorem. If the matrix we were decomposing was a covariance matrix, we would say that the columns of U were "orthogonal," making an angle of $90°$ in the space of linear combinations. But in this application, the interpretation is quite a bit easier. We have singular vectors

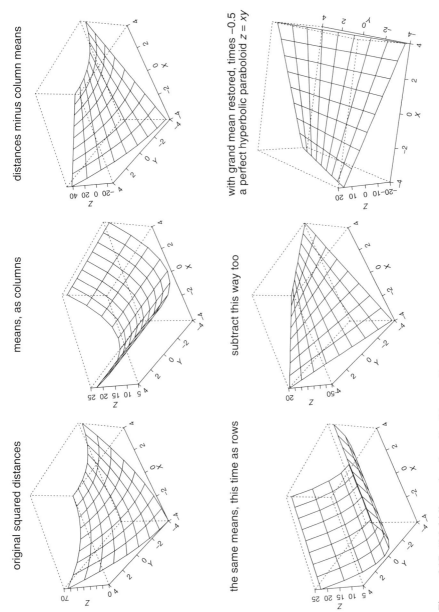

Figure 6.17. "Multidimensional scaling" in one dimension.

U_i that have one entry for each case of the original data set (remember the matrix D is $k \times k$), and all rows and columns have mean zero (an effect of the pre–post multiplication by H), so that the word we use for orthogonality in this application is **noncorrelation**. The principal coordinates that the SVD of the matrix $-HDH/2$ produces are correlated zero.

But we already know about some quantities that are correlated zero: the principal components of the original data considered as a covariance structure (Section 6.3.1). If the distances you started with arose as the sums of squared differences on two variables $(x_1, \ldots, x_k)$ and $(y_1, \ldots, y_k)$, such that the x's and the y's average zero and are uncorrelated as variables, then the principal coordinates of the composite Euclidean squared distance matrix, computed as eigenvectors of $-HDH/2$, will just be the vectors $(x_1, \ldots, x_k)$ and $(y_1, \ldots, y_k)$ with which you began. (In the algebra we just finished, the ijth entry of the matrix $-HDH/2$ ends up equal to $x_i x_j + y_i y_j$, which is clearly a matrix of rank two with singular vectors $(x_1 \ldots x_k)$ and $(y_1 \ldots y_k)$ on both sides of D.)

Principal components, because they are simply a rotation of your original data vectors, don't change interspecimen distances. Then the principal component scores of any data set **are** the principal coordinates of the corresponding distance matrix. In other words:

the principal coordinates of a squared intersample Euclidean distance matrix are exactly the same as the principal component scores of the covariance matrix among the variables contributing to the Euclidean distance formula.

However many dimensions are required to explain or approximate the covariance structure or the distance structure, we can still start with a matrix of squared Euclidean distances between the cases instead of a matrix of covariances among the original coordinates, and arrive at the same ordination from the SVD of the doubly centered distance matrix $-HDH/2$, singular vector by singular vector, that we get from the covariance matrix of the original coordinates now treated as variables one by one.

Unless the original data really were two uncorrelated variables, $x_1 \ldots x_k$ and $y_1 \ldots y_k$ centered and with $\Sigma x_i y_i = 0$, of course we don't get back these actual x's and y's from either technique. We get a set of scores that are centered at $(0, 0)$ and rotated to put the principal coordinates (or principal components) along the Cartesian axes. Otherwise both PCA and PCO remove information about centering from the raw data, and PCO further loses the information about rotation, because it was not actually coded in the distances.

Bibliographical remark. A good technical reference for this material is Chapter 14 of Mardia et al. (1979), who refer to the approach by its original 20th-century label, "multidimensional scaling." Mardia et al. trace the original idea back to a brief remark published in 1935. The most commonly cited modern source, even more often than the technical paper by Gower (1966), is Torgerson (1958), which is actually a textbook of psychometrics. Reyment and Jöreskog (1993) briefly mention an extension to asymmetric matrices, but this seems not to have caught on. Though the approach here is envisioned for dissimilarity data arising from physical measurements, there is an equally powerful methodological toolkit for the dissimilarity data that arise by direct introspection of irreproducible reports by individuals of perceived dissimilarities on an ordinal scale. For these methods, which are usually called "unfolding techniques," see Coombs (1956).

6.5.2 Classification

A principal coordinates analysis is an *ordination*. This means that, as in Figure 6.7, it puts your set of entities-with-distances down on a statistical display where you can look for clusters, gaps, or other suggested explanations for the pattern of their distances. It is an exploratory technique, one looking for structure. As a data analysis, it is a projected version of cluster analysis; the purpose is to observe *either* clusters *or* curves/lines *or* separations *or* other spatially striking patterns that deviate from the homogeneous (disorganized) models of Gaussian ellipsoids and the like. In this way PCO functions like the Hertzsprung–Russell plot or the Ramachandran plot for domains in which we haven't guessed at the appropriate control space a priori (perhaps because theory can't tell us where to expect the empty space in these diagrams).

It appears as if you might be able to apply these plots for a different purpose entirely, to wit, to decide whether groups that you knew about in advance are "separate enough" in the plane in which you are plotting them. Phrased this way, the intention, while forgivable, is not interpretable in terms of the principles of numerical inference we are exploring, as there is no hypothesis stated, let alone the set of two or more that our method (strong inference) requires. But we can convert the situation into something of the proper flavor if we go on to introduce a *probe*, an additional point whose location is known but whose group is unknown, and inquire as to the relative weight of evidence for *two* hypotheses: that the probe is in fact a member of group 1, versus that it is really a member of group 2.

Even though this is a chapter on multivariate data, to set the problem of classification on firm intuitive ground let us assume, first, that we measured only one feature. Also, we presume that within each group our measure is distributed as a Gaussian variable. (This assumption can be waived, but the algebra and the narrative become more complicated – the topic becomes the discipline of *machine learning*, see, e.g., Hastie et al. [2009].) Then membership of the probe in either group is a hypothesis to which there is a corresponding likelihood, which is the probability of that probe value within the statistical distribution characterizing the group. The situation is as in the panels of Figure 6.18 labeled "densities": for each group there is a distribution function, in thin continuous line for group 1 and in thick dashed line for group 2. For any probe value, the likelihood ratio that we use to test whether one hypothesis dominates the other (whether the data support a claim of membership in one group over the other) is the ratio of heights (probability densities) between these two curves observed at the measured value.

There is one particularly simple case, the improbable situation in the upper left panel of the figure, where the two groups have different averages but the same standard

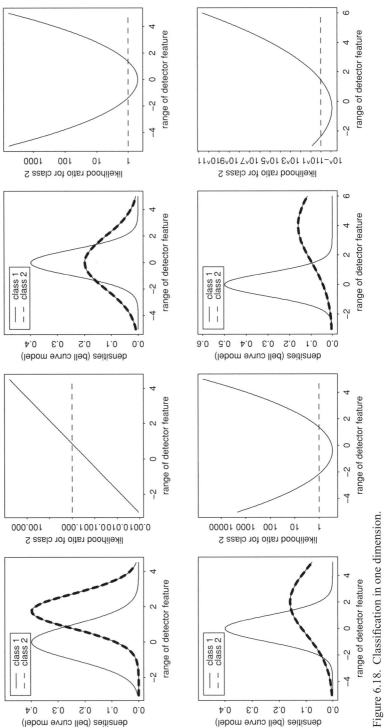

Figure 6.18. Classification in one dimension.

deviation. From the formula for Gaussian distributions, Section L4.2, it can be shown that the logarithm of the likelihood ratio driving our classification rule will be a linear function of the measurement for the probe. This logarithm is the straight line shown to the right of the corresponding paired density plot. Its value is zero (i.e., the likelihood ratio is 1, equivocal as regards group membership) midway between the averages of the two groups, where the probability densities of the two distributions are the same.

But such a situation is totally unrealistic – groups with different averages should certainly be expected to have different variances as well. An equally unrealistic scenario is the situation in the two upper right panels. Here the means are identical but the variances are different. An equally powerful classification rule becomes feasible here, whereby all probes near the common mean are assigned to the group with the smaller variance and probes far enough away from the common mean are assigned to the group with the larger variance. Again the changeover points (there are two of them) are where the density curves cross. From the formula for the Gaussian distribution, the shape of the likelihood ratio curve is a parabola with vertex at the average of the groups.

This functional form, the parabola, turns out not only to characterize this unrealistic case of equal means, but all the *realistic* cases, the cases of unequal means *and* unequal variances, as well. (The sum of a linear function and a parabola is just a different parabola.) The bottom row shows two examples that differ only slightly in detail. At lower left, the range of probe values classified into group 2 arises in two disconnected parts, one corresponding to the mean shift for group 2 and one corresponding to the fact that, while small values are improbable on either hypothesis, they are less improbable on the hypothesis of the greater variance. At lower right, the shift of means is so great that this second region no longer generates probe values with any realistic probability, and so the classification rule again reduces, more or less, to a simple threshold.

The same situation persists into the context of two (or more) measured values. Here the model of a Gaussian distribution is that of a bell-shaped *surface*, of circular or elliptical cross-section. In Figures 6.19–6.21 these are represented by their "waists" only, the curves around the bell corresponding to the inflection of the corresponding univariate sections (see Figure 4.12). Figure 6.19 shows several examples analogous to the two upper left panels of Figure 6.18, the analysis in the unrealistic case that variances are equal between the groups. Here in two (or more) dimensions this assumption is even more implausible, as not only do variances need to be equal for every variable, but also covariances (correlations) – a total of $p(p+1)/2$ demands, where p is the count of variables. Nevertheless we plot the response of our hypothesis test to the various probe values by lines of equal log likelihood. The thick line corresponds to equal support of the data for either hypothesis.

At upper left in the figure is a scenario of two circular normal distributions with different averages. When variances are equal, the logarithm of the likelihood ratio characterizing the differential support of the data for one of our group membership hypotheses over the other is a plane; here it is drawn by its contour lines of equal values, which are of course parallel lines at equal spacing (on this log scale).

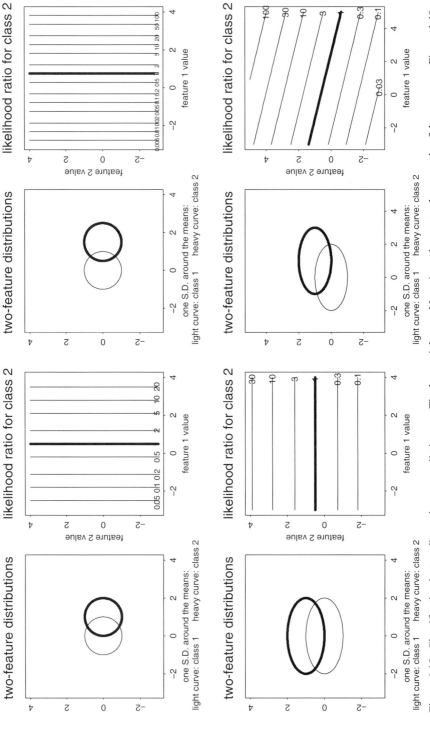

Figure 6.19. Classification in two dimensions, unrealistic case. The lower right panel here is analogous to the example of the sparrows, Figure 4.40.

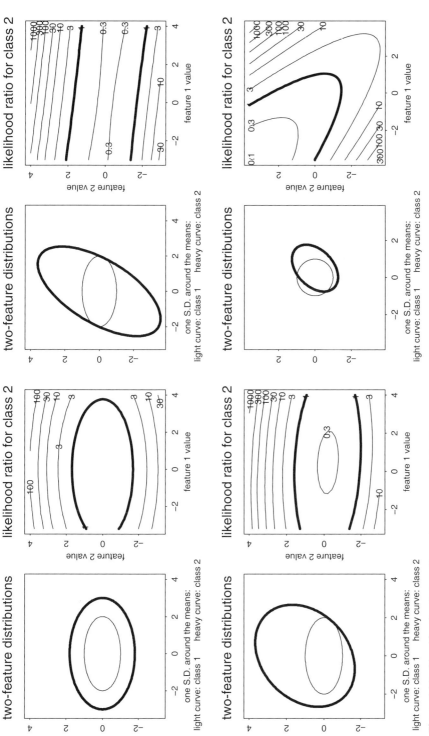

Figure 6.20. Classification in two dimensions, more realistically.

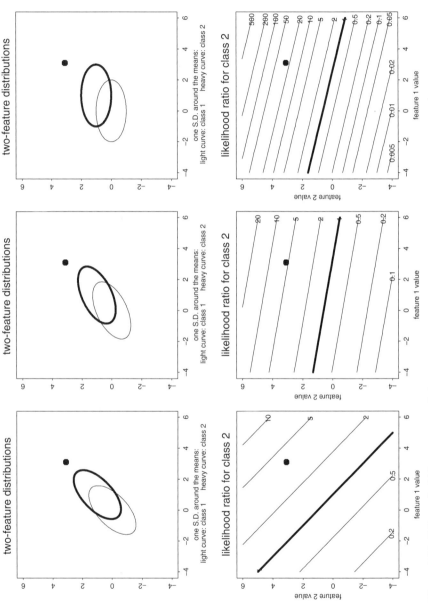

Figure 6.21. Effect of within-group correlation.

If the means are close (upper left), the contour lines are widely spaced; if the means are farther apart (upper right), the contours are more closely spaced. The same linearity is true if the within-group variances are not equal over the directions in which the probe can vary, lower left. If the group mean difference is not aligned with the coordinate axes here, neither are the contours along which the plane of log likelihood ratio intersects parallels to the plane of probe values (lower right), but the contours remain parallel lines equally spaced on the log scale. We have already seen an example of this scenario in Section L4.5.3.3 at Figure 4.40.

Everything changes in this simple visualization when the unrealistic restriction to equal covariance structures is removed, Figure 6.20. The two-dimensional equivalents of the parabolas in Figure 6.18 are now always quadric surfaces (recall Section 6.1), variously ellipsoids, paraboloids, or hyperboloids, and again the plane is divided into regions within which one group or the other is preferred by a probe. Figure 6.20 shows a collection of possibilities. You can see that the likelihood ratio in support of one classification or the other can be remarkably high, in the 100's, for probes that can arise quite realistically from the group of larger variance. By contrast, the data never support classification to the group of smaller variance too strongly. (This same figure shows, in passing, the range of geometric forms that the separatrix between regions of likelihood favoring one class versus likelihood favoring the other can take. In its full n-dimensional geometry this surface is a generalized quadratic form, with some planes slicing it in ellipses, some in hyperbolas, and some perhaps in pairs of straight lines.) The case of the hyperboloid has already been shown for an example with more convenient centering in the middle row of Figure 5.9.

Even when covariance structures are equal, this classification machinery is strongly dependent on the interaction between the mean difference vector and the covariance structure. Figure 6.21 shows three different instances of the same probe being subject to classification by two variables with the same group means (hence, same mean difference) but varying within-group correlations. As you can see, the likelihood ratio test is least informative when the groups have the greatest variance in the direction of the displacement to the probe, and is steadily more and more informative as that direction of greatest variance rotates away from the direction of displacement toward the probe.

There is an ostensibly competing technique, *canonical variates analysis* (CVA), that purports to achieve one part of this same purpose, the visualization of separated subgroups, when starting from flat data matrices (cases by variables). CVA is related to the SVD's of this chapter, as it is a PCA of the data after the pooled within-group covariance structures has been made precisely spherical. These PCs are the *relative eigenvectors* of the matrix of sums of squares and cross products of the group mean locations with respect to the within-group covariances. They bear the extremes and stationary points of *ratios of squared length*, or ratios of variance, between the two matrices. Geometrically, they are the principal axes of either ellipsoid after the transformation of the dual space that sends the other ellipsoid to a sphere (a set of uncorrelated variables all of the same variance), and, for ellipses, they are the directions that are *conjugate* in both forms, meaning that for both forms the tangent direction to the curve at the piercing point of either is the direction of the other. See the left panel of Figure 6.22. (By contrast, the PLS version of this computation is just a PCA of the group means alone, making no reference to those covariances.) The *relative eigenvalues* corresponding to these eigenvectors will concern us later in this chapter.

CVA is not aimed at the same goal as PCO – it is not actually a technique suited to the framework of numerical inference that is the subject of this book, nor does it belong in the company of PCA in the ordinary multivariate textbooks, as it cannot cope with organized systems. The most familiar version of CVA, Fisher's linear discriminant analysis (LDA), reduces to a regression of a two-valued variable for group upon an arbitrary list of measured variables. Instead of seeing the situation as a pair of hypotheses pertaining to a single probe value, we consider the assignment of finding the best direction in which to take the probe. This is surely not the same scientific question (we have seen an example in Section L4.5.3). In this regression setting, the "model" is of predicting a grouping variable by measurements: a version of the left-hand panel in Figure 6.10 where the variable Y is not a quantitative outcome but instead a binary decision rule, such as alive/dead, chimpanzee/gorilla, or healthy/diseased.

Applying the formalism of normal equations (Section L4.5.3, especially Figure 4.33) to such a diagram leads to nonsense. The decomposition of paths into direct and indirect makes no sense for such an outcome, as the idea of adding up their arithmetic values no longer corresponds to any empirical process of superposition. The only interpretation is in terms of the likelihood ratio of the corresponding hypotheses, as already explained in Chapter 4.

This regression-based approach fails to meet the standards of this book for two more fundamental reasons as well. First, in the numerical-inference framework we have adopted, the LDA or CVA step is playing the logical role of some preliminary *abduction* – some surprise is to be observed (the groups separate in the prediction space, or the accuracy is high) that awaits reduction to something routine. But why should the success of any such classification, with groups identified in advance, *ever* come as a surprise? It is as if we were attempting to inform animals of their scientific names. **Of course** the group name can be predicted by our measurements to some extent – otherwise, why would the groups have different names? And the LDA/CVA machinery makes no use of any information about the probes, which are the actual subject of the hypotheses that the classification task is considering. The second fundamental difficulty is that the vectors generated by these methods have a meaning for numerical inference only when the within-group covariance matrices are identical, and this requires $p(p+1)/2$ separate equalities, a collection which is quite unlikely to obtain if the groups are really different enough to deserve different names to begin with. Note, for instance, that in none of the situations shown in Figure 6.20 does a vector formalism or a linear path model such as that of Figure 4.33 apply meaningfully.

The logical difficulties become even worse when variables come with the sort of a priori structure that qualifies them for explanatory roles in organized systems

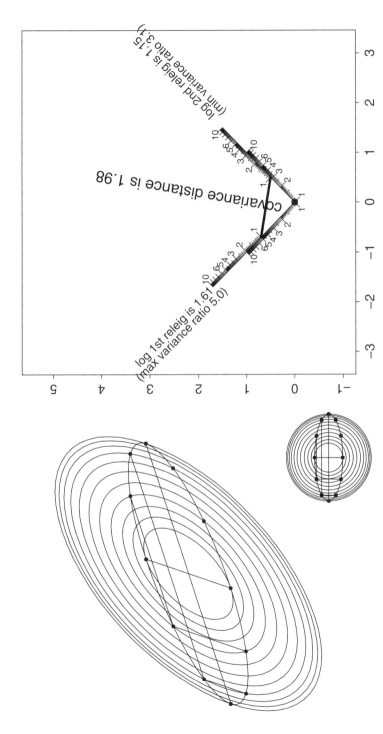

Figure 6.22. The natural distance function for 2×2 covariance matrices. (Left) Geometry of relative eigenanalysis, here for the case of two 2×2 covariance matrices. The relative eigenvalues of the oblong ellipse bearing the dots with respect to the outermost of the nested ellipses are the same as the ordinary eigenvalues of the corresponding ellipse in a coordinate system sheared so that the ellipses of the nested system have been circularized. (Right) The corresponding distance function (for a different example), square root of the sum of squares of the logarithms of the relative eigenvalues.

393

(Section 6.3.2). In that case, the list of actual variables going into the "regression" is relatively arbitrary – there could be an indefinitely large number of these – and thus the competent investigator is forced into either an artificial termination of the list of predictors (so-called "stepwise regression," which is always unwise; see D. R. Anderson [2008]) or instead a replacement of the predictor set as a whole by its principal components. But in that case, as these predictors are themselves uncorrelated, the "regression" reduces to a sum of direct paths without any indirect paths, and thus to a Partial Least Squares analysis, not a regression at all.

By contrast, the findings coming in Figures 6.25 and 6.26 will indeed be surprising. The revealed structure here is not built into the computations, but emerges as an empirical finding. In particular, the pattern of straight segments separated by sharp turns, corresponding to a succession of single-factor models, is quite surprising on any model of disorganization, but would be expected if the regulation of form is actually calibrated by a sequence of factors within a sequence of distinct endocrinological regimes. The sharp turns correspond to the discovery of discrete times of systematic *reorganization* of these very highly organized systems as they balance the lifelong sequence of contending physiological and biomechanical demands.

CVA is a visualization technique, not an explanatory one, and the cognitions to which it corresponds are quite primitive, epistemologically speaking: simple categorizations, not even forecasting, let alone any of the more sophisticated aspects of organized system description. Also, the assumptions it must enforce, notably, equality of within-group covariance structures, are highly unrealistic in most of the examples I have encountered. In fact, differences in those covariance structures are typically more informative than the mean differences themselves over which CVA obsesses, and lead both to greater surprise and to more important understandings of these organized systems. Early in Part III I mentioned that the sciences exploiting multivariate numerical data seem to divide into two regimes, one interested in numerical explanations and the other interested in human actions related to management of social systems (null-hypothesis "significance tests," predictions, classifications [supervised learning]). CVA falls wholly in the second category, and is not mentioned further here.

6.5.3 Covariance Distance

The methodology explained in this section was originally the idea of my Vienna colleague Philipp Mitteroecker, and the basic argument here is adapted from the original proposal in Mitteroecker and Bookstein (2009a).

Suppose there are p variables with a $p \times p$ covariance structure $S_1, S_2, \ldots$ in each of two (or more) subsamples, at more than one observation time, etc. Then there are very good reasons to rely on the following formula as the *unique, natural* **distance between two covariance matrices** S_1, S_2 on the same variables:

$$\|S_1, S_2\|_{\text{cov}} = \sqrt{\sum_{1}^{p}(\log \lambda_i)^2}, \tag{6.1}$$

6.5 Another Tableau: Dissimilarities as Primary Data

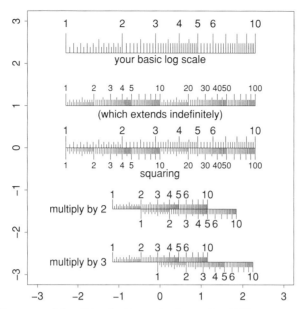

Figure 6.23. Geometry of the slide rule, for readers born after 1970.

where the λ_i are the relative eigenvalues of either S_i with respect to the other one, defined as in Section 6.5.2. Figure 6.22 (left), deferred from that earlier discussion, is the illustration of a two-dimensional case. For the formula to make sense, none of the eigenvalues of either matrix can be zero. That is, each matrix S_i must be of full rank separately.

Applied to a 2×2 relative eigenanalysis that multiplies variances by a factor of 5 in one direction and a factor of 3.15 in the perpendicular direction, there results a covariance distance of **1.98**. The right-hand panel of Figure 6.22 shows where that number comes from in this specific application.

Why this funny-looking formula? Why logarithms? Consider the humble slide rule (Figure 6.23) familiar to readers of a certain age. A slide rule is a mechanical analogue computer for multiplication – a 17th-century invention that multiplies numbers by adding their logarithms. The scaling in terms of the logarithm is unique in the sense that the log (to any base) is the only function f such that $f(xy) = f(x) + f(y)$ for all positive x, y.

Think of multiplication by a number as a mathematical *operator* on the positive real line, so that the number 3, for instance, is both a dot on the line and the action of tripling, and the number 2 is both a dot and the action of doubling. We have $\log\ 2 + \log\ 3 = \log\ 6$. In terms of distances along the slide rule, the distance between 1 and 2 is the same as the distance between 3 and 6.

In this way, we assign a measure ($\log 2$) to the operation of doubling, a distance between x and $2x$ that is the same no matter what the positive number x may be, and likewise a measure $\log 3$ to the operation of tripling, such that the distance that corresponds to the combination of the operations (first do the one, then do the other) is the sum of the distances of the two factors each from the identity map (the number 1). We emphasize that the logarithm is the *unique* distance function with this property. (We have already seen this claim in connection with the Shannon measure H of information, Section 5.2, that underlies the Akaike Information Criterion for strong inference among multiple statistical hypotheses.)

The logarithms in equation (6.1) are there for precisely the same reason as the logarithms built into the scales of a slide rule. For a covariance matrix is both an object and an operator in the same sense that a number like 2 or 3 is both a value and an operator. Covariance matrices multiply vectors to give other vectors, for example. If we establish a distance measure (write it using a pair of double vertical bars, $\|X, Y\|$) between two matrices X and Y, and we re-express both of them using a new set of basis vectors T, we would like to leave the distance unchanged: we want $\|X, Y\| = \|T^{-1}XT, T^{-1}YT\|$ for all invertible transformations (changes of basis) T. In biometrical terms, this means that we want our distance measure to be invariant against changes of the factor structure of the measurements in which we express it. The formula (6.1) is the unique formula with this property.

To begin understanding why we care about being independent of changes of factor structure, let us consider a simple set of covariance matrices: the $k \times k$ matrices

$$M_1(x) = \begin{pmatrix} x & 0 & \cdots & 0 \\ 0 & 1 & \cdots & 0 \\ & & \cdots & \\ 0 & 0 & \cdots & 1 \end{pmatrix}$$

Each of these is the operation that multiplies the first element of a k-vector by x and leaves all the rest alone. (For instance, $M_1(1)$ is just the identity matrix.) The dimension k of the vector isn't stated explicitly in the notation because we won't need to refer to it for this demonstration.

Suppose we want an expression for the distance between two of these matrices that meets the "slide rule requirement," the requirement that the distance between $M_1(x)$ and $M_1(1)$ is the same as the distance between $M_1(xy)$ and $M_1(y)$. Then it follows that the distance of $M_1(x)$ from $M_1(1)$ has to be $|\log x|$, and this is the same as the distance $|\log x|$ of $M_1(xy)$ from $M_1(y)$. The absolute value $|\cdot|$ comes from the fact that "distance" on the slide rule is the same in both directions. The distance between 1 and 2 is the same as the distance between 1 and 0.5; both are equal to log 2.

If you insert the two matrices $M_1(y)$, $M_1(xy)$ in formula (6.1), the relative eigenvalues you get are just $\lambda_1 = xy/y = x$, $\lambda_2 = \lambda_3 = \cdots = 1$. Then (because log $1 = 0$) formula (6.1) reduces to just

$$\|M_1(xy) - M_1(y)\| = |\log x|,$$

which is the same as the slide rule formula we had before.

You can see that this argument would be the same whatever single diagonal entry of M we were tinkering with. In fact, it is the same for any *single-factor model*

$$M_2(1 + \lambda, v) = I + \lambda v v^t$$

where v is any vector whose variance is being inflated by a factor λ while leaving the variance of every other direction of the measurement space unchanged. The distance between $M_2(1 + \lambda_1, v)$ and $M_2(1 + \lambda_2, v)$ has to be $|\log(1 + \lambda_1) - \log(1 + \lambda_2)|$ for exactly the same reason that $\|M_1(x) - M_1(y)\|$ had to be exactly $|(\log x - \log y)|$.

There is another way to think about exactly the same construction. We want the distance of $M_2(1 + \lambda_2, v)$ from $M_2(1 + \lambda_1, v)$ to be the same as that of $M_2((1 + \lambda_2)/(1 + \lambda_1), v)$ from the identity because $M_2((1 + \lambda_2)/(1 + \lambda_1), v)$ is just the matrix $M_2(1 + \lambda_2, v)$ *reexpressed in the basis that makes* $M_2(1 + \lambda_1, v)$ *equal to the identity.* So we have arrived at the unique distance function that is invariant against changes of basis (for this limited set of matrices, anyway).

Let us play the same logic in reverse. Suppose we have a set of matrices $M_2(1 + \lambda, v)$ for which distances add exactly – for which

$$\|I, M_2(1 + \lambda_1, v)\| + \|I, M_2(1 + \lambda_2, v)\| = \|I, M_2((1 + \lambda_1) \times (1 + \lambda_2), v)\|.$$

It is tempting (and ultimately entirely appropriate) to interpret this as the statement that the matrices $M_2(1 + \lambda, v)$ *lie on a straight line* in the space we are building. (The previous sentence actually subsumes an entire undergraduate math course's worth of introduction to the construct of geodesics in Riemannian geometry, which we have agreed to skip over for the purposes of this book.)

We began with a slide rule, which is a physical object. We have arrived at a metric that makes sense of single-factor models as lying on straight lines – slide rule–like lines – in some sort of metric space suited for the covariance matrices they embody. We need only one more step to make the formula (6.1) quite general: the extension from one single-factor model to the superposition of two. To show this, at least informally, let us examine the covariance matrix with not one but two single factors, $\Sigma = I + \lambda_1 v_1 v_1^t + \lambda_2 v_2 v_2^t$, where v_1 and v_2 have dot product zero. For distances of this matrix from the identity I of the same dimension, the eigendecomposition gives us the relative eigenvalues $1 + \lambda_1, 1 + \lambda_2, 1, \ldots, 1$ and thus the squared distance $\log^2(1 + \lambda_1) + \log^2(1 + \lambda_2)$ from formula (6.1). But this is just the Pythagorean sum of the two squared distances $\log^2(1 + \lambda_1)$, $\log^2(1 + \lambda_2)$ of the components deriving from the factor models on v_1 and v_2 separately. Because *any* covariance matrix can be built up as the superposition of these – this is just another version of the singular-value theorem – we have actually shown that the formula (6.1) applies to *any* difference between a matrix and the identity, that is, any covariance matrix at all. By iterating the one-dimensional change of basis a few paragraphs above as many times as necessary – once for each dimension of the natural basis – we find that we have established the formula (6.1) as the *unique* formula for distance between covariance matrices that is invariant against changes of basis for the covariances, which is to say, invariant against changes of the true factor model underlying our data sets. That is what this section set out to argue.

This has been an informal demonstration. For more mathematical rigor of language, see Mitteroecker and Bookstein (2009a) or Helgason (1978).

For 2×2 covariances near a reference metric that is taken as circular (by a preliminary transformation), Figure 6.24 shows what the three perpendicular dimensions of

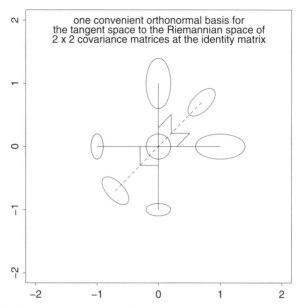

Figure 6.24. The three ways a 2 × 2 covariance matrix can vary. See text.

this space (three is all there are) look like. Drawn horizontally is the dimension for changing the relative eigenvalue in the horizontal direction of this circle of directions (linear combinations). Drawn vertically, with perpendicularity indicated, is the symmetrically construed version for the vertical direction through the circle. And drawn in foreshortened fashion in the figure is the third direction, perpendicular to both of the first two: change of ratio of variances of linear combinations at 45° to either of the first two directions. The first two of these three correspond to Flury's model of common principal components, Section 6.3.2, the third to the possibility of rotating the components (represented here in 2D as a shear of the originally perpendicular horizontal and vertical diameters of the circle). As the dimensionality k rises, the number of parameters to which Flury's method has access rises as the count k of variables but the degrees of freedom for shears rises as $k(k-1)/2$.

Example 1. This involves 18 genetically homogeneous male rats imaged via cephalogram at eight ages between 7 days and 150 days: the so-called "Vilmann data set," Figure 6.25 (left), imaged by Henning Vilmann, digitized by Melvin Moss, and used for demonstrations in the textbooks by Bookstein, Dryden and Mardia, and others. For a complete data listing see Bookstein (1991, Appendix 4).

In this example there are nearly as many shape coordinates as specimens, and in the next example there are actually more shape coordinates than specimens. In both settings, we reduce the morphometric descriptor space to the projection on some number of principal components before proceeding with the relative eigenanalysis. As part of that procedure, one must carefully check the results to ensure that they are not taking advantage of chance at any specific count of PCs, but are instead typical of the findings over a range of dimensionalities. That is the situation here.

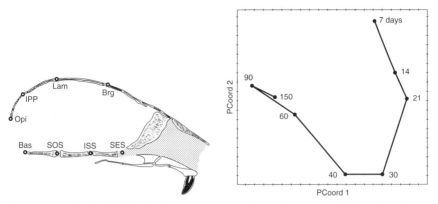

Figure 6.25. (Left) Eight landmarks on the rat midsagittal neural skull. (Right) First two principal coordinates for shape covariance at eight ages, Vilmann data. This and the next figure are from Mitteroecker and Bookstein (2009a). Landmarks: *Bas*, Basion, *Opi*, Opisthion; *IPP*, Interparietal Suture; *Lam*, Lambda; *Brg*, Bregma; *SES*, Spheno-ethmoid Synchondrosis; *ISS*, Intersphenoidal Suture; *SOS*, Spheno-occipital Synchondrosis. For operational definitions of these points, see Bookstein, 1991, Table 3.4.1.

We computed covariances of the Procrustes shape coordinates (a total of 16), reduced to the subspace of the five dominant principal components (for robustness), and produced the analysis at right in Figure 6.25 by the distance formula and the principal coordinate analysis explained above. **Finding:** the covariance structure of this growth process makes a turn of nearly 180° around the time of weaning. Closer investigation (Fig. 7.16) shows that the phenomenon involves a canalization (decrease in variance) of the vertical component of the uniform transformation of this shape between ages 7 and 30 days, and then a corresponding reinflation of this component of individual difference between 40 and 150 days. Consilience might be with the associated changes in craniofacial function, specifically, action of the muscles of mastication.

Example 2. This example involves 26 normal 20th-century American children growing from age about 3 years to 18 years of age and imaged cephalometrically (again) by 18 mostly conventional landmarks. The data, Figure 6.26 (left), are from the Denver Growth Study; digitizing was by Ekaterina Bulygina. See Bulygina et al. (2006).

We computed covariances of the 36 Procrustes shape coordinates, reduced to their 15 principal components, and again proceeded by the distance formula and the principal coordinate analysis explained above. **Finding** (Fig. 6.26, right): the covariance structure of this growth process makes a turn of nearly 90° at each of three crucial stages of craniofacial growth re-regulation. Consilience might pursue more detailed studies of the hormonal regulation of bone growth at its principal growth sites.

6.6 Concluding Comment

This chapter has explored the rhetoric of a considerable range of tableaux for pattern analysis of complex systems, always emphasizing the logic of the numerical

400   6 The Singular-Value Decomposition

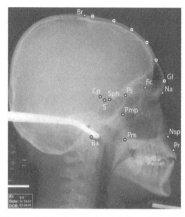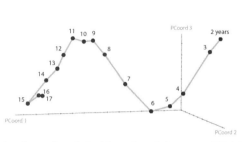

Figure 6.26. (Left) An 18-(semi)landmark scheme for clinical lateral cephalograms. (Right) First three principal coordinates, shape covariances at 13 ages, Denver data. Landmarks: *N*, Nasion; *Nsp*, Nasospinale; *Pr*, Prosthion; *Pns*, Posterior nasal spine; *Ba*, Basion; *CP*, Clival point; *S*, Sella; *Sph*, Sphenoidale; *Pmp*, Greater wings of the sphenoid; *PS*, Planum Sphenoideum point; *FC*, Foramen Caecum; *Br*, Bregma; *Gl*, Glabella. For operational definitions of these points, see Bulygina et al., 2006, Table 1.

inferences they might empower. I argued that for application to the science of a complex organized system, the geometry of multivariate patterns is *not* the conventional least-squares formalism of multiple regression – vertical distances to lines or planes – that was the explicit focus of the lectures in Chapter 4, but instead the far richer geometry of explicit patterns as conveyed by the fit to some entire table of explananda at once. From 19th-century analytic geometry we borrowed a structure more complex than planes, the hyperbolic paraboloid, whose symmetries are different and whose parameterization (by virtue of its double ruling) maps much better onto our intended explanatory context. The corresponding pattern algebra rests on the singular-value theorem and the associated decompositions, a mathematics considerably deeper than the sums of squares of linear combinations that drive multiple regression models.

From this expansion of our numerical toolkit follow pattern search engines in great variety. I selected three of these for particular attention: tables of raw data, units by attributes, Section 6.3; tables of associations of blocks of (related) attributes by other blocks, Section 6.4; and tables of pairwise dissimilarities between study units, Section 6.5. In their details, and also in their graphics, these styles overlap; likewise in the consiliences that sometimes help resolve the Peircean abductions launched by perceived surprises in the detected patterns.

In the technical literature of applied statistics, these methods ramify into an extensive library of treatises and handbooks customized to data patterns arising from specific machines and of interest to specific disciplines. If a book like this one chose to survey these and other praxes in adequate depth it would become too thick for binding in one single volume. So I have adopted a different strategy: to demonstrate the interweaving of most of these methods in one common domain of application. That domain, of which I am particularly fond, is *morphometrics*, the pattern analysis of biological shape and shape change. Enough of its data resources come from machines to meet the requisites of abduction and consilience on which we rely everywhere. By augmenting this core information with other sorts of observations of the same entities – measures of function, notes about development or disease, perceptions of dissimilarity or conspecificity – we can illustrate the interplay of the whole range of these pattern engines as they collaborate (literally: work together at one and the same time) to make sense of the associated scientific concerns. Organisms are the most interesting complex organized systems we've got, and so a methodology that handles patterns from this particularly rich domain of interrelated information resources may serve as a good prototype for the necessary development of statistical methods for organized complex systems more generally. The next chapter concerns itself with the passage from machine-derived data to numerical inference in this exemplary field.

7

Morphometrics, and Other Examples

Biological forms are particularly highly organized systems. Here in 2013, morphometrics is the careful construction of a framework for quantification of these systems that supports pattern analyses over pointlike, curvelike, or surfacelike image data represented in ways that highlight their organized aspects. Beyond all the geometry, its methods align well with the toolkit introduced in the previous chapter, the singular-value decomposition (SVD) in its three different guises. But the applications context for these tools has been optimized by massive intellectual investment in the structure of variables a priori, a strategy aligned with the criteria of Section 6.3.2 about where good sets of measurements come from. Morphometrics is my favorite example of the real work actually involved in setting up that numerical information from images of living or once-living creatures, so as to be fully open to both abduction (surprise) and consilience (confirmation by measurements via other quantitative protocols).

Right at the core of morphometrics is a particularly careful construal of what we *mean* by "shape measurement" – how raw information from images of living or once-living creatures, modulated by the installed knowledge base of comparative anatomy, is transformed into the geometry and algebra that allow us to pursue the theories and explanations that govern this domain of organized form. This process, the rearrangement of intuitive turn-of-the-20th-century ideas of biometrics so as to suit modern approaches to numerical inference in organized systems, culminated only recently, from 1980 through 2005 or so, in the hands of the present author in collaboration with a range of colleagues including W. D. K. Green, James Rohlf, Kanti Mardia, John Kent, Dennis Slice, and Paul O'Higgins. The present chapter is not a vademecum of morphometrics, but instead a survey of its styles of knowledge representation and reasoning in terms of the principles driving this book as a whole.

In a manner oddly reminiscent of the completely different discipline of psychometrics, morphometrics springs from a primitive notion of dissimilarity as fully quantified *distance* in a very specific geometry, one where the measure of distance between any two specimens is the celebrated *Procrustes distance*. This most basic quantification pertains to two forms, not one. It is a lot of work, mathematically speaking, to get back to the cases-by-variables setup of standard statistical computations. The mathematics of this formulation then reveals a pleasant surprise: the existence of a canonical set of *shape coordinates* for which Procrustes distance is a distance of the ordinary sort, square root of a sum of squares that the SVD and its family of techniques will find convenient to examine. We will see that principal coordinates analysis (PCO) of Procrustes distances is equivalent to principal components analysis (PCA) of these shape variables – a circumstance that is unprecedented across other applications of PCO to primary data that are pairwise dissimilarities.

Corresponding to these new coordinates is a new style of diagram, the *thin-plate spline*, that rigorously attaches multivariate pattern descriptor vectors to the picture of the underlying living form in a startlingly evocative way going well beyond D'Arcy Thompson's suggestions of a century ago. The diagram covers the import of the numerical inference for *every* landmark that might have been taken in-between the ones that were actually located: covers, so to speak, the whole set of variables that could have been observed, not just the arbitrary set that were actually made available by the overworked graduate student doing the digitizing. The spline brings with it, too, a further candidate representation of ignorance to match our Gaussian and other random models: ignorance of the pattern of deviations from uniformity in shape changes – ignorance of "where we will point," and with how broadly tipped a pointer, to call the reader's attention to the morphological patterns that inform us about growth, evolution, or disease.

Whether an abduction or a consilience, the core morphometric data flow goes in four steps:

- Design of a system for archiving the forms of interest by locating conventionally named components of their drawings or images
- Production of a square matrix of interspecimen distances, using one of two alternative formulas, and manipulation of those distances to generate the Procrustes average form and the shape coordinates in terms of which pattern computations go forward
- Extraction of surprising or consilient patterns of those shape coordinates as they vary over organisms or over other attributes of the same organisms
- Visualization of the patterns back in the picture or the space of the original data, in order to adumbrate whatever theories of system function might govern the organized forms under study

This version of morphometrics is the approach I find best suited to the specific purposes of numerical inference as this book construes it: settling arguments among competing scientific claims arising from contemplation of pictures of living or previously living creatures. Other methodologies may work better for applications of a different sort, for instance, predicting the locations at which bones will break under stress, or staging a cancer, or putting a name to the skull of a war crimes victim. When the techniques appear not to be answering the right questions, it may be that the questions do not suit the framework of numerical inference of explanations for which this version of morphometrics was designed, or it may be that the representation of ignorance (and its symmetries) that underlies the distances used here overlooks some actual a-priori knowledge about differences that should be allowed to supplement the generic formalisms.

Let us call this approach **geometric morphometrics**, which is the name by which it is known in the literature and on the Web. Of the three computational themes driving geometric morphometrics, Chapter 6 has already reviewed one, the SVD-based family of pattern engines that are useful across quite a broad range of applications. The other two, which are specific to morphometrics, are the main focus of our attention in this chapter. One is the core data representation of landmark locations, which is how we formalize all that prior biological knowledge for interactions with data, Section 7.1, and the other is the core diagram style, the thin-plate spline, that induces viewers to speculate about potential interpretations of these coordinate rearrangements as traces of biological processes, Section 7.5. The information that is the subject of the morphometric conversion is summarized in one of two quantities, shape distance or form distance, that stand for differences represented as *geometrical discrepancies* among two or more forms that arrive with attached *labels at landmark structures.* This prior information is craft knowledge, the lore of how you point reliably to places inside an organism. From the locations inside a volume, on a volume section, or along a surface curve or curving surface patch arise the Cartesian coordinates and thereafter the Procrustes distance representation and the shape coordinates that drive pattern analyses.

Morphometrics "works" to the extent that this unitary construal of geometric discrepancy, ramified into regionalized descriptors by manipulations of the thin-plate spline, conveys the important embedded information about the biological processes underlying the resemblances. It works, in other words, to the extent that this particular quantitative image of one specific cognitive dissimilarity, a formalism in the mind of the trained biologist, captures information relevant to the understanding of the particularly complicated organized systems we call organisms. The formulas do not know any biology, and there is no claim that Procrustes distance itself makes any biological sense. Nevertheless the raw data are intended to convey all the prior knowledge needed when the pattern engines we've just reviewed are applied to the shape coordinates that reify this single a-priori quantification of difference.

A brief historical comment. The morphometrics reviewed here combines pattern-centered multivariate statistics with a specific biomathematical approach to image geometry. Because syncretic methodologies like this only rarely spring forth Athena-like from the brow of just one creator, readers may wish to consult a range of oddly inconsistent essays that review the potential history of this method from a variety of perspectives (e.g., Bookstein, 1978, 1996b, 1997b; Slice, 2004). In 1907 Francis Galton invented the method of two-point shape coordinates we shall introduce presently, and suggested them as a biometric method for identifying persons of interest to the police. Even earlier, in 1905, Franz Boas almost (but not quite) stumbled over the Procrustes superposition (he overlooked only the rescaling aspect). And there must have had to be an intuitive sense of shape feature spaces in the mind of every 18th-century caricaturist, at least, those who were able to make a living at this. By contrast, two books with the explicit word "morphometrics" in their titles (Blackith and Reyment, 1971; Pimentel, 1979) have virtually no overlap with the syllabus of this chapter in spite of being only 30 or 40 years old. (By the time of the second edition of the Blackith volume, Reyment et al. [1984], this lacuna was remedied.)

Omitted from all these anticipations is the context emphasized in this chapter: abduction and consilience in complex systems as exemplified by morphometrics as *rhetoric*, one component of a narrative strategy for persuading from data. The ability of morphometrics to supply a coherent multivariate praxis consistent with the knowledge base of classical comparative anatomy lay unformalized even in the early review articles and textbook chapters (e.g., Reyment and Jöreskog, 1993). In retrospect, what was lacking was the "consilience" part – the use of morphometrics not just to compute stuff but also to *establish the truth of hypotheses* that were then (1) replicated by others and also (2) wielded to generate new hypotheses leading to new discoveries, particularly revelations of mechanism, in turn.

Other reasons for the lag might be the failure of any of these techniques to appear in the standard "statistical packages for biologists" – their dissemination had to await the emergence of Internet-shared code, such as R libraries, well after the turn of the present century – and the (related) explosion of personal computer technology that made possible the management of large archives of 2D and even 3D data resources in one's lap, along with powerful software running locally for satisfyingly responsive visual interaction (see Weber and Bookstein, 2011).

The obsession of the established curriculum with p-values, likewise, was a major obstacle to acceptance of this style of morphometrics until just a few years ago.[1] P-values cannot be reformulated to suit the statistics of the organized system sciences. The persuasive power of rendered pattern discoveries has vastly outrun any facility for the corresponding probability calculus. As Wilson explained in 1998 and as we have already seen in Chapter 2, it is now consilience with multilevel explanations, not significance testing, that drives the engine of collaborative research. Put another way, it has taken more than a hundred years for Peirce's elementary syllogism about surprise and its reduction to replace the malformed alternative that is classic null-hypothesis significance testing. The well-received monograph by Dryden and Mardia (1998) can be seen as transitional in this respect. It wholly embraces the multivariate algebra and most of the pattern engines I am putting forward here, but nevertheless continues to emphasize the context of null models and significance testing that characterized earlier incarnations of multivariate analysis for the psychological and social sciences. Regarding the concept of consilience, and the way morphometric findings are to be woven with other arguments into a persuasive paper, the morphometrics of the 1980's and 1990's had hardly any biologically sensible suggestions to offer.

Yet today's morphometrics may be the first applied statistical technology to fuse these approaches so effectively. Part IV comments about one possible future, in which other interdisciplinary methodologies follow its lead.

---

[1] And this battle itself is not new. Twenty years ago Reyment and Jöreskog's *Applied Factor Analysis in the Natural Sciences* attempted such a transition – the conversion of a confirmation-centered methodology into a method for discovery, and just before that Martens and Næs had attempted the same for chemometrics in their magisterial *Multivariate Calibration*. Neither book was properly appreciated for its depth of philosophical engagement until long after publication.

7.1 Description by Landmark Configurations

If morphometrics is driven by carefully validated differences among forms, then what counts as valid for this purpose must arise from the domain that generates the explanations ultimately arrived at. Correspondingly, for more than a century the raw data of morphometrics have arisen from the meaningfully labeled anatomical images and diagrams produced by organismal biologists. With only a few exceptions (e.g., applications to docking sites of enzymes), morphometrics applies to the structures of the *old* biology, the early 20th-century biology, rather than the -omes and other multilevel constructs that make up today's "bioinformatics." In the more classical application domains, concerned with the extended forms of organisms, what makes the work craftworthy is not measurement precision but careful restriction of the scientist's freedom of data representation – the management not of noise but of human arbitrariness. (For a provocative recent meditation on this topic, see Oxnard and O'Higgins, 2009.)

Such representations differ from domain to domain of biological form. For instance, the table of contents of Rudolf Martin's great classical treatise *Lehrbuch der Anthropologie in Systematischer Darstellung*, 1928, volume 2, "Kraniologie, Osteologie," highlights the main technical terms in craniometric praxis: planes and lines, standardized views (verticalis, basilaris, lateralis, frontalis, occipitalis, sagittalis), and a great diversity of points appearing in from one to four of those standardized views. In the absence of methods such as shape coordinates and the thin-plate spline, the principal formalisms for the pattern analyses of this chapter, Martin needed to proceed with a long list of instructions for scalars, and it was *these* instructions that conveyed the craft knowledge of his time. There were lengths and breadths and also a variety of other distances taken variously between points, lines, and curves. Besides distances, there are angles and even curvatures. "Indices" (combinations of other measurements) usually arose as ratios of distances. Martin also offers clear anticipations of some of the more mathematical craniometrics of today: three-dimensional reconstruction, the "truss method" for triangulating points from distances and angles, and even fractals, like his system of classifying sutures.

None of these measurements were under control of any theory. All were designed simply to quantify the anthropologist's preexisting intuitions about global or regional differences of the forms on the table before him. It is not obvious that a methodology superordinate to a specialized toolkit like his should exist in any more generally useful form. Morphometrics can supply the missing generality – can fuse the toolkits of the separate divisions of organismal biology – to the extent that it concentrates on the theorems that model information flow across the protocols of these analyses: that is, to the extent that it sharpens the application of quantifications of form to abductions or consiliences deriving from other domains of biological insight.

Biological form is three-dimensional even when our images are not, and crucial information can inhere in any physical or chemical property "at a point," meaning, in practice, over a subregion of any position, orientation, shape, and extent (see Cook et al., 2011). But once we leave the solid world of morphogenesis for the less informative domains of surface images or pictorial densities, information about three-dimensional biological forms is best imagined as packaged not in discrete points but in *curves*. These come in three types:

- *Observed curves*, such as intersections of two actual biological bounding surfaces, centerlines of tubular structures, or intersections of one smooth surface with a standardized plane
- *Ridge curves*,[2] which are curves of points at each of which the *perpendicular* curvature is a local maximum along that perpendicular direction
- *Symmetry curves*, like the corpus callosum midcurve that is discussed in the fetal alcohol example, Section 7.7. A symmetry curve is an ordinary smooth curve together with a perpendicular direction at every point such that every plane through that direction looks (approximately) symmetric in the vicinity of the curve with respect to reflection in that direction.

These types of curves characterize many of the most helpful landmark points by their intersections, while other points arise as extrema of local properties of curves (e.g., vertices) or surfaces (boss points [centers of bumps], saddle points). Such location data acquire scientific meaning only to the extent that they encode the morphogenetic processes that formed them or the biomechanical processes whose explanations they expedite. Thus even at this stage of data design the origin of morphometric questions in the phenomena of a morphology regulated throughout its lifetime (i.e., a highly organized system persisting in time) bears extensive implications for its metrology. We arrive at

The First Rule of Morphometrics. The configuration of raw location data underlying the pattern analyses must be a *Gesamtkunstwerk* (a lovely German term meaning "collective work of craftsmanship," such as the Kunsthistorische Museum in Vienna), a coherent scheme that is balanced in its coverage of spatial details while still managing to incorporate enough global information to be used in support of a wide range of explanations, not just the group differences or growth gradients that you think are your principal topic of analysis. Information omitted at the stage of landmark charting cannot be retrieved later. But not all information about organic form is reducible to landmark configurations. Huge domains, including many aspects of botanical form (branches, leaves) and, alas, the cortex of the human brain, remain out of reach of these formalisms.

Examples. We have seen several examples of this First Rule already, at Figure 6.8 (landmarks of the midsagittal human skull), 6.25 (rat skull), and 6.26 (lateral cephalogram). Others will be the image of the callosal midcurve *in situ*, Figure 7.21, and the layout of the landmarks for the schizophrenia cortical localization example, Figure 7.19. For one particularly thorough "standard" set this one for the anthropologist, see Figure 7.1.

---

[2] Also known as *crest lines* in the literature of medical computer vision. Technically, these are the preimages of cuspidal edges of the surface of centers. See Koenderink (1990).

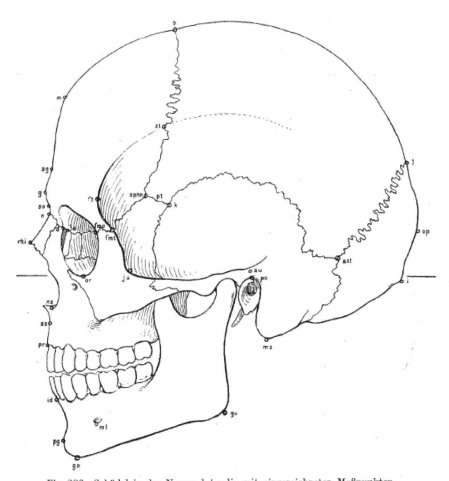

Fig. 286. Schädel in der Norma lateralis mit eingezeichneten Meßpunkten.

ast Asterion, *au* Auriculare, *b* Bregma, *d* Dakryon, *fmo* Frontomalare orbitale, *fmt* Frontomalare temporale, *ft* Frontotemporale, *g* Glabella, *gn* Gnathion, *go* Gonion, *id* Infradentale, *i* Inion, *ju* Jugale, *k* Krotaphion, *la* Lakrimale, *l* Lambda, *ms* Mastoideale, *ml* Mentale, *n* Metopion, *n* Nasion, *ns* Nasospinale, *op* Opisthokranion, *or* Orbitale, *pg* Pogonion, *po* Porion, *pr* Prosthion, *pt* Pterion, *rhi* Rhinion, *sphn* Sphenion, *st* Stephanion, *ss* Subspinale, *sg* Supraglabellare, *so* Supraorbitale.

Figure 7.1. Types of curves and landmarks, exemplified in Figures 286 and 290 of Martin, 1928. I sketch some operational definitions that exemplify options for the relation between these points and the curves on which they lie: *g*, glabella, point of sharpest curvature on a symmetry curve (the midline); *au*, auriculare, displaced representation of the opening of the aural canal (a centerline intersecting a virtual surface); *eu*, right euryon, one end of an extreme diameter; *go*, left gonion, locus of extreme curvature of a ridge curve; *st*, stephanion, intersection of a suture and a muscle insertion on a curving surface; *b*, bregma, intersection of three sutures; *n*, nasion, intersection of three sutures; *ml*, mentale, standardized representation

7.1 Description by Landmark Configurations

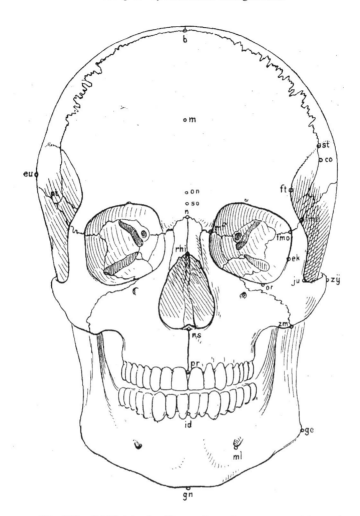

Fig. 290. Schädel in der Norma frontalis mit eingezeichneten Meßpunkten.

b Bregma, *co* Coronale, *ek* Ektokonchion, *eu* Euryon, *fmo* Frontomalare orbitale, *fmt* Frontomalare temporale, *ft* Frontotemporale, *gn* Gnathion, *mf* Maxillofrontale, *ml* Mentale, *m* Metopion, *go* Gonion, *id* Infradentale, *ju* Jugale, *n* Nasion, *ns* Nasospinale, *on* Ophryon, *or* Orbitale, *pr* Prosthion, *pt* Pterion, *rhi* Rhinion, *st* Stephanion, *so* Supraorbitale, *zy* Zygion, *zm* Zygomaxillare.

Figure 7.1 (*cont.*) of the aperture of the mandibular canal; *m*, metopion, point of greatest bulge on the midline of the forehead; *pr*, prosthion, contact point of the upper anterior alveolar ridge with the two central incisors; *or*, orbitale, lowest point of the orbital rim (which is not actually a well-defined curve); etc. Most of these points can be seen in both views.

7.2 Procrustes Shape Distance

Where "shape distances" come from turns out to be the same question as where "shape variables" come from. While the origin of landmarks is craft knowledge, morphometrics offers an unusually sophisticated response to the generation and algebraic representation of the differences among correspondingly labeled configurations. The match of Procrustes shape coordinates to Procrustes shape distance or Procrustes form distance lies at the core of the methodology and likewise the core of the claim that sometimes it strengthens scientific arguments.

A language for "shape coordinates," not for "shape." There is a way of talking about morphometrics in which one refers to an organism as "having a shape," and even draws pictures of "shapes," but that is not how data arise from the scanners or how they get attached to biological explanations (which, as we have known since Galileo, are usually not scale-free). The praxis here refers not to "shapes" but, following the language originally published by Mosimann (1970), to shape *variables* or, when data come from Cartesian coordinates, to "shape coordinates," which, it will turn out, no longer come in centimeters. It is very difficult to convey in words what any specific shape coordinate is actually measuring. Rather, they need to be regarded as a collection, a complete alternative set of Cartesian coordinates for which the scientist has altered the geometry of the digitizing device in a way that is different from speciment to specimen. *Shape coordinates leave the form alone but retroactively change the way it was digitized.*

What differentiates shape variables from other kinds of variables (or, what differentiates shape coordinates from other kinds of coordinates) is their invariance under some kinds of transformation of the raw data. For shape variables, following Mosimann (1970), the invariance is against scaling (changing the ruler). For shape coordinates, now following Kendall (1984) or Dryden and Mardia (1998), the invariance is against everything that is arbitrary when you put an object into a device that scans it: where, exactly, it should be put and in what direction, exactly, it should be facing. Our pattern analyses of coordinates must have both of these properties of positional invariance – translation (where the specimen is placed) and rotation (which way it faces) and, optionally, Mosimann's scale-free property as well.

In other words, the quantities we are calling "shape coordinates," *considered as functions of the original Cartesian data*, need to be immune to the same nuisances of positioning (or scaling) as those to which the shape differences themselves will ultimately prove immune. This is arranged by a clever trick due originally to John Kent and me independently. *Define* shape distance as the minimum of familiar (Euclidean) distance over all possible choices of of position and orientation, and then *define* shape coordinates as a set of Cartesian coordinates for which ordinary Euclidean (sum-of-squares) distance is almost exactly identical with this minimum *for all pairs of specimens in the original sample.* For a characterization like this to be useful there needs to be a *constructive* method – an easy way to produce the coordinates in question.

7.2 Procrustes Shape Distance

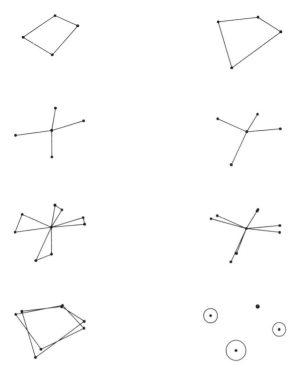

Figure 7.2. Procrustes distance between two quadrilaterals. (Top) Two quadrilaterals. (Second row) Center and scale each one separately. (Third row) Superimpose the centroids, then rotate to best fit. (Bottom) The resulting superposition has minimum summed area of the little circles shown.

The conventional Procrustes shape distance. How, then, *do* we work with landmark configurations independent of accidents of positioning the specimen? We begin with a distance measure that has the independence we seek by virtue of its explicit construction. The underlying formulation concerns two forms at a time that have the same k landmarks: for instance, the two quadrilaterals ($k = 4$) in the top row of Figure 7.2. In a preliminary step, compute the *centroid* of each set of four points (average of the points, not center-of-gravity of the area). Then the two-form Procrustes fit[3] proceeds in three steps. *First*, rescale the forms separately so that for each one the sum of squared distances of the landmarks from their centroid is 1. *Second*, place the centroids on top of one another. *Third*, rotate one of the forms (it doesn't matter which) over the other until the sum of the squared lengths of the little vectors on the right – the summed squared distances between corresponding landmarks of this superposition – is as small as any rotation can make it. In one formulation (but see

---

[3] Procrustes was not the inventor of this technique, but a Greek bandit who specialized in changing the size of his victims – the name of the technique preserves a little statistical joke. Who says statisticians have no sense of humor?

footnote 4 below), the squared Procrustes distance between the two landmark shapes is taken as the sum of squares of all those little distances, landmark to homologous landmark, in this superposition. This sum of squares is the same as the total area of the little circles drawn, divided by π.

There are other ways of arriving at this same quantity. For two-dimensional data, one approach is via a regression formula using complex numbers (Bookstein, 1991), and for data in either two or three dimensions there is a formulation (Rohlf and Slice, 1990) based on a modification of the singular-value decomposition we have already introduced in Section 6.2 for an entirely different purpose. (The modification consists in producing a matrix of all the cross-products of Cartesian coordinates in one form by the same coordinates in the other form, producing the SVD UDV^t of this cross-product matrix, and then setting all diagonal entries of the diagonal matrix D to 1.0. The resulting rotation UV^t is now the solution of this new least-squares problem, not the one for which the SVD was imported in the previous chapter.)

It is not only the last step, the rotation, that minimizes the summed area of those little circles. The first step, the joint centering of both forms at the same spot, must reduce that sum of areas, and, to a very good approximation, the second step as well, the setting of the sums of squared landmark distances from their centroid to be the same for the two landmark sets, must also reduce the sum of squares. The quantity being standardized at this step, which will prove crucial to studies of allometry, is called the squared **Centroid Size** of the landmark configuration. A physicist would recognize it as the second central moment of the set of landmarks treated as point masses. (In our statistics it usually appears in the form of its logarithm for reasons tied to the construction of form space, Section 7.4.) Of course we could make the areas of all of those circles as small as we wanted just by shrinking the figure. In the interests of mathematical precision all this needs to be phrased just a little more carefully: once one configuration has its scale fixed, the minimum of summed circle areas is achieved when the second configuration has (almost exactly) the same Centroid Size as the first.

7.3 Procrustes Shape Coordinates and Their Subspaces

Generalized Procrustes Analysis for samples. We can get a sample average form for free at the same time we fit each form of a sample *to* that average.

This is not as paradoxical as it sounds. The Procrustes fit is defined by a least-squares criterion in sums of squares, and sample averages are already defined by such a criterion. (Recall from Section L4.1 how the average $\Sigma_1^n x_i / n$ of any set of n numbers x_i is the minimizer v of $\Sigma_1^n (v - x_i)^2$ over v.) The central reassurance of the Procrustes approach is the assertion that for reasonably similar landmark configurations, no matter how you start off the algorithm in Figure 7.3, by the time it converges it must have constructed the shape that has the least summed squared Procrustes distances to all the configurations of any sample labeled by the same landmarks.

7.3 Procrustes Shape Coordinates and Their Subspaces

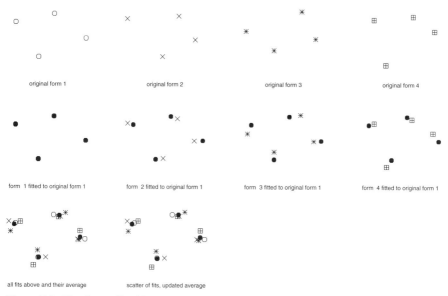

Figure 7.3. The Generalized Procrustes Algorithm for averages. See text.

Step 1. Suggest any specimen in your sample, for instance, the first specimen, as the initial guess at the average shape.

Step 2. Superimpose all the specimens over this candidate average in the least-squares position (row 2 of the figure).

Step 3. Take the ordinary Euclidean (center-of-gravity) average, landmark by landmark and coordinate by coordinate, of all of the Procrustes-superimposed specimens (row 3, left).

Step 4. Replace the candidate average by this new average, scaled up to Centroid Size 1, and loop over the combination of Step 2 and Step 3 until the candidate average stops changing from cycle to cycle.

This is **Generalized Procrustes Analysis** (GPA). For any competently designed landmark configuration, GPA converges to the same average form no matter which specimen you chose in Step 1. Each step reduced the same sum of squares, the sum of the areas of all the circles we could have drawn in every frame of Figure 7.3, and this series of minimizations has to converge to the smallest possible positive value.[4]

Now think again about the situation in the final panel of Figure 7.3 as it looks when the iterations have converged. Figure 7.4 enlarges this image and enhances it

---

[4] I have omitted some details that slightly affect matters if the forms of a sample vary wildly around the average we are computing here (in which case you should probably not be using tools like averages anyway). Procrustes distance pertains to a curved space (something like the surface of the earth, not like the interior of a room). So sometimes it is useful to state the measurement as a length on a certain abstract sphere, as the angle subtended by that length out of the center of the sphere, as the sine of that angle, or as double the chord of the half-angle. Also, there is a weighting factor proportional to the squared cosine of the same angle. In all good applications the sines are all much less than 0.1 and the squared cosines correspondingly bigger than 0.99, and the variation among all these choices can simply be ignored except in the punctilio of theorems.

414　　　7 Morphometrics, and Other Examples

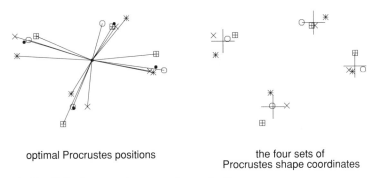

optimal Procrustes positions　　　　the four sets of
　　　　　　　　　　　　　　　　Procrustes shape coordinates

Figure 7.4. The GPA gives you shape coordinates for free.

a bit. There is a set of point locations for the average (the big black dots) and for each specimen of the sample a set of point locations, one in the vicinity of each black dot. The little subdiagrams here have two properties at the same time. For each landmark, the average of the four dots for the four specimens is the big black dot standing for the average shape, and also for each specimen, the squared Procrustes distance from the average is (almost exactly) the sum of squares of the coordinatewise differences between the specimen's dot and the black dot over all the landmarks of the configuration. In fact, this distance property (sums of squared interlandmark distances in the plot equaling the squared Procrustes distances we started with) obtains not only for distances from the average (the black dots) but also, to a very good approximation, for the distances between every pair of specimens.

Even better: for small variations of shape around an average, these shape coordinates accommodate *every* conventional shape variable – "small" variations in *every* ratio of interlandmark distances, *every* angle (or sine or cosine), and *every* index that combines these in further algebraic expressions. That means there is some linear combination of the shape coordinates that, in the limit of small shape variation, correlates perfectly ($r \sim 1.0$) with any distance, ratio, or index the ghost of Rudolf Martin might care to specify. This is just another **theorem** – for a more careful statement with proof, see Bookstein (1991).

So we have arrived at the variable values whose summed squared differences are the squared distances of the matrices that go into the principal-coordinates analysis of Section 6.5. Hence these **Procrustes shape coordinates** are exactly what we wanted: coordinates for the variations of shape within or between samples that match the Procrustes distances that are the real quantities driving the pattern analysis.[5] It is also

---

[5] *Warning.* These coordinates are not of full rank as a set of variables. Exactly or approximately, they satisfy four linear constraints (2D data), or seven (3D data), that follow from the standardizations carried out in the course of the Procrustes fits to the average. (For two-dimensional data, the coefficients in these constraint equations are just the first four rows of the matrix J, equation (7.1) in the next section.) So do not use the Procrustes shape coordinates themselves as predictors in regressions, canonical correlations or canonical variates analyses, or multivariate analyses of covariance. But anyway these techniques, which involve inversion of covariance matrices, are not of much use in toolkits like this one, which are designed for applications to organized systems where matrix inversion plays no role in fitting the more plausible models of how the system is actually behaving.

7.3 Procrustes Shape Coordinates and Their Subspaces

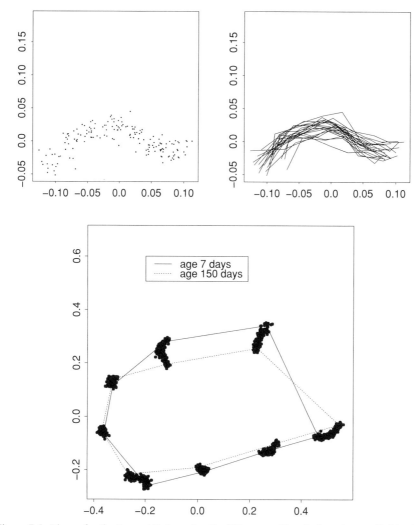

Figure 7.5. Figure for the Second Rule, using the Vilmann midsagittal rat data as digitized by Melvin Moss.

true that the principal components of these shape coordinates are the same as the principal coordinates of the distances. Hence,

The Second Rule of Morphometrics. Once you have the Procrustes shape coordinates, immediately draw *both* of the crucial scatterplots: the plot of the first two principal coordinates, Figure 7.5 (top), and the plot of the shape coordinates each around its own location in the average shape, Figure 7.5 (bottom). Your statements about the morphometrics of your sample have no authority until you have produced, labeled or colored, and examined these two central scatterplots. Both of these plots, by the way, need to be "squared," which means they have to have the same scale on the horizontal axis as on the vertical axis.

We exemplify this **Second Rule** in Figure 7.5. For this example, the First Rule was fulfilled at Figure 6.25. The design of the data set, as reviewed in Bookstein (1991), is a careful serial study of 21 male laboratory rats observed at eight ages that, like the dates of observation of viremia in the example in Section 6.2, are spaced roughly evenly on a scale of developmental time that appears nearly exponential when converted to calendar dates: ages 7 days after birth, 14, 21, 30, 40, 60, 90, and 150. The study is *longitudinal*, meaning that the same animals were observed at every occasion (and the structure is almost perfect: of the 168 combinations of animal by age, only 4 had any missing data at all; those four images were omitted from the analysis here). The data were digitized in a fixed coordinate system with lambda at (0, 0) and bregma directly to its left along the x-axis, but this makes no difference for any of the Procrustes steps.

Figure 7.5 is typical of the application of the Second Rule in that it affords the opportunity to annotate the standard images by the information beyond actual biological form that accounts for their role in the study design. At upper left in the figure is the standard scatter of the first two principal coordinates of Procrustes distance. There can be no doubt that these two dimensions are meaningful: the eigenvalues of the data set are 0.686, 0.065, 0.021, 0.015,..., meaning that the first two are distinct from one another and from the third and following. The scatter is evidently not elliptical, which would be in keeping with models for disorganized systems, but shows a structure that will concern us in several further figures. At upper right, the same figure is enhanced by 21 lines, one for each animal whose skull form is being tracked. We see three apparently outlying forms that represent individual states of three different animals, and thus no hint of a subtyping. At the bottom of the figure is the Second Rule's other mandatory display, the scatter of all the Procrustes shape coordinates separately. We see that at nearly all landmarks this scatter, too, is far from elliptical – indeed, most of the little scatters are reminiscent in their form of the single scatter of the first two principal coordinates at the upper left. This plot, too, can be enhanced by a simple overlay. As the information at the upper right probed the organization of the data set by animal, so the annotation here at the bottom displays the information by animal's age: the average form for the 7-day-olds, in the solid line, compared to the form for the 150-day-olds, dashed line. The effect of age is clear and systematic, and nearly exhausts all the information in the data set. (The end-to-end aspects of this age effect, straight from age 7 to age 150 days, were already drawn in Figure 5.9.) For biomathematical purposes, the animals in this classic data resource are nearly interchangeable. (They were not genetically homozygous, but they did arise from a purebred laboratory strain.)

Terminology. For essentially historical reasons (priority of the name in Bookstein, 1991), in the application to Procrustes shape analysis, the principal components of the shape coordinates are often called *relative warps*, and the scores (equivalently, the principal coordinates of the Procrustes distances) are sometimes called *relative warp scores.*

Is the Procrustes average form ever worth looking at? The sample Procrustes average, the configuration of black dots in Figure 7.4, might itself be of interest – maybe according to some hypothesis about an "optimal" form it should be seen to have some exact geometric characteristic, such as symmetry. In most applications there is no such a-priori geometric prediction. Instead the average is itself just another nuisance variable. It is the variations *around* the average, along with their dependence on the labels of the scatterplots (group, age, latitude and longitude, experimental conditions, etc.), that interest the student of organized systems. For both kinds of questions the variation around the average contributes to scientific explanations, but only for hypotheses about optimal form do the parameters of the average shape enter directly (the extent to which they support various hypotheses can be weighed by the likelihood-surface-based χ^2's of Chapter 5). Besides parameterizing the construal of "variations from an average," the average also affects the entries of the matrix J to be introduced below, and thereby, though only to a tiny extent, the information that is available for modeling shape after the standardization of position, orientation, and scale.

The isotropic Mardia–Dryden distribution. Corresponding to the geometry of the preceding three figures there is a statistical distribution, the **offset isotropic Gaussian distribution**, for which the probability density at any specimen is purely a function of its Procrustes distance from the population mean. This elegant shape distribution arises when each landmark point of your data varies around its own average location by noise that is of the same variance in every direction for every landmark independent of every other. Formulas for this distribution, which is also called *the Mardia–Dryden distribution with isotropic covariance matrix*, can be found in the textbook by Dryden and Mardia. If this model for probability is true, then Procrustes distance must be the best statistical quantity for conventional statistical inference questions, such as comparisons of group mean shape or correlations of shape or form with some claimed cause or effect of form.

I have never seen a real data set that is characterized by the extreme symmetry that this null (isotropic) Mardia–Dryden distribution demands. All real data have differences of variance by location or direction and local shape factors of greater or lesser scope that contradict the stark symmetry of the distribution *when stated as an empirical fact.* But that is not why we use Procrustes distance in a statistical context. Instead, we use this distribution to model our **ignorance of underlying biological processes prior to the data analysis**, rather as the Gaussian distribution was the best representation of ignorance for ordinary quantitative variables (Section 5.2). There will be another principle of ignorance, the one pertinent to the thin-plate spline, coming in at Section 7.5. The only exception to this a-priori ignorance of factor structure that is encountered at all often is the presence of *bilateral symmetry*, a topic we take up at Section 7.7.5.

In this statistical geometry of shape coordinates, zero correlation stands for a geometric property, orthogonality (perpendicularity). The following table, intended for the reader who has had some multivariate algebra, is the geometric decomposition of the vector space tangent to the non-Euclidean geometry underlying all the Procrustes machinery here. *Only in this subsection*, we refer to a **mean shape** μ on k landmarks $(x_1, y_1), (x_2, y_2), \ldots, (x_k, y_k)$ vectorized as $(x_1, y_1, x_2, y_2, \ldots, x_k, y_k)$

with $\Sigma x_i = \Sigma y_i = \Sigma x_i y_i = 0$, $\Sigma(x_i^2 + y_i^2) = 1$ (meaning: μ is centered, its Centroid Size is 1, and it has been rotated to principal axes horizontal and vertical). Write $\alpha = \Sigma x_i^2$ and $\gamma = 1 - \alpha = \Sigma y_i^2$ – the central moments of the mean configuration in its principal directions – and define $\delta = 1/\sqrt{k}$, $c = \alpha/\sqrt{\alpha\gamma}$, $d = \gamma/\sqrt{\alpha\gamma}$.
Consider the matrix

$$J = \begin{pmatrix} \delta & 0 & \delta & 0 & \cdots & \delta & 0 \\ 0 & \delta & 0 & \delta & \cdots & 0 & \delta \\ -y_1 & x_1 & -y_2 & x_2 & \cdots & -y_k & x_k \\ x_1 & y_1 & x_2 & y_2 & \cdots & x_k & y_k \\ cy_1 & dx_1 & cy_2 & dx_2 & \cdots & cy_k & dx_k \\ -dx_1 & cy_1 & -dy_2 & cy_2 & \cdots & -dx_k & cy_k \end{pmatrix}, \quad 6 \times 2k, \qquad (7.1)$$

and write J_i, $i = 1, \ldots, 6$, for the ith row of J.

- J is orthonormal by rows. (By the constraints on μ and the definitions of α, γ, and δ, JJ^t is a 6×6 identity matrix.)

For any small σ, sample configurations C of k landmark locations from the distribution $N(\mu, \sigma^2 I)$, treated as column vectors, and consider the various derived configurations

$$C^j = C - \Sigma_{i=1}^{j} J_i^t(J_i C), \quad j = 2, 3, 4, 6.$$

Geometrically, we have projected out 2, 3, 4, or 6 dimensions of the spherical distribution $N(\mu, \sigma^2 I)$ with which we began. Because J is orthonormal, each C^i is spherically symmetric within its subspace.

- C^2 is centered (has mean zero) in both even-numbered (vertical) and odd-numbered (horizontal) coordinates.
- C^3 is centered and has been rotated to a position of zero torque against μ.
- The fourth row of J scales each C to centroid size 1 (approximately, to second order in σ). (For dy small with respect to y, $(y + dy)^2 \sim y^2 + 2y \cdot dy$; y^2 is a constant, and the coefficient 2 goes away when we revert from squared Centroid Size to just plain Centroid Size, because $\sqrt{1 + \epsilon} \sim 1 + \frac{1}{2}\epsilon$ for small ϵ.) Hence C^4 is, approximately, the vector of Procrustes shape coordinates for the distribution C, written out as deviations from the sample average form.
- For small σ, because $\log(1 + \epsilon) \sim \epsilon$, C^3 is a rotation of the matrix that augments C^4 by log Centroid Size; hence it is a representation of the form space of C to be introduced in the next subsection.
- The fifth and sixth rows of J correspond to the shear and dilation terms in the uniform component of shape around μ. Hence $(J_5 C, J_6 C)$ is the uniform component of any configuration C, while C^6 is, approximately, the vector of nonaffine shape coordinates for the distribution C. The distribution C^6 has rank $2k - 6$. (The uniform component is often a useful component of explanations; it is introduced at Section 7.7.)

We can keep track of these projections row by row. Starting from $2k$ Cartesian coordinates in 2D, we lose two for centering (the first two rows of J), one for rotation (the third row), and one for scaling (the fourth row). Stopping here leaves us with a linear approximation to our familiar Procrustes shape coordinates. Much of the craft of a numerical inference consists in a judicious approach to standardizations of this sort, analogous to the physicist's craft knowledge of experiment design that Krieger (1992) talks about: to remove sources of irrelevant variation while preserving the signal that will be the ultimate subject of explanation. Projecting out the first four rows of J results in residuals optimal for searching over the descriptor space of *any* signal involving shape, with all patterns of shape change that have the same Procrustes length weighted equally. Other descriptors, such as ratios to a variety of standard lengths, can be proved to be distinctly less efficient than this reduction to shape coordinates. For the formalisms by which those proofs go forward, see, in general, Bookstein (1991).

The entries of the matrix J explicitly depend on the Procrustes average shape. The effect of this average on the covariance structure of the shape coordinates under the Mardia–Dryden model is the subject of the theorem in Bookstein (2009d).

7.4 Procrustes Form Distance

This is the most recent component (Mitteroecker et al., 2004) of the standard Vienna syllabus I'm reviewing here. Its principal applications are in studies of growth and in studies of group differences or time trends for which allometry (the dependence of shape upon size) is a confounding factor. Let us return for a moment to the equations in the previous section. The rows of J are mutually perpendicular, and so we can leave one in without affecting the consequences of projecting out the others. In particular, let us consider *omitting* the step that takes out Centroid Size, row 4 of the matrix J, equation (7.1).

It is not actual Centroid Size we were projecting out, because the average form has been scaled to unit size as the very first step in the notation. The projection along the direction of μ is (approximately) equivalent to dividing each Centroid Size by the grand mean Centroid Size before considering variation. But for any positive variable y of small variation, the derived variable $y/\bar{y}$ is nearly identical to the variable $\log y$, where logarithms, as always, are to base e. After this transformation we can discuss the relative amounts of variation of size vs. shape regardless of the actual centroid size of the configuration about which we are distributing these increments in x and y.

Instead of "not taking out Centroid Size," then, we can *put it back into the data*, in the form of deviations from the mean, by writing it back in, or rather its logarithm, as a column at the right of the matrix of Procrustes shape coordinates mentioned in Section 7.2. All the SVD-based statistics then go forward as before, except that the distance being interpreted now includes one more component that will let us properly accommodate size differences and size–shape correlations in our pattern analyses. The "presumption of ignorance" reviewed at Section 5.2 now extends to one additional clause, the absence of any prior knowledge about allometry (the dependence of shape

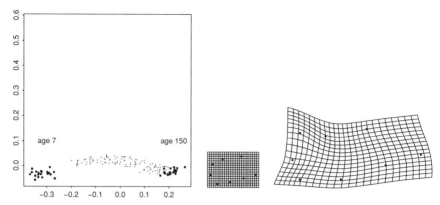
Figure 7.6. Figure for the Third Rule.

on size). *(Important reminder:* size is "put back in" by appending this extra column to the Procrustes shape coordinates, not by multiplying it into the shapes.)

After all its columns are centered again, the SVD of this augmented matrix is equivalent to a principal coordinates analysis of a squared *Procrustes form distance* that starts with squared Procrustes shape distance and adds in the square of the logarithm of the ratio of Centroid Sizes. Hence:

The Third Rule of Morphometrics. If size could possibly be an explanatory factor for your data set, then each principal coordinates plot like Figure 6.10 or Figure 7.5 needs to be augmented by a second principal coordinates plot in which Procrustes shape distance is replaced by Procrustes form distance, like the example in Figure 7.6.

Our illustration of the **Third Rule** of morphometrics (Figure 7.6) continues with the same Vilmann data resource we used in Figure 7.5. The inclusion of size as an additional descriptor of form further increases the domination of the first eigenvector of the descriptor system (eigenvalues are now 0.168, 0.023, 0.015, . . .) but only subtly alters the shape of the principal coordinate scatterplot, left panel. For instance, there is now a considerable gap separating the 7-day-old data from the 14-day-old data – even though the shape of the octagon here hasn't changed much, the size has apparently altered by about 15% (judging from the increment of about 0.15 on the horizontal coordinate, which is dominated by the natural log of Centroid Size). Otherwise the shape is the same, including the orthogonal drift in the direction of the second principal coordinate for the first half of the record followed by its restoration in the second half. To the right of this scatterplot I have drawn an added bonus figure pair, the thin-plate spline (to be introduced in the next section) corresponding to the first principal coordinate, at correct relative scale. Over the change from age 14 days to age 150 days, Centroid Size has changed in a ratio of 1.89. You can see how that compromises between a somewhat larger ratio in the horizontal direction and a somewhat smaller ratio in the vertical – we show later (at Figures 7.16 and 7.18) how this change is indeed directional at the largest two geometrical scales.

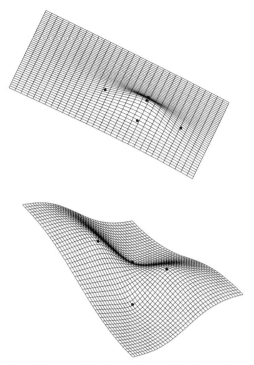

Figure 7.7. Two simple and interpretable shapes that a thin-plate spline might take.

7.5 The Thin-Plate Spline in 2D and 3D

The third major theme that makes geometric morphometrics work, complementing landmark selection and Procrustes distances and their principal coordinates, is a tool for making the geometry of its multivariate patterns explicitly visible in the picture plane (or space) of the typical form. The thin-plate spline interpolant we use for shapes was originally developed a few years earlier as an interpolant for surfaces over scattered points at which the height of the surface was known. In that context, the thin-plate spline is an interpolant that is linear in those "heights" and that minimizes a global figure of energy, *bending energy*, that is a quadratic form in the vector of "heights" with coefficients determined by the spacing of the landmarks in the starting form.

Prototypes. The thin-plate spline starts as a mathematical model for an actual metal plate, and so its mathematics is close to one's intuitive appreciation of the bending of an originally flat metal plate into the (real) third dimension (Figure 7.7). The figure shows two of the typical ways these grids can behave as if they applied to *real* metal plates with a square grid painted on beforehand. Both panels are perspective views of the grids that describe the movements of some landmarks against a background of others fixed in position. In the top panel, only the middle landmark of a quincunx (the five-spot of a die) is lifted, and the form of the surface is that of a bulge in the original plane. As a grid, the lines are moved on the page in a graded fashion

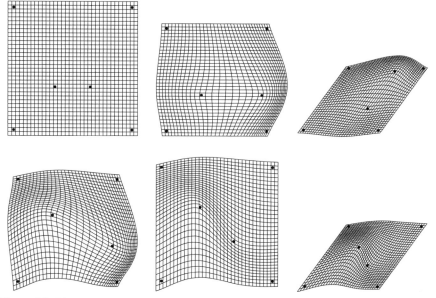

Figure 7.8. The TPS as a surface in four dimensions.

in the direction of displacement of the driving landmark. The bottom panel shows a fundamentally different configuration, the one called a "purely inhomogeneous transformation" in Bookstein (1991). Here one diagonal of a (projected) square has been lifted and the other depressed (on the page, but the illusion of a third dimension is irresistible). The surface here has alternating symmetry when rotated by 90° around its central "normal" (recall that the hyperbolic paraboloid, Section 6.1, that we use to diagram SVDs has the same property). The general spline combines features like these spanning every pair of landmarks using intentionally incompatible definitions of the direction that is "up" (Figure 7.8).

Formulas for the two-dimensional case (from Bookstein, 1989). From mathematical physics we borrow the idealized equation of a thin metal plate infinite in extent, infinitely thin, and uniform in its properties, displaced by small distances perpendicular to itself at a finite number of distinct points. Let U be the function $U(r) = r^2 \log r$, and let $P_i = (x_i, y_i)$, $i = 1, \ldots, k$, be k points in the plane. Writing $U_{ij} = U(P_i - P_j)$, build up matrices

$$K = \begin{pmatrix} 0 & U_{12} & \cdots & U_{1k} \\ U_{21} & 0 & \cdots & U_{2k} \\ \vdots & \vdots & \ddots & \vdots \\ U_{k1} & U_{k2} & \cdots & 0 \end{pmatrix}, \quad Q = \begin{pmatrix} 1 & x_1 & y_1 \\ 1 & x_2 & y_2 \\ \vdots & \vdots & \vdots \\ 1 & x_k & y_k \end{pmatrix},$$

and

$$L = \begin{pmatrix} K & Q \\ Q^t & O \end{pmatrix}, \quad (k+3) \times (k+3) \quad (7.2)$$

7.5 The Thin-Plate Spline in 2D and 3D

where O is a 3×3 matrix of zeros. Write $H = (h_1 \ldots h_k \; 0 \; 0 \; 0)^t$ and set $W = (w_1 \ldots w_k \; a_0 \; a_x \; a_y)^t = L^{-1}H$. Then the thin-plate spline $f(P)$ having heights (values) h_i at points $P_i = (x_i, y_i)$, $i = 1, \ldots, k$, is the function

$$f(P) = \sum_{i=1}^{k} w_i U(P - P_i) + a_0 + a_x x + a_y y. \quad (7.3)$$

This function $f(P)$ has three crucial properties:

1. $f(P_i) = h_i$, all i: f interpolates the heights h_i at the landmarks P_i.
2. The function f has minimum **bending energy** of all functions that interpolate the heights h_i in that way: the minimum of

$$\iint_{\mathbf{R}^2} \left(\left(\frac{\partial^2 f}{\partial x^2}\right)^2 + 2\left(\frac{\partial^2 f}{\partial x \partial y}\right)^2 + \left(\frac{\partial^2 f}{\partial y^2}\right)^2 \right) \quad (7.4)$$

where the integral is taken over the entire picture plane.
3. The value of this bending energy is

$$\frac{1}{8\pi} W^t K W = \frac{1}{8\pi} W^t \cdot H = \frac{1}{8\pi} H_k^t L_k^{-1} H_k, \quad (7.5)$$

where L_k^{-1}, the *bending energy matrix*, is the $k \times k$ upper left submatrix of L^{-1}, and H_k is the initial k-vector of H, the vector of k heights.

Partial warps are helpful visual interpretation of the nonzero eigenvectors of this matrix $H_k^t L_k^{-1} H_k$ that use them as coefficients for the displacements of the points P_i of the starting landmark configuration in any convenient direction on their plane. For k landmarks there are $k - 3$ of these warps, usually numbered in increasing order of their eigenvalue. Thus the "first" partial warp is the one of smallest eigenvalue, the one corresponding to the largest scale of bending. In Figure 7.7, the upper panel is one perspective view of the last (smallest-scale) partial warp of the quincunx configuration, while the lower panel shows is an analogous perspective view of the first partial warp here, which is also the sole partial warp of a simpler landmark configuration, a starting square. There will be another example of a first partial warp in Figure 7.18 of Section 7.7. The $k - 3$ partial warps span, in total, $2k - 6$ dimensions of the full space of the landmark configurations $P_i, i = 1, \ldots, k$. Two more dimensions of shape space are spanned by the fifth and sixth rows of the matrix J at equation (7.1), while the final four dimensions (the first four rows of J) are the four dimensions of the landmark data that are nullified by the Procrustes operations that project them down into this linearized version of shape space. All of these counts apply to two-dimensional data. In three dimensions, there are $k - 4$ partial warps that together span $3k - 12$ dimensions of shape space, and the uniform transformations span another five dimensions, while the Procrustes fit has approximately obliterated the remaining seven (three for centering, three for rotation, and one for Centroid Size).

In the application to two-dimensional landmark data, we compute two of these splined surfaces, one (f_x) in which the vector H of heights is loaded with the x-coordinate of the landmarks in a second form, another (f_y) for the y-coordinate. Then the first of these spline functions supplies the interpolated x-coordinate of the map we seek, and the second the interpolated y-coordinate. The resulting map ($f_x(P)$, $f_y(P)$) is now a deformation of one picture plane onto the other which maps landmarks onto their homologues and has the minimum bending energy of any such interpolant. In this context one may think of the bending energy as localization, the extent to which the affine derivative of the interpolant varies from location to location. Out of all the maps that do actually involve the right pairing of corresponding landmark locations, this particular spline is the map having the least possible amount of "local information" in this sense; this is the model of ignorance that the spline is aggrandizing (rather as a Normal distribution maximizes entropy subject to constraints of mean and variance). The affine part of the map is now an ordinary shear, which, in the metaphor of the lofted plate, can be thought of as the shadow of the original gridded plate after it has been only tilted and rescaled but not bent.

Figure 7.8 deconstructs all this diagrammatically. At upper left is a starting configuration, a square with two points eccentrically placed inside. At lower left is the final configuration, in which one of the inner landmarks has moved upward and the other has moved rightward. In the rest of the top row is the spline that handles the horizontal coordinate here: at upper right, as a surface in space (in perspective view) with the rightmost inner landmark *raised* – up is in the x-direction – and in upper center as a component of the final transformation (notice that the lines of equal y-value have not bent). In the rest of the lower row is the corresponding treatment of the vertical coordinate: at right, the perspective view of the surface with the leftmost inner landmark raised – now up is in the y direction – and at lower center the corresponding grid back in the picture plane; now it is the lines of constant x-coordinate that have not been bent. The final graphic, lower left, combines the x-coordinate from the upper center panel with the y-coordinate from the lower center panel. In fact it is a projected view of a surface in four dimensions. (By the way, it is easy to show that you get the same spline even if you rotate coordinates from a starting (x, y) system to one that is oblique.)

Because $W = L^{-1}H$ at equations (7.2–7.3), the spline is linear in the coordinates H of the right-hand form. In the application to the Procrustes construction for shape, the x- and y-coordinates we are using are actually displacements of shape coordinates after the Procrustes fit of the previous section. When the starting form of the spline is the sample average shape, then we have extended the linear machinery of shape space in $2k - 4$ dimensions to a system of exactly equivalent diagrams in two dimensions. For form space, we restore Centroid Size simply by scaling the grid appropriately. The relation of any specimen shape to the average, and the relation of any two shapes (such as group means) that are both near the average, are thereby shown as deformations with coefficients that are linear in the shape coordinates. In this way Cartesian grids deforming the average shape have been directly incorporated into the biometric framework, a formally linear function space of nonlinear maps. So we have arrived at a coordinate-free visualization of pairwise comparisons, shape coordinate singular vectors, or *any* pair of vectors of Cartesian coordinates. Because the spline is linear in the coordinates of the target form, furthermore, we can extrapolate it until its features are readable, or even reverse its sign. Practical applications of this possibility include Figure 7.9 and the example explored in Section 7.7.3.

Why this model of deformation? The thin-plate spline is far more helpful than anyone could have expected prior to its promulgation (by Bookstein [1989], borrowing from the literature of computer graphics in which it had been installed a few years before). It may be that minimizing bending energy may actually be a principle of biomathematical design. It is more likely, in my view, that our visual systems, which are so attuned to navigating over a landscape, are appropriately processing the gradients and other inhomogeneities that are communicated here for reasons probably buried deep in human brain evolution. If you doubt that these cognitive processes are hard-wired, try following Figure 7.7 with your eye as you rotate this whole book until the figure is upside-down. You will see each of the plates change its "actual" (three-dimensional) form as you turn, passing through a position that is difficult to interpret in 3D as any form at all. This may be a gridded version of the Helmholtz Paradox, a similarly noninvertible recovery of shape from shading in 3D. There is an extensive consideration of these and related speculations in Bookstein (2011).

There is no basis for claiming that these diagrams are truthful in any reductionist sense. Biological form need not, and probably does not, deform as a thin plate in four or six dimensions, and the spline needs to be replaced by an appropriately validated physical operator whenever real deformations are under the control of that operator in the application at hand. This means, in particular, that the thin-plate spline is not suited for diagramming real deformations, as of bone under strain, that are actually elastic. Its energy term has no component quadratic in the first-order directional derivatives of a map, which is the standard formalism of energy for modeling material elasticity. For processes like those, the thin-plate spline is minimizing the wrong quantity; it is not a substitute for a finite-element analysis. The spline's bending energy expresses *inhomogeneities* of strain, not strain itself – those are the squared *second* derivatives, not the first derivatives, that appear in the definition of bending energy, equation (7.4). But most comparisons of organismal form are across specimens or over long intervals of time and hence cannot benefit from the realistic physical models. For nonphysical comparisons, the spline is as good a graphic as we have. Also, it is more symmetrical in its algebra than schemes like finite-element analysis that require the form to be divided up into cells prior to the description and that assert identity of material properties of the contents of these cells across all pairs of forms in the data set. For general discussions of the relation between morphometrics and finite-element computations, see Weber et al. (2011), or Bookstein (2013a).

In 3D, the kernel function U of equation (7.2) is $|r|$, not $r^2 \log r$, the matrix Q has one more column for the z-coordinates of the landmarks in the target form, the integral has six terms instead of three and is over $\mathbf{R}^3$ (all space) instead of $\mathbf{R}^2$ (just a plane), and the constant in the bending energy equation has a minus sign. The corresponding algorithms are built into packages such as Edgewarp and the

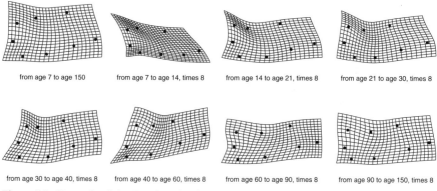

Figure 7.9. Example of the Fourth Rule of Morphometrics.

new *EVAN Toolkit* released in 2010 by a team headed by Paul O'Higgins, Hull–York Medical School, and the late Roger Phillips, University of Hull, UK, working with the present author and W. D. K. Green.

Hence, finally,

The Fourth Rule of Morphometrics: Use the thin-plate spline to draw all between-group contrasts of interest, each of the first few relative warps, and any within-group vectors of interest, such as the vectors produced by a Partial Least Squares (PLS) analysis of shape or form against its exogenous causes or effects. Carefully inspect these splines at a visually helpful extrapolation for hints of meaningful global or local processes.

Most examples in this chapter culminate in a display according with this Fourth Rule. The demonstration in Figure 7.9 exploits the longitudinal structure of the Vilmann data set to decompose the net shape change over 143 days, upper left, into its seven component segments, changes from age 7 to 14, age 14 to age 21,..., age 90 to age 150 days. (Each segmental transformation, but not the net change, has been extrapolated eightfold for legibility.) Clearly these are not all the same transformation. From age 7 to 14 the transformation is nearly uniform except for some local deformations near landmark IPP, the interparietal point. From 14 days to 60, over five consecutive waves of observation, the changes seem relatively consistent – the upper border is shrinking with respect to the lower – and then this process, which the literature called "orthocephalization," ceases. Over the entire 143-day period, furthermore, there is a steady reduction in the ratio of height to width of this structure.

7.6 Semilandmarks: Representation and Information Content of Curving Form

There were some additional dots in Figure 6.11 beyond the landmark coordinates. It is time to explain them.

Figure 7.10 shows the rest of the raw data for the forms in Figure 6.10: curves connecting various pairs of the landmarks of the earlier analysis along outlines of the sections made by the imaging plane within the original CT volumes. These total five:

7.6 Semilandmarks: Representation and Information Content of Curving Form 427

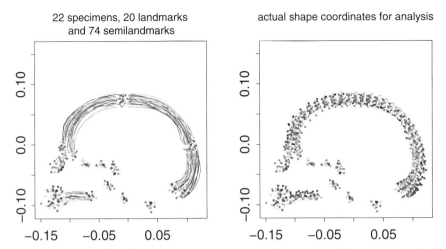

Figure 7.10. Semilandmarks represent curves in a scene for purposes of Procrustes analysis. Each specimen is in a randomly selected color (see Plate 7.10) or gray level. (After Weber and Bookstein, 2011.)

long arcs subdividing the outer and inner vault, an arc along the palate, and very short segments between Alv and ANS and between Rhi and Gla. The curves are made up of *semilandmarks*, a total of 74 of them. Semilandmarks are points that minimize bending energy (in the sense of the thin-plate spline, equation (7.5)) for deviations from the average while remaining along the curves upon which they were originally digitized. There is a general formula driving the computations here, the formula

$$T = -(U^t Q U)^{-1} U^t Q Y_0, \qquad (7.6)$$

which in most expositions, even primary peer-reviewed articles, appears as a naked dictum, a formula without any proof, without even any attempt to make it seem comprehensible. But it *is* comprehensible, and this section attempts to place it in an appropriate pedagogical context.

Like the thin-plate spline itself, this formalism, first introduced in Bookstein (1997a) and applied to data in Bookstein et al. (1999), has proved far more useful than anyone expected at the time of original promulgation. Thinking only in terms of the biomathematics of "shape measurement," we have radically altered the ground metaphor by extending the list of ignored variables well past the stereotyped "position, orientation, and scale" of the Procrustes methods to incorporate the full range of possible *relabelings* of points on curves as well. Over in the organized system we are studying, the biological form that is growing, evolving, or just getting on with the business of living, the semilandmark method bridges from the biomathematics of the Procrustes method, which deals with isolated points incapable of bounding biological extents, to the languages of morphogenesis and biomechanics, which have always dealt with bounded regions and the corresponding integrals of physiology or force generation that arise there. It is quite a surprise that the addition of one single elementary geometric operation so enormously expands the visual information resources available to the systems biologist who would retrieve as much information as possible from these ubiquitous images of organized form.

7.6.1 Sliding One Single Landmark

The method of semilandmarks rests on one principal algebraic fact the proof of which would distract from the main theme and is thus omitted here. Suppose you have two configurations of k landmarks in any two relative positions. The thin-plate spline from one to the other has some bending energy. If you add a new landmark, a $(k + 1)$st landmark, to the starting form, and place it down on the target form at exactly the point that the spline map on the first k landmarks would put it, then there is no change in the bending energy of the new $(k + 1)$-landmark spline. If, thereafter, you move this new $(k + 1)$st landmark in the target form, then the bending energy of the spline that takes it to any new position is proportional to the square of the distance of the new position from the original position that that landmark was assigned by the spline on the first k. The proportionality constant is a complicated function of the positions of all $k + 1$ landmarks in the starting form: high if the $(k + 1)$st is near any of the other k, and lower if it is far from all of them. For the sketch of a proof, see Bookstein and Green (1993).

We can apply this fact to geometrize the problem of placing that new landmark somewhere on some *line* in the picture: it translates into the straightforward task of finding the closest approach of a line to a fixed point not on it. In Figure 7.11, suppose that the location where the spline on the original k landmarks, whatever their configuration, puts the new landmark is $(0, 0)$ in this coordinate system, that this new landmark has actually been located at the point $(c, 0)$ (where $c < 0$ in the figure as drawn), and that it is free to slide on the line making angle θ with the x-axis, as shown. (In reality it will be sliding on a curve, not a line, but if the original location was a reasonably good guess, the line approximation will do, and we can always iterate the algorithm after projecting the line down to match the observed boundary curve or surface and repeating everything.)

We write the points on this line as $Y_t = (c, 0) + t(\cos\theta, \sin\theta)$, where t is an unknown adjustment parameter we have to estimate, and, because bending energy for one point at a time is circular, we are looking for the value of t at which the point Y_t is closest to the rest point $(0, 0)$. That is the point for which the distance d between the line of Y_t's and $(0, 0)$ is minimum. The square of the distance d is $d^2 = (c + t\cos\theta)^2 + (t\sin\theta)^2$, and the derivative of this with respect to t is $2\cos\theta(c + t\cos\theta) + 2\sin\theta(t\sin\theta) = 2c\cos\theta + 2t$, which we set to zero. (Remember that $\cos^2 + \sin^2 = 1$.) Solving for t, we get $t = -c\cos\theta$ and hence

$$Y_t = (c, 0) - c\cos\theta(\cos\theta, \sin\theta) = c(1 - \cos^2\theta, -\sin\theta\cos\theta)$$
$$= -c\sin\theta(-\sin\theta, \cos\theta),$$

a point in the direction $\theta + 90°$ out of $(0, 0)$ at signed distance $-c\sin\theta$ in that direction.

As θ rotates, the line through Y^0 in its direction rotates too, and the result of the basic sliding geometry in Figure 7.11 is to locate Y_t where the constraint line

7.6 Semilandmarks: Representation and Information Content of Curving Form

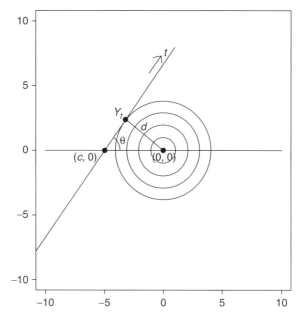

Figure 7.11. Geometry for one single semilandmark restricted to a line.

is tangent to the appropriate circle in each case: the one centered on $(0, 0)$, the landmark's rest position, and having radius $-c \sin \theta$. To be tangent to such a circle is to be perpendicular to the radius through the touching point, which makes the triangle on $(0, 0)$, $(c, 0)$, and the touching point a right triangle with hypotenuse from $(c, 0)$ to $(0, 0)$. By a theorem of Euclid, this locus (all the touching points that are generated as θ varies while $Y^0 = (c, 0)$ stays the same) is a circle on the segment from $(c, 0)$ to $(0, 0)$ as diameter.

This is formula (7.6) for the case of $T = t$, a vector with only one element. In Vienna's Department of Anthropology, the rule is that nobody can get a degree based on analysis of a data set involving semilandmarks who can't reproduce this specific derivation, verbatim, at the time of the General Exam.

7.6.2 Sliding Any Number of Coordinates

From this setup you can imagine an approach to the problem of sliding more than one landmark by sliding each of them separately in turn, over and over. This would be ludicrously inefficient, but it would actually work eventually, as in Figure 7.12. We are sliding two points now on two lines with two parameters t_1, t_2 that both need estimating. In the figure this bending energy is drawn by its contours of equal value – you can see that it has a minimum somewhere, and it is this point we are trying to find. We proceed by cycling over the pair (in general, the whole list) of sliding landmarks. At every step, we fix all the landmarks except one at their current positions, and apply the circle construction to the one that remains free to move. That construction will find the location of least bending for the parameter in question given the current

430 7 Morphometrics, and Other Examples

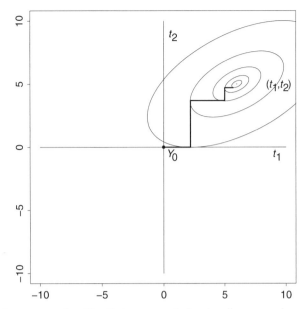

Figure 7.12. A very poor algorithm that can nevertheless handle any number of SLs.

positions of all the other semilandmarks. (In the figure, we optimized t_1 first while t_2 was fixed at its starting value of 0.) We then fix this landmark at its new position and free the next one to slide; in the figure that is t_2, which gets optimized for the just-computed value of t_1.

With only two sliding landmarks, the first cycle is complete. We fix the second one now and free the first to slide to a new value of t_1, as shown; then again t_2, back to t_1, and so on. Evidently the algorithm converges to the pair (t_1, t_2) that together have the lowest value of bending energy. (It *must* converge, because the bending energy drops at every step and there is only one minimum for the actual quadratic form we are using.)

The formula $T = -(U^t Q U)^{-1} U^t Q Y_0$, equation (7.6), is a shortcut that computes this minimum value without the iteration, by sliding on all landmarks at once to arrive at the (unique) position of tangency $U^t Q(Y^0 + UT) = 0$ to some ellipsoidal hypercylinder of constant bending energy $Y^t Q Y$. The formalism is now of many sliding points instead of just one, so what was distance from $(0, 0)$ is now bending energy $Y^t Q Y$ (for some bending energy matrix Q) over the set of target forms $Y = Y^0 + UT$ where Y^0 stands for the starting form in a notation where a form is not a matrix of landmark numbers by Cartesian coordinates but instead a single vector: not

$$Y^0 = \begin{pmatrix} x_1 & y_1 \\ x_2 & y_2 \\ \cdot & \cdot \\ \cdot & \cdot \\ x_k & y_k \end{pmatrix} \quad \text{or} \quad Y^0 = \begin{pmatrix} x_1 & y_1 & z_1 \\ x_2 & y_2 & z_2 \\ \cdot & \cdot & \cdot \\ \cdot & \cdot & \cdot \\ x_k & y_k & z_k \end{pmatrix}$$

7.6 Semilandmarks: Representation and Information Content of Curving Form 431

but

$$Y^0 = (x_1 \, x_2 \, \ldots \, x_k \, y_1 \, y_2 \, \ldots \, y_k)^t \quad \text{or} \quad Y^0 = (x_1 \, x_2 \, \ldots \, x_k \, y_1 \, y_2 \, \ldots \, y_k \, z_1 \, z_2 \, \ldots \, z_k)^t.$$

UT needs to be a vector conformable with Y^0 (meaning: capable of being added to Y^0 – having the same shape), and T has one entry per sliding landmark, so U must be a matrix of $2k$ or $3k$ rows by one column for each landmark that is going to slide, a total count, say, of j of them. The first column of U talks about the effect of sliding the first landmark, so it probably has entries only in positions 1 and $k + 1$, which together specify the x- and y-coordinates of that first landmark (for 3D data, there is an entry in position $2k + 1$, for z_1, as well). The second column of U probably has entries only in the second and $(k + 2)$nd positions, as it is responsible mostly for x_2 and y_2 (but for 3D also position $2k + 2$ associated with coordinate z_2); and so on.

In this notation, $Y = Y^0 + UT$ is the configuration of all the semilandmarks *after* sliding, and we want to know how far to slide each one. Each semilandmark slides by its own parameter t_i times the unit vector $(U_{i,i}, U_{i+k,i})$ or $(U_{i,i}, U_{i+k,i}, U_{i+2k,i})$ that gives us the direction it's sliding in 2D or 3D. (The letter U comes from the word "unit" – in one version of this algorithm, the U are direction cosines [cosines of angles with the two or three Cartesian axes], the squares of which total 1.0.)

To work in the same spirit as for the circle, we need to use the statisticians' notation of a derivative with respect to a vector parameter like T. This means the vector made up of the derivatives with respect to every single entry of T. These need to all be 0 at the same time. The semilandmarks must slide to the points $Y^0 + UT$ at which

$$\frac{d}{dT}(Y^0 + UT)^t Q (Y^0 + UT) = 2 U^t Q (Y^0 + UT) = 0.$$

That last 0 was a *vector*, a list of one 0 derivative for each sliding parameter t_i in T. So this is really a set of j equations, one for each sliding parameter even though each equation involves all the sliding parameters simultaneously. As we have already seen for the similar optimization task that arises in multiple regression, the matrix equation can be verified one slot at a time just by exhaustive enumeration and collection of terms.

But we can solve this equation for T: $U^t QUT$ must equal $-U^t Q Y^0$, so **the sliding parameter vector T we seek is, again,**

$$T = -(U^t QU)^{-1} U^t Q Y^0, \tag{7.6}.$$

This is the "magic formula" that puts all the sliding landmarks into their appropriate places in one fell swoop, as my mother used to say (though not in this context).

For the circle, Figure 7.11, we had $Y^0 = \binom{c}{0}$, $Q = \begin{pmatrix} 1 & 0 \\ 0 & 1 \end{pmatrix}$, that is, $(x \; y) Q \binom{x}{y} = x^2 + y^2$, $U = \binom{\cos \theta}{\sin \theta}$, and so $U^t QU = \cos^2 \theta + \sin^2 \theta = 1$ and thus $T = t_1 = -(U^t QU)^{-1} U^t Q Y_0 = -(\cos \theta, \sin \theta) \cdot (c, 0) = -c \, \cos \theta$, which is the right result.

The **interpretation** of the formula $T = -(U^t QU)^{-1} U^t Q Y^0$ flows likewise from the previous version $U^t Q (Y^0 + UT) = 0$. However many elements the vector T has, this is the statement of the same geometry that we immediately understood in Figure 7.11 when it applied to one sliding landmark at a time. The equation $U^t Q(Y^0 + UT) = 0$ states that at the point $Y^0 + UT$ at which we have arrived after the sliding, none of the allowable changes (shifts of the sliding parameters T applied to the point locations via the geometry coded in U) have any effect on the bending energy (for small changes – the concept is of the *gradient* of bending energy, its deviations near the optimum). That $U^t Q(Y^0 + UT) = 0$ means that at the correct T the whole space of all possible slidings is tangent to the ellipsoidal hypercylinder representing the locus of constant bending energy (that's a mouthful; just think "circle"), and *that* means that we must have arrived at the global minimum we were seeking.

This single formula $T = -(U^t Q U)^{-1} U^t Q Y^0$ applies to the curving data of the left side in Figure 7.10 to convert each form from its previous 20-point configuration to the new 94-point configuration shown on the right. (Each form is in a different, randomly selected color.) In this way we have brought in much of the rest of the information in each image: not the information having to do with gray levels or colors or textures, but at least the information pertaining to the locations of boundary structures and curves that display the interations of form with function much more evocatively than landmarks can.

For a great deal more on semilandmarks, including more specific formulas for the 3D case, see Gunz et al. (2004). That reference explains, in particular, how straightforward it is to slide points on surfaces – one simply assigns two of these parameters t to each point, one for each of any pair of directions it can slide within the tangent plane of the surface to which it is bound. Once they have been located by this algebra, semilandmarks, whether from curves or from surfaces, are treated in exactly the same way as landmarks are for all the manipulations of the morphometric synthesis, and that is just what we are about to do here.

The display in Figure 7.10 has considerable graphical density, and a slightly different superposition than that for the landmark points only. But the statistical summary diagrams of the sample of 22 specimens are not much affected – all those additional points have not actually added much information to our understanding of the system under study here. The upper left panel in Figure 7.13 plots the first two principal coordinates (relative warp scores); this result is close to the plot in Figure 6.7, and similarly for form space, lower left, versus the previous ordination on landmarks only (lower right). At upper right in the figure is the grid for the first principal coordinate, the horizontal axis in the plot to its left; it closely resembles the one in Figure 6.9 based on the 20 landmarks only. This is often the case for analyses of data in either 2D or 3D, and should be taken as quite reassuring. For instance, in terms of raw Procrustes distances, the two versions of those distances share $(0.976)^2$ of their summed squared variation around 0. The classic comparative anatomists did pretty well at placing the tips of their calipers on points that, collectively, extracted most of the information needed for at least a qualitative understanding of the vagaries of form in the taxa they were studying. What we are using the SLs to visualize here may have been in the mind of the expert all along. In the presence of enough point landmarks, the semilandmarks add a great deal of verisimilitude but evidently not much actual information.

7.6 Semilandmarks: Representation and Information Content of Curving Form

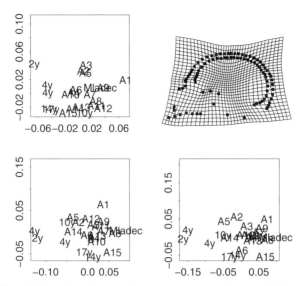

Figure 7.13. The addition of semilandmarks does not much alter our SVD-based descriptors. (Upper left) First two principal coordinates of shape for the data in Figure 7.10. (Upper right) The first principal coordinate as a thin-plate spline. (Lower left) Principal coordinates of form. (Lower right) The same for just the 20 landmark points. Specimen labels are as in Figure 6.7.

7.6.3 Organizing the Semilandmarks in Practice: Working from a Template

The method of sliding semilandmarks may well account for the bulk of computer cycles used and entropy generated by the entire "Vienna school of morphometrics" considered as a human community. Our students have grown accustomed to challenging the R statistical package with data sets of upwards of a thousand coordinates. These could not possibly represent conscientious individual measurements form by form; they must have been generated automatically by a system such as what has just been reviewed. The key preparatory step is to set up an initial digitizing scheme, or *template*, with extreme care.

The formalism we are using here is algorithmic, not algebraic. The implementation of any point–curve–surface semilandmark scheme involves orchestrated sequences of a modest variety of image matching/warping operations. These basic steps are construction of privileged planes (symmetry plane, nongeneric tangent planes), intersections with planes, location of points, location of ridge curves, warping by partial information, and pointwise projection post-warp orthogonally onto curves or surfaces.

The example here is a template-driven configuration of 15 point landmarks, 127 curve semilandmarks, and 273 surface semilandmarks from a study of the surface of the growing human mandible as seen in CT scans of living specimens. Originally published as Bookstein, 2008, it relies on data belonging to Dr. Michael Coquerelle and was discussed at greater length in its appropriate biological context in Coquerelle et al. (2010). The implementation was in Bill Green's program package `edgewarp`.

Otherwise explicit formulas become implicit via a "digitizing flowchart," a sequential set of instructions – a constructive morphometric geometry – according to which a template is very carefully constructed on a single a-priori "typical" specimen. The same constructions control the task of digitizing every additional specimen. In this setting, the result of a digitization is a carefully controlled warp of the entire template into the space of each target specimen in turn. Figure 7.14 shows a variety of views of this one single template. In the color figure, the surface extracted from the original CT scan is colored red. The green crosses with labels are conventional landmark points. The white polylines are curves, and the green crosses without labels are curve semilandmarks or surface semilandmarks. Landmarks are or are not bound to curves, which in turn may or may not be bound to surfaces, according to incidence properties to be reviewed later.

The constructive geometry begins with two specific planes:

P1. A subjectively located "midsagittal plane" around which the surface is most nearly symmetric as viewed by eye. For instance, this plane passes through the saddle in the alveolar ridge between the two lower front incisors. It is *not* the mirroring plane generated by Procrustes superposition of the form on its own mirror image, because the average normal mandible is not quite bilaterally symmetric.

P2. A best-fit plane tangent to the upper margin of the mandible. This plane is typically in approximate contact with the mandibular surface at four separate points CorL, CorR, and the two condylar tops ConT shown. These structures typically lie very close to a plane owing to the symmetry required for efficient biomechanics of chewing. They comprise our first four landmarks.

From the interaction of P1 with the surface we extract one particular curve:

C1. The *symphysis* (drawn as a closed loop in the figure) is the plane curve cut by the midsagittal plane P1 upon the surface.

This curve includes two further landmark points extracted by hand once this structure is visible: Inf1, Inf2, lingual and buccal vertices of the curve C1 near the incisors.

It is convenient at this time to augment the list of landmarks by four additional points MenL, MenR, MandL, MandR, the paired mandibular foramina, which float in space over the apertures of those foramina where the corresponding tangent plane is actually "missing." Also produced are the points ConRL, ConRM, ConLM, ConLL where the surface normal of the condyle is perpendicular to the symmetry plane P1.

Following the general constructive theory of ridge curves (cf. Koenderink, 1990) we extract three curves of this type by inspection of the corresponding normal sections.

C2. The *mandibular border* is the ridge curve along the bottom of the mandibular bone, from the midsagittal outward in both directions as far as it can be extended. On the *template*, production of this polyline is by careful handwork spacing forward along the tangent by distances inversely proportional to curvature and inspecting normal sections to ensure that the candidate point is a vertex of every curve of normal section.

7.6 Semilandmarks: Representation and Information Content of Curving Form   435

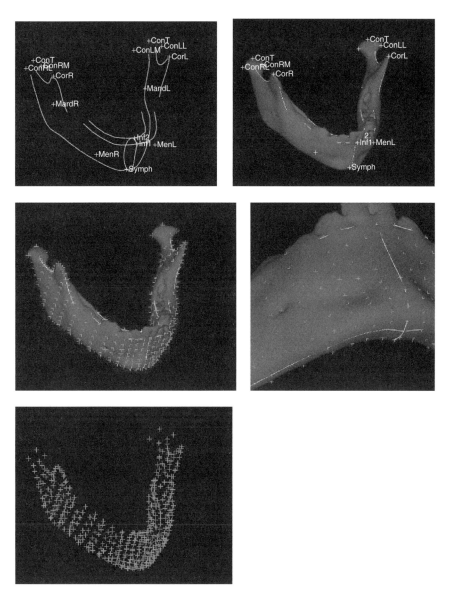

Figure 7.14. Example of an extended semilandmark scheme. (Top left) Curves and landmark points. Plane P1 of the text is implied by its closed curve of intersection with the symphysis; plane P2, spanning the tops of coronoids and condyles, is not shown. (Top right) Curves on the anatomical surface. (Middle left) Adding surface semilandmarks. (Middle right) Detail at Menton. (Lower left) The data as received by the Generalized Procrustes algorithm that would follow. (Figure from Bookstein, 2008; its data courtesy of Michael Coquerelle, see Coquerelle et al., 2010.) Figures like these are typically published in color, as in Plate 7.14.

Because the bilateral asymmetry of the mandible is one frequent topic of investigation, the morphometric geometry requires symmetric rosters of semilandmarks. In practice, one places them sensibly on one half of the form, then warps the form onto itself via reflected relabeling of landmarks and finally projects from the warp to the nearest point of the corresponding (sub)manifold on the other half. In effect, the "real" template consists only of one hemimandible (including the unpaired curve, C1); the other half is digitized as a warp of the first, hence at constrained spacing.

C3, C4. Typically each coronoid point is quite nearly upon the *anterior ramus ridge curve* that runs from just behind the last molar to just in front of the condyle. We extract these curves, left and right. Again, for the template, the second of these curves is created by pointwise projection of the reflected relabeled warp of the first, the one that had been digitized at free spacing.

The remaining landmark point Symph is produced as the intersection of the curves C1 and C2, a plane curve and a ridge curve. It is close to the point that orthodontists and craniofacial biologists often name "Menton."

There remain two more curves, the *alveolar curves* C5, C6, which run along the rim of the alveolus where the erupting teeth cut it (always near the ridges of the alveolar surface itself, structures that are visible only where the teeth are absent). These are thus not ridge curves, but ordinary polylines made up of variable numbers of discretely sampled touching points. They will eventually be resampled like any other set of curve-bound semilandmarks.

Finally, the surface points of one side of the template, spaced pleasantly in a quasi-grid arrangement, are paired to their antimeres by projection downward from the warp driven by the reflected relabeling of all the landmark points and curves.

Thus the template. The application of this scheme to any other form proceeds in the same conceptual order interweaving planes, points, curves, more points, more curves, and surfaces. To digitize is to interweave point specification, warping, and orthogonal projection in arbitrarily complicated but fully specified sequences. Data digitized at any step constrain the forward warp of the template that applies to all subsequent steps. The digitizing is not order-independent, and so invariance of the scripted order is mandatory. At any stage, ridge curves are warped in from the template and then projected "down" to the visible ridges of the target surface (this is a simple operation given access to arbitrary surface sections, in this case, normal sections along the tangent to the ridge curve in question). Ridge curves can also be digitized by subdivided updating of the warp, beginning with the midpoints, then the quarter-points, and so on. Finally, after the surface semilandmarks are warped into range (so that projection variability is attenuated) by a spline driven by landmark points and curves, they too are projected all at once, down onto the target surface. The normal components of these projections are evaluated independently, so we are no longer precisely optimizing any global figure such as bending energy.

In this way the constraints underlying any particular semilandmark scheme remain implicit in the digitizing sequence. All of our advance tacit knowledge – the symmetry of the form, the four-point plane of upper paired structures, the reliable ridge curves, the special treatment of the alveolar ridges – must be explicitly coded in two sets of instructions, one for the template and one for its warping onto the target. Symmetry is ubiquitous. It is handled by explicit construction of a midsagittal plane (that is for orientation, not for mirroring, as already mentioned), by explicit indexing of paired versus unpaired landmarks, and by explicit symmetrizing of semilandmarks that are themselves paired. The forward warp is constructed over and over: first from the landmarks that are easiest to locate without the warped template, then from the foramina and the semilandmarks on the condyles, then from the last landmark on C1, and finally from all the other ridge curves. The constructions keep carefully separate the incidence relations of points, curves, and surfaces. Ridge curves, of course, are necessarily on the surface. The alveolar curves are not ridge curves, but polylines.

In this way the scientist has almost unlimited freedom to adapt the formal schemes of morphometrics to prior knowledge of what aspects of the morphology under study are reliable, problematical, salient, or just plain interesting.

7.7 Putting Everything Together: Examples

This extraordinarily flexible mensuration scheme nevertheless submits to the singular-value approach to signal extraction reviewed in the preceding chapter. The four Rules of Morphometrics combine with all the pattern analysis possibilities to underwrite a remarkably wide variety of scientific explorations. Each analysis combines a choice of a Procrustes distance (shape-only or size-and-shape), a list of landmark/semilandmark locations, a scatterplot of Procrustes shape coordinates, another scatter of principal coordinates labeled or colored by everything else you know in advance from textbooks, specimen labels, or other measurements, and some thin-plate splines visualizing the interesting vectors that apply to the Procrustes coordinates as they emerge from principal coordinates or principal components analysis, comparisons by experimental group or taxonomic group or any other labeling variable, or Partial Least Squares (PLS) analysis against nonmorphometric quantities, like function, that happen to be correlated with form. We're looking for clusters, clouds, submanifolds, trends, or outliers in the ordinations produced by the pattern engine; at the same time, we are looking for spatially coherent patterns and scales in the deformation grids.

Here are five intentionally diverse examples combining the specific morphometrics formalisms of this section with the pattern analyses preceding it in this chapter into persuasive grounds for numerical inference in a variety of contexts.

7.7.1 Allometric Growth

This example deals with the data set of Figure 6.10, extended to semilandmarks as in Figure 7.10. Because the context is one of a growth study (recall that there are six children in the reduced data set, as well as 16 adults), we choose to work in form space, where squared distance equals ordinary squared Procrustes shape distance plus the square of the logarithm of the ratio of the pair's Centroid Sizes (square roots of second central moments). The principal coordinate analysis finds eigenvalues of 0.0526, 0.0279, 0.0206,... and hence one dominant component, which can be

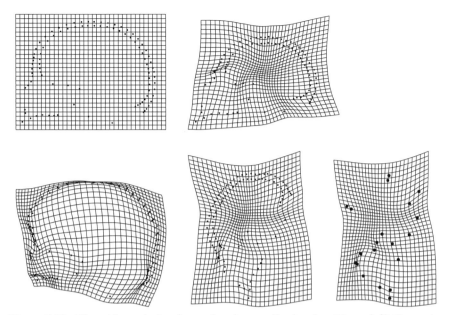

Figure 7.15. Allometric analysis of growth using semilandmarks. (Upper left) Procrustes average of the 94-point representation (20 landmarks, 74 semilandmarks). (Lower left) In the direction along form space PC1 given by smaller size. (Upper center) In the direction of larger size. (Lower center) The same, rotated to the craniofacial auxologists' preferred orientation. (Lower right) The same for the form space on the 20 landmarks only.

interpreted as a size allometry inasmuch as its loading for log Centroid Size is 0.901. Figure 7.15 represents allometric shape as the shape coordinate part of this same first principal coordinate – in effect, shift of each shape coordinate in proportion to its own covariance with log Centroid Size. At lower left is a deformation in the direction of one extreme on this coordinate, evidently corresponding to an exaggerated embryonic stage; at upper center is the opposite deformation, which indicates relative shrinkage of the braincase against the maxillary apparatus. The analyst is free to set the orientation of a grid like this in any convenient way. At lower center, the same deformation is drawn to a different grid aligned with the orthodontist's typical *growth axis* from Sella turcica toward Prosthion. This figuration of growth allometry based on the additional 74 semilandmarks differs hardly at all from the corresponding diagram from the information in the 20 landmark points only, lower right. Chapter 4 of Weber and Bookstein, 2011, explains why this is the most suggestive orientation for interpreting this growth at the largest scales.

7.7.2 Simplifications of Grids

Figure 7.5 showed the first two principal coordinates of the Vilmann rat skull data, and Figure 7.9 presented thin-plate splines for various age intervals, but to this point we have not visualized the actual axes of the scatterplots in Figure 7.5, the principal coordinates themselves. These are shown in the left column of Figure 7.16. Relative warp 1, which increases steadily with age, looks quite a bit like the simple contrast of average 7-day-old against average 150-day-old already shown in Figure 7.5. Relative warp 2 rises from age 7 to age 30 days and then falls back.

The pattern at upper left in Figure 7.16 is strikingly similar to the pattern in Figure 7.17. This is the first *principal warp* of the bending-energy matrix for this configuration of landmarks, the eigenvector of smallest bending per unit Procrustes distance in the space C^6 of purely nonuniform transformations mentioned in Section 7.3. Figure 7.17 shows this in two orientations, first with the bending "straight up," a particularly suggestive pose, and second in the orientation that corresponds to the application here. In fact this pattern accounts for more than half of the nonuniform signal all by itself.

Given this coincidence, it is instructive to plot the individual configurations of the data set in this simplified space of two descriptors only, the uniform term (completely nonlocalizable deformation) and this largest-scale feature. We see this in the center column of Figure 7.16 for all 164 radiographs and in the right column just for those at ages 7, 30, and 150 days. The display is quite suggestive. The pattern of largest-scale bending, top row, is indeed quite systematic from end to end of this growth sequence, and the trend and then reversal of the uniform term is not only nearly perfect but also shows a striking decrease in variance, followed by increase, that accounts for the hairpin in Figure 6.25, the PCO of covariance structures for the same data. At right in Figure 7.17 are the two components of that uniform term, the shear (upper panel) that progresses from age 7 to age 30 days and then reverses, and the horizontal dilation, which steadily decreases relative height throughout the entire growth period.

We can look at these two terms in detail over the two successive changes scattered there. At left in Figure 7.18 are the uniform terms for the mean changes from 7 to 30 days (top) and from 30 to 150 days (bottom). Evidently the shear component is reversed while the horizontal extension persists. By comparison (center column), the component of the net change along this largest-scale partial warp is nearly identical for the two subdivisions of growth. (For the definition of the partial warps, see at equation (7.5), or consult Bookstein, 1991.)

In my judgment, this is the context in which currently fashionable studies of "integration" or "modularity" should be pursued: not as matters of correlation among shape coordinates, but as a matter of the scale of the patterns extracted by the more sophisticated engine, the SVD.

Two-point shape coordinates. The uniform term just introduced is a bridge between the current technology of Procrustes shape coordinates and a somewhat different version that I introduced a few years earlier: the method of *two-point shape coordinates* (Bookstein, 1986) that combined with thin-plate splines as the central formalism of my monograph on the subject, Bookstein (1991). These coordinates are named "Bookstein coordinates" (see Stuart and Ord, 1994, p. 279) because, after Stigler's Law of Eponymy, Bookstein was the second person to discover them. The first was Sir Francis Galton, also discoverer of weather maps, fingerprints, and the correlation

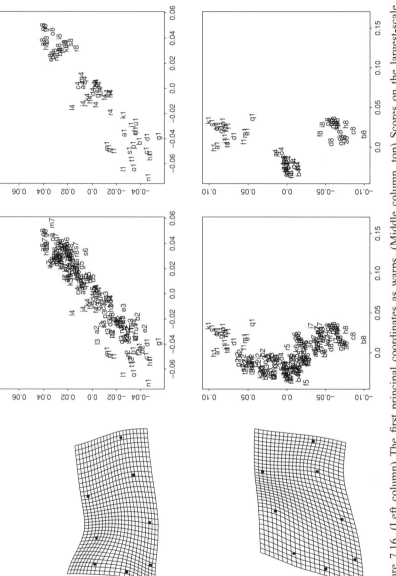

Figure 7.16. (Left column) The first principal coordinates as warps. (Middle column, top) Scores on the largest-scale nonlinear feature of this shape space, the first partial warp, for all 164 configurations. (Upper right) The same for ages 7, 30, and 150 days only. Note the momentum. (Lower row, left to right) The second principal coordinate as a warp appears mainly uniform; scores for all 164 configurations; for ages 7, 30, and 150 days only. Note the reversal of the horizontal coordinate, which the next figure will show to be a shear. Individual specimens are labeled by a letter (a through u for animal numbers

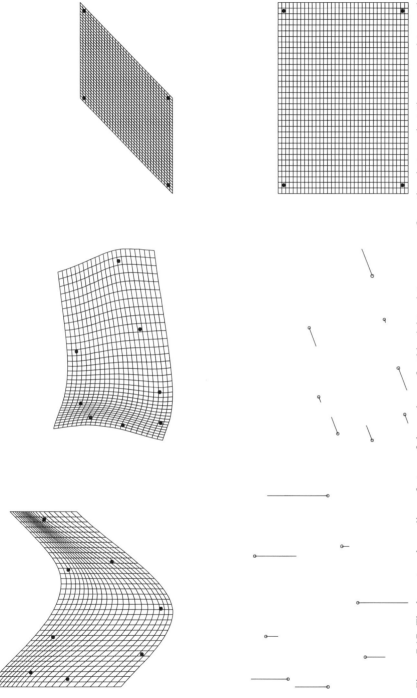

Figure 7.17. The largest-scale nonlinear feature of shape change for this eight-point average configuration, in two orientations: upward, top and bottom left; aligned with the data, top and bottom center. (Right column) The two coordinates of the uniform term, rows 5 and 6 of the matrix J, equation (7.1). (Top) x-coordinate (shear). (Bottom) y-coordinate (vertical–horizontal reproportioning).

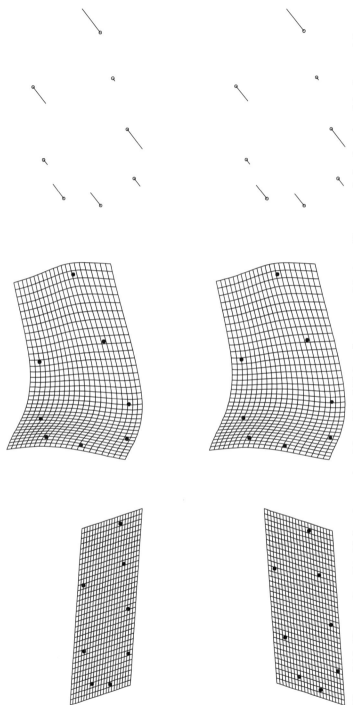

Figure 7.18. Changes from 7 to 30 days (top row) and from 30 to 150 days (bottom row) in the uniform term (left column) and the first partial warp (middle and right columns) are respectively reversing and continuing.

coefficient, in a *Nature* note of 1907 that is most easily viewed on the corresponding pages of Karl Pearson's hagiographic four-volume biography *The Life and Labours of Francis Galton*, at volume 2, pages 325–326. Galton meant them for use in biometric identification (as it was in those days), the recognition of criminals by matching their profiles across photographs.

The two-point shape coordinates do not conform to the interplay between Procrustes distance and Procrustes shape coordinates. Their principal components are not meaningful, and they do not serve as principal coordinates of any meaningful distance function. Of all the uses reviewed in chapter 6 of Bookstein (1991), only one survives into the era of Procrustes: their role in simplifying reports of uniform or nearly uniform transformations. In this special case, the uniform term can be translated into a *tensor*, a representation by an ellipse having precisely the same number of free parameters (three, for two axis lengths and an orientation; compare the three for two uniform components and a Centroid Size). See Rohlf and Bookstein (2003). The tensor formalism has a direct verbalization in terms of "directions over the form that show the greatest and the least ratios of change," directions that must be perpendicular, and sustains further suggestive interpretations in terms of interlandmark segments, perpendiculars, angles, or angle bisectors with which these principal directions align. There is an example of the reduction to these descriptors at Section 4.5.2.3 of Weber and Bookstein (2011).

7.7.3 Focusing on the Interesting Region

Figure 5.16 in Section 5.2 introduced a configuration of 13 landmarks for the purpose of exemplifying a model of self-similar noise. In the course of that discussion, Figure 5.17 showed a first principal component against that unusual model of disorganization. We noted that the phenomenon here was, more or less, "pinpointed" – it is at very small scale – but we did not talk about what it might have represented in terms of the biology of the data under analysis.

Actually the data here came from a two-group design, doctors against schizophrenic patients (although not the data set used for the study of the callosal outline, Section 5.3.3). At upper left in Figure 7.19 are the Procrustes shape coordinates for these 28 configurations, and at upper right the plot of group average shape coordinates that conveyed the finding in the original publication. At lower left is the thin-plate spline corresponding to this finding. You probably didn't see it before, but you can see now, that there is a visually apparent region of light inking (a bulge) in the grid as presented there. This exemplifies one of the principal advantages of the spline: you see that bulge without looking for it – you cannot avoid attenting to signals generated by your visual cortex that evolved eons ago for the preservation of small binocular primates wandering in a dangerous landscape.

The processing of this grid deformation is a hybrid method, part algorithmic (the spline) and part cognitive (detection of that bulge). It would be an advance to be able to bundle these together – to build a version of the thin-plate spline that functioned as a pattern engine for detecting focal intensifications of shape change like these.

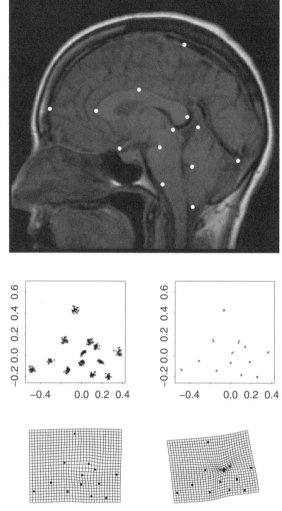

Figure 7.19. Adapted from Bookstein (2000). (Top) Layout of landmarks on the midsagittal brain image (quality circa 1994). (Center left) 28 sets of 13 shape coordinates. (Center right) group means for doctors (+) and patients (×). (Lower left) Thin-plate spline from doctors to patients. The image is suggestive. (Lower right) -2.303 times the transformation at left, in a grid rotated to simplify the detected feature even further. The focus – a *crease* where one derivative is precisely zero – is now obvious, as is the absence of any other features in the pattern.

Such a method was published at the turn of the century as Bookstein's (2000) method of *creases*. Local extremes of deformation grids like these can be characterized by a specific *catastrophe*, the word being used in a specific mathematical sense, and a specific grid orientation that renders their interpretation clearest. For the case at hand, that preferred presentation corresponds to a *reversal* of the transformation followed by an intensification in a ratio of 2.303.[6] This makes the extreme of that directional stretch fall precisely to zero, the overwriting of two consecutive grid lines, with no such pinching anywhere else in the image. There results the scene at lower right in Figure 7.19, which unequivocally shows where the signal is that corresponds to the group difference under study (and, in passing, suggests that there is basically no other information to be had anywhere else in this diagram).

This grid is a startlingly close match to the grid in Figure 5.17, which was produced under entirely different assumptions back in Section 5.2. There, we were simply looking for the smallest-scale feature of a data set, in the sense of the feature that showed the most excess shape variation regardless of scale. Here, we were seeking a specific description of the difference between the average over the schizophrenic subgroup and the average over the others. That these are so similar is a powerful abduction indeed, one leading to a range of speculations (why is the group difference so focused? is it a disease effect or instead a drug effect?) and thereby a great range of potential follow-up studies.

An analogous investigation of Vilmann's rat data shows that, when one searches for the region of *slowest* growth over the age interval (14–40 days) of most homogeneous growth according to the grids in Figure 7.9, every one of the 20 rats with data at those two ages shows a minimum in very nearly the same location. This description (Figure 7.20) is, again, startling. It is a far more precise signal than anything in the ordinary biometrics of these growth trajectories had led us to expect, and it is certainly capable of driving an abduction. My preference would be an explanation in terms of the overlengthening of the cranial base (drawn at the right edge of these little diagrams), together with the likelihood of functional constraints on the anatomically anterior and posterior edges that make the management of the calvarial shortfall less costly if it falls right in the middle of the anatomically superior arc, where it involves proportions to which no specific function is allocated.

7.7.4 A Symmetry Curve in 3D, and a Classification It Drives

The discussion of landmarks, Section 7.1, mentioned three kinds of curves, of which one was relatively novel: the *symmetry curve* that runs up the middle of a nominally bilaterally symmetric biological structure. Symmetry curves are often an efficient way of representing form for systems explanations, especially of systems that themselves are partly regulated by aspects of the same symmetry. Such systems include most processes of mammalian neurogenesis, and so the following study, originating around the year 1997, is very much in sympathy with the systems principles of this book.

We are studying the *corpus callosum* of the human brain, a bundle of nerve fibers (axons, not cell bodies) laid down fairly early in neurogenesis to connect the two

---

[6] The formula for the extrapolation of a transformation $S \to T$ by any factor α is the thin-plate spline $S \to S + \alpha(T - S)$, where the subtraction is taken in the sense of shape coordinates in the Procrustes representation. The value $\alpha = 1$ leaves the original transformation $S \to T$ unchanged; the grid for $\alpha = -1$ is the transformation $S \to 2S - T$, which approximates the transformation $T \to S$, the inverse of the original grid.

446 7 Morphometrics, and Other Examples

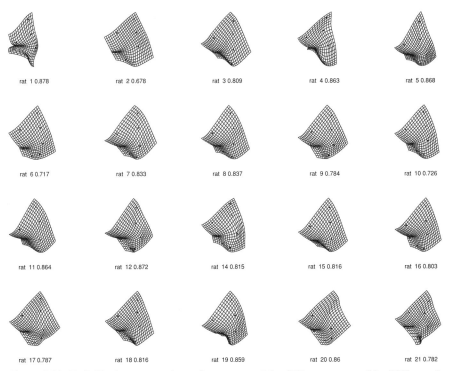

Figure 7.20. Individual crease analyses for neurocranial midline octagons, 20 of Vilmann's rats, from age 14 to age 40 days. (From Bookstein, 2004).

halves of the cerebral cortex. The callosum is important inasmuch as many core cognitive processes involve both sides of the brain and transfer information from one side to the other (cf. DeHaene, 2009) but also, for purposes of the study about to be reviewed, because the maps these fibers follow in order to be laid down across the middle of the brain are greatly confounded by the effects of the ethanol (alcohol) molecule, so that the callosum is the single most sensitive structure of the body with respect to detection of prenatal alcohol damage, more sensitive even than the facial signs that gave the fetal alcohol syndrome its definition (see Figure 5.6).

In full 3D the callosum is a very complex structure that branches and twists as it runs outward from the paleocortex into the neocortex. Near its midsection, however, it becomes quite a bit more tractable, especially in normal human brains, and likewise a good deal easier to reduce to data for our purposes of abduction and consilience. To this end we designed a simple "two-and-a-half-dimensional" template (a nearly but not quite flat closed curve), that traced completely around this structure along a path of greatest local symmetry. Every point was digitized (by hand) to lie on a local mirroring plane for its vicinity, as in Figure 7.21, and then each data set of 40-gons was treated as a chain of semilandmarks and slipped according to the algebra of equation (7.6), Section 7.6, as applied to curves in space.

7.7 Putting Everything Together: Examples 447

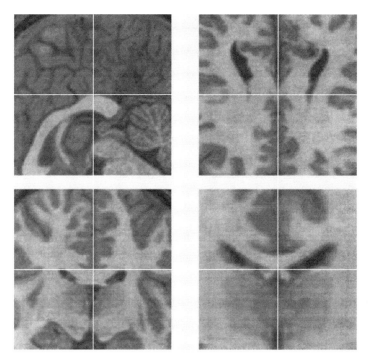

Figure 7.21. How to digitize a symmetry curve. From Bookstein, Sampson et al. (2002). (Upper left) Section of a volumetric MR brain image by one estimated symmetry plane touching the callosal midcurve at the intersection of the crosshairs in the picture. The horizontal crosshair is along the tangent to the midcurve there, while the local axis of symmetry is perpendicular to the picture through the same point. (Upper right) Intersection of this brain image by a plane along the horizontal crosshair of the upper left panel and perpendicular to that panel. Note the approximate left-right symmetry. (Lower left) Intersection of the same brain image by the plane along the vertical crosshair from the upper left panel and perpendicular to the upper left image. Note, again, the near-symmetry across the vertical crosshair here. (Lower right) A plane just a bit anterior to the plane at lower left but now centered on a point of the lower (rather than the upper) arc of this midcurve. Images produced by W. D. K. Green's `edgewarp` program package.

There resulted the data set shown in Figure 7.22 pursuant to the First Rule and part of the Second.

Figure 7.23 fulfills the Fourth Rule for this same data set. We notice a striking hypervariation along the first principal coordinate here: the variance of this vector in the FASD's is about 2.25 times as large as that in the unexposed.

Back in the 19th century, there was a "science" called *phrenology*, now discredited, whose practitioners claimed to be able to predict aspects of character (especially the conventional middle-class virtues and vices) from the bumps on the outside of the skull. The flavor of research being reviewed here could perhaps be referred to as *endophrenology*, the study of the bumps on the head "from the inside," from the shape of brain structures instead. Witticisms aside, we can pursue this possibility because

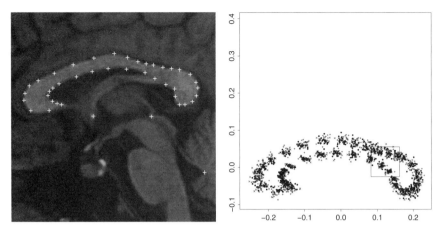

Figure 7.22. A typical callosal midcurve as digitized (First Rule), and the Procrustes shape coordinates (one component of the Second Rule; the rest is in the next figure) for all 45 adult males of the Seattle study, 30 of whom had been diagnosed with an FASD.

the same data resource that includes the callosal outlines in Figure 7.22 also includes measurements of a variety of psychometric test scores, a domain that also meets the requisites of Section 6.3.2. The SVD-based technique that we use for searching for commonalities of structure in two otherwise unrelated measurement domains is, of course, Partial Least Squares (PLS). PLS analysis of the Procrustes shape coordinates of 45 callosal outlines against the hundreds of behavioral scores resulted in two pairs of latent variables, of which that in Figure 7.24 is the *second*. (The first pair, not shown here, corresponded to the a-priori knowledge that people with FASD diagnoses have much lower average scores on nearly every test of cognitive function, as we already showed in Figure 5.7.) When PLS is applied to shape coordinates, as it is here, the latent variables it produces are usually called *singular warps.*

Because singular warp 1 necessarily must separate the mean of the normals from the mean of the FASD's, SW 2, shown here, must leave the means in the same place; and this it does. Nothing about the method, however, suggests the striking pattern of additional variation shown in Figure 7.24. The FASDs show variation far beyond that of the normals in behavioral profiles that deviate from normal in two different ways, corresponding to the two ends of the corresponding deformation, the second singular warp, sketched along the corresponding horizontal axis. The structure here suggests the existence of two different cognitive variants of FASD, associated with two different spatial patterns of brain damage. Weaknesses of executive function, the behaviors that get you in trouble with the law, are associated with a callosum that backs off from the normal close juxtaposition with the frontal lobe, which is beyond the left little grid to its upper left, while weaknesses of motor control, which get you into much less trouble, are associated with general thinning of this structure and a shearing forward. To be a useful stimulus to any novel scientific observation, such a profile score must be approximately uncorrelated with IQ, and that is clear from the figure, where the two- or three-digit integers are in fact the IQ scores of these same

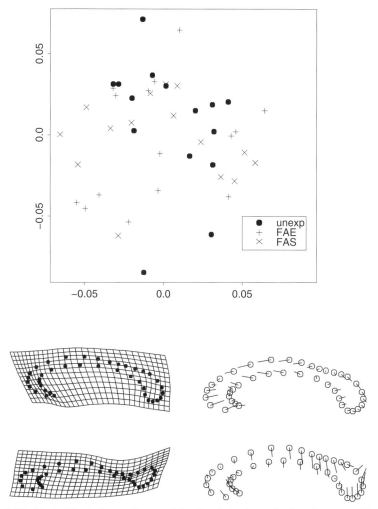

Figure 7.23. (Top) Principal coordinates of the Seattle brain study data, by group. Note two outlying unexposed cases. (Bottom) Axes of the principal coordinate plot as warps (left) and joint shape coordinate displacements (right).

45 adult males. For more information on all of these considerations, see Bookstein, Streissguth et al. (2002).

You may be wondering why there is a little square superimposed upon the Rule Two scatterplot at the right in Figure 7.22. It is a reminder that the literature of fetal alcohol syndrome (e.g., Riley et al., 1995, or NCBDDD, 2004) advises us to look carefully just about here, in the vicinity of the *isthmus* of the callosum, for possible clinically important information. This is where the callosal midcurve is typically thinnest, but also where the callosum typically ends in the birth defects where it is incomplete or disconnects in the birth defects where it is disconnected (the two situations collectively referred to as *partial callosal agenesis*). It also happens

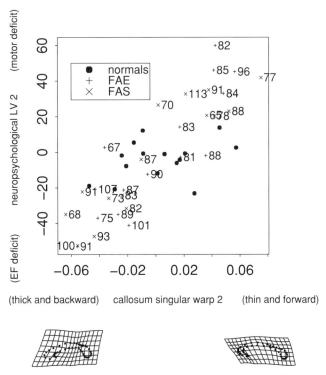

Figure 7.24. The second singular warp and second PLS latent variable score for callosal shape against executive function find a very strong hypervariation characterizing both and independent of measured IQ. See text. (After Bookstein, Streissguth et al., 2002.) Note the resemblance of this SW2 to PC1 in Figure 7.23.

to be where our Seattle subjects with FASD showed the greatest variability with respect to the unexposed. A typical analytic plot is laid down in Figure 7.25, roughly corresponding to that boxed region in Figure 7.22. The figure is set as if every instance of semilandmark 29 (at lower center in the box in the guide figure) was placed in precisely the same spot on the graph. Then the corresponding arcs of the *opposite* side of the arch, the upper side, can be drawn as the bouquet of lines in the upper two-thirds of this figure. The heavy solid lines correspond to the 15 adult male subjects without prenatal alcohol exposure; the light dashed lines, to the other 30 adult males in the study, 15 with an FAE diagnosis and 15 with FAS. Obviously (I think the word is permitted in connection with this figure) the normal forms are radically more regular here than the forms of the FASD patients. So we have passed the Interoptical Trauma Test, Epigraph 1, at this step of our study. For the conversion of this contrast to a formal likelihood-ratio test, see the center right panel of Figure 5.9.

The recentering at one selected landmark is, in a sense, the replacement of Figure 7.22's Procrustes superposition at (0, 0) (the centroid of the shape) by a superposition chosen to be much more informative for the explanatory task at hand, which might, for instance, be the referral of the callosal outline of a new subject, a person in

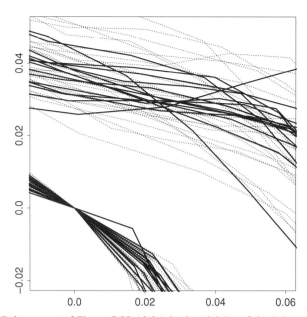

Figure 7.25. Enlargement of Figure 7.22 (right) in the vinicity of the isthmus, showing the enormous excess of shape variation in the fetal-alcohol–affected group when the Procrustes superposition is replaced by one more relevant to the scientific context. Solid lines: repositioned callosal midcurves for the 15 Seattle unexposed adult males. Dotted lines: the same for the 30 Seattle adult males diagnosed with an FASD.

legal difficulties for whom a diagnosis of fetal alcohol spectrum disorder might be materially relevant to the outcome of his criminal proceedings. If this "46th subject" was known to have been exposed to high levels of alcohol before birth, then we can plot his outline in the same fashion as these 45 and examine whether the relevant arc of his recentered isthmus border lies within the normal bundle here, within the FASD bundle, or outside both. There results an odds ratio for the likelihood of brain damage, a rhetoric that can be quite helpful to the judge or jury trying to assign criminal responsibility (*mens rea*, literally, "guilty mind"). See Bookstein and Kowell (2010). For the data pertinent to the person plotted at the XX in the center right panel of Figure 5.9, this odds ratio is about 800 to 1 (log-odds about 6.7) in favor of the hypothesis of prenatal damage from alcohol.

Permutation testing. We make a momentary detour from the main theme of this chapter to deal with a technical consideration that readers have perhaps encountered before under the heading of "data dredging" (Feinstein, 1988) or "Bonferroni correction." The issue is one of protecting oneself against overenthusiasm arising from ordinary levels of meaningless disorder that have been misunderstood as meaningful coincidences, that is, nonrandom patterns. The following protocol serves as a prophylaxis against this risk applicable to a wide variety of studies of organized complex systems.

Every abduction begins with a claim of surprise. But whenever such a claim is based on looking over a distributed pattern for a heightened focus of the signal in

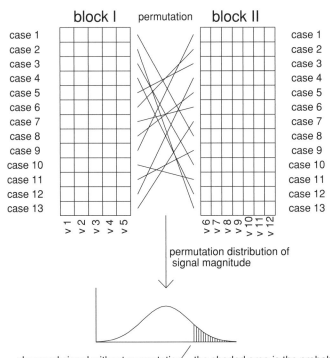

Figure 7.26. Diagram for the logic of the permutation test. See text.

some subdomain, that claim itself must be subjected to some numerical scrutiny. The scene in Figure 7.25 certainly wasn't expected by the original researchers responsible for the callosum study – it was not, for instance, mentioned even as a possibility in the grant proposal of 1997 that underwrote the study from which this portion of the data is extracted – and so perhaps they were naïve in embracing it with relative eagerness in their publications. After all, it is conceivable that had they considered all possible paths across the typical callosum, they would typically find some transects apparently manifesting overdispersion of one group or another even if group actually had nothing to do with the shape of the callosum. We need to check the subjective response of surprise, then, against some objective evidence of actual implausibility. In other words, the surprise driving the abduction needs some justification of its own.

The basic logic of this approach arose in the 1930s, long before the computers existed that cycle nowadays through the thousands of random reorderings of the sample that are entailed. (For this history, along with a technical exposition setting out exact definitions and the accompanying theorems, see Good, 2000.) The phrase "even if group actually had nothing to do with the shape of the callosum" from the previous paragraph is taken to mean *even if the actual matching between a particular group designation and an actual callosal shape were broken by explicit randomization*, and this is the proposition that is actually checked. Then the computation goes as in Figure 7.26.

One must choose a descriptive statistic corresponding to the ultimate verbal report. Here, since the claim is that the FASD subset is overly disparate (or, equivalently, that the unexposed subset is overly concentrated), we turn to an ordinary scalar, the ratio of the variances of the distances between pairs of semilandmarks that face each other across the callosal midcurve. For the transect shown in Figure 7.25, which is from semilandmark 16 to semilandmark 29, this variance ratio is 9.405. On an ordinary significance test (an F on 29 and 14 degrees of freedom) this corresponds to a p-value of 0.000035, but that is not the question at hand. We wish to know, instead, how often we would get this ratio of variances across the callosal outline *anywhere* if in reality there was no relationship between the prenatal alcohol diagnosis and this callosal form. In the abstract, the signal that was surprising is formalized as the claim of a numerical association between two *blocks* of measurements on the same cases, and the randomization explicitly destroys the matching of these paired measurement blocks, as the figure indicates by those crosshatched lines across the middle of the upper panel.

The scientist begins by designing a signal that corresponds to the intended report: in this case, the variance-ratio, FASD cases to unexposed cases. What is computed needs to match the actual process of discovery. So what we use is not the contents of Figure 7.25 – the ratio at precisely the locus already discovered to bear a surprise – but the maximum over all the possibilities we examined before focusing on Figure 7.25. The signal, in other words, is the *maximum* of those variance-ratios across a list of 15 transects going more or less straight across the callosal midcurve arch. Then, over and over for a very large number of iterations, an algorithm randomizes the matching of one set of measurements (here, just one measure, the diagnostic group) to the other set of measures (here, the actual polygon of Procrustes coordinates for each of 45 callosal outlines). The reported surprise is then the frequency with which any variance ratio from the randomized data exceeds the number we are choosing to anchor our abduction, generate our new hypothesis (here, the concentration of excess variation right at the isthmus), and launch our consilience search.

In the data set here, the 45 adult males of the Seattle study as published in 2002, the variance-ratio reported in Figure 7.25 is exceeded only once in 1000 permutations over 15 transects. (The exceedance is very modest – 9.52 instead of 9.41 – and it is for a different transect, one considerably more anterior along the arch.) For the editor who demands a "significance test," this can be converted to a p-value of 0.001, one greater excursion in 1000 permutation trials. For our purposes, it is instead merely permission to continue with the science: a reassurance, not a finding in its own right. The finding was already present: the abduction that follows from Figure 7.25. The permutation test result is one assessment of the strength of that abduction, to be replaced as the focus of our attention by consiliences (e.g., the analogous finding for babies; see Figure 8.4) as soon as those consiliences are available.

A similar test was used to establish the reality of the pattern claimed to represent growth allometry, Figure 7.15. As there was no claimed anatomical focus here, but instead a pattern distributed over the whole, the pivotal statistic was just the sum of squared residuals for all of the regressions of the shape coordinates on log

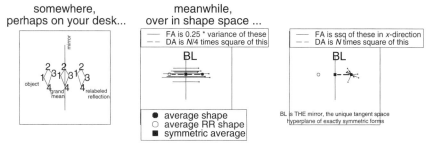

Figure 7.27. Procrustes construction of bilateral symmetry.

Centroid Size, a simpler computation in that no maximization over options was involved.

7.7.5 Bilateral Symmetry

Section 7.3 noted one common research context in which properties of the average shape are no longer a "nuisance variable" but rather a substantial concern of a scientist working with shape coordinates: the application to *bilaterally symmetric* structures, such as the vertebrate body. In this context, the Procrustes method offers a modification of all the standard SVD-based approaches that applies directly to the comparison at issue. It comes in two versions, one for paired objects (left hand, right hand) and one for rigid objects that have both a left side and a right side. We model this as a set of landmarks most of which are paired – left eye corner, right eye corner – along with a possible list of unpaired points up a nominal "midline" or "midplane."

Typical approaches (e.g., Palmer and Strobeck, 1986) would measure the two sides of the form separately, or else measure distances from the paired landmarks separately to some approximate midline. In the Procrustes approach, we instead measure the distance of the entire form from its own mirror image. Because of the independence of rotation that characterizes Procrustes distance, this measurement is independent of any specific choice of a midplane for the reflection. It is, instead, reflection in the Procrustes geometry itself, Figure 7.27: reflection in the hyperplane *BL* of all possible perfectly symmetric shapes with the same counts of paired and unpaired landmarks. As part of the mirroring procedure, we have to renumber the landmarks so that left and right sides are restored; this is the reason for the term "relabeled reflection" in the figure, along with its acronym RR.

As the figure's upper left panel indicates, the average of a form and its own relabeled reflection is a perfectly symmetrical form, from which the original form has a Procrustes distance (which is half its distance from its own relabeled reflection). A sample of forms yields a sample of these relabeled reflections, a sample average, and a sample average of the averages; this is the point plotted at the solid square in either right-hand panel of the figure. Mardia et al. (2000) and Schaefer et al. (2006) show how to extract quantities standing for the various standard quantifications of asymmetry (fluctuating, directional, total) from a scheme like this – it reduces to an

7.7 Putting Everything Together: Examples   455

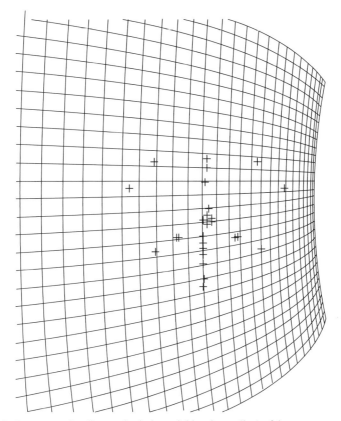

Figure 7.28. Representation (by quadratic-kernel thin-plate spline) of the component of human craniofacial asymmetry for bilateral difference in the uniform component.

ordinary analysis of variance – but, in keeping with the theme of this chapter, it is more interesting to see how the descriptors we have adopted can be used to visualize *types of asymmetry*. These include differences in size or shape of the left and right halves with respect to one another, differences of their positions with respect to the midline, rotation of the unpaired structures with respect to the mirroring line between the sides, and unflatness of the midline.

Figure 7.28, from Bookstein (2004), shows one of these components for another classic craniofacial data set, the "Broadbent–Bolton standards" for craniofacial growth in three dimensions. This is a sample of about thirty Cleveland children selected for the pleasantness of their appearance (i.e., not a definable sample of any population), who were imaged by head X-rays in two directions, which permits a 3D reconstruction of some of the characteristic points from Figure 7.1. Analysis here is of a frontally projected view. The thin-plate spline pictured is the deformation of the true average form to its own relabeled reflection, but it is not the spline introduced in Section 7.4. That spline is linear at infinity, and to see how the sides differ at large scale we want instead a spline that is quadratic at infinity. This is arranged by replacing the kernel function $|r|$ of the standard spline with the function $|r^3|$ and augmenting the bordering matrix Q in the formulas of Section 7.4 by the products $x^2, xy, \ldots, z^2$ of the Cartesian coordinates as well as their actual values. (For the mathematics that allows us to do this, see Wahba, 1990.)

In the display of Figure 7.28, the intended focus is not on the average landmark locations themselves (the little + signs – if you look hard, you can see all the pairs, and also the unpaired stack along the midline) but on the largest-scale features of the grid. The left side of the figure is larger than the right side everywhere, and the dilation is directional: it has become wider faster than it is becoming taller. As a result, the cusp where the spline dwindles to a zero derivative is aligned vertically (at the right margin) rather than horizontally, as it would have had the opposite gradient applied to the raw data. The left–right difference has been exaggerated by a factor of ten so that this feature of the grid will appear near enough to the data to be printed at the same scale.

7.7.6 Combinations of Linear Constraints

While the sliding of semilandmarks introduced in Section 7.6 feels intuitively like an addition of information, from the point of view of the algebra it is a restriction of information, an imposition of one additional linear equation per sliding point. Because the maneuvers of assessing Procrustes symmetry and asymmetry just explained are likewise linear in the shape coordinates, the two approaches can be combined in a technique useful for studies that require the use of both these tools. Suppose, for instance, that we want to study the asymmetry of a form that includes a bilateral pair of curves, such as a face viewed from the front and having a cheek outline on both sides. If the ultimate concern of the study is the asymmetry of the form, we would do well to insist that the sliding of semilandmarks, whatever else it accomplishes, does not alter the composite assessment of asymmetry.

This is not difficult to do. In the following algebra, suppose we we wish to restrict the sliding by an additional j_1 linear constraints, where j_1 is some number of degrees of freedom fewer than the j to which we actually have access. Write those additional constraints in the form of an equation $FT = 0$, where F is some new matrix that is j_1 by j. Now we seek the parameter vector $\hat{T}$ that minimizes the bending energy $(Y^0 + UT)^t \mathbf{L}_k^{-1} (Y^0 + UT)$ not for all T but only for T that happen to satisfy the additional condition $FT = 0$.

The easiest way to notate the solution to this problem is to rewrite the matrix F as the combination $(F_f | F_b)$ of a "front part," F_f, made up of the first $j - j_1$ columns of F, followed by a "back part" F_b comprising the last j_1 columns. The matrix F_b is square of order j_1; we assume that it is invertible. Similarly, break up the j-vector T into a front $(j - j_1)$-vector T_f and a back j_1-vector T_b. (Actually, these are "top" and "bottom," as T is a column vector.) The equation $FT = 0$ can then be expanded as $F_f T_f + F_b T_b = 0$ or $T_b = -F_b^{-1} F_f T_f$. We then have $T = \begin{pmatrix} I_{j-j_1} \\ -F_b^{-1} F_f \end{pmatrix} T_f$, where I_{j-j_1} is the identity matrix of order $j - j_1$.

Replace the unconstrained sliding parameter matrix U with a new version

$$U_F = U \begin{pmatrix} I_{j-j_1} \\ -F_b^{-1} F_f \end{pmatrix},$$

which is now $2k$ by $j - j_1$ instead of $2k$ by j. Under the conditions $FT = 0$ we have $UT = U_F T_f$, so that the bending energy of $Y^0 + UT$ is the bending energy of $Y + U_F T_f$. Then one solves this F-constrained problem by minimizing the bending energy over just the first $j - j_1$ sliding parameters (elements of T_f), which remain unconstrained but which (by virtue of what they entail for T_b) correspond to the displacements conveyed by U_F instead of U. In other words, we seek the minimization over T_f of the form $(Y^0 + U_F T_f)^t \mathbf{L}_k^{-1} (Y^0 + U_F T_f)$ (equation (7.5)). But by the same formula as before, the minimizer is $\hat{T}_f = -(U_F^t \mathbf{L}_k^{-1} U_F)^{-1} U_F^t \mathbf{L}_k^{-1} Y^0$. From this $\hat{T}_f$, compute $\hat{T}_b = -F_b^{-1} F_f \hat{T}_f$. The sliding vector $\hat{T}$ that moves all j of the landmarks so as to minimize bending energy consistent with the constraint $FT = 0$ is then the reassembly $\hat{T} = (\hat{T}_f, \hat{T}_b)$.

For an application preserving symmetry, we want to leave unchanged the separation vector between the form and its reflected relabeling in Procrustes space. This requirement corresponds to a matrix F that takes the form $F = \begin{pmatrix} |x_1| & |x_2| & \dots & |x_k| & 0 & 0 & \dots & 0 \\ 0 & 0 & \dots & 0 & x_1 & x_2 & \dots & x_k \end{pmatrix} U$, 2 rows by j columns. This is two additional equations, the first for the horizontal uniform component of the sliding and the second for the vertical. The horizontal equation asserts that the net weighted horizontal shift of the left paired landmarks is equal to the negative of that for the right paired landmarks; the vertical equation, that the net weighted vertical shifts left and right are equal (not opposite). It is here that we needed to assume that the starting form S is exactly symmetric and is in Procrustes pose, so that the horizontal coordinates x_i of its unpaired landmark loci are 0 and those of its bilateral pairs sum to 0. (Software implementing the algorithm here can guarantee this most simply by symmetrizing the starting form beform any sliding is performed.) The weighting factors here correspond to the **uniform factor estimate** of Bookstein, 1991, pp. 277–280, in contrast to the more familiar least-squares (Procrustes) uniform term of Bookstein (1996) or Dryden and Mardia (1998), and also in contrast to the affine part of the thin-plate spline formula. We use the coefficients from S, the starting configuration (which is bilaterally symmetric, you will recall) so that shifts don't alter the assessment of asymmetry around this representative form; that is the whole point of the exercise. The meaning of the two-dimensional constraint, simply put, is that the sliding is prohibited from introducing, altering, or attenuating **either** a general inconsistency of expansions of width, left versus right of the midline, **or** a general shear parallel to the midline. If either of these features is patent in the original data it is preserved in the course of the sliding, presumably for later analysis along with all the other interesting features of the data set at hand.

An example of this, unrealistically simple so that its essential features can be understood visually, is shown in Figure 7.29. A symmetric six-landmark template, upper left, is to be related to the lopsided form at upper right. The lateral displacement of the landmark at the right "cheekbone" represents a substantial asymmetry; we

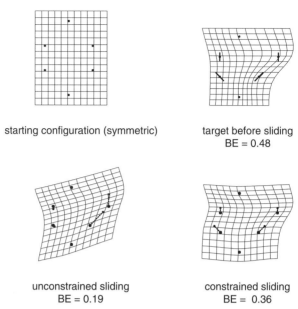

Figure 7.29. Further constraining a sliding operation: the application to preserving asymmetry. (Upper left) A simplistic symmetric template of two midline landmarks and a bilateral pair of curves represented by two semilandmarks each. (Upper right) The raw data as digitized, with the direction of sliding indicated for each bilateral point. (Lower left) Result of the original sliding algorithm, Section 7.6, adds a shear along the midline that was not present in the original data. (Lower right) Constraint by the additional pair of equations represented by the matrix F in the text replaces this with a more bent solution that induces no shear along this midline.

wish to preserve the information it conveys. But because of the dependence between sliding displacements for multiple points on the same curve, this one-sided lateral shift introduces a one-sided vertical shift (lower left) that would be interpreted as an additional asymmetry in the configuration as a whole. The additional restriction by the two equations conveyed by the matrix F, lower right, in effect constrains the two "cheekbones" to slide in a nearly symmetric fashion, and likewise the two "corners of the jaw." There is a cost to this procedure, the diminution of the reduction in bending from the original 60% (from 0.48 to 0.19) to only 25% (from 0.48 to 0.36). You can see that the effect of the additional constraint was to considerably increase the extent of sliding for the two landmarks on the left side, those that "didn't need it" according to the original, unconstrained manipulation.

Other reasonable constraints can be placed on the global aspect of the sliding process by similar reductions or approximations to linear expressions. One might, for instance, require that the sliding not affect the centroids of the specimens (two equations), or the Procrustes fits of the specimens (four equations). Or one might require that parts of the form (e.g., the eye sockets of the skull) not shift their centers with respect to the rest of the form (four equations). In this way, one protects the explanatory power of the analyses from interference by side-effects of this otherwise quite useful algebra.

For an entirely different approach to this problem, one that substitutes the presence of a preferred direction (a line of symmetry) for the location of one or more landmark points, see Bookstein and Ward (2013).

7.7.7 "Form and Function": An Example from the Morphometrics of Strain

Another context for active research into shape phenomena is the experimentally induced *reversible shape changes* induced by the application of mechanical loads. The mathematical modeling of these experiments has been understood in principle for nearly 200 years and in practice since at least the 1950s, when extended computations of the corresponding finite element approximations became practical. The following example, extracted from Bookstein (2013a), adapts the method of Partial Least Squares, Section 6.4, to the problem.

As Figure 7.30(a) shows, the example leverages a convenient analytic solution (Nguyen, 2007) of an exact problem in mechanical engineering. Imagine a population of uniformly tapered three-dimensional *beams* made of some uniform, stiff material and with a uniform thickness in the third dimension (into the printed page). The left end is presumed rigidly affixed to some structure (perhaps a building, perhaps a skull) and the right end is loaded vertically by a weight that is the same for all the experiments. The thicker the beam, the less it bends under this load. There are two canonical scalar descriptions of the resulting deformation (strain). One is our standard Procrustes distance (from a 42-point polygonal approximation of the outline), and the other is the engineer's standard quantity of *strain energy*, the work done by the weight that is stored in the form of elastic energy of the deformed beam. According to the figure, the beams come in three families having thicknesses at the fixed end equal to 5%, 7%, or 10% of the length of the undeformed beam. Each family varies by a thickness at the right (free) end that varies over 41 steps from no height at all (a sharp edge) to twice the thickness at the left end. The figure goes on to show the same unrealistic multiple of the actual deformation computed by Nguyen's approach for each of the $41 \times 3 = 123$ beams in the simulation.

Shape analysis and deformation analysis to a constant load use precisely the same geometrical data and hence must encode the same information. The plot verifies this by the near-linear relation between one Procrustes summary statistic, the bending energy of Section 7.5, equation (7.5), and the net strain energy just mentioned. The relation between these two summaries is log-linear. But this is not a form-function relation as the scientists usually interpret that phrase, an estimation of functional response (to strain) from the description of unstressed form. Figure 7.30(b) shows the oversimplified analysis (a version of the method of O'Higgins et al., 2011) that simply predicts strain energy from the relative warps of those 123 starting forms separately. (Note the three curves, corresponding to the three families of different fixed-end thickness.) So defective a prediction is empirically useless.

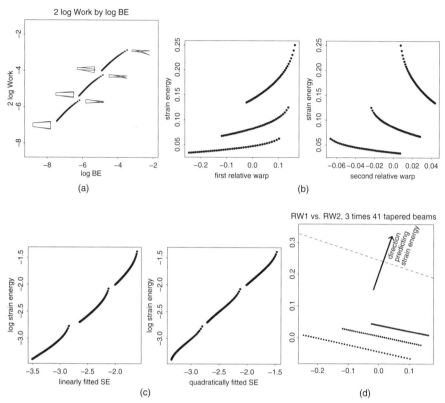

Figure 7.30. Four views of a deterministic simulation of biomechanical bending. (a) Design of the example: three families of 41 tapered beams each, fixed at the left end and subjected to a constant simulated load at the free (right) end. In each family the thinnest form tapers to a point. The little sketches show unstrained and strained forms for the thinnest and thickest members of each of the three families. (b) Faulty choice of a "form–function analysis": strain energy against the principal components of the unstrained beams. The analysis is not useful. (c) Correct Partial Least Squares analysis of the same configuration, showing the essentially perfect prediction of function by form (unsurprising, as they are based on exactly the same geometric information). (Left) Linear prediction. (Right) Prediction by quadratic terms and product term as well. (d) Key to the success of the prediction in panel (c): there is a direction that predicts strain energy best, but it is oblique to the ordination of the three families that determined the orientation of those Procrustes principal components.

As we learned from the description of the PLS method, Section 6.4, the principal components that are most effective at predicting the values of a block of measurements are not necessarily the most effective predictors of causes or consequences of the measures of that block. Figure 7.30(c) shows a prediction that is nearly perfect for a multiple regression (Section 4.5.3) of log strain energy against the whole *list* of principal components of the starting shapes in Figure 7.30(a). The underlying geometry is as in Figure 7.30(d): the direction of prediction is oblique to the orientation of those principal components in precisely the direction that lines up the sample by its bending as in Figure 7.30(a).

Even though couched in some details of our approach to morphometrics, this artificial example makes an important, more general point about the central topic of Part III, the statistics of organized systems. Measurements (in this case, the geometrical locations of corresponding points) can serve multiple purposes at one and the same time, and conventional styles of explanation (in this instance, "form versus function") might arise from different computations applied to precisely the same data base. To the experimentalist's simple intuition in Figure 7.30(c) – to wit, "there must be some easy way to extract a predictor of function from our systematic descriptions of form" – corresponds the simple diagram in Figure 7.30(d), which *displays* the linear part of that *predictor* as a new morphometric *descriptor.* For a more extensive discussion of all these matters, see Bookstein (2013a).

7.8 Other Examples

Morphometrics makes powerful use of aspects of human vision (the grids, the direct apprehension of shape) that are not normally exploited in scientific visualization. When these possibilities are unavailable, the same bravado of analysis can lead to a variety of paradoxes and pitfalls. This section surveys four examples exploring the limits of the reasoning recommended in this chapter and the preceding.

7.8.1 It Wasn't Just Alcohol

Our demonstration of PLS at Figures 6.13 and 6.14 showed how the pattern of correlations between a collection of measures of alcohol exposure and a psychometrically calibrated set of IQ probes resulted in a single pair of latent variables, one for prenatal alcohol dose and the other for IQ status. It would have been tempting to interpret that association as evidence for "the effect of prenatal alcohol exposure on IQ," and that is exactly how this finding was described in many earlier publications by the team that pursued the study. But that description was wrong.[7]

We exploit the immediacy of the printed page to augment the displays in Figure 6.14 by additional data measured at the same time as the alcohol exposure. For conciseness, the IQ score is now not the latent variable score from the analysis recounted in Section 6.4, but instead the usual psychometric total score as standardized to mean 100 and standard deviation 15 in representative earlier populations.

From a range of possible additional antenatal predictors of IQ we selected a particularly potent one, the *Hollingshead stratum* of the child's father as observed back in 1974–1975. This is a conventional score of social class derived from New Haven in the 1950s (the same population, oddly enough, from which Milgram took his subjects for the study reported in Section E4.7). A. B. Hollingshead's two-factor index of social position (see Hollingshead and Redlich, 1958) was the most common indicator of socioeconomic status at the time of the prenatal interviews in 1974–1975. Our analysis below uses Hollingshead's seven-level occupation coding according to a hierarchy ranging from the "low evaluation of unskilled physical labor toward the

---

[7] The critique that follows has not been published previously.

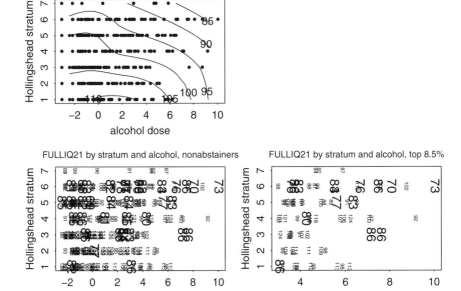

Figure 7.31. The prediction of IQ deficit shown in Figure 6.14 is misleading. See text.

more prestigeful use of skill, through the creative talents, ideas, and the management of men *[sic]*" (Hollingshead and Redlich, 1958, Appendix 2, p. 391).

The main finding is shown at upper left in Figure 7.31. When the data for the full sample (not just the 250 most highly exposed as in Figure 6.14) are analyzed by alcohol exposure and birth social status jointly, the principal factor explaining IQ shortfall is evidently the factor going from bottom to top of the figure rather than the factor running from left to right. *This is social class, not alcohol.* The contours of equal predicted IQ do not ever rotate to be insensitive to social class; by contrast, throughout the bulk of the underlying scatter they are in fact insensitive to alcohol.

We can sharpen the report by concentrating on the practical aspects of this study: where, on this chart, are the 42 subjects who rated in the lowest 10% of this sample on measured IQ 21 years later? These unfortunate subjects, those who would be thought to be "manifesting the consequences of their mother's drinking during pregnancy," in fact concentrate not along the high-alcohol region of this plot, but in the low-social-class rows. The figure shows this twice, once for the children of roughly the top 250 drinkers in our original sample (lower left), and again for the top 8.5% of drinkers (lower right). In both of these charts, each subject is represented by his or her actual IQ score, and the lowest 10% of those scores are printed in a larger font. At left, one sees that the highest density of these low IQ scores is evidently in the sixth row of the chart, for fathers who were "working-class" in 1975. If we zoom in on the group of highest alcohol exposure, we see only one subject of rich parents manifesting a low IQ (the single occurrence of IQ 86 at the lower left corner of the plot), and only 5 others from anywhere in the middle class, but a majority (8 out of the 14) again from the working class. It is true that within the stratum-6 row of the chart the apparent density of these persons in deficit appears to increase from left to right (from moderate up to high alcohol dose along this transect), but the sample sizes here are too small for that inference to bear any information according to a formal modeling along AIC lines.

The abduction from Figure 6.14, then, *cannot* be converted into a consilience. It was wrongly inferred, and requires support by consiliences at lower levels of mechanism, that is to say, much more emphasis on reductionism and the neural causes of lowered IQ. Current research efforts regarding the causes of mild intellectual deficits such as these (e.g., Yumoto et al., 2008) emphasize the presence of multiple social challenges, not just alcoholism but also social disorganization and a variety of alternative childhood stressors both prenatal and postnatal, along with other medical causes, such as pregnancy diabetes. While the findings in respect of the diagnosed FASD patients, Section 7.7.4, are consilient with earlier findings about the human corpus callosum (Riley et al., 1995) and with extensive findings in monkeys (Miller et al., 1999) and rodents (Guerri, 2006), the findings in respect of these cognitive defects in the undiagnosed children of social drinkers have proven resistant to all attempts at consilience across levels of measurement. Here in 2013, the case against mild levels of fetal alcohol exposure as responsible for adult performance deficits in broadly representative Western populations (as distinct from populations that are multiply disadvantaged) has proven to be strikingly weak, and the question no longer seems a promising domain for further research.

You may have noticed that this is not an SVD-type computation. In fact, it is not even a linear model (a least-squares procedure), although the graphical smoothing engine we are using, the `lowess` function in `Splus`, is based on local manipulations of least-squares fits to part of the data (Cleveland, 1993). While the roster of alcohol variables meets the criteria of Section 6.3.2 regarding properly organized sets of variables, the other predictor possibilities, the list from which Hollingshead stratum was drawn, are not such a list, but include instead a measure or two corresponding to every single theory of delayed child development that was in the air at the time this study was designed: variables such as maternal parity, prenatal nutrition, parents' marital status, mother's and father's education, and many others. An SVD is inappropriate in a setting like this, because hardly any linear combination of variables can be given a meaningful explanatory interpretation as a coherent composite, one having what the psychologists call "construct validity."

7.8.2 Karl Pearson, Right or Wrong

Pearson and morphometrics. I have said many nice things about Karl Pearson over the course of this volume. Here is another: he *almost* invented morphometrics, in the 1930's, when he examined quite a considerable range of ratios of interlandmark distances (in three dimensions) in order to demonstrate, "to a moral certainty," that the so-called "Wilkinson skull" was in fact the skull of Oliver Cromwell (Pearson and Morant, 1935). This article was physically too large and had too many figures to appear in the bound volumes of *Biometrika* for that year. Distributed separately, it has become rare as a codex volume, although it is available from the academic journal resource service JSTOR as a .pdf.

This near-miss of a late-life creative explosion arose in the course of a forensic investigation of remarkable eccentricity (matched, perhaps, by that of Fred Mosteller and David Wallace in their 1964 book, *Inference and Disputed Authorship*, a formal probabilistic investigation of who wrote some of the Federalist papers that steered the discussions leading to the American constitution). Pearson and Morant were fascinated by a single physical object (Figure 7.32), a severely weathered mummified head from 17th-century England. The history of the object is fascinating, although macabre. In the mid-17th century the body of Oliver Cromwell was buried, disinterred, hanged, and beheaded, in that order. For about 30 years after that, his head was mounted on a pike at the south end of Westminster Hall, *pour encourager les autres* – don't kill a British monarch. At some time thereafter it was lost to general view. Late in the 18th century, a skull accompanied by a claim that it was the skull of Oliver Cromwell, as salvaged by a "sentinel on guard at Westminster" after it fell off its pole during a thunderstorm, turned up in the possession of a family named Wilkinson. It is this object that concerned our authors. The Wilkinson skull was "that of a person who has been embalmed and decapitated after embalmment, and... exposed long enough on an iron spike fastened to an oak pole for the pole to rot and the worms to penetrate pole and head." (Pearson and Morant, 1935, p. 108). What can one say about the claim that the head is Cromwell's?

To supply an answer, Pearson and Morant had to anticipate much of the multivariate toolkit that Sections 7.1–7.7 just reviewed: tools that exploited the full information content of data from solid objects in the space of their workroom. They inspected all the extant death masks of Cromwell, the single life mask, and all the busts, and on each located a variety of what we are now calling landmarks: the deepest point of the nasal bridge, external orbital corners, diameter of mouth, height of lipline, tip of chin, and, most cleverly, the center of Cromwell's prominent wart over the right eyebrow. (The Wilkinson head sported a wart socket at a very similar location.) These configurations were archived as sets of interlandmark distances and then converted to a series of size-standardized ratios (*indices*, Rudolf Martin would have called them). On the resulting quantities one could then carry out a sturdy statistical inference regarding a hypothesis of identity that actually meets all the requisites of the book you are holding in your hand.

The issue is whether the Wilkinson skull, as measured in the 20th century, matches this compendium of quantities as averaged over the various representations from three centuries before, each first painstakingly adjusted for factors known to have intervened on the living Cromwell, such as aging, or on the postmortem head, such as desiccation. Their analysis is summarized in the table I've reproduced here in Figure 7.33. Pearson (for surely this part of the paper is his alone) actually took enough distance-ratios to cover the space of shape coordinates we have been discussing, and his finding, the "moral certainty" that the Wilkinson head is Cromwell's, inheres in the agreement between the two lines of scaled distances (one the Wilkinson's, one the Cromwell average) in the table. Notice that the entries here are not vectors of

7.8 Other Examples 465

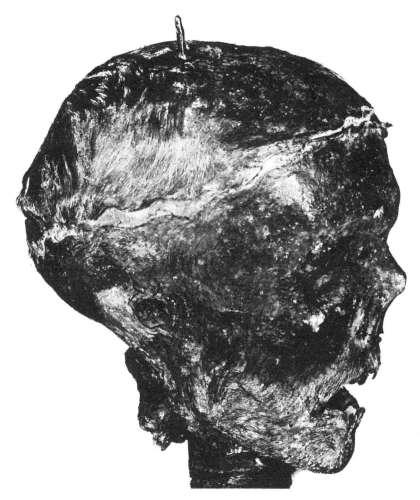

The Wilkinson Head in Right Profile, showing the oak pole and the corroded tip of the iron prong, and the cincture marking the removal of the skull-cap to take out the brain. Note flowing moustache and hair on chin.

Figure 7.32. Photograph of the Wilkinson skull (Pearson and Morant, 1935, facing page 2). Republished with permission of Oxford University Press, from *Biometrika* (1934) 26 (3) *The Wilkinson Head of Oliver Cromwell and Its Relationship to Busts, Masks and Painted Portraits*, Karl Pearson and G.M. Morant, 1934; permission conveyed through Copyright Clearance Center, Inc.

shape variables *sensu stricto*, instead, values of certain hypothetical size variables contingent on the values of others; but, still, to compare them is exactly the same as comparing the shape variables of which they are multiples. Had Pearson gone on to supplement this splendid analysis with a distribution theory for such comparisons of vectors, instead of merely asserting their identity, he would have invented the entire core of contemporary morphometrics in 1935, and I would be out of a job. Put another way, if the Wilkinson head had *not* been Cromwell's, Pearson would have

TABLE II.
Final Comparison of the Wilkinson Head with Masks and Busts.

Characters	External Ocular Distance	External Orbital Distance	Length of Month	Nasion to Lip-line	Nasion to Wart Centre	Nasion to Gnathion	Nasion to lowest point of "beardlet" root	Nasion to Subnasal Point	Wart Centre to Right External Lid-meet	Wart Centre to Right External Orbital Margin
Mean Masks and Busts	98.7	113.75	58.9	75.2	28.05	122.3	95.9	53.8	48.9	52.4
Wilkinson Head	96.6	112.3	59.5	76.2	27.2	117.7	98.8	53.3	49.3	51.1

Characters	Wart Centre to Left External Lid-meet	Wart Centre to Left External Orbital Margin	Glabella to Subnasal Point	Glabella to Lip-line	Glabella to lowest point of "beardlet" root	Glabella to Gnathion	Subnasal Point to lowest point of "beardlet" root	Breadth of Nose, without also	Interpupillary Distance
Mean Masks and Busts	74.9	81.2	71.9	96.65	121.9	139.6	48.25	32.2	73.55
Wilkinson Head	74.4	81.3	71.7	98.1	125.8	135.7	48.4	32.0	70.7

Figure 7.33. Pearson's summary table of the scaled distances on the Wilkinson skull and the average of his exemplars of Oliver Cromwell. (From Pearson and Morant, 1935, p. 105.) Republished with permission of Oxford University Press, from *Biometrika* (1934) 26 (3) *The Wilkinson Head of Oliver Cromwell and Its Relationship to Busts, Masks and Painted Portraits*, Karl Pearson and G.M. Morant, 1934; permission conveyed through Copyright Clearance Center, Inc.

had to develop all the methods to show that it was *not*; but as it *was*, there was no abductive surprise necessary, only the consilience that is conveyed by the table.

In 1960, pursuant to the wishes of a Wilkinson who had just inherited the head from his father, the Wilkinson skull – or, we would now say, the Cromwell skull – was buried in a secret location on the grounds of Sidney Sussex College, University of Cambridge, England.

Pearson and "civic virtue." Ironically, the same determined self-confidence that drove Pearson to such great perspicacity in matters morphometrical completely betrayed him in the matters he considered his greatest achievements: the continuation of Francis Galton's work on "civic virtue," a somewhat Utopian combination of racism and self-deluding smugness. The following comments are adapted from Bookstein (1995).

Modern statistics appeared on the intellectual scene with remarkable suddenness. A field that even as late as 1880 was a miscellany of probability theory and least squares was, by Pearson's death half a century later, fully sketched out with scholarly paraphernalia like textbooks and academic curricula. Pearson was the principal early figure in this great surge. He invented the test for the correlation coefficient, the

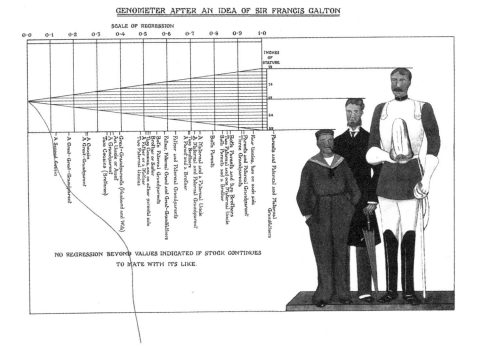

Figure 7.34. Pearson's "genometer" for graphical regressions of height. (From Pearson, 1930, facing page III:30.)

chi square statistic, and principal components; he produced three editions of *The Grammar of Science*; he wrote on socialism, national identity, and (in four volumes) the life of his mentor Francis Galton; and he ran a laboratory, a museum, and a curriculum, all at a level of energy that seems preternatural in view of the absence of any computing equipment beyond mechanical calculators.

Yet, incomprehensibly, Pearson spent the best years of his professional life mired in the hopeless attempt to demonstrate that human worth is hereditary. This is, of course, the theme of *eugenics*, a word coined by Galton in 1883 that did not acquire its fully pejorative connotations until after Pearson's death.

The pervasiveness of this conflation of biological and civic is well-illustrated by Figure 7.34, originally in color, from Pearson's biography of Galton. In the publication, the curlicue was an actual piece of red string for readout of the slopes here. We have a machine for predicting heights of adult males, calibrated to Galton's then-new method of regression, but . . . look at those cartoons at the right: a short sailor, a schoolmaster of middling height, and a towering cuirassier in fancy dress uniform.

By 1901 Galton was explicitly conflating biometric variables such as height with "civic worth," arguing that

> . . . the natural character and faculties of human beings differ as least as widely as those of the domesticated animals, such as dogs and horses, with whom we are familiar. In disposition some are gentle and good-tempered, others surly and vicious; some are courteous, others timid; some are eager, others sluggish; . . . some are intelligent, others

stupid. So it is with the various natural qualities that go toward the making of civic worth in man. Whether it be in character, disposition, energy, intellect, or physical power, we each receive at our birth a definite endowment. (Galton, 1901, p. 659)

This was from a Huxley Lecture (a series of anthropological lectures named after Thomas Henry Huxley, the great defender of Darwin). Galton reviews the Normal distribution and shows how to assign Normal equivalents to Charles Booth's classification of the "grades of civic worth" of London, from criminals and loafers through high-class labour, the lower middle classes, and even beyond. The lower classes are poor "from shiftlessness, idleness, or drink," while the lower middle classes "as a rule are hard-working, energetic, and sober." The threshold of the upper classes is analogous to a height of 6 feet $1\frac{1}{4}$ inches, "tall enough to overlook a hatless mob." Galton concludes, without any evidence, that "civic worth is distributed in fair approximation to the normal law of frequency." It is impossible to read this stuff today without wincing.

The example I wish to spotlight comes two years later, in 1903, when Pearson was offered the podium at the same lecture series. His title, "On the inheritance of the mental and moral characters in man, and its comparison with the inheritance of the physical characters," was part of an attempt to be more persuasive on these topics than Galton had been. Pearson designed a study on the data sheet shown in my Figure 7.35.

The "psychical" characteristics here are an interesting list: vivacity, assertiveness, introspection, popularity, conscientiousness, temper. For the finely categorized ratings, like "intelligence," Pearson worked out a statistic to show that the regression of one sib's score on the others looked linear, in accordance with the prescriptions of Galton's theory. (The method is still in use today, under the name of "polychoric correlation.") From these data Pearson computed the summary tables and the diagram I have collected in Figure 7.36.

The correlations of these variables between sibs were all approximately the same. From this he draws a stunning conclusion:

> How much of that physical resemblance [Table III] is due to home environment? You might at once assert that size of head and size of body are influenced by nurture, food, and exercise. [But] can any possible home influence be brought to bear on cephalic index, hair colour, or eye colour? I fancy not, and yet these characters are within broad lines inherited exactly like the quantities directly capable of being influenced by nurture and exercise. I am compelled to conclude that the environmental influence on physical characters... is not a great disturbing factor when we consider coefficients of fraternal resemblance in man.... Now turn to [Table IV] of the degree of resemblance in the mental and moral characters. What do we find?... Why, that the values of the coefficient again cluster round .5. I have illustrated the whole result in [Figure 7.36].... We are forced absolutely to the conclusion that the degree of resemblance of the physical and mental characters in children is one and the same.... [There is an invocation of Occam's razor, and then:] We are forced, I think literally forced, to the general conclusion that the physical and psychical characters in man are inherited within broad lines in the same manner, and with the same intensity.... That sameness surely involves something additional. It involves a like heritage from parents.... We inherit our parents' tempers, our parents' conscientiousness, shyness and ability, even as we inherit their stature, forearm and span. (Pearson, 1903, 202–204)

7.8 Other Examples

APPENDIX IB.

DATA PAPER FOR COLLATERAL HEREDITY INVESTIGATIONS.

B. SISTER-SISTER SERIES.
(Whole, not half sisters.)

No. in whole series.
(Not to be filled in.)

Please return this Paper to Professor KARL PEARSON, F.R.S., University College, London.

School:
Observer: No. in School Series
Date:

Place a cross against the class of each sister under as many headings as possible, except under III and VIII. Please read first the General Directions.

	ELDER SISTER.	YOUNGER SISTER.
Name		
Age		
District of Home		

I. PHYSIQUE:

	Very Strong.	Strong.	Normally Healthy.	Rather Delicate.	Very Delicate.	Athletic.	Non-Athletic.
ELDER SISTER							
YOUNGER SISTER							

II. ABILITY: (a) *General Scale.*

	Quick Intelligent.	Intelligent.	Slow Intelligent.	Slow.	Slow Dull.	Very Dull.	Inaccurate-Erratic.
ELDER SISTER							
YOUNGER SISTER							

(b) HANDWRITING: (See Back.)

	Very Good.	Good.	Moderate.	Poor.	Bad.	Very Bad.
ELDER SISTER						
YOUNGER SISTER						

(c) WORK:

		Classics.	Modern Languages.	History.	Mathematics.	Descriptive Science.	Drawing.	Singing, Music.
ELDER SISTER	Good at...							
	Best at...							
	Likes best							
YOUNGER SISTER	Good at...							
	Best at...							
	Likes best							

(d) GAMES OR PASTIMES:

	ELDER SISTER.	YOUNGER SISTER.
Likes		
Good at...		

III. HEAD MEASUREMENTS.:

	Length.	Breadth.	Height.	a.	b.	c.	(a), (b), (c), Indices (not to be filled in).
ELDER SISTER							
YOUNGER SISTER							

IV. HAIR:

	Red.	Fair.	Brown.	Dark.	Jet Black.	Smooth.	Wavy.	Curly.
ELDER SISTER								
YOUNGER SISTER								

V. EYES:

	Light.	Medium.	Dark.
ELDER SISTER			
YOUNGER SISTER			

VI. RELATIVE CAPABILITIES: This is *only* to be filled in in those cases wherein the two sisters fall into the *same* class.

	Physique, stronger in	More Athletic.	Ability, greater in	Handwriting, better in	Hair, darker in	Eyes, darker in
ELDER SISTER						
YOUNGER SISTER						

VII. CHARACTER, ETC.:

	Noisy.	Quiet.	Self-conscious.	Unself-conscious.	Self-assertive.	Shy.	Conscientiousness. Keen.	Dull.	Popular.	Unpopular.	Quick.	Temper. Good-natured.	Sullen.
ELDER SISTER													
YOUNGER SISTER													

VIII. GENERAL REMARKS. Add here any striking features of resemblance or dissimilarity in the sisters.

ELDER SISTER	
YOUNGER SISTER	

[On the back of the Schedule spaces were arranged for samples of the handwriting.]

Figure 7.35. Pearson (1903), data sheet, page 210.

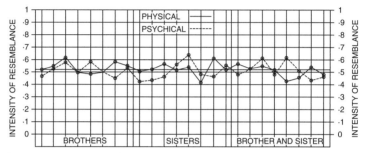

Figure 7.36. Pearson's (1903) evidence for "inheritance of civic virtue": the summary charts.

Note the triple repetition of the trope of physical force, to conceal the lack of all inferential force. The breadth and bravado of Pearsons' peroration is remarkable:

> If the views I have put before you tonight be even approximately correct, the remedy lies beyond the reach of revised educational systems: we have failed to realize that the psychical characters, which are, in the modern struggle of nations, the backbone of a state, are not manufactured by home and school and college; they are bred in the bone; and for the last forty years the intellectual classes of the nation, enervated by wealth or by love of pleasure, or following an erroneous standard of life, have ceased to give us in due proportion the men we want to carry on the ever-growing work of our empire, to battle in the fore-rank of the ever intensified struggle of nations.... The real source of an evil is halfway to finding a remedy.... Intelligence can be aided and can be trained, but no training or education can create it. You must breed it, that is the broad result for statecraft which flows from the equality in inheritance of the psychical and the physical characters in man. (loc. cit., p. 207)

For this to be the transcript of a major public lecture by the founder of my own field, my own intellectual grandfather,[8] is inexplicable and nearly unbearable even at this distance of more than a century. I need not point out to you how grave is the fallacy here, or how attenuated the reasoning. *The abduction Pearson is attempting*

---

[8] Biometric statistics came from Galton and Pearson down to Fisher and Wright and in the next generation to me and my coevals.

just does not work. Eye color notwithstanding, all these results are just as consistent with the effects of the environment, combined with stimulus generalization on the part of the raters, as with effects of heredity. Pearson, who more or less singlehandedly invented the method of significance testing, has here completely failed to consider any hypothesis beyond Galton's. He knew how to do this sort of thing – he invented the method of partial correlations, for instance, specifically to adjust one cause for others acting simultaneously. But he simply refused to consider the relevance of those explicitly skeptical tests to the present circumstance, owing to the enormous ideological centrality of the issues involved. His own textbook of philosophy of science, the *Grammar*, notes, as the fourth canon of "legitimate inference,"

> While it is reasonable in the minor actions of life, where rapidity of decision is important, to infer on slight evidence and believe on small balances of probability, it is opposed to the true interests of society to take as a permanent standard of conduct a belief based on inadequate testimony. (Pearson, 1911, p. 60)

Recall our epigraph from William Kingdon Clifford (whose book-length essay *The Common Sense of the Exact Sciences* Pearson himself had edited):

> It is wrong, always, everywhere, and for anyone, to believe anything upon insufficient evidence.

Pearson continued to the end of his life to behave in this paradoxical way. Otherwise a superb statistician scrupulously conscientious regarding assumptions and formulas, and wise and wary of measurement errors, he would in his eugenic work – and only there – grossly mold the data analysis to suit this predetermined ideology. His work in other sciences, including physical anthropology, is generally first-rate (for instance, the identification of the Wilkinson skull as Cromwell's that I just reviewed), but his work on eugenics and "racialism" is worthless. His advances in measurement and biometric analysis are limited to the earlier work on strictly physical characters, ironically, the themes of his early collaboration with Weldon cut short by the latter's death in 1906. Please notice also that the variables in Pearson's study are completely unorganized. The equivalent for measurement of the "hatless mob" that Galton abjured in humanity, they illustrate the logical difficulties of applying multivariate statistical techniques to unstructured variables, whether in the social sciences or the natural sciences.

A folly of this extent is indeed a terrible cost to have been paid by a great scholar otherwise aware of the "determination to sacrifice all minor matters to one great end." In Bookstein (1995) I traced this to a *folie á deux* between Pearson and his great late-life father figure Galton, a possibility consistent with a tremendous amount of detailed evidence packed into Pearson's extravagant four-volume biography of his mentor, whom he believed to be the greatest scientist England ever produced.

7.8.3 Structured Factor Analysis, Guttman-Style, and the "Horseshoe"

The preceding is a story of intellectual failure, but outside of morphometrics the standard methods I am recommending here can also fail on grounds that are instead purely algebraic. To exemplify this, I turn to the work of one of my teachers who deserves to be better remembered than he is: the psychometrician Louis Guttman (1916–1987). The closest I have seen to the morphometric approach in any other field of applied statistics is Guttman's inspired and original theory of the structure of psychometric scale strategies.

Perhaps the best-known of these is the *Guttman scale*, an ordered set of yes–no questions such that the total count perfectly accounts for the actual pattern of yes answers for a "rational respondent." These might be attitude questions, tests of achievement, or suites of measures in any other domain. For achievement tests, there is a probabilistic version of this approach, *Rasch scaling* (Rasch, 1980), in which each question is characterized by a difficulty, and each individual by an ability, and the probability of a correct answer to the question is some function of the difference between difficulty and ability, perhaps $\frac{e^{\text{abil}-\text{diff}}}{1+e^{\text{abil}-\text{diff}}}$. Inasmuch as it places fairly stringent requirements on the domain of measures as well as the domain of respondents, the Rasch scaling model fits well under the rules for organized variable sets listed in Section 6.3.2.

The Guttman scale model is one of a family of variously organized sets of "scale items" characterized by a range of different geometries, not just a single ordered underlying dimension. For a good review of these concerns, see the chapter by Guttman in Cattell's 1966 compendium *Handbook of Multivariate Experimental Psychology*. In this approach, variables have quantitative attributes prior to any measurement, rather as the perception of color names would be based on the prior quantification of their dominant frequency. Guttman's approach is to retrieve diverse sorts of these a-priori arrangements from patterns in a correlation matrix of a sort different from what the SVD is attuned to. *Simplex patterns* have full rank according to the SVD but are nevertheless easily reduced to a lower-dimensional explanation in terms of a-priori variable orderings. For instance, the correlation matrix of highly symmetric structure

$$\begin{pmatrix} 100 & 75 & 50 & 25 & 0 & 25 & 50 & 75 \\ 75 & 100 & 75 & 50 & 25 & 0 & 25 & 50 \\ 50 & 75 & 100 & 75 & 50 & 25 & 0 & 25 \\ 25 & 50 & 75 & 100 & 75 & 50 & 25 & 0 \\ 0 & 25 & 50 & 75 & 100 & 75 & 50 & 25 \\ 25 & 0 & 25 & 50 & 75 & 100 & 75 & 50 \\ 50 & 25 & 0 & 25 & 50 & 75 & 100 & 75 \\ 75 & 50 & 25 & 0 & 25 & 50 & 75 & 100 \end{pmatrix} / 100$$

has singular values 4, 1.707, 1.707, 0.29, ... and thus requires three rank-one terms to arrive at any competent approximation. (The dominant eigenvalue is the exact integer 4, corresponding to the average entry, 4/8, of this matrix.) Patterns of association like this can arise from circadian data or other measurements that have the logic of an angle or "phase." Another simple pattern for a correlation matrix drops away evenly from the diagonal in both directions; this corresponds, roughly speaking, to a set of variables numbered $i = 1, \ldots, k$ such that the ith is measuring "distance from the value i." For both of these, the SVD completely fails to grasp the simplicity of the structure; the lowered rank it is looking for does not match the actual geometry by which measurements map onto cases. If we replace "distance from the value i" by "being at least i in value," we arrive at the classic Guttman scale. The very simple 21×20 data set with $X_{ij} = 0$, $i \le j$; $X_{ij} = 1$, $i > j$, has singular values $1.425, 0.715, 0.479, 0.362, \ldots$, and again the SVD utterly fails to grasp the joint structuring of variables and cases, in spite of the fact that the raw data matrix, printed out, is just a triangle of 1's under another triangle of 0's. In matrix form, the data model lines up like

$$G = \begin{pmatrix} 0 & 0 & \ldots & 0 & 0 \\ 1 & 0 & \ldots & 0 & 0 \\ 1 & 1 & \ldots & 0 & 0 \\ . & . & \ldots & . & . \\ 1 & 1 & \ldots & 1 & 0 \\ 1 & 1 & \ldots & 1 & 1 \end{pmatrix}, \quad (k+1) \times k,$$

where k is the maximum scale score.

Recall the discussion of Section 6.5 on "discriminant function analysis." If we divide the Guttman scale data set into two groups by cutting the total scale score at any value, say, k, there is a perfect univariate predictor of this classification: the kth original variable, which is 0 for scale scores less than k and 1 for scale scores that are k or greater. But manipulations in a least-squares context such as the SVD nevertheless cannot formalize the fact that these variables are ordered as strongly as they are. If we treat each subsample of distinct scale score as a separate subgroup, then the sum of squared distances between any two of them, say the ith and the jth, is just $|i - j|$.

This is the same pattern of distances that arises from simple random walk in a very high-dimensional space, Section 5.2. It seriously confuses our pattern engine, inasmuch as if the groups were evenly spread out along a line the distances would be of the form $(i - j)^2$ instead. (For this set of d's, the corresponding $-HDH/2$ is of rank 1 – it is simply the one-dimensional scale of integers already shown in the metric Figure 6.17.) For the complete mathematical analysis of this specific matrix, see Bookstein (2013b). Its principal coordinates are approximately the functions $\cos\theta$, $\cos 2\theta$, $\cos 3\theta$, ... where θ rescales the index numbers of the specimens to a range from 0 to π, and the eigenvalues (variances of those principal coordinates) are approximately proportional to the series $1, \frac{1}{4}, \frac{1}{9}$, and so on. Principal coordinates analysis of distance patterns like those a Guttman scale generates, distances that are growing "too slowly" in this manner, are encountered frequently enough to have a name of their own, the *horseshoe* (cf. Mardia et al., 1979, Section 14.5.2). As Figure 7.37 shows, the principal coordinates analysis for this situation shows enough special features (specifically, the exactly parabolic relationship between PC1 and PC2) that we can suspect the underlying scaling; but this suspicion emerges only from familiarity with its graphics, not from its conventional numerical reports. The fraction of total variation accounted for by this parabolic projection is $\frac{1.25}{\pi^2/6} = 0.76$, because

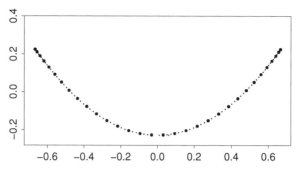

Figure 7.37. The "horseshoe" for data that is precisely linear in distance-squared against an underlying ordination. Shown is the plot of principal coordinate 1 against principal coordinate 2; both are exact trigonometric functions of the specimen numbering. The solid circles indicate the spacing of these points with respect to row or column index of the distance matrix. This plot encompasses 76% of the variation of the full sample in its setting of much higher dimensionality.

the sum of the reciprocals of the squares of the integers, $1 + \frac{1}{2^2} + \frac{1}{3^2} + \cdots$, equals $\pi^2/6$. For much more on this paradoxical situation, see, again, Bookstein (2013b).

7.8.3.1 Mathematical Note 7.1: A Wonderful Identity of Euler's
The preceding discussion touched, in passing, on a mathematical identity as wholly unexpected and beautiful as any of those dealing with the Gaussian distribution (the bell curve) that we encountered in Chapter 4. It demonstrates the mathematician's "ruthless"[9] exploitation of the ambiguity of the written setting (are the symbols just meaningless letters, or are they real?) as well as anything else in the undergraduate canon. This example is from the work of the great Swiss mathematician Leonhard Euler (1707–1783). The problem was to find a formula for the sum of the reciprocals of the squares of the positive integers. What is the value of

$$1 + \frac{1}{2^2} + \frac{1}{3^2} + \cdots = \sum_{i=1}^{\infty} \frac{1}{i^2} \,?$$

We will see that its value is $\pi^2/6$: another of those startling appearances of the constant π in branches of mathematics that seem absurdly far from the context of circles and diameters where you first encountered it. Identities such as this one are useful in all sorts of applied realms, and part of the mathematician's craft is to keep a reasonably long list of them active in memory and an even longer list in retrievable form (formerly, in thick compendia; nowadays, on the computer's desktop). Regardless of its role in that example, this derivation serves as yet another wonderful example of how mathematicians actually go about their business.

Euler approaches this question entirely indirectly. You may remember from high school or Gymnasium that if you have a polynomial (an expression $F(x) = \sum_{j=0}^{k} a_j x^j$, where the a's, for the purposes of this discussion, are $k+1$ real

---

[9] The adjective is from Wigner (1960).

numbers, $a_k \neq 0$) then you can rewrite it as $F(x) = a_k(x - r_1)(x - r_2)\ldots(x - r_k)$ where the $r's$ are the *roots* of the polynomial, the arguments (guaranteed to number k when you count them correctly[10]) for which $F(r) = 0$. Because a_0 is $F(0)$ and a_k/a_0 is the product of all the roots, then assuming that no root is zero we can divide out their product termwise: if F is a polynomial of degree k with $F(0) = 1$ and roots $r_1, r_2, \ldots, r_k$, then

$$F(x) = \left(1 - \frac{x}{r_1}\right)\left(1 - \frac{x}{r_2}\right)\cdots\left(1 - \frac{x}{r_k}\right).$$

Write this using the symbol $\prod$ for an extended product (analogue to Σ for sums):

$$F(x) = \prod_{j=1}^{k}\left(1 - \frac{x}{r_j}\right).$$

As a special case, if the polynomial has only terms in powers of x^2, its roots come in pairs $\pm r$, and then the formula for F can group as well: since $(1 - \frac{x}{r})(1 - \frac{x}{-r}) = (1 - \frac{x}{r})(1 + \frac{x}{r}) = 1 - \frac{x^2}{r^2}$, then for $F(x) = 1 + a_2 x^2 + a_4 x^4 + \cdots + a_{2k} x^{2k}$, we must be able to factor F as $F(x) = \left(1 - \frac{x^2}{r_1^2}\right)\left(1 - \frac{x^2}{r_2^2}\right)\cdots\left(1 - \frac{x^2}{r_k^2}\right) = \prod_{j=1}^{k}\left(1 - \frac{x^2}{r_j^2}\right)$ in terms of its roots $\pm r_j$, $j = 1, \ldots, k$.

You might also remember from high school what the Taylor series is for the function $\sin(x)$: because $\sin(0) = 0$ with first, second,... derivatives at 0 equal to $\cos(0), -\sin(0), -\cos(0), \sin(0), \ldots = 1, 0, -1, 0, 1, \ldots$, we have the identity

$$\sin x = x - \frac{x^3}{3!} + \frac{x^5}{5!} - \cdots.$$

Now comes the creative moment, the cognitive leap for which there was no justification until well after Euler's death. Assume that the product formula $F = \prod(1 - \frac{x^2}{r^2})$ applies to functions F that have the value 1 at argument 0 *even when F is not a polynomial* – indeed, even when F has infinitely many zeroes. Such a function is the expression $F(x) = \sin(x)/x$. It is 0 wherever the sine function is 0, except that for $x \to 0$ this expression goes to 1. We can immediately write down its Taylor series by lowering every exponent of x by 1 in the corresponding formula for $\sin x$: we must have

$$\frac{\sin x}{x} = 1 - \frac{x^2}{3!} + \frac{x^4}{5!} - \cdots.$$

Pretending that the formula $F(x) = \left(1 - \frac{x^2}{r_1^2}\right)\left(1 - \frac{x^2}{r_2^2}\right)\ldots\left(1 - \frac{x^2}{r_k^2}\right)\ldots$ *still applies,* even with that second set of dots... in there, we explore the implications of the possibly nonsensical expression equivalencing the two entirely different versions of

---

[10] This is the *Fundamental Theorem of Algebra*, which was suspected in the 17th century, which Euler himself had tried to prove, which Gauss believed he *had* proved in 1799, and which was finally proven in 1806. For an enlightening discussion, see the Wikipedia page http://en.wikipedia.org/wiki/Fundamental_theorem_of_algebra, and, more generally, Latakos (1976).

the same function $\frac{\sin x}{x}$. Consider the possibility that

$$1 - \frac{x^2}{3!} + \frac{x^4}{5!} - \cdots$$

must equal

$$\prod\left(1 - \frac{x^2}{r^2}\right)$$

where the product is over the r's that are the zeroes $k\pi$, $k > 0$ of the function $\frac{\sin x}{x}$. In other words, can we learn something from the symbolic equivalence

$$1 - \frac{x^2}{3!} + \frac{x^5}{5!} - \cdots = \left(1 - \frac{x^2}{\pi^2}\right)\left(1 - \frac{x^2}{(2\pi)^2}\right)\left(1 - \frac{x^2}{(3\pi)^2}\right)\cdots$$

$$= \left(1 - x^2\frac{1}{\pi^2 1^2}\right)\left(1 - x^2\frac{1}{\pi^2 2^2}\right)\left(1 - x^2\frac{1}{\pi^2 3^2}\right)\cdots ?$$

Yes, we can, Euler claimed. Consider the term in x^2, for instance. The coefficient on the left is just $-\frac{1}{6}$. The coefficient on the right comes from using the term 1 in every binomial of the infinite product except for each factor in turn, for which you take the coefficient $-\frac{1}{\pi^2 k^2}$ and add these up for all k. You must get the same coefficient on both sides. Then

$$-\frac{1}{6} = -\frac{1}{\pi^2}\left(1 + \frac{1}{2^2} + \frac{1}{3^2} + \cdots\right),$$

or, in one final rearrangement,

$$1 + \frac{1}{2^2} + \frac{1}{3^2} + \cdots = \frac{\pi^2}{6},$$

which is the correct formula.

The relation between $\frac{\sin x}{x}$ and its partial products can be illustrated by terminating the infinite product at any convenient term, for instance, the 150th, as shown in Figure 7.38. Even though every approximating polynomial must go to ∞ as the argument x becomes arbitrarily large, whereas the Euler curve goes to 0 instead, still the curves here are indistinguishable within the range of $(-10, 10)$ or so. The first omitted term in the product formula was $\left(1 - \frac{x^2}{474.38^2}\right)$.

7.8.4 A Cautionary Closing Remark: Beware Epigraph 4

One of the epigraphs of this volume is Pearson's, from his eulogy on the premature death of his great colleague W. F. R. Weldon. The contents of this subchapter were meant to lead you to the suspicion that the epigraph may be meant ironically in the present context, however sincerely it was originally meant. The passion to which Pearson refers is, in practice, nearly independent of the "sympathy with Nature" that must accompany it, and, especially in contexts where public welfare is a possible rhetorical trope (from global warming through civic virtue to birth defects), the rules

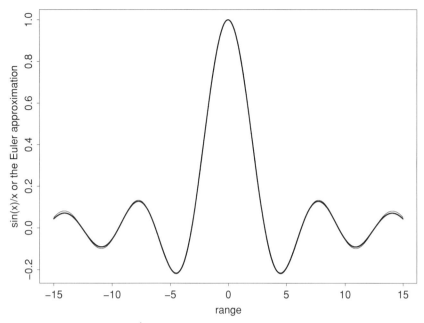

Figure 7.38. Euler's function $\frac{\sin x}{x}$ and the 150-term version of the polynomial approximation from which Euler derived the formula for the sum of the reciprocals of the squares of the integers. See text.

of abduction and consilience are among the only prophylactics we have against being misled by one's own hopes combined with the ambiguity of statistical graphics and multivariate methods.

Karl Popper was wrong about falsificationism – it is consilience, not falsification, that combines with abduction to drive the fundamental engine of scientific progress – but an awareness of the ubiquity of mistaken judgments is part of the fundamental ethos by which that progress should be communicated to colleagues and to the public. The Wilkinson affair shows how the careful, conscientious matching of hypothesis to data can sometimes settle an argument; but the next example, eugenics, showed Pearson succumbing to the temptation to prematurely proclaim an issue in a particularly complex, particularly organized human system to be "settled." And the example in which the origin of adult performance deficits in antenatal social class was overlooked shows how their own hypothesis blinded a whole group of investigators, including me, to the meaning of their own data. In this last setting, an organized system (a birth defect) was being studied by a disorganized system of variables in place of the organized system that characterizes contemporary morphometrics, among other general systems. A desire to have something plangent to say overruled the skepticism appropriate to a devolution of the corresponding SVD class of methods.

Clearly things are not yet perfectly aligned in this domain of numerical inference in organized complex systems. In the final part of this book, I turn to some summary comparisons and recommendations that might reduce somewhat the chances of errors this spectacular continuing to appear in the hands of otherwise quite shrewd and competent investigators. I will argue that the key is to remain focused on the main themes of this book – abduction and consilience – but also to be fully cognizant of the difference between the simple systems (Part II) and the systems of organized complexity (Part III), which require rhetorical sophistication of an altogether different level to effectively convert arithmetic into understanding.

Part IV

What Is to Be Done?

We need toolbuilders who can bridge the cognitive divides of abduction and consilience between the applications to simple systems and the applications to complex systems, between the more familiar or conventional and the less familiar. *What is to be done?* To build, and teach, those tools.

8

Retrospect and Prospect

This last chapter summarizes the implications of all that has preceded for the praxis of statistical science in the near future. It is divided into three sections. Section 8.1 introduces one final example: another study of fetal alcohol exposure, this time based on ultrasound brain images of infants. With its aid I review the notions of abduction and consilience from a point of view emphasizing the psychosocial structures of science rather than the logic of these forms of inference per se. If a scientific fact is "a socially imposed constraint on speculative thought" (Ludwik Fleck's main theme), then abduction and consilience work in somewhat contrasting ways to effect that constraint depending on whether the scientific context is one of simple measurement or the probing of a complex system. Section 8.2 shows how all these procedures depend on prior consensus as to what constitutes agreement or disagreement between a numerical representation of some pattern and an expectation about a scientific regularity. In the final section I step back to examine the whole protocol by which forms of numerical reasoning are communicated across the academic and technological generations. The chapter concludes with some recommendations for major changes in the way we teach statistics. If this praxis of abduction *cum* consilience toward the understanding of complex systems is to keep up with the requirements for science and for the public understanding of science over the next few decades, special attention must be paid to the curricula by which all these strategies are taught to the next generation of our colleagues.

8.1 Abduction, Consilience, and the Psychosocial Structures of Science

This is not a book on multivariate analysis in general, still less on morphometrics in particular, and yet you have just finished studying, or at least browsing, Part III, some 190 pages, that dwelt at length on these two themes. You have seen how quantifying the most powerful contemporary forms of scientific reasoning, abduction and consilience, remains possible in a multivariate context, but involves far more subtlety and creativity than we typically recognize. For complex systems, the construction of the derived quantitative structures that support grounded inference is a matter of high craftmanship. Consequently, the sophistication of the search for quantitative patterns in these multivariate contexts is a far more stubborn and independent component of a research program than was the routine when we were measuring only two or three quantities at a time. The protean character of the singular-value decomposition and the analytic tools (PCA, PCO, PLS) that derive from it is couched in a more realistic rhetoric than that of the linear models that most of our methodology teaching is centered on, and is capable of a great deal more nuance, now that we can refer to parametric properties of variable definitions as well as data values. Especially in morphometrics, to revisit well-quantified and familiar data sets, such as the Vilmann rat skull data published more than twenty years ago (Bookstein, 1991; Dryden and Mardia, 1998), often results in discoveries of new patterns that are genuinely surprising – see, for instance, Bookstein (2004).

To go forward effectively we must weaken the authority of prior hypotheses over the empirical setting. Perhaps the single most pernicious aspect of the current system of scientific review, as it applies in the domain of complex organized systems measured by organized systems of variables, is the utterly inappropriate requirement that investigations be phrased in terms of positivistic "hypotheses": "We hypothesize that...." As Platt (1964) said so pointedly, no work in support of a hypothesis is good work unless at the same time it weakens support for all the equally reasonable competing hypotheses. Scientific inference is a matter of deciding *between* plausible hypotheses, not pursuing the implications of one single beloved option. You have far greater freedom to discover something important, and to persuade others of its importance, if you play by the rules of evidence (Section 5.3) instead of the rules of advocacy.

Much of the power of morphometrics, Chapter 7, derives from its flexibility – its freedom to pursue so many of these schemes of competing hypotheses jointly in the same geometrical setting. Even when restricted to the same raw data resources, the shape coordinates of landmarks and semilandmarks, our tools are capable of confronting more than one model of noise (isotropic diffusion, random timing, self-similar random fields) and many different vocabularies of biological interpretation (group difference, multiple form factors, excessive variation, global and local features, classification and diagnosis). We have seen how descriptors at one level of abstraction (relative warps) can be taken apart and reassembled in different combinations (for example, the partial warps) to hint at analogous recombinations of the associated scientific explanations. And we have seen how longstanding vocabularies of process-based interpretation can be enormously enlarged when the new tools are used to explore patterns in organized data about organized systems.

This flexibility is not unique to morphometrics. Much of it was borrowed from the SVD family of pattern engines, which are themselves universally open to arguments from multiple orderings of variables, multiple scales for the expression of answers, and the like. What is special about morphometrics is the *geometric origin* of those multiple orderings – around curves or surfaces, inside regions – alongside the remarkable cognitive power of the thin-plate spline to reuse our existing brain mechanisms for visualization of these quite different formalisms. This cognitive mechanism is not novel to the spline – it was already tacitly present in the art form known as *caricature* that we have exploited for half a millennium, if not since the time of the Venus of Willendorf – but it is nevertheless novel as a praxis of multivariate data analysis.[1]

To that extent, the multivariate pattern engines we have been reviewing are innovative on their own. Yet part of their intended role is to downgrade the claimed prominence of the less effective alternatives. Much of what we are doing with SVDs in general, and with morphometrics in particular, ends up by showing the vacuity of several common clichés of quantitative science, including some that are not yet widely recognized to *be* cliches. The idea of a null-hypothesis significance test is one such deleterious cliché long overdue for eradication (see Section L4.3.4). Others include the assumption of equal covariance structures at the root of most unthinking approaches to the problem of classification in the biological sciences (the workaround is the likelihood-ratio approach sketched in Section 6.5); the language of "independent" versus "dependent" variables and the distortions inculcated by the resulting restriction to linear models of dependence in place of singular-value decompositions of interdependent systems; the general preference for memory-free stochastic structures such as covariances over memory-dependent models even for interpretation of processes that encode contingent histories (Bookstein, 2013b); and the even more general preference for indexes that are guessed in advance of a data analysis over indexes derived from actual pattern engines. As our empirical world turns to concerns that focus more and more on *organized* systems, statistical vocabularies restricted to the language of the disorganized systems based in bell curves, linear mechanisms, and simple thresholds become more and more ineffectual. Perhaps one reason that physicists are doing so well in finance just lately (Mantegna and Stanley, 2007; Szpiro, 2011), a migration that began years ago with an invitation from Georgescu-Roegen (1971), is that their respect for quantitative anomalies, combined with their professional understanding of long-range order (one mathematization of this notion of organized complexity), leads to a healthy skepticism toward nearly all the claimed findings and theoretical deductions of macroeconomics. Yet it is also the case that physical models and statistical models may well agree with one another to the extent they are drawing on the same information from the real world. In this connection, recall the demonstration at Figure 7.30 that under carefully contrived circumstances a simulated physical measurement of a system undergoing a simulated deformation can produce the same quantitative summary as a Procrustes (i.e., physics-free) assessment of the same simulation.

Abduction. The multivariate approach based on seemingly theory-free pattern extraction from carefully organized data representations, the underlying theme of Part III, offers the careful scientist discoveries at multiple scales, multiple interpretations, and, most important of all, *the possibility of surprise*, whenever a pattern appears in a display that was not among those she imagined she would see. Abduction,

---

[1] On the facility that human culture has for reuse of brain wiring, the best example I know is the emergence of reading over historical time. See the brilliant discussion in DeHaene, 2009. But I have always believed that our processing of transformation grids is a reuse of comparable scope, and recently delivered an hour-long public lecture on that theme, Bookstein (2011).

the emergence of a surprise seemingly at just the same moment as the explanation that renders it comprehensible (at least if you credit Peirce's psychology), takes on heightened importance in these multivariate contexts. Reviewing the examples of the early sections of this book, we can see now how many of them are actually multivariate in character. Their multivariate essence was concealed somewhat by our familiarity with the ordinary two-dimensional page. In the episode in Chapter 1, about the missing submarine *Scorpion*, the inferences were actually in a three-dimensional space of latitude by longitude by time, within which the presence of a sequence of events going in the wrong direction is a "hairpin" of a geometry similar to that of Figure 6.26, the far more abstract graphic of human craniofacial growth variability as a function of hormonal regime. The principal examples in Chapters 2 and 3, as well – continental drift, climate change – are likewise multivariate of low dimensionality (rate of seafloor spreading, or regional rates of global warming). One way to think of multivariate analysis as a whole is as a generalization of *paper*, the paper on which we draw our simpler charts and diagrams without any thought to *its* geometry. (On this point, the primacy of diagrams in the process of scientific persuasion, see Latour [1987, 1990].)

We move away from abduction in this cognitive sense – the surprise of *inscriptions* – mainly when we work within the exact sciences, such as Newtonian mechanics, statistical thermodynamics, or the biophysics of the double helix, where the excellence of the numerical fit to an a-priori mathematical structure becomes itself a feature of the surprise. Elsewhere, as in the "historical natural sciences" for which Friedrich's *Wanderer* of my cover serves so elegantly as metaphor, the richness of the modern pattern engines far exceeds anything that John Snow or Rudolf Martin might have imagined, and serves to replace the organizing genius for facts of a Darwin or a Wegener with the more reliable but dogged examination of computational outputs that makes competent scientists of the rest of us.

Consilience. In this domain, too, consilience takes on a different persona than in the elementary settings of numerical verification ("Did you measure it as 5.38 volts per cm? Let me try"). Consider, in this context, a wonderfully wry comment from Feynman (1985). He notes that everybody teaches Galileo's measurement of the acceleration of gravity, but Galileo could not *possibly* have gotten the correct value, because he ignored the rotational kinetic energy of the ball he was rolling down the inclined plane. The acceleration of a rolling ball due to gravity is only $5/7$ of the acceleration of a free particle downward. Feynman notes that this was one of the epiphanies convincing him he was destined to be a very unusual physicist. Galileo, like Perrin, obtained the wrong value but used it in support of exactly the right theory. Kuhn (1961) offers a similar comment about Laplace's confirming a mismeasured value for the speed of sound in air as an implication from a hopelessly nonphysical theory. In a partial analogy, some of the morphometric examples of Chapter 7 involve abductions that lead to assertions for which no consilience has yet proved possible. I do not mean that these assertions have been contravened by new data sets; I mean that no experimental study of animals, no histology, no epigenetic bioinformatics has yet been set in play to determine whether in fact the isthmus *is* demonstrably

8.1 Abduction, Consilience, and the Psychosocial Structures of Science

hypersensitive to alcohol at the level of individual cell processes and fate maps. The laboratory scientists of fetal alcohol are pursuing molecular studies in accordance with their own training, not in accordance with the hypotheses produced by this distantly rumbling morphometric pattern engine. As for the necessary ties with the underlying biophysics, we are still far from the stage at which the language of such ties can even be assigned a grammar: see Cook et al. (2011).

What differentiates the consiliences of Part III from the consiliences reported in Chapters 1 through 4 is that consiliences in the setting of complexity embody community decisions to *vary the level of observation* of a process that has spun off an abductive surprise at one (generally aggregated) level of pattern analysis. The statistician cannot force any disciplinary community to cooperate in a search for consiliences such as these. Do not confuse the search for consilience with the confirmation or disconfirmation of a finding at the same level of analysis. Consilience involves, rather, a *reordering* or *rereading* of others' research programmes that is required for further confirmation or explanation. In the running example of fetal alcohol damage that has concerned some of the examples in Part III, for instance, the discovery of those patterns of shape of the adult corpus callosum has been embraced by the committees (e.g., NCBDDD, 2004) that explain to nonscientists how to use the findings of scientists. The issue thus becomes the pursuit of an explanation *of* these findings (rather as in the case of ulcers, Section E4.4, the question changed from the possibly infectious origin of ulcers to the construction of a treatment for the infection that causes them) instead of the elementary challenge to their validity that characterized the community response early on. Even if the current fashion is to pursue molecular explanations based in the genome or proteome, it remains an empirical fact that in the environment of a human embryo or fetus, alcohol is only one among a multitude of potential stressors of the same imperfectly buffered neuroanatomy. The example in Section 7.8, showing how alcohol combines with social stress to jointly enhance the probability of IQ deficits in a manner not attributable to alcohol alone, is intended to be combined with additional explanations in terms of prenatal nutrition, pregnancy diabetes, child abuse, and other consequences of social disorganization. As it happens, no one seems to be pursuing such consiliences of the callosal findings at this time (merely using them, after the fashion of Figure 7.25), but the scientific context remains multilevel.

A final example: fetal alcohol effects in the brains of human infants. Still, some attempts at consilience lead to new abductions. In the attempt to extend the findings about the FASD callosum to childhood (a time when the provenance of a diagnosis is most useful for remediation of later real-life difficulties), we attempted a replication in terms of another mode of imaging, transfontanelle ultrasound, at the earliest possible postnatal age. There resulted a paper (Bookstein, Connor et al., 2007) that required the invention of another new methodology (ultrasound image averaging, Figure 8.1) and, with its aid, a finding from otherwise standard Chapter-7 morphometrics that, back in 2007, was best conveyed in the form of two classes of averaged images, for the unexposed vs. for the half of the exposed that show a particular angle feature (Figure 8.2). The finding as published back then shows mainly one systematic change

486   8 Retrospect and Prospect

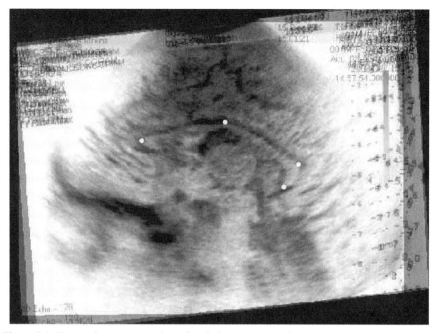

Figure 8.1. Four-point representation of arch form, unwarped averaged ultrasound images of an infant's corpus callosum (imaged through the fontanelle). Clockwise around arch from left: tip of genu (a landmark); top of arch (a semilandmark taken exactly halfway between tip of genu and tip of splenium); "splenium intercept point" (a semilandmark taken where the long axis of splenium intersects the upper arch margin); tip of splenium (a landmark). The exemplar here is typical of the alcohol-affected forms. (From Bookstein, Connor, et al., 2007.)

in the mean form of this little quadrilateral: the reorientation of the axis of the splenium so as to open the angle it makes with the diameter of the form. In this graphical setting, the report is an image, or, rather, an image prototype, something that the consulting physician should "keep in mind" when looking at any particular clinical ultrasound.

But that was not an adequate summary of the signal the data actually conveyed. Authors named Alvarez notwithstanding, the rest of us are permitted only one innovation of method per empirical peer-reviewed publication, and so we could not go on to demonstrate the power of shape coordinates for improving on clinical intuition about how classifiers arise from measurements of complex systems.

The example exploits four separate tools from Part III for numerical inference into organized systems in morphometrics. In brief: (1) The analytic technique driving those intersubject averages relies on the Procrustes shape coordinates shown in Figure 8.3. These embody the usual transform of discrete point configurations by translation, rotation, and scaling of axes that was introduced in Section 7.3 at Figures 7.3 and 7.4. (2) We restrict our attention to the scatterplot of the first two relative eigenvectors, Figure 6.22. The decision to stop at two is confirmed by the Anderson test graphed at Figure 5.12. (3) The technique by which the axes of a sample scatter are labeled with

8.1 Abduction, Consilience, and the Psychosocial Structures of Science   487

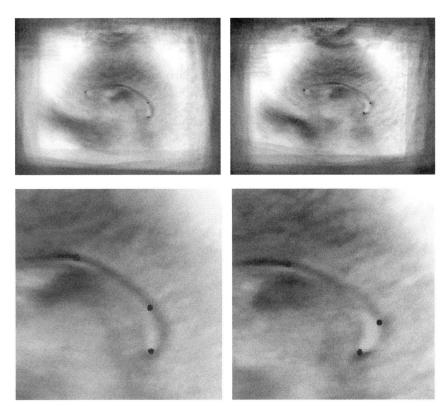

Figure 8.2. Further unwarped averaged images from the four-point representations of 45 babies as in Figure 8.1. (Left) Averaged both within and between subjects, for the 23 unexposed babies. (Right) For 11 of the 22 exposed. Above, full frames; below, zooming into the region of arch, isthmus, and splenium of the callosal midline. The shape difference in the vicinity of splenium is visually obvious; so is the hypervariation of the outline to its left. Both were confirmed by ordinary statistical test of a mean difference (required by the reviewers of the paper from which this figure and the previous were taken), even though the phenomenon is not described very well in terms of a mean shift alone. (From Bookstein, Connor, et al., 2007.)

deformation grids takes advantage of the convenient thin-plate spline characterized in Section 7.5 (cf. Figures 7.7 and 7.8). (4) In this specific projection plane, the likelihood-based discriminator is quadratic (Figure 6.20), not linear (Figure 4.40), and has the topology of the elliptical case at lower left in Figure 6.20. In fact, it is even simpler than that. When the ellipses of constant likelihood ratio are circles (which will be the case whenever the group-specific covariance matrices are proportionate), the discriminator can be phrased simply in terms of distance from their common center.

With all of these criteria satisfied, we are at last permitted to trust our visual impression of the corresponding ordination diagram, the subject of Figure 8.4. This one reminds us that there can be far better ways to describe a group difference than simply noting the mean shift of one single indicator. In keeping with our real theme, numerical inference from systems of organized complexity, we can assert a

488   8 Retrospect and Prospect

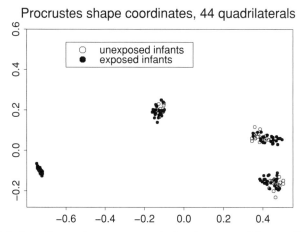

Figure 8.3. Underlying Procrustes structure producing the averaged figures of Figure 8.2.

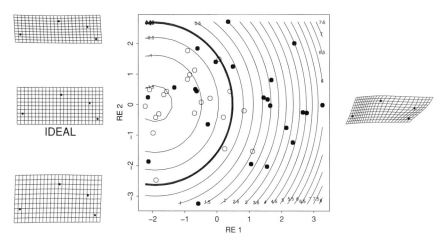

Figure 8.4. The data of Figure 8.3 support a clinically useful report of considerably higher complexity than the static prototype images in Figure 8.2: this quadratic discriminant function for detecting the infant callosal shapes at highest odds of having been alcohol-exposed. In the central display here, the horizontal axis is the relative tilt of the midsplenium segment (in effect, the same index reported in Bookstein, Connor et al. [2007]), and the vertical is effectively an affine term, height over width of this structure. The "ideal" pediatric form, average of the unexposed children at far left center in the scatter, is the quadrilateral in the grid in the left central margin. Discrimination is by distance from this ideal, which combines the tilt of the midsplenium, a unidirectional indicator, with the deviation of the overall height/width ratio from the average, a bidirectional indicator. Discrimination by the net dissimilarity from this ideal thus involves two dimensions of morphological variability, not just one. Solid disks, children prenatally exposed to alcohol; open circles, children not so exposed. Numbers printed on contours are natural logarithms of the likelihood ratio for the hypothesis of exposure. For instance, the line with the printed "3" represents a likelihood ratio of $e^3 \sim 20$ in favor of the hypothesis of exposure.

great deal more about the information in these configurations than just their averages and the average pictures that they reconstruct. Indeed Figure 8.4 has converted the information resources of Figure 8.2 into a clinically useful rule for detecting individual damaged children. The first two relative eigenvalues are nearly enough equal that a relative eigenanalysis of the type already exemplified in Figure 5.9 leads to a narrative summary that is tidy indeed: classification by **net similarity to an apparent "ideal form,"** the quadrilateral labeled as such in the center left margin. What justifies calling this particular quadrilateral "ideal" is its location at the maximum of likelihood for the hypothesis of normality – the centers of all those circles of equal likelihood ratio. Deviation from this ideal is assessed using a combination of the tilt of the midsplenium axis (here, along the horizontal) with the deviation of the aspect ratio of the quadrilateral from the average (along the vertical).

What all this implies is that some of the babies of alcoholic pregnant women can be shown to be likelier than not to have been damaged by prenatal alcohol in precisely the same region that is typically abnormal in adults diagnosed at some later age as prenatally damaged by this same agent. And the report of the detection criterion can go forward in the same rhetoric of mechanism, a "dysmorphy" – a disregulation of development, which is operationalized as a dissimilarity to forms in the normal range. Such a conclusion is far more congenial to the general theory of alcohol's effect on the fetal brain than the single index announced in the 2007 publication. In other words, the switch from simple-system methods to complex-system methods, without any change at all in the data, has translated the language of the infant finding into a version consilient with findings from earlier research into adolescents and adults. Further consilience must await further developments in pediatric ultrasound or early-life MRI.

The statistical manipulation here does not change the nature of the surprise, the massive shape difference shown via averaged pictures in Figure 8.2, but it makes it a more *useful* assertion, one that ameliorates the comfort with which the relevant community might respond to a finding of this somewhat untypical structure. Some communities are relatively open to succession of methodologies in this wise – the serial polishing of individual data sets by successive reanalyses. Others typically do not give much weight to reanalyses of old data, only to analyses of new data. But in today's world of complex systems, this latter strategy is inferior – it places too heavy a burden on the consilience step of a numerical inference, a burden that typically retards progress by comparison with the program of strong inference recommended by observers like Platt (1964) and emphasized in this book.

The combination of consilience and abduction can work to change the opinion of people whose opinions matter. John Snow (Section E4.1) altered the public health policies of a great city. Robert Millikan (Section E4.2) established the truth of Einstein's model of the photoelectric effect, and Jean Perrin (Section E4.3) that of Einstein's model of diffusion. That there exist health effects of passive smoking (Section E4.6) is no longer gainsaid even by tobacco companies. Ulcers were acknowledged to be infectious within 10 years of Barry Marshall's original insight

(Section E4.4). More than 30 years after his death, Alfred Wegener's hypothesis (Section 2.1) carried instantly as soon as a plausible mechanism was unearthed by which the effects he had suspected could be explained consiliently.

The differences between these trajectories of acceptance and those of findings that are still being negotiated, such as global warming, the dramaturgy of obedience, or, for that matter, my continuing effort to bring adequate statistical sophistication into studies of the fetal alcohol spectrum, may be sociological, or else a matter of cognitive style, the sort of willingness to be compelled by a quantification that Kuhn was reviewing in his essay on the role of measurement in the physical sciences. The exact sciences, and sometimes the public health sciences, defer to the rhetorical force of persistent anomalies, but the historical natural sciences seem not to. We cannot, for instance, force the physical anthropologist to concede that the observed signal from his measurements is too weak to support his theory in the face of reasonable alternatives. If he instead implacably asserts statistical significance over and over, he is behaving like an advocate, not a scientist. In contrast, Livingstone explained both the trend and the residual in his model for frequency of the sickle cell gene; an innovator like Milgram uses the full range of a scale to guarantee that the effect size justifies a reader's attention; a great teacher like Perrin runs the same experiment hundreds of times, varying every physical parameter he can manage; the Alvarez team invented new machines specifically to take seriously the hypotheses of colleagues with whom they did not agree. The best ironies in the history of science are often those that follow from abductions resisted or misunderstood as inconsilient with earlier observations. These often lead to sudden improvements in the rhetoric of numerical inference based on demonstrations of novel quantitative matches not anticipated until the resistance of a community made them necessary.

Recall the argument of Fleck (1935/1979) that led to Epigraph 14: a scientific fact is recognized by the stylized, instrument-driven way that what would otherwise be creative thought is constrained. Abductions are *likewise* recognized by the ways they constraint speculative thought (specifically, the thoughts that contribute to the confirmations/disconfirmations that must follow), and the less familiar role of new methodologies is to constrain the language in which those speculations go forward: those new languages must *incorporate* a rhetoric capable of confronting the abduced patterns with new data. We saw this in the example of ulcers (Section E4.4), for example, where initially the finding was unthinkable (or at least unpublishable) and where it took several years for a community to figure out how to attempt a consilience: not by checking that the bacteria cause ulcers (the test of Koch's postulates) but by checking that eliminating the bacteria eliminated the *recurrence* of ulcers. The rest of the argument then flowed easily through the customary media (*New England Journal of Medicine*, etc.) within a few years.[2]

---

[2] The first paper I ever submitted about shape coordinates, way back in 1982, was rejected by the flagship journal *Biometrics* on the grounds that its core methodological assertion – that you could measure shape exhaustively by a system of triangles – was obviously absurd inasmuch as nobody was currently measuring anything that way.

8.2 Implications for Inference Across the Sciences

Example by example, it can be shown that what unifies Parts II and III is more profound than what divides them. To every style of quantitative scientific research, there is a corresponding style of *representation of uncertainty* interposed between an explanation and its modes of quantitative confirmation or disconfirmation. Whether the context is the abduction this book recommends, or instead the more classical modes of induction and deduction that concern other approaches to scientific method, there is always a component whose job it is to bridge the measurements (the instrument readings) to the formal symbolic representations (regressions, exact equations, singular vectors) that notate the desired inferences most tersely. Recollect Platt's (1964) principle of strong inference, Section 3.4: data are used effectively only when they eliminate competing theories, quantitative or otherwise, at the same time that they help us argue the validity of our own. The task before us – to decide when data are *not* consistent with prior expectation – evidently must precede every form of quantitative inference, from the highly nuanced likelihood approaches to the equally computational searches in multivariate spaces arising from SVDs that have concerned us in Chapters 6 and 7.

Competence in this task is inseparable from expertise in the ways numerical patterns can relate to verbal summaries. These depend, in turn, on cognitive states of the receiving scientist's mind, and thus as much on psychological processes as on empirical facts. Extending this reasoning, one might aver that the foundation of numerical science must be an interdiscipline between cognitive psychology and probability theory; but this assigns too little weight to the role of embodied human knowledge and to the inborn pattern perception skills on which Peirce grounded his speculations about the mysterious capacity of abduction to arrive at truths from time to time. The division between the examples of Part II and those of Part III may be traced in the level of conceptual difficulty of associated claims about the rules of correspondence between measurement and symbol that bridge cognition to arithmetic. Consider the contrast between the immediacy of Watson's "sudden awareness," Section E4.5, about the shape match between the two nucleotide pairings at the core of the double helix, and the complexity of the summary rendition of the fetal alcohol corpus callosum finding (Figure 7.25) that permits its conversion into a practical detection rule for forensic applications (Bookstein and Kowell, 2010). Or consider what is required to confirm the peaks of the Ramachandran plot, Figure 5.3, by computations about the secondary structure of folded chains of amino acids when they are in their energetically stable configurations (Tramontano, 2005).

The reliability of computationally extracted summary descriptors is ordinarily given the label of "statistical inference," by contrast with the *numerical inference*, the construction of abductive and consilient justifications of numerical propositions, that is the topic of this book. (See, for instance, Cox 2006, who notes on p. 7 that statistical inference "takes the family of models as given and aims to give intervals or in general sets of values within which ψ [the parameter] is likely to lie, [and] to assess the consistency of the data with a particular parameter value ψ_0.") Broadly speaking, we have reviewed three different ways of dealing with the challenges it conveys. In Part I we relied on the Interocular Trauma Test, in keeping with which we report only the patterns most outrageously improbable a priori. For these, challenges to the detected pattern seem fatuous. The only rational response, at least among communities who care about the content domain, is to turn directly to explorations of possible consilience at other levels. In Part II, where summaries of patterns generally took the form of single quantities, we dismissed the possibility of "null-hypothesis statistical significance testing" in favor of a Bayesian approach simply setting out the likelihood of each hypothesis within a numerical range given the data at hand.

In Part III, the very idea of that null as a reference value of zero for some quantity is replaced by the notion of a wholly symmetrical (and thus uninterpretable, intuitively inaccessible) distribution of some geometrically more extended summary like a principal component or an underlying deformation grid. If an abduction or a consilience is to concern some a-posteriori feature of such a multivariate summary, that feature, too, needs to be challenged for stability. The earliest of these concerns, the stability of a principal component, was reduced to a formula in Section 5.2.3.4 using an approach modified from T. W. Anderson's work of the 1960s, while most of the other versions could have been demonstrated as information-based tests assessing the sharpening of the likelihood of an entire data set when the explanatory model is granted access to additional explanatory factors. The distinction between these two general approaches becomes much more salient for models that permit local findings (see Section 6.5 for a morphometric example). When data models do not postulate independence of cases, though, often the necessary likelihoods cannot be calculated by a formula. For numerical insights into data records that could be envisioned as the traces of random walks, for instance, the appropriate tests must be against explicit simulations of the descriptor actually intended for reporting.[3]

One converse of complexity is *clarity*. In all the examples of Part I and Part II it is eventually clear what a match between theory and explanation would *mean* for the numbers or the graphs. For the examples of Part III, the argument is more difficult, more nuanced. Before we claim a match between theory and data – the main sequence as seen in the Hertzsprung–Russell diagram (Figure 5.1), the beta sheet as seen in the Ramachandran plot (Figure 5.3), the effect of prenatal alcohol as seen in average images of corpus callosum form (Figures 7.22 and 8.3) – it is necessary to have argued that the named phenomenon *is* captured in the quantitative pattern analysis referred to, and that whatever metrological chain of instrument settings and environmental controls has been put in place is sufficiently powerful to disambiguate what would be the otherwise hopeless cognitive task of selecting from this "blooming, buzzing confusion" the message from Nature that corresponds to the explanatory intent.

As Wigner argued half a century ago, this is the most difficult task ever faced by the physicist: to guess what quantities might actually be the desired invariants, and to

---

[3] The topic is too technical for this book; see, instead, Bookstein (2013b).

show that they are, indeed, invariable against reasonable estimates of the measurement error inherent in their assessment – within what Kuhn (1961) calls "the limits of reasonable agreement." Perrin (1923) is particularly informative about this logic, and Watson (1968) clarifies what Watson and Crick (1953) left completely tacit on the same theme. This need for selection of particular quantities for discussion out of a wider range of choices is shared among all the sciences that use numerical inference in any form. Likewise shared between Part II and Part III are the limitations of the machines used for measuring. The precision of Rosalind Franklin's diffraction images is computed using some of the same formal mathematics as the reliability of the three-dimensional medical images driving the assessment of the Visible Female's anatomy (Figure 2.8) and likewise that of the alcohol-exposed newborns in Figure 8.1. We have seen examples of the failure of this sort of intuition, univariately (the claim of infinite dilution, Section L4.4) or multivariately (the misidentification of the psychosocial effects of social class as the effects of prenatal alcohol, Figure 7.29). Mistakes like these are not errors of numerical inference, but something much more fundamental: the snap judgment, the "jump to conclusions" – a built-in aspect of human cognition, antithetical to the skepticism on which normal science is founded, that is nevertheless of high value in other contexts than the scientific (see Kahneman, 2011).

There is a biographical/historical aspect to these concerns as well whenever the scholar is writing as participant in an ongoing scientific debate not yet settled. The very idea of "accuracy" requires prior agreement regarding the spread of predictions or repetitions that is to be assessed. Even in the presence of a newly precise pattern engine producing abductions that are broadly consistent with earlier qualitative understandings, there will necessarily be a time lag before issues of consilience are raised, a time lag over which a community must come to realize that the new abductions *are* comparable with earlier understandings, however different the words or the graphical iconography. Where communicating fields are not accustomed to this degree of precision, and so cannot keep up with the language of quantification, these lags may endure over entire academic generations. In Ernst Mach's metaphor, it is not only the believers in the old *theories* who need to die for a new theory to be established, but likewise the expert propagators of the old *methods*.

8.3 What, Then, IS to Be Done?

Most of our earlier examples began with an abduction, the perception of a surprising pattern in data together with a proposition that, if true, would reduce the observation to "a matter of course." One hopes that an abduction is followed sooner or later by a consilience, the match of some numerical summary of the data set at hand by analogous, but not strictly parallel, numerical summaries from other data sets generated in different samples or at different levels of measurement. Here arises an issue that becomes steadily more problematic as the volume of available data grows seemingly without limits: *how do we decide whether a complex data record matches an "expectation"?* The examples of Part II presume that the object of discussion is a very short vector (perhaps as short as one scalar, plus, say, a grouping label), and that all of the rhetorical concern is with the accuracy of reports of this one quantity. What renders the concerns of Part III more problematic and open-ended (indeed, in most fields, incapable of closure for any length of time) is that we have as yet no canonical methods for situations in which the complexity of the *data* exceeds our ability to provide persuasive summary rhetorical representations.

The collective task, then, is twofold. The first assignment is to make the various communities studying complex systems aware that simple methods of numerical inference (meaning, in practice, the simple methods of data summary and the associated statistical formularies that we teach our undergraduates) are not adequate. This is a task of *un*learning. The second, more advanced task is to begin injecting into the curriculum the methods of advanced pattern analysis that can be shared across the comparably many disciplines and interdisciplines that use complex data designs. The requisite methods are intended to be those that support actual human scientists: the topic is theory-grounded explanation, not just similarity detection.

The principles driving that curriculum can be distilled from the epigraphs of this book. In that form, they apply to the pedagogy of statistical science, regardless of discipline, at any level from the introductory course upward:

- The interocular trauma test, ITT (Epigraph 1, Edwards et al., 1963)
- Abduction, the scientific inference designed for numbers (Epigraph 9, Peirce, 1903)
- Consilience: measures should agree over machines (Epigraph 2, Wilson, 1998)
- The unreasonable effectiveness of mathematics (Epigraph 5, Wigner, 1960)
- Science as cognitive (Epigraphs 7 and 12, Kuhn, 1961, and Bloor, 2011) or social (Epigraph 14, Fleck 1969) rather than empirical
- Passion and the "deep sympathy with nature" (Epigraphs 3, 4: Russell, 1967, and Pearson, 1906)

These apply whether one's data resource is simple or complex, whether one's field is well-plowed or instead being cleared of brush for the first time. They do not merely describe my idiosyncratic approach to the rhetoric of statistical science per se; they also describe the ways in which that science can function to critique the rhetoric of the communicating disciplines that supply its data and its justifications. After all, one must find hypotheses that fit one's data sets, and one's noise models must match prior knowledge of the instruments that helped us gather those data. It is obligatory to show that no other explanations fit the data as well as our preferred mechanisms or causal narratives. And it is necessary to defend the role that human cognition and intuition have played in our choosing to pay attention to these particular hypotheses, these particular quantities or quantitative summaries – the issue of "where your list of variables came from." If these are overlooked, it is the scientist's loss, not the statistician's.

I said at the very outset of this book that my topic was the conversion of arithmetic into understanding. One who is zealously optimistic about the future of "artificial intelligence" or "machine learning" might argue that the human is not necessarily a part of this construction. (The most zealous presentation of all is probably

Ray Kurzweil's, from 2006; he actually forecasts a specific year by which the human's presence in this dataflow will be optional.) The kind of pattern analyses referred to here could conceivably be carried out purely *in silico*, with any subsequent "understanding" construed as, perhaps, a sort of optimized intervention in contexts like human medicine: robot surgery, perhaps, without a surgeon. This is one theme of a newly emergent field, bioinformatics, that has not been treated here except for a brief mention in Section 6.4.3.4. The reason is that the context of those pattern analyses does not capture my meaning, at least, not for the kinds of general systems queries that are raised across the range of sciences that this book has touched on. Rather, the task combines information processing with actual human understanding and with the dissemination of summaries of the emerging knowledge outward into the social systems that can make use of it.

In other words, the content of those qualitative summaries is a negotiation between one's "arithmetic" and one's "understanding." Between Part II and Part III there has been a substantial reinterpretation of a crucial metaphor, again from Thomas Kuhn's seminal 1961 essay. Kuhn wrote, on pages 189–190:

> *The road from scientific law to scientific measurement can rarely be traveled in the reverse direction.* To discover quantitative regularity one must normally know what regularity one is seeking and one's instruments must be designed accordingly; even then nature may not yield consistent or generalizable results without a struggle. [emphasis in original]

The metaphor to which I wish to attend, here at the close, is the notion of a path being traveled in one direction or "the reverse." We have seen that the relation between arithmetic and understanding is not like that. It is not a "road" to be "traveled," but the construction of a joint interpretation of natural or social reality that is held in common between two conceptual frameworks, the one of measurement with all its details of instrumentation and (un)reliability, the other of explanation, particularly consilience among explanations at different levels.

Kuhn is talking about physics, an exact science where explanations take the form of "laws" that, in turn, take the symbolic form of equations. But the equations of the inexact sciences are not like those of physics. Their epistemology is wholly different: they are not equalities in which *observed quantities* appear, but distributions in which the *uncertainties of observed quantities* or relationships appear. The governing metaphor for numerical inference in complex systems (systems with variability, systems with biography, systems governed by regularities at many levels at the same time) is not a "road" between data and theory but, instead, a woven intellectual *fabric*. The anatomical sciences have a better metaphor to offer us. A *rete mirabile* is an interpenetration of two physical branching networks allowing for exchange or mutual influence of their separate contents by virtue of both systems' approaching arbitrarily close to every point within the same organismal volume. In its original anatomical context the term pertains to vertebrate circulatory systems. As a metaphor it refers to the interconnection of data with their interpretations or explanations at all loci of an extended system, not only on some numerical boundary. The rhetoric by which we explore these extended interplays, as reviewed in Part III, alternates between assertion and skepticism, following Feynman's advice or Platt's, conscientiously challenging both sides of an argument in the same presentation.

The difference between Part II and Part III, then, is that in Part III, the setting of complex systems, there is no real possibility of "confirmatory" statistical analysis at all, not if by "confirmation" one means some specific further computation that settles an argument at the community level. All that exist are the procedures that explore pattern spaces; the data control the reports of these spaces far more directly than prior expectations no matter how strongly structured. Only such a rhetoric can combine the results of a pattern engine computation with its implications for the larger disciplinary or interdisciplinary community. The findings of pattern engines are *always* surprising, at least, at their level of observation. Consider, again, the three patterns illustrated in Section 5.1: the Hertzsprung–Russell diagram, explained by astrophysics and stellar nucleochemistry; the Ramachandran plot, explained by energetics of protein folding at small and smaller scales; and the face of fetal alcohol syndrome, explained by processes of cell migration ultimately modulated by the receptors on their surface membranes. The existence of the pattern would be of little interest were it not for the possibility of these cross-level consiliences. Indeed, the purpose of the pattern analysis is, precisely, to *motivate* (i.e., instigate) the investigations that reduce studies of the Part III type to studies of the Part II type: studies whose quantifications become explanatorily more cogent by virtue of being at a different level of measurement (for the Hertzsprung–Russell, not color and brightness but mass and hydrogen:helium balance, etc.).

Studies in the Part III style, in other words, are intrinsically predisposed to the context of the consilient multilevel narrative in a way that studies of the Part II style are not. And nearly all of the important investigations of applied quantitative science here in the early 21st century are indeed studies of this Part III form. The sense of closure that ends a Part II investigation – "Thus we have proved" or "So we have shown" or the other tropes of Section 1.1 – seems unsuited to studies of complex systems, where any pattern detected is intrinsically ambiguous, bearing a meaning that is best confirmed or challenged at other levels of analysis.

Today's important questions of quantitative science are Part III questions, questions about patterns in complex systems redundantly measured, and the standard statistical curricula cover these very badly when they cover them at all. The message of this book is thus ultimately a pedagogical one. **We need statistical curricula that center on complex systems, the Part III questions**, instead of the curricula of today that center on questions of the Part II style, where it is assumed you have measured the correct few quantities, and your grouping variables are the correct ones, and the relation of theory to data is just the matter of a summary sentence or two stating a comparison between two numbers.

Scientifically speaking, today's typical undergraduate curriculum in applied statistics is a strikingly unpromising investment of human capital – a waste of the students' and the professors' time. *No important question of today will be answered by the consideration of a single correlation* – if the question can be answered by examination of just one correlation, it is very unlikely to be of any importance. Likewise, *no important question can be answered by a multiple regression*, not only for the reasons given in Section L4.5.3 regarding the uncertainty of the resulting coefficients but because of uncertainty in the roster of variables doing the predicting and in the roster of indicators aggregated in the predictand, whatever it may be. The quantitative understanding of important questions cannot be disentangled, in practice, from the question of how to measure the specific quantities that are the crux of the corresponding explanations and, if the data were selected from measured spectra, how that selection from the manifold of all the other possibilities was managed. (This was the central philosophical point of Herman Wold's (1982) original methodology of Partial Least Squares, a method specifically crafted for application to "complex systems indirectly measured." For further thoughts on this underappreciated theme, see Martens and Kohler, 2009, and more generally Martens and Næs, 1989 and Munck, 2005.)

The points here are not specific to the techniques and examples I selected for either Part II or Part III. Instead the examples were selected mainly for the limited range of mathematical notation required for their effective understanding, so that the underlying logical and rhetorical points could be visible through as transparent as possible a symbolic superstructure. There are other kinds of organized variables than the spectra, images, and so forth reviewed in Section 6.3.2: there are, for instance, graphs and networks. These are not "beyond the scope of this volume," in the usual apologia, but instead just differently organized, along lines I have not the space to introduce properly. Event series (as in Atwater's example concerned with earthquake prediction, Section 3.2), multiple densely observed time series such as electrocardiograms or electroencephalograms that combine periodic signals with noise or chaos, and many, many other modern examples involve different algebraic formalisms than those used in the examples here, yet all rely ultimately on the same logic of numerical inference, combining abduction and consilience with an appreciation (shrewd, it is hoped) of the fallibility of the original data and the samples (both of specimens and of measures) on which they are based.

Across all of these applications, some exemplified in these pages and others not, the underlying rhetorical resources are the same: the cognitive origins of abduction in surprise and in pattern-matching, and the origins of consilience in scientific creativity, the perception of formal analogies, and the correct intuition of Nature's mechanisms. Numerical inference is not so much about formulas as about persuasion, and our statistical methodology and especially our statistical pedagogy need to follow, rather than lead, the needs of this rhetoric.

Here at the end of this long essay, then, I invite you to revisit the examples at the very outset of Chapter 1. How do we decide that data have shown us something, or not shown? How do we find the hints in numerical records that lead us to new hypotheses about the meaning of those records? How, in short, do we turn arithmetic into understanding across the full range of 21st-century scientific contexts and purposes, both the natural sciences and the social sciences?

For the simple classical questions, those for which issues of instrumentation and uncertainty have been mastered in centuries past, the lectures of Part II – averaging and least squares, bell curves, inverse probability, and path analysis – may be sufficient. But for the contemporary questions on which depend our present and future quality of life, questions of the Part III type, we have too little experience in converting pattern analyses into either stable explanations or stable rules for generating new hypotheses. It is the generalizations of the techniques of Chapter 6 – the singular-value decomposition and its variants, with applications from fetal alcohol damage to Netflix marketing strategies – that should be the toolkit of today's graduate students, not the generalizations of the techniques of Chapter 4. And it is the generalizations of Chapter 5 that should constitute the interesting data structures – not flat matrices but structures of much more complex geometry, and not Gaussian models but far more subtle ways of generating dependencies among quantities.

Chapter 5 briefly reviewed Kenneth Boulding's (1956) construction of a "general systems theory" in the context of the curriculum sketched by Bode et al. (1949) for the training of the "scientific generalist." Regarding statistics per se the Bode team's recommendation was somewhat blindered, limited to elementary mathematical statistics, some multivariate methods, and experimental design. Boulding's approach, though published only a few years later, is much more in keeping with the modern systems-based methodology exemplified by SVDs or morphometrics in this book, and so his syllabus might serve as a useful closing chart for our discussion here. Boulding suggests that the generalist should first be on the lookout for the kinds of models found useful in more than one discipline: for instance (page 200), models of "populations – aggregates of individuals conforming to a common definition, to which individuals are added (born) and subtracted (die) and in which the age of the individual is a relevant and identifiable variable." Models of population change, he notes, cut across a huge range of disciplines; likewise, models of the interaction of individuals with their environment; likewise, models of growth. These correspond, in their general logic, to the kinds of models we considered in Part II.

There is a second approach, though, that Boulding wrote about on pages 202–205 of the same essay and that he actually taught in the course of his that I took in 1966. This alternative did not pursue any list of commonalities among diverse discipline-specific models, but rather the "arrangement of theoretical systems and constructs in a hierarchy of complexity," in a spirit quite similar to that of Part III of this book. The work is not retrospective, looking back on what individual disciplines do well, but prospective, searching in the direction of what they currently cannot do very well or indeed cannot do at all. This second mode of Boulding's comprises a structured *sequence of examinations* of steadily greater sophistication and abstraction, beginning with the "static structures" of level 1 – geography and anatomy, patterns and arrangements – and continuing through clockwork dynamics (level 2), cybernetics and the maintenance of equilibria (level 3), open systems (level 4), botanical systems with division of labor (level 5), zoological systems with mobility and internal goal-setting (level 6), humans as conscious, symbol-processing individuals (level 7), and human social and historical structures, including the sciences themselves (level 8).

Each stage is more complicated than any preceding stage or any combination of preceding stages. In this context, the statistical curriculum that Bode et al. (1949) recommended for their scientific generalist corresponds well to Boulding's levels 1 and 2, but even in 1956 he had already anticipated today's emphasis on the sciences of organized complexity, along with the corresponding need for the much more powerful languages of pattern that communicate our understanding of how they work and how they can serve as analogies of one another. This multilevel *general systems theory*, he concludes (page 208), is "the skeleton of science," the framework on which to hang the understanding of the interrelationships among the disciplines. At the same time, the paucity of vocabulary for general systems theory was a sign of how far we have to go even now regarding analysis of "the problems and subject matters which do not fit easily into simple mechanical schemes," and, in that sense, "*where* we have to go."

In the contrast between the familiar statistical methods of Part II and the less familiar methods of Part III is a possible germ for a curriculum that translates the rhetoric of abduction and consilience, surprise and discovery, up Boulding's ladder of emergent structures. Beyond the key terms already introduced in Chapters 5 and 6, central concerns of such a curriculum would emphasize the origins of data in output of calibrated machines whenever possible, alternatives to the SVD-derived toolkit of pattern engines when they are applicable in specific fields, more specific formalisms for the challenge of fusing results across levels of measurement that reifies Wilson's fundamental vision of consilience, and modifications to all of these results when data arise not from machines but instead as reports filtered through human nervous systems. Those reports can take the form of instrumented images, verbal narratives, or tacit human choices from a range of options proferred by the environment, alone or in any combination. Among the sciences to be added to the coverage already exemplified in this book would surely be examples from such emerging interdisciplines as animal cognition, neuroethics, contemporary evolution of microörganisms, social contagion processes, and global monitoring of ecology and urban conditions. But such a rebuilding of a major component of our modern liberal-arts curriculum is a topic deserving of a full-length essay on its own, one drafted by a committee, not a short speculative comment late in the text of a single-authored essay like this one.

Assume that a curricular reform aligned with the principles of abduction and consilience is put in place sometime in the near future. To maintain the relevance of numerical inference as 21st-century science evolves in a social context that is likewise evolving, numerical inference itself has to evolve, in order to remain relevant to the data flows that comprise the information that needs to be converted into understanding. Without those new modes of inference, there is no point, scientifically speaking, to all of that wonderful data. It is too frustrating to go on analyzing 21st-century data records, in all of their rich and reticulated complexity, by the statistical methods envisioned by Karl Pearson more than a century ago. If quantitative reasoning is to remain relevant to the public face of science at all – if it is not to degenerate into the quasipolitical style of argument that characterizes today's arena for discussions of global warming, for example – then the relation between the arithmetic and the understanding has to be recast to accommodate the sea change in data complexity over the last century and continuing into the next.

This essay has summarized my view of where we are now along this trajectory. Driven by the computational environment of the near future – Google mash-ups, medical data federations, summaries of cell-phone connections, global environmental databases, and the like – an exposition dated 2030 will surely involve a completely different balance of techniques applied to a completely different series of examples. The success of this book is entwined with the contrast between the content here and the content of some successor volume, if successor there proves to be. I look forward with indescribable anticipation to the obsolescence of all the examples here, and their replacement by equally abductive, equally consilient instances drawn from the scientific successes of the near future. The succession will testify to the continuing relevance of numerical inference, which will remain one of the fundamental tools, however rarely foregrounded, by which we make individual and social sense of all these scientific worlds new to the new century.

References

Akaike, H. A new look at the statistical model identification. *IEEE Transactions on Automatic Control* AC19:716–723, 1974.
Alexander, C. *A Pattern Language: Towns, Buildings, Construction*. Oxford University Press, 1977.
Allais, M., and O. Hagen, eds. *Expected Utility Hypotheses and the Allais Paradox: Contemporary Discussions of Decisions under Uncertainty, with Allais' Rejoinder*. D. Reidel, 1979.
Alvarez, L. W., W. Alvarez, F. Asaro, and H. V. Michel. Extraterrestrial cause for the Cretaceous-Tertiary extinction. *Science* 208:1095–1108, 1980.
Andersen, P. K., and L. T. Skovgaard. *Regression with Linear Predictors*. Springer, 2010.
Anderson, D. R. *Model Based Inference in the Life Sciences: A Primer on Evidence*. Springer, 2008.
Anderson, T. W. Asymptotic theory for principal component analysis. *Annals of Mathematical Statistics* 34:122–148, 1963.
Anson, B. J. *An Atlas of Human Anatomy*. W. B. Saunders, 1950. 2nd edition, 1963.
Arrow, K. J. *Social Choice and Individual Values*. John Wiley & Sons, 1951.
Astley, S. J., E. H. Aylward, H. C. Olson, K. Kerns, A. Brooks, T. E. Coggins, J. Davies, S. Dorn, B. Gendler, T. Jirikovic, P. Kraegel, K. Maravilla, and T. Richards. Magnetic resonance imaging outcomes from a comprehensive magnetic resonance study of children with fetal alcohol spectrum disorders. *Alcoholism: Clinical and Experimental Research* 33:1–19, 2009a.
Astley, S. J., H. C. Olson, K. Kerns, A. Brooks, E. H. Aylward, T. E. Coggins, J. Davies, S. Dorn, B. Gendler, T. Jirikovic, P. Kraegel, K. Maravilla, and T. Richards. Neuropsychological and behavioral outcomes from a comprehensive magnetic resonance study of children with fetal alcohol spectrum disorders. *Canadian Journal of Clinical Pharmacology* 16:e178–201, 2009b.
Atwater, B. F. Evidence for great holocene earthquakes along the outer coast of Washington State. *Science* 236:942–944, 1987.
Atwater, B. F., M.-R. Satoko, S. Kenji, T. Yoshinobu, U. Kazue, and D. K. Yamaguchi. *The Orphan Tsunami of 1700: Japanese Clues to a Parent Earthquake in North America*. U. S. Geological Survey and University of Washington Press, 2005.
von Baeyer, H. C. *Taming the Atom: the Emergence of the Visible Microworld*. Random House, 1992.
Ball, P. *Nature's Patterns: A Tapestry in Three Parts*. Part 1, Shapes. Part II, Flow. Part III, Branches. Oxford University Press, 2009.
Barabási, A.-L. *Linked: How Everything is Connected to Everything Else and What it Means*. Plume, 2003.
Bauer, H. H. *Scientific Literacy and the Myth of the Scientific Method*. University of Illinois Press, 1994.

Baumrind, D. Some thoughts on ethics of research: After reading Milgram's "Behavioral Study of Obedience." *American Psychologist* 19:421–423, 1964.

Baxter, R. J. *Exactly Solved Models in Statistical Mechanics*, 2nd ed. Dover Books, 2007.

["Belmont Report."] *Ethical Principles and Guidelines for the Protection of Human Subjects of Research*. National Commission for the Protection of Human Subjects in Biomedical and Behavioral Research, U. S. Department of Health, Education, and Welfare, 1979.

Bem, D. J. Feeling the future: Experimental evidence for anomalous retroactive influences on cognition and affect. *Journal of Personality and Social Psychology* 100:407–425, 2011.

Benfey, O. T. [Spiral periodic table.] 1960. Retrieved August 5, 2011 from http://en.wikipedia.org/wiki/File:Elementspiral.svg.

Benjamini, Y., and Y. Hochberg. Controlling the False Discovery Rate: a practical and powerful approach to multiple testing. *Journal of the Royal Statistical Society* B57:289–300, 1995.

Benzécri, J.-P. *L'Analyse des Données*. Vol. 2, *Analyse des Correspondances*. Dunod, 1973.

Berry, W. *Life Is a Miracle: An Essay against Modern Superstition*. Counterpoint, 2000.

Bertin, J. *Semiology of Graphics: Diagrams, Networks, Maps*. ESRI Press, 2010.

Blackith, R. E., and R. A. Reyment. *Multivariate Morphometrics*. Academic Press, 1971.

Bloor, D. *The Enigma of the Aerofoil: Rival Theories of Aerodynamics, 1909–1930*. University of Chicago Press, 2011.

Blum, H. Biological shape and visual science, part 1. *Journal of Theoretical Biology* 38:205–287, 1973.

Boas, F. The horizontal plane of the skull and the general problem of the comparison of variable forms. *Science* 21:862–863, 1905.

Bode, H., F. Mosteller, J. Tukey, and C. Winsor. The education of a scientific generalist. *Science* 109:553–558, 1949.

Bolhuis, J. J., G. R. Brown, R. C. Richardson, and K. N. Laland. Darwin in mind: New opportunities for evolutionary psychology. PLoS Biology 9:e1001109, 2011.

Bookstein, F. L. *The Measurement of Biological Shape and Shape Change*. Lecture Notes in Biomathematics, Vol. 24. Springer-Verlag, 1978.

Bookstein, F. L. The line-skeleton. *Computer Graphics and Image Processing* 11:123–137, 1979.

Bookstein, F. L. Size and shape spaces for landmark data in two dimensions. *Statistical Science* 1:181–242, 1986.

Bookstein, F. L. Random walk and the existence of evolutionary rates. *Paleobiology* 13:446–464, 1987.

Bookstein, F. L. Principal warps: Thin-plate splines and the decomposition of deformations. *I.E.E.E. Transactions on Pattern Analysis and Machine Intelligence* 11:567–585, 1989.

Bookstein, F. L. *Morphometric Tools for Landmark Data: Geometry and Biology*. Cambridge University Press, 1991.

Bookstein, F. L. Utopian skeletons in the biometric closet. *Occasional Papers of the Institute for the Humanities, the University of Michigan*, number 2, 1995.

Bookstein, F. L. A standard formula for the uniform shape component in landmark data. In L. F. Marcus, M. Corti, A. Loy, G. J. P. Naylor, and D. E. Slice (Eds.), *Advances in Morphometrics* pp. 153–168. NATO ASI Series A: Life Sciences, Vol. 284. Plenum Press, 1996a.

Bookstein, F. L. Biometrics, biomathematics, and the morphometric synthesis. *Bulletin of Mathematical Biology* 58:313–365, 1996b.

Bookstein, F. L. Landmark methods for forms without landmarks: Localizing group differences in outline shape. *Medical Image Analysis* 1:225–243, 1997a.

Bookstein, F. L. Shape and the information in medical images: A decade of the morphometric synthesis. *Computer Vision and Image Understanding* 66:97–118, 1997b.

Bookstein, F. L. Creases as local features of deformation grids. *Medical Image Analysis* 4:93–110, 2000.
Bookstein, F. L. After landmarks. In D. E. Slice (Ed.), *Modern Morphometrics in Physical Anthropology* (pp. 49–71). Kluwer Academic, 2004.
Bookstein, F. L. My unexpected journey in applied biomathematics. *Biological Theory* 1:67–77, 2006.
Bookstein, F. L. Morphometrics and computed homology: An old theme revisited. In N. MacLeod (Ed.), *Proceedings of a Symposium on Algorithmic Approaches to the Identification Problem in Systematics* (pp. 69–81). Museum of Natural History, London, 2007.
Bookstein, F. L. Templates: A complex implicit parameter for morphometric analyses takes the form of an algorithmic flow. In S. Barber, P.D. Baxter, A. Gusnanto, and K. V. Mardia (Eds.), *The Art and Science of Statistical Bioinformatics* (pp. 58–63). University of Leeds, 2008.
Bookstein, F. L. Measurement, explanation, and biology: Lessons from a long century. *Biological Theory* 4:6–20, 2009a.
Bookstein, F. L. How quantification persuades when it persuades. *Biological Theory* 4:132–147, 2009b.
Bookstein, F. L. Was there information in my data? Really? [Essay review of Anderson, 2008.] *Biological Theory* 4:302–308, 2009c.
Bookstein, F. L. For isotropic offset normal shape distributions, covariance distance is proportional to Procrustes distance. In A. Guznanto, K. V. Mardia, and C. Fallaize (Eds.), *Proceedings of the 2009 Leeds Annual Statistical Research Workshop* (pp. 47–51). University of Leeds, 2009d.
Bookstein, F. L. Biology and mathematical imagination: The meaning of morphometrics. Rohlf Prize Lecture, Stony Brook University, October 24, 2011. http://youtu.be/LvwF3fYv7Ys.
Bookstein, F. L. Allometry for the twenty-first century. *Biological Theory* 7:10–25, 2013a.
Bookstein, F. L. Random walk as a null model for high-dimensional morphometrics of fossil series: Geometrical considerations. *Paleobiology* 39:52–74, 2013b.
Bookstein, F. L., B. Chernoff, R. Elder, J. Humphries, G. Smith, and R. Strauss. *Morphometrics in Evolutionary Biology. The Geometry of Size and Shape Change, with Examples from Fishes.* Academy of Natural Sciences of Philadelphia, 1985.
Bookstein, F. L., P. D. Connor, J. E. Huggins, H. M. Barr, K. D. Covell, and A. P. Streissguth. Many infants prenatally exposed to high levels of alcohol show one particular anomaly of the corpus callosum. *Alcoholism: Clinical and Experimental Research* 31:868–879, 2007.
Bookstein, F. L., and W. D. K. Green. A feature space for edgels in images with landmarks. *Journal of Mathematical Imaging and Vision* 3:231–261, 1993.
Bookstein, F. L., P. Gunz, P. Mitteroecker, H. Prossinger, K. Schaefer, and H. Seidler. Cranial integration in *Homo*: Singular warps analysis of the midsagittal plane in ontogeny and evolution. *Journal of Human Evolution* 44:167–187, 2003.
Bookstein, F. L., and A. Kowell. Bringing morphometrics into the fetal alcohol report: Statistical language for the forensic neurologist or psychiatrist. *Journal of Psychiatry and Law* 38:449–474, 2010.
Bookstein, F. L., and P. D. Sampson. Suggested statistical standards for NTT manuscripts: Notes from two of your reviewers. *Neurotoxicology and Teratology* 27:207–415, 2005.
Bookstein, F. L., P. D. Sampson, P. D. Connor, and A. P. Streissguth. The midline corpus callosum is a neuroanatomical focus of fetal alcohol damage. *The Anatomical Record – The New Anatomist* 269:162–174, 2002.
Bookstein, F. L., K. Schaefer, H. Prossinger, H. Seidler, M. Fieder, C. Stringer, G. Weber, J. Arsuaga, D. Slice, F. J. Rohlf, W. Recheis, A. Mariam, and L. Marcus. Comparing frontal cranial profiles in archaic and modern *Homo* by morphometric analysis. *The Anatomical Record – The New Anatomist* 257:217–224, 1999.

Bookstein, F. L., A. P. Streissguth, P. D. Sampson, and H. M. Barr. Exploiting redundant measurement of dose and behavioral outcome: New methods from the behavioral teratology of alcohol. *Developmental Psychology* 32:404–415, 1996.

Bookstein, F. L., A. P. Streissguth, P. D. Sampson, P. D. Connor, and H. M. Barr. Corpus callosum shape and neuropsychological deficits in adult males with heavy fetal alcohol exposure. *NeuroImage* 15:233–251, 2002.

Bookstein, F. L., and P. D. Ward. A modified Procrustes analysis for bilaterally symmetrical outlines, with an application to microevolution in *Baculites*. *Paleobiology* 39:213–234, 2013.

Boulding, K. E. General systems theory – the skeleton of science. *Management Science* 2:197–208, 1956.

Bowden, D. M., G. A. Johnson, L. Zaborsky, W. D. K. Green, E. Moore, A. Badea, M. Dubach, and F. L. Bookstein. A symmetrical Waxholm canonical mouse brain for NeuroMaps. *Journal of Neuroscience Methods* 195:170–175, 2011.

Bowker, G. C., and S. L. Star. *Sorting Things Out: Classification and Its Consequences*. MIT Press, 1999.

Bowman, C., O. Delrieu, and J. Roger. Filtering pharmacogenomic signals. In S. Barber, P. D. Boxter, K. V. Mardia, and R. E. Walls (Eds.), *Interdisciplinary Statistics and Bioinformatics: The 25th Leeds Annual Statistical Workshop* (pp. 41–47). University of Leeds, 2006.

Bowman, C., and C. I. Jones. Genetic evidence-of-no-interest, pathway Sudoku, and platelet function. In A. Gusnanto, K. M. Mardia, C. Fallaize, and J. Voss (Eds.), *High-throughput Sequencing, Proteins, and Statistics: The 29th Leeds Annual Statistical Research Workshop* (pp. 53–58). University of Leeds, 2010.

Brent, J. *Charles Sanders Peirce: A Life*, 2nd ed. Indiana University Press, 1998.

Brillouin, F. *Science and Information Theory*. Academic Press, 1962.

Brinton, W. C. *Graphical Methods for Presenting Facts*. New York: The Engineering Magazine Company, 1914.

Brumfiel, G. Academy affirms hockey-stick graph. *Nature* 441:1032–1033, 2006.

Buchler, J., ed. *The Philosophy of Peirce: Selected Writings*. Harcourt Brace, 1940.

Bullard, E., J. E. Everett, and A. G. Smith. The fit of the continents around the Atlantic. *Philosophical Transactions of the Royal Society of London A* 258:41–51, 1965.

Bulygina, E., P. Mitteroecker, and L. C. Aiello. Ontogeny of facial dimonphism and pattern of individual development within one human population. *American Journal of Physical Anthropology* 131:432–443, 2006.

Bumpus, H. C. The elimination of the unfit as illustrated by the introduced sparrow, *Passer domesticus*. *Biological Lectures of Woods Hole Marine Biological Laboratory*, 209–225, 1898.

Burnham, K. P., and D. R. Anderson. *Model Selection and Multimodel Inference: A Practical Information-Theoretic Approach*, 2nd ed. Springer-Verlag, 2002.

Buss, D. M. *Evolutionary Psychology: The New Science of the Mind*, 3rd ed. Allyn and Bacon, 2009.

Buss, D. M., M. Abbott, A. Angleitner, A. Asherian, A. Biaggio, A. Blanco-Villasenor, L. van den Brande, M. Bruchon-Schweitzer, H.-Y. Ch'u, J. Czapinski, B. Deraad, B. Ekehammer, N. el Lohamy, M. Fioravanti, J. Georgas, P. Gjerde, R. Guttman, F. Hazan, G. van Heck, S. Iwawaki, N. Janakiramaiah, F. Khosroshani, S. Kreitler, L. Lachenicht, L. van Langenhove, M. Lee, K. Liik, B. Little, S. Mika, M. Moadel-Shahid, G. Moane, M. Montero, A. C. Mundy-Castle, T. Niit, E. Nsenduluka, R. Pienkowski, A.-M. Pirtila-Backman, J. Ponce de Leon, J. Rousseau, M. A. Runco, M. P. Safir, C. Samuels, R. Sanitioso, R. Serpell, N. Smid, C. Spencer, M. Tadinac, E. N. Todorova, K. Troland, and K.-S. Yang. International preferences in selecting mates: A study of 37 cultures. *Journal of Cross-Cultural Psychology* 21:5–47, 1990.

Buss, D. M., and D. P. Schmitt. Sexual strategies theory: An evolutionary perspective on human mating. *Psychological Review* 100:204–232, 1993.

Byers, W. *How Mathematicians Think: Using Ambiguity, Contradiction, and Paradox to Create Mathematics*. Princeton University Press, 2007.
Campbell, D. T., and H. L. Ross. The Connecticut crackdown on speeding: Time-series data in quasi-experimental analysis. *Law and Society Review* 3:33–54, 1968.
Campbell, D. T., and J. C. Stanley. *Experimental and Quasi-Experimental Designs for Research*. Rand-McNally, 1966. Originally published in Gage, N. L. (Ed.), *Handbook of Research on Teaching*. Rand-McNally, 1963.
Ceccarelli, L. *Shaping Science with Rhetoric: The Cases of Dobzhansky, Schrödinger, and Wilson*. University of Chicago Press, 2001.
Challenor, P. G., R. K. S. Hankin, and R. Marsh. Toward the probability of rapid climate change. In H. Schellnhuber, W. Cramer, N. Nakicenkovic, and T. Wigley (Eds.), *Avoiding Dangerous Climate Change*. Cambridge University Press, 2006.
Challenor, P. G., D. McNeall, and J. Gattiker. Assessing the probability of rare climate events. In T. O'Hagan and M. West (Eds.), *The Oxford Handbook of Applied Bayesian Analysis*. Oxford University Press, 2010.
Chamberlin, T. C. The method of multiple working hypotheses. *Science*, old series, 15:92, 1890; reprinted as *Science* 148:754–759, 1965.
Cleveland, W. S. *Visualizing Data*. Hobart, 1993.
Clifford, W. K. The ethics of belief. *Contemporary Review*, 1877. Variously reprinted, including in *The Ethics of Belief and Other Essays*, Prometheus Books, 1999.
Cohen, J. The earth is round ($p < .05$). *American Psychologist* 49:997–1003, 1994.
Colby, A., and L. Kohlberg. *The Measurement of Moral Judgment*. Vol. 1, *Theoretical Foundations and Research Validation*. Cambridge University Press, 1987.
Collins, H. M. *Changing Order: Replication and Induction in Scientific Practice*. Sage, 1985.
Collins, M. Ensembles and probabilities: A new era in the prediction of climate change. *Philosophical Transactions of the Royal Society A* 365:1957–1970, 2007.
Conway Morris, S. *Life's Solution: Inevitable Humans in a Lonely Universe*. Cambridge University Press, 2004.
Cook, D. L., F. L. Bookstein, and J. H. Gennari. Physical properties of biological entities: An introduction to the Ontology of Physics for Biology. PLoS One, 6(12): e28708, 2011.
Cook, T. D., and D. T. Campbell. *Quasi-Experimentation: Design and Analysis Issues for Field Settings*. Houghton Mifflin, 1979.
Coombs, C. H. *A Theory of Data*. John Wiley & Sons, 1956.
Cooper, H., L. V. Hedges, and J. C. Valentine, eds. *The Handbook of Research Synthesis and Meta-Analysis*, 2nd ed. Russell Sage, 2009.
Coquerelle, M., F. L. Bookstein, J. Braga, D. J. Halazonetis, and G. W. Weber. Fetal and infant growth patterns of the mandibular symphysis in modern humans and chimpanzees. *Journal of Anatomy* 217:507–520, 2010.
Coquerelle, M., F. L. Bookstein, S. Braga, D. J. Halazonetis, G. W. Weber, and P. Mitteroecker. Sexual dimorphism of the human mandible and its association with dental development. *American Journal of Physical Anthropology* 145:192–202, 2011.
Courtois, H. M., D. Pomarède, R. B. Tully, Y. Hoffman, and D. Courtois. Cosmography of the local universe. *arXiv*: 1306.0091v4, accessed 6/13/2013.
Cox, D. R. *Principles of Statistical Inference*. Cambridge University Press, 2006.
Cox, D. R., and N. Wermuth. *Multivariate Dependencies: Models, Analysis, and Interpretation*. Chapman & Hall, 1996.
Craven, J. P. *The Silent War: The Cold War Battle Beneath the Sea*. Simon & Schuster, 2002.
Crick, F. H. C. *What Mad Pursuit: A Personal View of Scientific Discovery*. Basic Books, 1988.

Davenas, E., F. Beauvais, J. Amara, M. Oberbaum, A. Miadonna, A. Tedeschi, B. Pomeranz, P. Fortner, P. Belon, J. Sainte-Laudy, B. Poitevin and J. Benveniste. Human basophil degranulation triggered by very dilute antiserum against IgE. *Nature* 333:816–818, 1988.

DeHaene, S. *Reading in the Brain: The Science and Evolution of a Human Invention.* Viking, 2009.

Delrieu, O., and C. Bowman. Visualizing gene determinants of disease in drug discovery. *Pharmacogenomics* 7:311–329, 2006.

DeQuardo, J. R., F. L. Bookstein, W. D. K. Green, J. Brumberg, and R. Tandon. Spatial relationships of neuroanatomic landmarks in schizophrenia. *Psychiatry Research: Neuroimaging* 67:81–95, 1996.

DeQuardo, J. R., M. S. Keshavan, F. L. Bookstein, W. W. Bagwell, W. D. K. Green, J. A. Sweeney, G. L. Haas, R. Tandon, and J. W. Pettegrew. Landmark-based morphometric analysis of first-break schizophrenia. *Biological Psychiatry* 45:1321–1328, 1999.

Deutsch, K. W., A. S. Markovits, and J. R. Platt. *Advances in the Social Sciences, 1900–1980.* University Press of America, 1986.

Deutsch, K. W., J. Platt, and D. Senghaas. Conditions favoring major advances in social science. *Science* 171:450–459, 1971.

Devlin, K. *Mathematics: The Science of Patterns.* Scientific American Library, 1994.

Diaconis, P., S. Holmes, and R. Montgomery. Dynamical bias in the coin toss. *SIAM Review* 49:211–235, 2007.

Diaconis, P., and F. Mosteller. Methods for studying coincidence. *Journal of the American Statistical Association* 84:853–861, 1989.

Doll, R., and A. B. Hill. Lung cancer and other causes of death in relation to smoking. *British Medical Journal* 1071–1081, November 10, 1956.

Dragoset, R. A., A. Musgrove, C. W. Clark, and W. C. Martin. Periodic table: Atomic properties of the elements. National Institute of Standards and Technology [NIST]. Retrieved August 1, 2011 from http://www.nist.gov/pml/data/periodic.cfm

Dryden, I. V., and K. V. Mardia. *Statistical Shape Analysis.* Wiley, 1998.

Duncan, O. D. *Notes on Social Measurement, Historical and Critical.* Russell Sage, 1984.

Eckart, C., and G. Young. The approximation of one matrix by another of lower rank. *Psychometrika* 1:211–218, 1936.

Edwards, W., H. Lindman, and L. J. Savage. Bayesian statistical inference for psychological research. *Psychological Review* 70:193–242, 1963. (This essay is reprinted in the collected essays of Jimmie Savage, in Vol. 1 of *Breakthroughs in Statistics*, and elsewhere.)

Einstein, A. [The three immortal papers from *Annalen der Physik*, 1905.] "On a Heuristic Viewpoint Concerning the Production and Transformation of Light" [the photoelectric effect], June 9. "On the Motion of Small Particles Suspended in a Stationary Liquid, as Required by the Molecular Kinetic Theory of Heat" [Brownian motion], July 18. "On the Electrodynamics of Moving Bodies" [special relativity theory], September 26.

Elliott, L. P., and B. W. Brook. Revisiting Chamberlin: Working hypotheses for the 21st century. *BioScience* 57:608–614, 2007.

Ellis, B. *Basic Concepts of Measurement.* Cambridge University Press, 1966.

Elsasser, W. M. *The Chief Abstractions of Biology.* Amsterdam: North-Holland, 1975.

Elwood, J. M. *Critical Appraisal of Epidemiological Studies and Clinical Trials.* Oxford University Press, 1998.

Emerson, F. *The North American Arithmetic. Part Third for Advanced Scholars.* Jenks, Palmer, & Co., 1850.

Endler, J. *Natural Selection in the Wild.* Princeton University Press, 1986.

Environmental Protection Agency (EPA). *Respiratory Health Effects of Passive Smoking: Lung Cancer and Other Disorders.* U.S. Environmental Protection Agency, 1992.

Ewald, P. W. *The Evolution of Infectious Disease*. Oxford University Press, 1994.
Feinstein, A. R. Scientific standards in epidemiologic studies of the menace of everyday life. *Science* 242:1257–1263, 1988.
Feller, W. *An Introduction to Probability Theory and Its Applications*, Vol. 1, 2nd ed. John Wiley & Sons, 1957.
Felsenstein, J. *Inferring Phylogenies*. Sinauer Associates, 2004.
Feynman, R. P. *"Surely You're Joking, Mr. Feynman!" Adventures of a Curious Character*. W. W. Norton, 1985.
Feynman, R. P. *Feynman Lectures on Computation*. Addison-Wesley, 1996.
Feynman, R. P. Cargo cult science. *Engineering and Science*, June 1974, 10–13. Reprinted as pp. 205–216 of J. Robbins (Ed.), *The Pleasure of Finding Things Out: The Best Short Works of Richard P. Feynman*, Basic Books, 1999.
Feynman, R. P., R. B. Leighton, and M. Sands. *The Feynman Lectures on Physics*, three volumes. Addison-Wesley, 1963–1965.
Fisher, R. A. Cigarettes, cancer, and statistics. *Centennial Review* 2:151–166, 1958.
Fleck, L. *Entstehung und Entwicklung einer wissenschaftlichen Tatsache*. Schwabe, 1935. *Genesis and Development of a Scientific Fact*, tr. F. Bradley and T. J. University of Chicago Press, 1979.
Floud, R., R. W. Fogel, B. Harris, and S. C. Hong. *The Changing Body: Health, Nutrition, and Human Development in the Western World Since 1700*. Cambridge University Press, 2011.
Flury, B. *Common Principal Components and Related Multivariate Models*. John Wiley & Sons, 1988.
Fowler, B. *Iceman: Uncovering the Life and Times of a Prehistoric Man Found in an Alpine Glacier*. University of Chicago Press, 2001.
Frame, D. J., N. E. Faull, M. M. Joshi, and M. R. Allen. Probabilistic climate forecasts and inductive problems. *Philosophical Transactions of the Royal Society A* 365:1971–1992, 2007.
Franklin, R. K., and R. G. Gosling. Molecular configuration in sodium thymonucleate. *Nature* 71:740–741, 1953.
Freedman, D. A. Statistical models and shoe leather. *Sociological Methodology* 21:291–313, 1991.
Freedman, D. A. *Statistical Models: Theory and Practice*, 2nd ed. Cambridge University Press, 2009.
Freedman, D. A., R. Pisani, and R. Purves. *Statistics*. W. W. Norton, 1980 and subsequent editions.
Freedman, D. A., and H. Zeisel. From mouse to man. *Statistical Science* 3:3–56, 1988.
Fuentes, A. *Evolution of Human Behavior*. Oxford University Press, 2009.
Fuller, S. *Thomas Kuhn: a Philosophical History for Our Time*. University of Chicago Press, 2001.
Gaddis, J. L. *The Landscape of History*. Oxford University Press, 2002.
Galison, P. *How Experiments End*. University of Chicago Press, 1987.
Galton, F. The possible improvement of the human breed undeer the existing conditions of law and sentiment. *Nature* 64:659–665, 1901.
Galton, F. Classification of portraits. *Nature* 76:617–618, 1907.
Geller, M. J., and J. P. Huchra. Mapping the universe. *Science* 246:897–903, 1989.
Georgescu-Roegen, N. *The Entropy Law and the Economic Process*. Harvard University Press, 1971.
Gerard, R. W. Units and concepts of biology. *Science* 125:429–433, 1957.
Gerard, R. W. (Ed.). *Concepts of Biology*. Publication 560. National Academy of Sciences, 1958.
Gerard, R. W. Quantification in biology. In H. Woolf (Ed.), *Quantification: A History of the Meaning of Measurement in the Natural and Social Sciences* (pp. 204–222). Bobbs-Merrill, 1961.
Gleick, J. *The Information: A History, a Theory, a Flood*. Pantheon, 2011.
Glimcher, P. W., E. Fehr, C. Camerer, and A. Rangel, eds. *Neuroeconomics: Decision Making and the Brain*. Academic Press, 2008.

Goldstein, I. F., and M. Goldstein. *How Much Risk? A Guide to Understanding Environmental Health Hazards*. Oxford University Press, 2002.

Good, P. I. *Permutation Tests: A Practical Guide to Resampling Methods for Testing Hypotheses*, 2nd ed. Springer, 2000.

Goodstein, D. In defense of Robert Andrews Millikan. *Engineering and Science* 4:31–39, 2000.

Gould, S. J. *Wonderful Life: The Burgess Shale and the Nature of History*. Norton, 1989.

Gower, J. C. Some distance properties of latent root and vector methods used in multivariate analysis. *Biometrika* 53:325–338, 1966.

Gower, J. C., and D. J. Hand. *Biplots*. Chapman and Hall, 1995.

Gower, J. C., S. G. Lubbe, and N. Le Roux. *Understanding Biplots*. John Wiley & Sons, 2011.

Greene, J. Emotion and cognition in moral judgment: Evidence from neuroimaging. In J.-P. P. Changeux, A. Damasio, and W. Singer (Eds.), *Neurobiology of Human Values* (pp. 57–66). Springer Verlag, 2005.

Greenwald, A. G., M. R. Leippe, A. R. Pratkanis, and M. H. Baumgardner. Under what conditions does theory obstruct research progress? *Psychological Review* 93:216–229, 1986.

Grenander, U. *General Pattern Theory: A Mathematical Study of Regular Structures*. Oxford University Press, 1994.

Grenander, U. *Elements of Pattern Theory*. Johns Hopkins University Press, 1996.

Grenander, U., and M. Miller. *Pattern Theory: From Representation to Inference*. Oxford University Press, 2007.

Guerri, C., G. Rubert, and M. Pasqual. Glial targets of developmental exposure to eathanol. In M. W. Miller (Ed.), *Brain Development: Normal Processes and the Effects of Alcohol and Nicotine* (pp. 295–312). Oxford University Press, 2006.

Guillemin, J. *Anthrax: The Investigation of a Deadly Outbreak*. University of California Press, 1999.

Gunz, P., P. Mitteroecker, and F. L. Bookstein. Semilandmarks in three dimensions. In D. E. Slice (Ed.), *Modern Morphometrics in Physical Anthropology* (pp. 73–98). Kluwer Academic, 2004.

Guttman, L. A note on the derivation of formulae for multiple and partial correlation. *Annals of Mathematical Statistics* 9:305–308, 1938.

Guttman, L. Order analysis of correlation matrices. In R. B. Cattell (Ed.), *Handbook of Multivariate Experimental Psychology* (pp. 438–458). Rand McNally, 1966.

Hackshaw, A. K., M. R. Law, and N. J. Law. The accumulated evidence on lung cancer and environmental tobacco smoke. *British Medical Journal* 315:980–988, 1997.

Hamming, R. W. The unreasonable effectiveness of mathematics. *American Mathematical Monthly* 87:81–90, 1980.

Hanahan, D., and R. A. Weinberg. Hallmarks of cancer: The next generation. *Cell* 144:646–674, 2011.

Hand, D. J. *Measurement Theory and Practice: The World through Quantification*. Arnold, 2004.

Hansen, J., M. Sato, and R. Ruedy. Perception of climate change. *PNAS*, early online posting. Retrieved August 6, 2012 from www.pnas.org/cgi/doi/10.1073/pnas.1205276109

Harpending, H., and G. Cochran. *The 10,000 Year Explosion: How Civilization Accelerated Human Evolution*. Basic Books, 2009.

Harwit, M. *Cosmic Discovery*. Basic Books, 1981.

Hastie, T., R. Tibshirani, and J. Friedman. *The Elements of Statistical Learning: Data Mining, Inference, and Prediction*, 2nd ed. Springer, 2009.

Hauser, M. *Moral Minds: How Nature Designed our Universal Sense of Right and Wrong*. Ecco, 2006.

Helgason, S. *Differential Geometry, Lie Groups, and Symmetric Spaces*. Academic Press, 1978.

Hentschel, E., S. Bandstatter, B. Drgosics, A. M. Hirschl, H. Nemec, K. Schutze, M. Taufer, and H. Wurzer. Effect of ranitidine and amoxicillin plus metronidazole on the eradication of

Helicobacter pylori and the recurrence of duodenal ulcer. *New England Journal of Medicine* 328:308–312, 1993.

Herrnstein, R. J., and C. Murray. *The Bell Curve: Intelligence and Class Structure in American Life.* The Free Press, 1994.

Hersh, R. *What Is Mathematics, Really?* Oxford University Press, 1997.

Hilbert, D., and S. Cohn-Vossen. *Anschauliche Geometrie.* Springer, 1932. Translated as *Geometry and the Imagination.* Chelsea, 1952.

Ho, C. Y., R. W. Powell, and P. E. Liley. Thermal conductivity of the elements: A comprehensive review. *Journal of Physical and Chemical Reference Data* 3, Supplement 1, 1–244, 1974.

Hogben, L. *Chance and Choice by Cardpack and Chessboard.* Chanticleer Press, 1950.

Hogben, L. *Statistical Theory: The Relationship of Probability, Credibility, and Error.* George Allen and Unwin, 1957.

Hollingshead, A. B., and F. C. Redlich. *Social Class and Mental Illness.* John Wiley & Sons, 1958.

Horton, R. The statin wars: why Astra Zeneca must retreat. *Lancet* 362:1341, 2003.

Houser, N., and C. J. W. Kloesel, Eds. *The Essential Peirce: Selected Philosophical Writings*, 2 vols. Indiana University Press, 1992, 1998.

Hovland, C. I., A. A. Lumsdaine, and F. D. Sheffield. *Experiments on Mass Communication.* Studies in Social Psychology in World War II, Vol. 3. Princeton University Press, 1949.

Hovmöller, S., T. Zhou, and T. Ohlson. Conformations of amino acids in proteins. *Acta Crystallographica* D58:768–776, 2002.

Huber, S., F. L. Bookstein, and M. Fieder. Socioeconomic status, education, and reproduction in modern women: An evolutionary perspective. *American Journal of Human Biology* 22:578–587, 2010.

Huff, D. *How to Lie with Statistics.* W. W. Norton, 1954.

Huizenga, J. R. *Cold Fusion: The Scientific Fiasco of the Century*, 2nd ed. University of Rochester Press, 1993.

Hyman, A. T. A simple Cartesian treatment of planetary motion. *European Journal of Physics* 14:145–147, 1993.

Intergovernmental Panel on Climate Change [IPCC]. *Climate Change 2007: The Physical Science Basis.* Cambridge University Press, 2007.

Jaynes, E. T. *Probability Theory: The Logic of Science*, ed. G. L. Bretthorst. Cambridge University Press, 2003.

Jeffreys, H. *Theory of Probability.* Clarendon Press, 1939, 2nd ed., 1961.

Jerison, H. J. *Evolution of the Brain and Intelligence.* Academic Press, 1973.

Jones, K. L., D. W. Smith, C. N. Ulleland, and A. P. Streissguth. Pattern of malformation in offspring of chronic alcoholic mothers. *Lancet* 301:1267–1269, 1973.

Josefson, D. US flight attendants win settlement over passive smoking. *BMJ* 315:968, 1997.

Josephson, J. R., and S. G. Josephson, Eds. *Abductive Inference: Computation, Philosophy, Technology.* Cambridge University Press, 1996.

Kahneman, D. *Thinking, Fast and Slow.* Farrar Straus Giroux, 2011.

Keay, J. *The Great Arc: The Dramatic Tale of How India was Mapped and Everest was Named.* Harper Collins, 2000.

Kendall, D. G. Shape manifolds, Procrustean metrics and complex projective spaces. *Bulletin of the London Mathematical Society* 16:81–121, 1984.

Kendall, D. G. A survey of the statistical theory of shape. *Statistical Science* 4:87–120, 1989.

Kerr, R. A. Huge impact tied to mass extinction. *Science* 257:878–880, 1992.

Knutti, R. Should we believe model predictions of future climate change? *Philosophical Transactions of the Royal Society of London A* 366:4647–4664, 2008.

Koenderink, J. *Solid Shape.* MIT Press, 1990.

Krantz, D. H., R. D. Luce, and P. Suppes. *Foundations of Measurement*, 3 vols. Academic Press, 1971–1990.
Krieger, M. H. *Doing Physics: How Physicists Take Hold of the World*. Indiana University Press, 1992.
Krzanowski, W. *Principles of Multivariate Analysis: A User's Perspective*. 1988. 2nd ed. 2000.
Kuhn, T. S. The function of measurement in modern physical science. *ISIS* 52:161–193, 1961. Also, in H. Woolf (Ed.), *Quantification: A History of the Meaning of Measurement in the Natural and Social Sciences* (pp. 31–63). Bobbs-Merrill, 1961.
Kuhn, T. S. *The Structure of Scientific Revolutions*. University of Chicago Press, 1962, 2nd ed., 1970. (Originally prepared for Neurath et al., Eds., 1970 in 1962.)
Kurzweil, R. *The Singularity is Near: When Humans Transcend Biology*. Penguin, 2006.
van de Lagemaat, R. *Theory of Knowledge for the IB Diploma*. Cambridge University Press, 2006.
Lakatos, I. *Proofs and Refutations*. Cambridge University Press, 1976.
Lakoff, G., and M. Johnson. *Metaphors We Live By*. University of Chicago Press, 1980.
Lande, R. Natural selection and random genetic drift in phenotypic evolution. *Evolution* 30:314–443, 1976.
Langmuir, I. Pathological Science. *Physics Today*, October 1989, 36–48. (Originally a lecture, 1953.)
Latour, B. *Science in Action: How to Follow Scientists and Engineers through Society*. Harvard University Press, 1987.
Latour, B. Drawing things together. In M. Lynch and S. Woolgar (Eds.), *Representation in Scientific Practice* (pp. 19–68). MIT Press, 1990.
Law, M. R., J. K. Morris, and N. J. Wald. Environmental tobacco smoke exposure and ischaemic heart disease: An evaluation of the evidence. *British Medical Journal* 315:973–980, 1997.
Ledley, R. S. *Use of Computers in Biology and Medicine*. McGraw-Hill, 1965.
LeGrand, H. E. *Drifting Continents and Shifting Theories: The Modern Revolution in Geology and Scientific Change*. Cambridge University Press, 1988.
Letvin, N. L., J. R. Mascola, Y. Sun, D. A. Gorgone, A. P. Buzby, L. Xu, Z. Yang, B. Chakrabarti, S. S. Rao, J. E. Schmitz, D. C. Montefiori, B. R. Barker, F. L. Bookstein, and G. J. Nabel. Preserved CD4+ central memory T cells and survival in vaccinated SIV-challenged monkeys. *Science* 312:1530–1533, 2006.
Lieberson, S. *Making It Count: The Improvement of Social Research and Theory*. University of California Press, 1985.
Lima, M. *Visual Complexity: Mapping Patterns of Information*. Princeton Architectural Press, 2011.
Lindsay, R. B. *Introduction to Physical Statistics*. John Wiley & Sons, 1941.
Lipton, P. *Inference to the Best Explanation*. Routledge/Taylor & Francis, 2004.
Livingstone, F. B. Anthropological implications of sickle cell gene distribution in West Africa. *American Anthropologist* 60:533–562, 1958.
Lynd, R. S. *Knowledge for What? The Place of Social Science in American Culture*. Princeton University Press, 1939.
Mach, E. Critique of the concept of temperature. (Translation of an extract from *Die Prinzipien der Wärmelehre*, 1896.) In B. Ellis, *Basic Concepts of Measurement* (pp. 183–196). Cambridge University Press, 1966.
Machlup, F. The problem of verification in economics. *Southern Economic Journal* 22:1–21, 1955.
Mackenzie, D. *Inventing Accuracy: A Historical Sociology of Nuclear Missile Guidance*. MIT Press, 1990.
Maddox, J. Waves caused by extreme dilution. *Nature* 335:760–763, 1988.
Maddox, J., J. Randi, and W. W. Stewart. "High-dilution" experiments a delusion. *Nature* 334:287–290, 1988.
Mandelbrot, B. *The Fractal Geometry of Nature*. W. H. Freeman, 1982.

Mann, M. E., R. S. Bradley, and M. K. Hughes, Global scale temperature patterns and climate forcing over the past six centuries. *Nature* 392:779–787, 1998.
Mantegna, R. N., and H. E. Stanley. *Introduction to Econophysics: Correlations and Complexity in Finance*. Cambridge University Press, 2007.
Marcus, P. M., and J. H. McFee. Velocity distributions in potassium molecular beams. In J. Esterman (Ed.), *Recent Research in Molecular Beams*. Academic Press, 1959.
Mardia, K. V. Statistical approaches to three key challenges in protein structural bioinformatics. *Applied Statistics* 62:487–514, 2013.
Mardia, K.V., F. L. Bookstein, and I. J. Moreton. Statistical assessment of bilateral symmetry of shapes. *Biometrika* 87:285–300, 2000.
Mardia, K. V., J. T. Kent, and J. Bibby. *Multivariate Analysis*. John Wiley & Sons, 1979.
Mardia, K. V., F. L. Bookstein, J. T. Kent, and C. R. Meyer. Intrinsic random fields and image deformations. *Journal of Mathematical Imaging and Vision* 26:59–71, 2006.
Mardia, K. V., C. C. Taylor, and G. K. Subramaniam. Protein bioinformatics and mixtures of bivariate von Mises distributions for angular data. *Biometrics* 63:505–512, 2007.
Margenau, H. *The Nature of Physical Reality: A Philosophy of Modern Physics*. McGraw-Hill, 1950.
Marshall, B. History of the discovery of *C. pylori*. In M. J. Blaser (Ed.), Campylobacter pylori *in Gastritis and Peptic Ulcer Disease* (pp. 7–23). Igaku-Shoin, 1988.
Marshall, B. J., and J. R. Warren. Unidentified curved bacillus on gastric epithelium in active chronic gastritis. *Lancet* 321:1273–1275, 1983.
Marshall, B. J., and J. R. Warren. Unidentified curved bacilli in the stomach of patients with gastritis and peptic ulceration. *Lancet* 323:1311–1315, 1984.
Martens, H. Domino PLS: A framework for multidirectional path modelling. In T. Aluja, J. Casanovas, V. Esposito Vinci, A. Morineau and M. Tenenhaus (Eds.), *PLS and Related Methods* (pp. 125–132), SPAD, 2005.
Martens, H., and A. Kohler. Mathematics and measurements for high-throughput quantitative biology. *Biological Theory* 4:29–43, 2009.
Martens, H., and T. Næs. *Multivariate Calibration*. John Wiley & Sons, 1989.
Martin, R. *Lehrbuch der Anthropologie in Systematischer Darstellung*, 2nd ed. Gustav Fischer Verlag, 1928.
Maxwell, J. C. On the cyclide. *Quarterly Journal of Pure and Applied Mathematics* 9:111–126, 1868.
May, P. A., J. P. Gossage, W. O. Kalberg, L. K. Robinson, D. Buckley, M. Manning, and H. E. Hoyme. Prevalence and epidemiologic characteristics of FASD from various research methods with an emphasis on recent in-school studies. *Developmental Disabilities Research Reviews* 15:176–192, 2009.
May, R. M. Tropical arthropod species, more or less? *Science* 329:41–42, 2010.
Mayer, E. *Introduction to Dynamic Morphology*. Academic Press, 1963.
Mazurs, E. G. *Graphic Representations of the Periodic System During One Hundred Years*, 2nd ed. University of Alabama Press, 1974.
McGrayne, S. B. *The Theory that Would Not Die*. Yale University Press, 2011.
McKay, B., D. Bar-Natan, M. Bar-Hillel, and G. Kalai. Solving the Bible Code puzzle. *Statistical Science* 14:150–173, 1999.
McKillop, T. The statin wars. *Lancet* 362:1498, 2003.
Meehl, P. E. Theory testing in psychology and physics: A methodological paradox. *Philosophy of Science* 34:103–115, 1967.
Meehl, P. E. Why summaries of research on psychological theories are often uninterpretable. *Psychological Reports* 66:195–244, 1990.
Meiland, J. W. *College Thinking: How to Get the Best Out of College*. New American Library, 1981.

Merton, R. K. Science and technology in a democratic order. *Journal of Legal and Political Sociology* 1:115–126, 1942. Reprinted in R. K. Merton, *The Sociology of Science: Theoretical and Empirical Investigations*. University of Chicago Press, 1973.

Meselson, M., J. Guillemin, M. Hugh-Jones, A. Langmuir, I. Popova, A. Shelokov, and O. Yampolskaya. The Sverdlovsk anthrax outbreak of 1979. *Science* 266:1202–1208, 1994.

Milgram, S. *Obedience to Authority: An Experimental View*. Harper & Row, 1974.

Miller, A. G. *The Obedience Experiments: A Case Study of Controversy in Social Science*. Praeger, 1986.

Miller, M. W., S. J. Astley, and S. K. Clarren. Number of axons in the corpus callosum of the mature *Macaca nemestrina*: Increases caused by prenatal exposure to ethanol. *Journal of Comparative Neurology* 412:123–131, 1999.

Millikan, R. A. A direct photoelectric determination of Planck's "h." *American Journal of Physics* 7:355–390, 1916.

Mills, C. W. *The Sociological Imagination*. Oxford University Press, 1959.

Mitchell, M. *Complexity: A Guided Tour*. Oxford University Press, 2009.

Mitteroecker, P., P. Gunz, M. Bernhard, K. Schaefer, and F. L. Bookstein. Comparison of cranial ontogenetic trajectories among great apes and humans. *Journal of Human Evolution* 46:679–697, 2004.

Mitteroecker, P. M., and F. L. Bookstein. The evolutionary role of modularity and integration in the hominoid cranium. *Evolution* 62:943–958, 2008.

Mitteroecker, P. M., and F. L. Bookstein. The ontogenetic trajectory of the phenotypic covariance matrix, with examples from craniofacial shape in rats and humans. *Evolution* 63:727–37, 2009a.

Mitteroecker, P. M., and F. L. Bookstein. Examining modularity by partial correlations: Rejoinder to a comment by Paul Magwene. *Systematic Biology* 58:346–347, 2009b.

Mitteroecker, P. M., and F. L. Bookstein. Linear discrimination, ordination, and the visualization of selection gradients in modern morphometrics. *Evolutionary Biology* 38:100–114, 2011.

Montford, A. M. *The Hockey Stick Illusion: Climategate and the Corruption of Science*. Stacey International, 2010.

Morgenstern, O. *On the Accuracy of Economic Observations*. Princeton University Press, 1950.

Morrison, D. E., and Henkel, R. E., eds. *The Significance Test Controversy*. Aldine, 1970.

Morrison, D. F. *Multivariate Statistical Methods*. McGraw-Hill, 1967. Fourth edition, Duxbury Press, 2004.

Mosimann, J. E. Size allometry: size and shape variables with characterizations of the log-normal and generalized gamma distributions. *Journal of the American Statistical Association* 65:930–945, 1970.

Mosteller, F., and J. W. Tukey. *Data Analysis and Regression: A Second Course in Statistics*. Addison-Wesley, 1977.

Mosteller, F., and D. L. Wallace. *Inference and Disputed Authorship*. Addison-Wesley, 1964. Republished as *Applied Bayesian and Classical Inference: the Case of the Federalist Papers*. Springer, 1984.

Moyers, R. E., F. L. Bookstein, and K. Guire. The concept of pattern in craniofacial growth. *American Journal of Orthodontics* 76:136–148, 1979.

Munck, L. *The Revolutionary Aspect of Exploratory Chemometric Technology*. Narayana Press, Gylling, Denmark, 2005.

Nash, L. K. *The Nature of the Natural Sciences*. Little-Brown, 1963.

NCBDDD [National Center on Birth Defects and Developmental Disabilities]. *Fetal Alcohol Syndrome: Guidelines for Referral and Diagnosis*. Centers for Disease Control and Prevention, Department of Health and Human Services, 2004.

Neel, J. V. Human hemoglobin types, their epidemiologic implications. *New England Journal of Medicine* 256:161–171, 1957.

Neurath, O., R. Carnap, and C. Morris, eds. *Foundations of the Unity of Science: Toward an International Encyclopedia of Unified Science*, Volume II. (Original title, *International Encyclopedia of Unified Science*.) University of Chicago Press, 1970.

Nguyen, C. T.-C. 2007. EE C245 – ME C218, Introduction to MEMS Design, Fall 2007, University of California at Berkeley. Lecture 16, Energy Methods. Retrieved February 22, 2012 from inst.eecs.berkeley.edu/~ee245/fa07/lectures/Lec16.EnergyMethods.pdf.

Nisbett, R. E. *The Geography of Thought: How Asians and Westerners Think Differently – and Why*. Free Press, 2004.

Nisbett, R. E., and D. Cohen. *Culture of Honor: The Psychology of Violence in the South*. Westview Press, 1996.

Nye, M. J. *Molecular Reality: A Perspective on the Scientific Work of Jean Perrin*. McDonald and American Elsevier, 1972.

Office of Research Integrity. Definition of research misconduct. http://ori.hhs.gov/misconduct/definition_misconduct.shtml, 2009.

Offit, P. A. *Autism's False Prophets: Bad Science, Risky Medicine, and the Search for a Cure*. Columbia University Press, 2008.

Ogle, R. *Smart World: Breakthrough Creativity and the New Science of Ideas*. Harvard Business School Press, 2007.

O'Higgins, P., S. N. Cobb, L. C. Fitton, F. Gröning, R. Phillips, J. Liu, and M. J. Fagan. Combining geometric morphometrics and functional simulation: an emerging toolkit for virtual functional analyses. *Journal of Anatomy* 218:3–15, 2011.

Olby, R. C. *The Path to the Double Helix*. University of Washington Press, 1974.

Olson, E. C., and R. L. Miller. *Morphological Integration*. University of Chicago Press, 1958.

Oxnard, C., and P. O'Higgins. Biology clearly needs morphometrics. Does morphometrics need biology? *Biological Theory* 4:84–97, 2009.

Palmer, A. R., and C. Strobeck. Fluctuating asymmetry: Measurement, analysis, patterns. *Annual Reviews of Ecology and Systematics* 17:391–421, 1986.

Parsons, T., E. Shils, K. D. Naegele, and J. R. Pitts, eds. *Theories of Society: Foundations of Modern Sociological Theory*. Free Press, 1961.

Pearl, J. *Causality: Reasoning, Models, and Inference*. Cambridge University Press, 2000. Second edition, 2009.

Pearson, K. On the inheritance of the mental and moral characters in man, and its comparison with the inheritance of the physical characters. *Journal of the Anthropological Institute of Great Britain and Ireland* 33:179–237, 1903.

Pearson, K. Walter Frank Raphael Weldon, 1860–1906. *Biometrika* 5:1–52, 1906.

Pearson, K. *The Grammar of Science*, 3rd edition. A. & C. Black, 1911.

Pearson, K. *The Life, Letters and Labours of Francis Galton*. Three volumes bound as four. Cambridge University Press, 1914–1930.

Pearson, K., and A. Lee. On the laws of inheritance in man. I: Inheritance of physical characters. *Biometrika* 2:357–462, 1903.

Pearson, K., and G. M. Morant. The portraiture of Oliver Cromwell, with special reference to the Wilkinson head. *Biometrika*, vol. XXVI. Cambridge University Press, 1935.

Peirce, C. S. Hume on miracles. In his *Collected Papers*, 6:356–369, 1901.

Peirce, C. S. *Collected Papers*, ed. C. Hartshorne and P. Weiss. Eight volumes. Harvard University Press, 1931–1958.

Peirce, C. S. A neglected argument for the reality of God. In his *Collected Papers*, 6:302–339, 1908.

Peirce, C. S. Lectures on Pragmatism, as delivered at Harvard, 1903. Cited on page 151 of Buchler, ed., 1940.

Perrin, J. *Atoms*, 2nd English edition, revised. London: Constable, 1923.

Perrow, C. *Normal Accidents: Living with High-Risk Technologies*. Basic Books, 1984.

Petley, B. W. *The Fundamental Physical Constants and the Frontier of Measurement*. Adam Hilger, 1985.

Pickering, A. *The Mangle of Practice: Time, Agency, and Science*. University of Chicago Press, 1995.

Pimentel, R. A. *Morphometrics: The Multivariate Analysis of Biological Data*. Kendall/Hunt, 1979.

Platt, J. R. Strong inference. *Science* 146:347–353, 1964.

Pólya, G. *Mathematics and Plausible Reasoning*, two volumes. Princeton University Press, 1954.

Popper, K. R. *Of Clouds and Clocks*. The Arthur Holly Compton Memorial Lecture. Washington University, 1965.

Provine, W. B. *Sewall Wright and Evolutionary Biology*. University of Chicago Press, 1986.

Raff, A. D., and R. G. Mason. Magnetic survey off the west coast of North America, 40° N. latitude to 52° N. latitude. *Geological Society of America Bulletin* 72:1267–1270, 1961.

Ramachandran, G. N., C. Ramakrishnan, and V. Sasisekharan. Stereochemistry of polypeptide chain configurations. *Molecular Biology* 7:95–99, 1963.

Ramsay, J., and B. W. Silverman. *Functional Data Analysis*, 2nd ed. Springer, 2010.

Rao, C. R. *Linear Statistical Inference and Its Applications*, 2nd ed. John Wiley & Sons, 1973.

Rasch, G. *Probabilistic Models for Some Intelligence and Attainment Tests*, expanded edition. University of Chicago Press, 1980.

Rendgen, S., and J. Wiedemann. *Information Graphics*. Köln: Taschen Books, 2012.

Reyment, R. A. *Multidimensional Palaeobiology*. Pergamon, 1991.

Reyment, R. A., R. E. Blackith, and N. Campbell. *Multivariate Morphometrics*, 2nd ed. Academic Press, 1984.

Reyment, R. A., and K. H. Jöreskog. *Applied Factor Analysis in the Natural Sciences*. Cambridge University Press, 1993.

Richardson, K. Introduction to reprint of Boulding, 1956. E:co, special double issue "General systems theory – the skeleton of science," 6:127, 2004.

Riggs, D. S. *The Mathematical Approach to Physiological Problems: A Critical Primer*. MIT Press, 1963.

Riley, E. P., S. N. Mattson, E. R. Sowell, T. L. Jernigan, D. F. Sobel, and K. L. Jones. Abnormalities of the corpus callosum in children prenatally exposed to alcohol. *Alcoholism: Clinical and Experimental Research*, 19:1198–1202, 1995.

Robbins, H. Statistical methods related to the law of the iterated logarithm. *Annals of Mathematical Statistics* 41:1397–1409, 1997.

Rohde, R., R. A. Muller, R. Jacobsen, E. Muller, S. Perlmutter, A. Rosenfeld, J. Wurtele, D. Groom, and C. Wickham. A new estimate of the average earth surface land temperature spanning 1753 to 2011. Manuscript submitted to *Journal of Geophysical Research*. Retrieved August 6, 2012 from www.berkeleyearth.org as results-paper-july-8.pdf

Rohlf, F. J., and F. L. Bookstein. Computing the uniform component of shape variation. *Systematic Biology* 52:66–69, 2003.

Rohlf, F. J., and D. E. Slice. Methods for comparison of sets of landmarks. *Systematic Zoology* 39:40–59, 1990.

Rosen, J. The brain on the stand: How neuroscience is transforming the legal system. *The New York Times Magazine*, March 11, 2007.

Rosenthal, R., and L. Jacobson. *Pygmalion in the Classroom: Teacher Expectation and Pupils' Intellectual Development*. Holt, Rinehart, and Winston, 1968.

Rothman, K. J., S. Greenland, and T. L. Lash. *Modern Epidemiology*. Lippincott, Williams & Wilkins, 2008.
Rousseau, D. L. Case studies in pathological science. *American Scientist* 80:54–63, 1992.
Royall, R. M. *Statistical Evidence: A Likelihood Paradigm*. Chapman & Hall, 1997.
Ruhla, C. *The Physics of Chance, from Blaise Pascal to Niels Bohr*. Oxford University Press, 1992.
Russell, B. *The Autobiography of Bertrand Russell*, 3 volumes. London: George Allen and Unwin, 1967–69.
Sataki, K., and B. F. Atwater. Long-term perspectives on giant earthquakes and tsunamis at subduction zones. *Annual Reviews of Earth and Planetary Sciences* 35:349–374, 2007.
Schaefer, K., and F. L. Bookstein. Does geometric morphometrics serve the needs of plasticity research? *Journal of Biosciences* 34:589–599, 2009.
Schaefer, K., T. Lauc, P, Mitteroecker, P. Gunz, and F. L. Bookstein. Dental arch asymmetry in an isolated Adriatic community. *American Journal of Physical Anthropology* 129:132–142, 2006.
Schaltenbrand, G. Darstellung des periodischen Systems der Elemente durch eine räumliche Spirale. *Z. anorg. allgem. Chem.* 112:221–224, 1920. Excerpted in Mazurs, 1974.
Schulte, P., L. Alegret, I. Arenillas, J. A. Arz, P. J. Barton, P. R. Bown, T. J. Bralower, G. L. Christeson, P. Claeys, C. S. Cockell, G. S. Collins, A. Deutsch, T. J. Goldin, K. Goto, J. M. Grajales-Nishimura, R. A. F. Grieve, S. P. S. Gulick, K. R. Johnson, W. Kiessling, C. Koeberl, D. A. Kring, K. G. MacLeod, T. Matsui, J. Melosh, A. Montanari, J. V. Morgan, C. R. Neal, D. J. Nichols, R. D. Norris, E. Pierazzo, G. Ravizza, M. Rebolledo-Vieyra, W. Uwe Reimold, E. Robin, T. Salge, R. P. Speijer, A. R. Sweet, J. Urrutia-Fucugauchi, V. Bajda, M. T. Whalen, and P. S. Willumsen. The Chicxulub asterioid impact and mass extinction at the Cretaceous – Paleogene boundary. *Science* 327:1214–1218, 2010.
Sebeok, T. A., and J. Umiker-Sebeok. "You know my method": A juxta position of Charles S. Peirce and Sherlock Holmes. *Semiotica* 26:203–250, 1979.
Shannon, C. E. A mathematical theory of communication. *Bell System Technical Journal* 27:379–423, 623–656, 1948.
Sheehan W. *Planets and Perception: Telescopic Views and Interpretation, 1609–1909*. University of Arizona Press, 1988.
Slice, D. E. Introduction. In D. E. Slice (Ed.), *Modern Morphometrics in Physical Anthropology* (pp. 1–47). Kluwer Academic, 2004.
Smolin, L. *The Life of the Universe*. Oxford University Press, 1997.
Snow, J. *On the Mode of Communication of Cholera*. London: John Churchill, 1855.
Sontag, S., and C. Drew. *Blind Man's Bluff: The Unfold Story of American Submarine Espiorage*. Public Affairs Press, 1998.
Spindler, K. *The Man in the Ice: the Discovery of a 5000-year-old Body Reveals the Secrets of the Stone Age*. Harmony Books, 1994.
Stent, G. S. *The Coming of the Golden Age: a View of the End of Progress*. American Museum of Natural History, 1969.
Stevens, P. S. *Patterns in Nature*. Little-Brown, 1974.
Stigler, S. M. *The History of Statistics: The Measurement of Uncertainty before 1900*. Harvard University Press, 1986.
Stolley, P. D. When genius errs: R. A. Fisher and the lung cancer controversy. *American Journal of Epidemiology* 133:416–425, 1991.
Stone, L. D., *Theory of Optimal Search*. Academic Press, 1975.
Streissguth, A. P. *Fetal Alcohol Sydrome: A Guide for Families and Communities*. Paul H. Brookes, 1997.

Streissguth, A., H. Barr, F. L. Bookstein, P. Sampson, and B. Darby. Neurobehavioral effects of prenatal alcohol. Part I. Literature review and research strategy. Part II. Partial least squares analyses. Part III. PLS analyses of neuropsychologic tests. *Neurotoxicology and Teratology* 11:461–507, 1989.

Streissguth, A. P., H. Barr, J. Kogan, and F. L. Bookstein. *Understanding the occurrence of secondary disabilities in clients with fetal alcohol syndrome (FAS) and fetal alcohol effects (FAE)*. Final report, CDC grant CCR008515. Fetal Alcohol and Drug Unit, University of Washington School of Medicine, August, 1996.

Streissguth, A. P., F. L. Bookstein, H. Barr, S. Press, and P. Sampson. A fetal alcohol behavior scale. *Alcoholism: Clinical and Experimental Research* 22:325–333, 1998.

Streissguth, A. P., F. L. Bookstein, P. Sampson, and H. Barr. *The Enduring Effects of Prenatal Alcohol Exposure on Child Development, Birth through Seven Years: A Partial Least Squares Solution*. University of Michigan Press, 1993.

Stuart, A., and J. K. Ord. *Kendall's Advanced Theory of Statistics*. Vol. 1, *Distribution Theory*, 6th ed. Edward Arnold, 1994.

[Surgeon General of the United States.] *Smoking and Health: Report of the Advisory Committee to the Surgeon General of the Public Health Service*. U.S. Department of Health, Education, and Welfare, 1964.

Szpiro, G. G. *Pricing the Future: Finance, Physics, and the 300-year Journey to the Black-Scholes Equation*. Basic Books, 2011.

Taleb, Nassim. *The Black Swan: The Impact of the Highly Improbable*. Penguin, 2008.

Taylor, J., A. Evans, and K. Friston. In memoriam: A tribute to Keith Worsley, 1951–2009. *NeuroImage* 46:891–894, 2009.

Thagard, P. *How Scientists Explain Disease*. Princeton University Press, 1999.

Thompson, D'A. W. *On Growth and Form*. Cambridge University Press, 1917. 2nd, enlarged edition, 1942. Abridged edition (J. T. Bonner, ed.), 1961.

Thurston, W. P. On proof and progress in mathematics. *Bulletin of the American Mathematical Society* 30:161–176, 1994.

Torgerson, W. S. *Theory and Methods of Scaling*. John Wiley & Sons, 1958.

Tramontano, A. *The Ten Most Wanted Solutions in Protein Bioinformatics*. Chapman & Hall, 2005.

Trueblood, K. N., and J. P. Glusker. *Crystal Structure Analysis: A Primer*. Oxford University Press, 1972.

Tuddenham, R. D., and M. M. Snyder. *Physical Growth of California Boys and Girls from Birth to Eighteen Years*. University of California Publications in Child Development 1:183–364, 1954.

Tufte, E. R. *The Visual Display of Quantitative Information*. Graphics Press, 1983.

Tukey, J. W. Analyzing data: Sanctification or detective work? *American Psychologist* 24:83–91, 1969.

Tyson, J. J., K. C. Chen, and B. Novak. Sniffers, buzzers, toggles and blinkers: Dynamics of regulatory and signaling pathways in the cell. *Current Opinion in Cell Biology* 15:221–231, 2003.

Varmus, H. *The Art and Politics of Science*. W. W. Norton, 2009.

Vedral, V. *Decoding Reality: The Universe and Quantum Information*. Oxford University Press, 2010.

Vine, F. J., and J. T. Wilson. Magnetic anomalies over a young oceanic ridge off Vancouver Island. *Science* 150:485–489, 1965.

Vossoughian, N. *Otto Neurath: The Language of the Global Polis*. NAI Publishers, 2011.

Wagenmakers, E.-J., R. Wetzels, D. Borsboom, and H. L. J. Maas. Why psychologists must change the way they analyze their data: The case of psi. Comment on Bem (2011). *Journal of Personality and Social Psychology* 100:426–432, 2011.

Wahba, G. *Spline Models for Observational Data*. SIAM, 1990.

Wakefield, A. J., S. H. Murch, A. Anthony, J. Linnell, D. M. Casson, M. Malik, M. Berelowitz, A. P. Dhillon, M. A. Thomson, P. Hafvey, A. Valentine, S. E. Davies, and J. A. Walker-Smith. Ileal-lymphoid-nodular hyperplasia, non-specific colitis, and pervasive developmental disorder in children. *Lancet* 351:637–641, 1998. Retracted, February 2, 2010.

Watson, J. *The Double Helix: A Personal Account of the Discovery of the Structure of DNA.* New York: Atheneum, 1968.

Watson, J., and F. H. C. Crick. A structure for deoxyribose nucleic acid. *Nature* 71:737–738, 1953.

Watts, D. J. *Six Degrees: The Science of a Connected Age.* W. W. Norton, 2004.

Weaver, W. Science and complexity. *American Scientist* 36:536–544, 1948.

Weber, G. W., and F. L. Bookstein. *Virtual Anthropology: A Guide to a New Interdisciplinary Field.* Springer-Verlag, 2011.

Weber, G. W., F. L. Bookstein, and D. S. Strait. Virtual anthropology meets biomechanics. *Journal of Biomechanics* 44:1429–1432, 2011.

Wegener, A. *Die Entstehung der Kontinente und Ozeane.* [The Origin of Continents and Oceans], 3rd ed., 1922.

Weiner, J. *The Beak of the Finch.* Vintage, 1995.

Weisberg, S. *Applied Linear Regression*, 3rd ed. John Wiley & Sons, 2005.

Weiss, P. A. [Comments.] In R. W. Gerard (Ed.), *Concepts of Biology* (p. 140). National Academy of Sciences, 1958.

West, G. B., J. H. Brown, and B. J. Enquist. A general model for the origin of allometric scaling laws in biology. *Science* 276:122–126, 1997.

West, G. B., J. H. Brown, and B. J. Enquist. The fourth dimension of life: Fractal geometry and allometric scaling of organisms. *Science* 284:1677–1679, 1999.

Whewell, W. *The Philosophy of the Inductive Sciences, Founded upon Their History.* 1840.

Wigner, E. The unreasonable effectiveness of mathematics in the natural sciences. *Communications in Pure and Applied Mathematics* 13:1–14, 1960.

Williams, R. J. *Biochemical Individuality: The Basis for the Genetotrophic Concept.* John Wiley & Sons, 1956.

Wilson, E. B. *An Introduction to Scientific Research.* McGraw-Hill, 1952.

Wilson, E. O. *Consilience: The Unity of Knowledge.* Knopf, 1998.

Wilson, J. T. Transform faults, oceanic ridges, and magnetic anomalies southwest of Vancover Island. *Science* 150:482–485, 1965.

Witztum, D., E. Rips, and Y. Rosenberg. Equidistant letter sequences in the book of Genesis. *Statistical Science* 9:429–438, 1994.

Wold, H., and K. Jöreskog, Eds. *Systems under Indirect Observation: Causality, Structure, Prediction.* 2 vols. North-Holland, 1982.

Wolfram, S. *A New Kind of Science.* Wolfram Media, 2002.

Woolf, H., ed. *Quantification: A History of the Meaning of Measurement in the Natural and Social Sciences.* Bobbs-Merrill, 1961.

Wright, S. Isolation by distance. *Genetics* 28:114–138, 1943a.

Wright, S. Analysis of local variability of flower color in *Linanthus parryae. Genetics* 28:139–156, 1943b.

Wright, S. Path coefficients and path regressions. *Biometrics* 16:189–202, 1960.

Wright, S. *Evolution and the Genetics of Populations.* Vol. 1, *Genetic and Biometric Foundations.* University of Chicago Press, 1968. Vol. 4, *Variability within and among Natural Populations.* University of Chicago Press, 1978.

Yablokov, A. V. *Variability of Mammals.* Amerind, Ltd., for the Smithsonian Institution and the National Science Foundation, 1974.

Yamaguchi, D. K., B. F. Atwater, D. E. Bunker, and B. E. Benson. Tree-ring dating the 1700 Cascadia earthquake. *Nature* 389:922–923, 1997.

Yumoto, C., S. W. Jacobson, and J. L. Jacobson. Fetal substance exposure and cumulative environmental risk in an African American cohort. *Child Development* 79:1761–1776, 2008.

Ziliak, S. T., and D. N. McCloskey. *The Cult of Statistical Significance: How the Standard Error Costs Us Jobs, Justice, and Lives*. University of Michigan Press, 2008.

Ziman, J. *Reliable Knowledge*. Cambridge University Press, 1978.

Zimbardo, P. G. *The Lucifer Effect: Understanding how Good People Turn Evil*. Random House, 2007.

Zimmer, C. How many species? A study says 8.7 million, but it's tricky. *New York Times*, August 23, 2011.

Index

The following index combines people, topics, techniques, and examples. Persons are cited by last name and one or two initials. References to figures are in **boldface**, preceded by **Pl.** if the figure also appears in the color insert. Entries pointing to footnotes go as page number followed by "n" and then footnote number, as with "74n1".

abduction (a principal mode of numerical
  inference), 13, 27, 85–100
  in combination with consilience, 115, 289–290,
    483–485
  in complex systems, 305
  definitions or characterizations of, xxviii, 5,
    86–88
  in discriminant analysis, as unlikely to arise, 392
  examples of
    anthropogenicity of global warming, 88–89,
      Pl.3.4
    in biology, 99–100
    continental drift, the graphical argument for,
      Pl.2.1
    faulty, 463, 470
    heart attacks in wives of smokers, 259
    lung cancer in wives of smokers, 262
    from maps, 122, 127
    in mathematics, 95
    in medicine, 204, 256–262
    and pathological science, 210–211
    in physics, 95–96
    in public health, 96–97
    Scorpion wreckage, discovery of, 8, 97–98
    in social psychology, 270
    in solar system science, 280–284
    and the fate of two space shuttles, 99
    in time-series data, 263–265
    see also under examples, main
  grammar of the word, 86
  intellectual precursors of, 86–87
  in linear regression

  for the intercept, 257–259
  for the slope, 260–262
  as the failure of the matrix formalism, 312
  emphasis on selection, 290
  in morphometrics, 405, **7.19**, **7.20**, **7.25**
  in multiple regression, as unlikely to arise, 244
  the two polarities of, 286
  psychological origin of, 91
  this argument is transcendental, 92
  surprise, the role of, 59–60, 88, 117, 259, 286
  see also under surprise; plausible rival
    hypotheses
added-variable plots, *see under* multiple
  regression, special cases
additive conjoint measurement, 379–380, **6.16**
agreement, reasonable, *see under* approximation,
  Kuhn's theory of
AIC, Akaike Information Criterion, 104, 334–338,
  5.19
  example, the corpus callosum in schizophrenia,
    338–344
  formulas for, 335, 337, 340n14
alcohol
  dose, measurement of, 373
  fetal exposure to, *see* fetal alcohol
allometry, 308, 437–438, **6.9**, **7.15**
Alvarez, W., study of the cause of the
  Cretaceous-Tertiary extinction, 4, 11,
    14, 34–35, 99, 117, 280–284, 285, 491
anatomical variants, 44
"and so on," *see* mathematical induction
Andersen, P., 234

519

Anderson, D., 244, 252, 331, 334, 337
Anderson, T., 322, 492
Anson, B., 44
anthrax, Sverdlovsk 1979 epidemic of, 3, 35, 127, **4.6**
    one-sentence summary of, 3
anthropogenic greenhouse gases, effect on global temperature of, **Pl.3.4**
apagoge, 86
approximations
    essential, 135–136
    Kuhn's theory of, 37–38, 178, 181, 262, 493
    mathematical notation for, 175
    in Perrin's work, 178
arc length, index of a variable type, 360
arc-sine law, 46–48, **2.10**
arithmetic into understanding (main theme of this book), xix, 497
Arrow, K., 279
asthma, epidemics of, in Barcelona, 129, **4.7**, 200, 211
asymmetry, human facial, 455–456, **7.28**
Atlantic Ocean, fit by E. Bullard to continents around, 19, **Pl.2.1**
atoms, existence and size of, *see under* Perrin
Atwater, B., 97–98
autism, as thimerosal poisoning, 207
average
    definition, 131
    as least-squares descriptor, 131
    as maximum-likelihood, under a Normal assumption, 191–193
    meaning of, as inseparable from theories about, 192
    of nonnumerical quantities, 142–143
    original sense of the word, 129
    precision of, 132–135
    weighted, 134–135
        Millikan's modification according to expert knowledge of error, 148
        regression slope as a weighted average of casewise slopes, 141
Avogadro's law, *see under* Perrin

Bayes, T., scholium (fundamental question) of, 188
Bayesian inference, *see under* inverse probability
Belmont Report, 271–272
bending energy, *see* thin-plate spline
Benveniste, J., *see under* dilution, infinite
Benzécri, B., 378–379
Berkeley Earth Study, analysis of global climate change by, 107–109, **3.9**
Berkeley Guidance Study, 240–241

Berry, W., xxviii
beta weights (β), *see* multiple regression, coefficients in
bilateral symmetry, *see* symmetry, bilateral
billiards table, as substrate for Bayesian inference, 188, **4.24**
binomial distribution, 151–152, **4.11**
    see also coin flips
biology
    abduction, biological examples of, 99–100
    numerical inference in, 13, 18
    rhetoric of unification arguments in, 30
    see also under anthrax, asthma, double helix, Fetal Alcohol Spectrum Disorder, Ramachandran plot, schizophrenia, Snow
biplot (of an SVD), 353–354
    in morphometrics, 362
Bloor, D., xxviii, 42
Bolhuis, J., 36
Boltzmann, L., 331
    see also Maxwell–Boltzmann distribution
Bookstein, F.
    email addresses of, xxiv
    and passive smoking trials, 254
    worked or published examples by
        on brain damage in fetal alcohol disorders, 445–449, **7.22**, **7.23**, **7.24**, **7.25**
        on brain damage in schizophrenia, 328–330, 338–343, **5.21**, 443–445, **7.19**
        on fetal alcohol effects upon adult IQ profile, 372–375, **6.13**, **6.14**, **7.31**, 461–463
        on hominoid evolution, 362–364, **6.7**
        on human facial asymmetry, 455–456, **7.28**
        on human skull growth, 364–366, **6.9**, 399, **6.26**, **7.13**, **7.15**
        on Pearson and his great folly, 466–471
        on rodent skull growth, 398–399, **6.25**, **7.5**, **7.6**, **7.9**, **7.17**, **7.18**, **7.20**
        on strain statistics, *see* strain, statistics of
        on viremia in monkeys, 354–357
    see also under thin-plate spline
Bookstein coordinates, *see under* shape coordinates, two-point
Boulding, K., 30–31, 71, 301–302, 498–499
Bowman, C., xxiv, 244, 377–378
"Boys have more" (Google search), 3–4
brain, human, damage to, from prenatal alcohol exposure, *see under* fetal alcohol
Broadbent-Bolton data, 455
Brownian motion, 38, 49–50, **2.12**, **4.19**, 116, 179–182
    certain origin of, *see under* Perrin
    maximal disorder of, **4.19**

orderly scaling of, 183
specific observable properties of, 176–177
Bullard, E., see under Atlantic Ocean
Bumpus, H., sparrow study of, 242–244, **4.39**
principal finding, **4.40**
Burnham, K., xxii, 331
Buss, D., 275–276, **4.52**

$_nC_k$ ("enn choose kay"), symbol of counts of combinations, 151
cameras, consiliences of, as machines, 32–33, **Pl.2.8**
Campbell, D., 263–266
cancer biology, as fundamentally heterogeneous, 44
canonical variates analysis (CVA), 391–394
carbon dioxide, trend and circannual pattern of, 76, **3.2**
Cargo Cult science, 212
caricature, 483
cases-by-variables matrix, 307
  see also principal components analysis, SVD
causation
  causal chains, correlations for, 223
  common causes, correlations for, 223
  Freedman and Weisberg on, 220–221
  as one grand theme of statistics, 220
  of multiple effects, 168
    see also Partial Least Squares
  normal equations for, 229–230, 233
  in the quincunx, 166, 220
  and regression lines, 167, **7.34**
  see also multiple regression
Ceccarelli, L., 30
centering matrix, 382
Central Limit Theorem, 159, 160
centrifugal force, 62
Centroid Size, 412, 418, 419–420, 424
  see also Procrustes form distance; Procrustes shape distance
ceteris paribus, see multiple regression, as counterfactual
challenges, in multiple regression modeling, 252–262
Challenor, P., see Gulf Stream
Chamberlin, T., 13, 101, 338
Chargaff, E., see under double helix
Chicxulub (meteor crater), 60, 117, 284
chimpanzees, as skull data, 364, **6.7**
χ^2 (chisquare), statistic, **4.20**, 253, 260, 315, 322
cholera, London epidemics of 1849, 1853, 1854, see under Snow
Christmas tree, see meta-analysis

civic virtue, as mensurand of a Pearson study, 3, 466–471
  one-sentence summary for, 3
classification, see discriminant analysis
Clifford, W., xxviii, 471
cognitive neuroscience, 36
  as having difficulties with consilience, 39
Cohen, J., 198–201
coin flips (model system), 132–135, 150–154
  as actual data, 132n1
  frequency falloff from its maximum, 153–154
    see also mathematics, notes, Note 2
  likelihood ratio tests for, 194–196
  tie between heads and tails, chances of, 152
cold fusion (pathological science), 207
Collins, H., 41–42
complex organized systems, 14–15, 34, 291–478
  advantages of, for both abduction and consilience, 483–485
  complexity scale of Boulding for, 302
  confirmatory statistical analyses for, as unlikely, 496
  heterogeneity as a problem in, see under heterogeneity
  measurement types in, 71
  morphometrics as a good example of analyses in, 405
  organisms as, 71, 160
  patterns in, 14, 289
  reductionism and, 34–35
  statistics of, as complement to the Gaussian model, 303, 316
computed tomography (CT)
  of the adult brain, **7.19**
  see also under Visible Female
conic sections, 347
Connecticut speeding crackdown, see plausible rival hypotheses
conservation equations, 162
consilience (a principal mode of numerical inference), 13, 27, 28, 29–37
  in combination with abduction, 115, 289–290, 483–485
  but temporally lags behind, 490, 493
  in complex systems, 304, 490
  emphasis on selection, 290
  definition of, 17
  examples of
    and anthrax, **4.6**
    and asthma, **4.7**
    and Avogadro's number, 178–179, 183–184
    and the Cretaceous–Tertiary extinction, 284
    and the effects of environmental tobacco smoke, 259, 262

consilience (*cont.*)
  and Einstein's law of the photoelectric effect, 149–150, **4.10**
  between Kepler's laws and Newton's law of gravitation, 66, 68
  and the Maxwell–Boltzmann distribution, **4.23**
  and Newton's apple, 28–30
  and obedience behavior, **4.50**
  and Planck's constant, 148
  in re the *Scorpion* sinking, 5–10
  in seafloor spreading, 22–27, **2.5**
  and ulcers, **4.25**
  see also under examples, main
  in morphometrics, 405
  in multiple regression, usually unavailable, 244
  numbers, special role of, 30–31, 39–40
  and pathological science, 210–211
  in physics, 40, 57; *see also* Kuhn, Wigner
  quantifying, 36
  requisites for visualization in, 28, 57–58
  the slide rule, as a machine for, 219
  trust in, as the foundation of the natural sciences, xxvii, 30, 186
Consilience (book), 29
continental drift, 18–27
  magnetometer data and, 22
  paleomagnetism and, 19–21
  one-sentence summary for, 3
contrastive method (J. S. Mill), 86
Coombs, C., 279
coordinates
  for the double helix, 218, **4.28**
  Cartesian, 33, 410
  polar, 62
copper, heat capacity, meta-analysis of, **4.41**
Coriolis force, 60, 62–63, **2.20**
corpus callosum, in human brain studies, **5.20, 7.21, 7.22, 7.23, 7.25, 8.1**
  details of permutation testing, 453
correlation
  formula for, 171, 222
  Pearson's data as exemplifying, **4.17, 7.36**
  as product of two regression coefficients, 222
  modified quincunx for, **4.29**, 221–222
  as a regression between standard scores, **4.29**
correlation matrix, 320, 369, 373
  not suited for principal components analysis of shape coordinates, 361, 362
  Wright's modification of, 369
correspondence analysis, *see under* singular value decomposition
covariance distance, **6.22**, 394–399
  pattern, for human skull growth, **6.26**
  pattern, for rodent skull growth, **6.25**
  explained by the uniform term of shape variation, 439
  see also relative eigenanalysis
covariance matrix, 357
  rank, in PLS, 372
crease (morphometric features), **7.20**, 444
Cretaceous-Tertiary extinction, *see under* Alvarez
Crick, F., *see under* double helix
Cromwell, O., skull of, 3, 464–466, **7.32**
crud factor, *see* Meehl
cudos ($\kappa \tilde{\upsilon} \delta o \varsigma$), "honor," acronym for a set of scientific norms, 41
curves, types of, 407
  see also ridge curve, symmetry curve

Darwin, C., 163, 185
DeHaene, S., studies of reading, 274
detective stories, as a model for numerical inference, 11, 91
determinantal rule, for intersections of lines, 140
diet, as confound in environmental tobacco studies, *see under* environmental tobacco smoke
dihedral angles, **5.2**
  see also Ramachandran plot
dilution, infinite, Benveniste study of, 208–210, **4.26**
direct effect, direct path, *see under* multiple regression, as path model; path analysis
discriminant analysis
  linear, 242–244, **6.18**, 387, **6.19, 6.21**, 387–391
  example of, *see under* Bumpus's sparrows
  quadratic, 387, **6.20**
  examples of, **5.9, 8.4**
  unlikely to lead to surprise, 392
disorder, 161
dissimilarity, *see* distance matrix
distance matrix
  example of, **5.9**
  Procrustes shape distance, Procrustes form distance, *see under* Procrustes shape distance; Procrustes form distance
  see also principal coordinates analysis
distances, commensurability of, as a principle of consilience, 28
distribution, *see* binomial distribution; Mardia–Dryden distribution; Normal distribution; prior distribution; random walk
DNA (deoxyribonucleic acid), *see under* double helix
"doctors study" of smoking and cancer, 96
domino PLS, 377

double-centered data, *see* principal coordinates analysis
double helix, 72, 214–220, 285
   Chargaff's rule for base-pair counts in DNA, 99, 216–217, 219
   discovery, as an abduction, 219
   discovery, as a consilience, 215, 219
   one-sentence summary for, 3
   X-ray crystallographic image of, 215, **4.27**
Double Helix, The (book), 116, 214
Duncan, O., xxii, 278

e, Euler's constant, the base of natural logarithms, *passim*
E, expectation operator, 168
earthquakes
   in Seattle, xxii, 97–98
   worldwide, **2.2**
   question posed by this figure, 22
economics, 278
`Edgewarp` (computer program), **Pl.2.8**, **Pl.7.14**, 425
Edgeworth, F., 141, 248
Edwards, A., 193–201
   see also under Interocular Trauma Test
effects, direct, indirect, and total, in multiple regression, 229, 233
effect sizes, in analysis of complex systems, 337
eigenanalysis, 309
Einstein, A., 3, 84, 96, 143–145, 179–180, 218n11, 490
elasticity, 45
ellipses
   in geometry, 60–61
   in gravitational physics, 69
   in statistical data analysis, 171–73, **4.18**
ellipsoids
   in geometry, **6.6**
   in principal-components analysis, 322–324
   variation of the principal axes of, 357–359
Elsasser, W., 13, 21, 42–43, 70, 160
empty space, as dominant feature of multivariate patterns for complex systems, 294–296, **5.1**, **5.3**
   abductions deriving from, 296
endophrenology, 447
entropy (physical concept)
   relative, Gibbs's inequality regarding, 331
   maximized by a Normal distribution, 161, 315
environmental tobacco smoke, 3, 116, 251, 253–262
   and heart attack risk, 256–259
   and lung cancer risk, 260–262
   one-sentence summary for, 3

epidemiology, role of multiple regression in, 244–252
equation, error in the, in linear regression, 137
   linear regression as the minimization of the mean square of these, 139–140, **4.8**
equipoise, 201
Euler, L., *see e*; mathematics, Note 7.1
EVAN (European Virtual Anthropology Network), xxiii, 426
Eve, *see* Visible Female
evolution, hominoid, 362–364, **6.7**
evolutionary psychology, limits of abduction and consilience in, 213, 274–276
Ewald, P., contemporary version of Koch's postulates by, 205–206
examples, main, of this book, *see under* Alvarez, anthrax, asthma, corpus callosum, double helix, environmental tobacco smoke, evolution, Fetal Alcohol Spectrum Disorder, growth, Hertzsprung–Russell diagram, IPCC, Livingstone, Millikan, obedience, Perrin, Ramachandran plot, schizophrenia, *Scorpion*, seafloor spreading, Snow, *Tetrahymena*, ulcers, viremia
   see also under Bookstein, examples by
excursion statistics, *see* random walk
expected value of log likelihood, *see* AIC
"explained" variance
   via correlations, 222–223
   see also under multiple regression
extrapolation, of a shape change, 445n6
   see also crease
extremes, rhetoric and risks of, 75–76

fact, scientific, *see under* Fleck
factor, 162, 396
   general
      role in epidemiological studies, 252
      Wright's method for extracting, 369
   special, 369
factor analysis, 368–370
   Guttman's models for, 472–473
factorial function ($n!$)
   definition of, 151
   Stirling's approximation for, 152, 173–175
FASD, *see under* fetal alcohol
Feller, W., 46
Felsenstein, J., xxii, xxiv
fetal alcohol
   effects of exposure to, at low doses, 15, 372–375, 461–463
   Fetal Alcohol Spectrum Disorder (FASD)

fetal alcohol (cont.)
   in the adult brain, 3, **7.21, 7.22, 7.23, 7.25,** 491
   legal and forensic aspects, 449–451, **5.9, 7.25**
   behavior in, **5.6, 5.7**
   corpus callosum in, **5.9, 7.21, 7.22, 7.23, 7.25**
   in relation to executive function, 448, **7.24**
   discovery of, 97
   the face in, 296, **5.4**
   in the infant brain, 3, 485–489
   one-sentence summary form for, 3
   test scores in, **5.5**
Feynman, R., 62–63, 484
   on pathological science, 212
Fisher, R., 96, 242, 310
Fleck, L., xxviii, 14, 42, 93, 185, 301, 481, 490
Flury, B., 362, 398
force
   gravitational, see under Newton's law of gravitation
   as metaphor, xxvii, xxviii, 272, 468–470
   see also centrifugal force, Coriolis force
forensic neurology, see fetal alcohol spectrum disorders, legal and forensic aspects
fractals, random walks as, **4.19,** 184
Franklin, R., 215, 219, 493
Freedman, D., xxi, 122, 220
   scale of practicality of regression of, 233–234
Friedrich, C., see under Wanderer
fruitcake, as metaphor for social measurement, xxii

Galton, F., 46, 404, 466–471
   averaged photography by, 34
   breadth of work by, 163
   Pearson and, 466–467
   quincunx machine of, 46, 162, **4.13**
      invention of, in holograph, **4.15**
      relation to coin flips, 163–164
      regression, discovery of, **4.14**
Gauss, K.
   the average of a Normal distribution is the likeliest mean, 193
   if the average is the likeliest value, the data must be Normal, 314
   see also Normal distribution
Gauss–Laplace synthesis, 191–193
general systems theory, 301–302, 499
Generalized Procrustes Analysis (GPA), 412–413, **7.3**
   the Procrustes shape coordinates it produces, 413–414, **7.4**

geodesy, 33
geologists, relative cognitive nonclosure of, 19
geometric morphometrics (GMM), see morphometrics
geometry, **1.1,** 11
   of covariance matrices, 397–398
   of multiplication, see hyperbolic paraboloid
   of scientific visualization, 57–58
Gerard, R., 13, 71
Gesamtkunstwerk, see rules of morphometrics, first
Gibbs, J., 315
Gibbs's inequality, 331
GMM, see morphometrics
Goldstein, I. and H., 129, **4.7**
Google, 3–4
Gould, S., 51
Gower, J., 382
Granger causality, 89n5
Green, W., xxiv, **Pl.2.8,** 338, **Pl.7.14,** 402
greenhouse gases, anthropogenic forcing of, see IPCC
Greenwald, A., propaganda study of, 272–273, **4.51**
grids
   examples, see rules of morphometrics, fourth
   random fields over, 326–328, **5.15**
   as representations of shape patterns, see thin-plate spline
   simplifications of, 439–443
   see also partial warps
   see also rules of morphometrics, fourth
growth, analysis of
   by added-variable plots, 240, **4.37**
   by grids, 366, **6.9**
   examples, see under Bookstein, examples, human skull growth and rodent skull growth
growth axis, of the human skull, 438
Guillemin, J., see anthrax
Gulf Stream, risk of collapse of, 110–111
Guttman, L., 472–473

Hagen's hypothesis (Normality of psychophysical variables), 139n2, 160–161
Hansen, J., approach to climate change of, 106–107
   see also NOAA
hat matrix, 233
heart attacks, risk of, from environmental tobacco smoke, 3, 256–259, 285
   platelet aggregation as explanation of, 259, **4.44**
Helicobacter pylori, see under ulcers
helix, double, see under double helix

Herrnstein, R., 276–277
Hertzsprung–Russell diagram (astronomy), 291–293, **5.1**, 492
heterogeneity
  of cancer biology, 44
  hidden, 160
  in high-exposure groups, 247
  interferes with consilience in the rhetoric of biology, 43
  as a mechanism for generating correlation, 227
  of random walks, when analyzed inappropriately, 51
  of strain, *see under* thin-plate spline, bending energy
"hockey stick" (overstated statistical model of global warming), 76–78
hominid evolution, as example of principal component analysis, 362–366
Huff, D., 4
hyperbolic paraboloid, 15, 346–349, **6.1**, **6.2**
  alternating fourfold rotational symmetry of, 349
  as the blend of two oppositely opening parabolas at 90°, 348–349
  for depicting individual terms of an SVD, *see* singular value decomposition
  as the finding, in a low-dose fetal alcohol study, **6.13**
  as the surface $z = xy$, 347
hypotheses, scientific
  a-priori choice of, as unhelpful for studies of complex systems, 482
  discriminant analysis as testing, 385
  the kinetic theory as, *see under* Perrin
  likelihood of, *see under* likelihood ratios
  multiplicity of, as a strength of morphometrics, 482
  null, as lacking a role in most competent numerical inferences, 200

ignorance, assumptions about
  in Bayes's scholium, 188
  as justification for the Normal distribution, 315
  in morphometrics
    as justification for the isotropic Mardia–Dryden distribution, 417
    as justification for the thin-plate spline, 403
indirect effect, indirect path
  bounds on, from a meta-analysis, 260–261
  Wright's equations
    example, 368
    and PLS, 368–369
  see also under multiple regression, as path model; path analysis
individuality, 16

  see also heterogeneity
inference, numerical
  abduction, *see under* abduction
  arithmetic and, 105, 289
  Bayesian, *see under* prior distributions
  and complex systems studies, 300–301, 494–498
  deduction, 86, 92
  detective stories and, 10, 91, 99
  the double helix, as the best biological example ever, 219–220
  examples of, in one sentence each, 3
  induction, 86, 92
    see also mathematical induction
  and Koch's postulates, 205–206
  norms of, 12
    apparently incongenial to the psychological and social sciences, with some exceptions, 272–280
  rhetorical form of, 3–5
  strong, *see under* strong inference
  as the subject of this book, xix, 16, 494, 499
  visualization and, **1.1**, 57–58, **3.3**, **Pl.3.4**, 85
    see also under Latour
inference, statistical, 491–492
information, Shannon's quantification of, 331–332, **5.18**
information matrix
  in bioinformatic PLS, 377–378
  Fisher's, 310
integers
  and continua, *see* mathematics, Note 4.5
  the Pythagorean epiphany, xxvii, 152
  sum of the squared reciprocals of the, *see* mathematics, Note 7.1
integration, 439
interaction, in causal reasoning, 252, 302, 311, 337, 462–463, **7.31**, 498
intercept, of a regression line, 135, 142, 145, 148, 238, 285, 335–336, **5.19**
Intergovernmental Panel on Climate Change, *see* IPCC
Interocular Trauma Test (ITT), xxvii, 76, 129, 260, 292
interrupted time series design, 263–265
inverse probability, 188–191
  prior distributions and, 178, 189
  see also likelihood ratios
Ionian enchantment, 30
IPCC (Intergovernmental Panel on Climate Change)
  Fifth Assessment Report, 74n1
  Fourth Assessment Report (FAR), 74–90
    anthropogenicity of global warming, **Pl.3.4**, 13, 74–85

IPCC (*cont.*)
  as a classic abduction, 88–90, 95
  global temperature change contributors to, 75
  treatment by, 76–77, **3.2**, **3.3**
  see also hockey stick
  greenhouse gases, treatment by, **3.1**
  methodology of not explaining events, 89–90
  refutation by Hansen, 106–107
  policy implications of, 84–85, **3.6**
  visualizations, quality of, 85
IQ, 370n5, 372–375, 461–463
iridium, anomalous concentrations of, 281–283
ischaemic heart disease, *see* heart attacks
Ising model (for binary images), 328

Jaynes, E., xxi, 289, 314, 315, 331
Jeffreys, H., 26, 193–201, 289, 314
  Jeffreys's paradox, with increasing sample size, 196–197
Josephson, B., on variations in the strength of abductions, 90
Juan de Fuca Ridge, *see* seafloor spreading

Kahneman, D., xx, 279, 493
Keeling, C., and the Keeling curve, 76, **3.1**, 94
  see also IPCC
Kepler's laws, 64–65
  as abductions from Newton's law of gravitation, 66, 68
  see also Newton's law of gravitation
kinematic quantities, 132
kinetic theory, use by Perrin, 176–177
Koch's postulates, 205; *see also* Ewald, P.
Konrad Lorenz Institute for Evolution and Cognition Research, xx, xxiii, 70
Krantz, D., 379
Krieger, M., on the craft of physics, 96
Krzanowski, W., xxi
Kuhn, T., xxvii, 12, 13, 15–16, 35, 38–40, 71, 93, 94, 101, 262, 286, 359, 484, 495
  argument for the centrality of strong inference, 100–101
  normal science, concept of, 39
  persistent anomalies, analysis of, 39–40
  as abductions, 93–94
  limits of, in the historical natural sciences, 491
Kullback-Leibler (K-L) divergence, 333–334, **5.18**

Lagemaat, R. van de, xxi
landmarks, in morphometrics, 404, 406–409, **7.1**, 464
  see also rules of morphometrics, first
Langmuir, I., 207

Laplace, P., *see under* likelihood ratios
Gauss–Laplace synthesis, 191–193
latent variable scores, *see* Partial Least Squares, two-block, interpretation in terms of scores
Latour, B., 31–33, 484
laws, physical
  auxiliary assumptions underlying, 136, 180, 200
  and consilience, 179
  see also under Perrin
  often are linear, 137, 181–182, **4.21**
  fit by linear regression, as weighted averaging, **4.8**
  significance testing in, 199
  see also under Newton's law of gravitation; Millikan; Wigner
least-squares methods
  average as, 131
  Gauss–Laplace synthesis of, 191–193
  straight-line fits (linear laws) as, 135–142
  see also Millikan; multiple regression
LeGrand, H., narrative of continental drift of, 19–26
Letvin, N., 354
lever, as example of consilience, 29
likelihood ratios, 116, 193–198
  and the AIC, 338, 340
  −2 times the log of, distributed as a chisquare, 315
  in Hansen's approach to global warming, 107
  in Jeffreys' approach to hypothesis testing, 193–196
  Laplace's theorem, 189
  maximum likelihood, principle of, 190
  average as maximum likelihood, under a Normal assumption, 191–193
  are not to be found in NHSST, 196, 198–201
  as corrected for number of parameters, *see* AIC
  on the quincunx, 166–167
  see also inverse probability
linear multivariate analysis, 290
  not conducive in general to abduction or consilience, 311–312
Lipton, P., on "inference to the best explanation," 87, 94, 96
Livingstone, F., 34, 212–213, 276, 491
logarithm
  as the function for which
    $f(xy) = f(x) + f(y)$, 159, 332, 395–396
  as the geometry of the slide rule, 395–396
  of the likelihood function
    in discriminant analysis, 242, 387
    see also likelihood ratios, AIC
  of a Normal distribution, **4.12**

as a useful transform for growth data, 237, **4.36**
log-likelihood, *see under* likehood ratios, AIC
London, England, cholera epidemics in, *see under* Snow
longitudinal analysis, *see under* Berkeley Growth Study; Vilmann
lung cancer, risk of
    from environmental tobacco smoke, 3, 260–262, 285
    from smoking, 96–97

machines
    Alvarez's, 281–283
    the astronomers', 19, 291
    Galton's, 163–165
    for the Maxwell-Boltzmann distribution, **4.22**
    Millikan's, 145
    and PLS analysis, 372
    the slide rule, as a machine for abduction, 219
    Wigner's characterization of, 6, 31, 33
Maddox, J., on the infinite dilution experiment, 208–210
magnetic resonance (MR) images of the brain, **7.19**
magnetic reversals, *see under* seafloor spreading
magnetometer data, *see under* continental drift
malaria, evolution of, 212–213
Mandelbrot, B., 46, 184
mandible, template for the morphometrics of, 433–436, **Pl.7.14**
Mardia, K., xxiii, 294, 402, 454
Mardia–Dryden distribution, of shapes, 417
Margenau, H., 55–57, **2.18**
Marshall, B., 116, 490; *see also under* ulcers
Martens, H., xxi, 377, 405n1
Martin, R., 406, 464, **7.1**
mathematical induction, 95, 175
mathematics
    appearance of brilliance in, 175; *see also under* Wigner
    obscurity of, in the reader's youth, xxii
    notes, mathematical
        Note 4.1, on extended equalities, 131–132, 141
        Note 4.2, on coin models, 134
        Note 4.3, on what a step in a derivation is, 151–152
        Note 4.4, on clever tricks, 156
        "Interlude," on the geometry of the regression line and the ellipse, 171–173
        Note 4.5, on the interplay between integers and continua, 174–175
        Note 4.6, maximum likelihood: symbols or reality?, 192

Note 5.1, derivation of the precision matrix, 321–322
Note 5.2, why does the arithmetic mean exceed the geometric mean?, 324–326
Note 7.1, A wonderful identity of Euler's, 474–476
power of, xxvii
relation to scientific understanding, xix, 4
role in natural philosophy, xxvii
role in scientific training, xxii
as a set of stratagems for symbol manipulation, 66
unreasonable effectiveness of, in the physical sciences, *see under* Wigner
matrix (mathematics)
    arithmetic of ($+$ and $\times$), 307
    labels of rows and columns, 15, 309–310
    general nature of, 306–307
    notation, failure of, as grounds for abduction, 312
    notation of multiple regression in, 233
    rank, 350, 372
    shift matrix, 310
    subordination of, in complex systems analysis, 15, 311
    incompatible with most forms of feature selection, 300
    adding more subscripts, 310
    varieties of, in statistical data analysis, **5.9**
    covariance structure, 307–308
    data structure, 308
    gridded surface, 307
    operator, 307
    pattern of distances, 309
    quadratic form, 308–309
    see also cases-by-variables matrix, correlation matrix, covariance matrix, distance matrix, hat matrix, image matrix, information matrix, precision matrix, singular value decomposition
Maxwell, J. (British physicist), 37, 144, 294
Maxwell–Boltzmann distribution, 37, 328
    derivation by Maxwell, 158
    as the most disordered distribution, 161
    experimental confirmation of, 186–188, **4.23**
mean centering, 141
mean square, 132; *see also* least squares
measurement, measurements
    additive conjoint, *see* additive conjoint measurement
    in biology, 43–44
    regarding the double helix, 215
    design of, 183, 184–185
    for complex adaptive systems, 360–362
    in developmental psychology, 37

measurement, measurements (*cont.*)
  foundations of, 36
  in physics, 37
  indirect, via regression, 140
  for questions with one principal explanandum, Chapter 4 *passim*
  repetition with variation, importance of
    in Alvarez's study, 281–283
    in Milgram's study, **4.50**
    in Perrin's study, 177–178, 183
    see also under Collins
  in science in general, xxviii, 40
  in the context of theory confirmation, 40, 101, 188
  cannot precede understanding, in Kuhn's view, 495
medial axis, **5.23**, 343–344
medical images
  as consiliences, 33
  multiple approaches for organizing, 343
  as provoking a renaissance in biometrics, 72
  as a variable type, 360
  see also computed tomography, magnetic resonance
Meehl, P., 199, 303n4
Meiland, J., xx
Merton, R., 12, 41
meta-analysis, 41, 256, 259, **4.45**
  Christmas tree (type of plot), 260, **4.45**
metrological chain, the Alvarez example of, 35
Michelson and Morley, experiment of, 96, 143
microstates, as unsuited for biological measurement, 44
microwave background radiation, **5.8**
midsagittal plane, 362, 434, **Pl.7.14**
Milgram, S., 14, 117, 491; *see also under* obedience
Mill, J., 85
Millikan, R., study of the photoelectric effect by, 14, 115, 143–150, 285, 490
  measurements, design of, 145, **4.9, 4.10**
  overruling of linear regression by, 145–148
  Planck's constant, estimates of, 148
  skepticism of, 144
    overruled by data, 150
  one-sentence summary of, 3
  as an example of the "unreasonable effectiveness of mathematics," 150
Mills, C., 279–280
Mitteroecker, P., 394–399, 419
Mladeč (archaic human skull), 364
modularity, 439
molecular magnitudes, *see under* Perrin
Moon, orbit of the, 27, **2.7**

moral reasoning, contemporary experimental approaches to, 272
morphometrics, 402–461
  as exemplar of analyses of complex organized systems, 15, 44, 482
  shape coordinates as structured variables, 360
  data flow in, 403; *see also* rules of morphometrics
  brief history of, 404–405
  the unusually broad role of hypotheses in, 482
  permutation testing and, 453
  principal components analysis in, *see* rules of morphometrics, second
  principal coordinates analysis in, *see* rules of morphometrics, second
  rules of, *see* rules of morphometrics
  templates and, **Pl.7.14**
  examples, *see under* human skull growth, rodent skull growth, schizophrenia
Mosteller, F., 239–241
"Mrs. Ples" (hominoid skull), 363
multiple regression
  assumptions of, in epidemiological studies, 226–227, 244–252
  plausibility of, Freedman's scale for, 233–234
  predictable failures of, 252
  adjustment for bias, 250
  biological plausibility, 247–248
  confounders, assumptions regarding, 251
  dose-response relationships, 248–249
  breadth of evidence, 249–250
  for challenging models that are otherwise plausible, 252
  coefficients of, as direct effects, 229
  examples
    artificial example: Success on IQ and SES, 224, **4.31, 4.32**
    in studies of environmental tobacco smoke, 116, 253–262
    indirect effects, meta-analytic estimates of, 257, **4.43**, 260–261
    as part of linear discriminant analysis, *see under* Bumpus
    malaria in West Africa, *see under* Livingstone
  as incapable (by itself) of driving numerical inferences, 251–252, 497
  the three interpretations of
    causal and least-squares analyses are arithmetically the same, 232–233
    as counterfactual, 227–231, **4.32**, 233
    "holding constant," 206, 230–231
    as path model, 231, **4.33**, 233, **6.10**
    direct and indirect effects, 229
    as least-squares fit, 231–233

see also normal equations
modeled as geometry, **4.31**
special cases of
 analysis of covariance, 234–239, **4.34**
 ecological fallacy in, **4.35**, 237
 added-variable plots, 239–240, **4.37**, **4.38**
 linear discriminant analysis, 242–244, **4.40**
 variance, "explained" and "unexplained,"
  language of, 223
 with uncorrelated predictors, 224–227, **4.32**
see also additive conjoint measurement
multiplication
 geometrization of, *see* hyperbolic paraboloid
 and rank-one matrices, *see* singular value
  decomposition
"multivariate statistics," *see* linear multivariate
 analysis

N (Avogadro's number), 179, 186
 see also under Perrin
Nash, L., 29
Nature, assumptions regarding, *see* ignorance; law,
 physical
Nature (journal), 208–210
Neanderthals, skull data of, 364, **6.7**
Nebelmeer, Der Wanderer über dem, *see* Wanderer
Netflix Prize, won by an SVD algorithm, 354
network diagrams, **5.11**
Newton's law of gravitation, 28, **2.7**, 58–70, 143
 as both mathematics and physics, 68–70
 relevance of the inverse-square formulation, 59
NHSST, *see* significance testing
Nisbett, R., studies of culture, 273–274
NOAA (National Oceanographic and Atmospheric
 Administration), approach to climate
 change of, 107, **3.8**
Nobel prizes, *see under* Alvarez, Arrow, Crick,
 Einstein, IPCC, Kahneman, Marshall,
 Millikan, Perrin, Warren, Watson
nomogram, 380
Normal distribution, 14, 46, **2.10**, 116, 154–163
 in biology, only an idealization, 160
 no longer a requisite for using textbook
  statistics, 162
 ignorance, as Jaynes's justification for its use,
  315
 in mathematical statistics, 161–163
 its two independently measureable and
  interpretable parameters, 183
 as characterized by the linearity of all
  regressions, 169
 the reproducing characterization, 314
 sums of Normals are again Normal, 161,
  162, 166

 as "most disorderly" for given mean and
  variance, 161, 315
 see also entropy
 notation and properties, 154
 "width" at the inflections is 2 standard
  deviations, 156
 "bell curve" as exponential of a parabola,
  4.12
 origin of the normalizing constant $1/\sqrt{2\pi}$ of,
  156
 origins
  distribution of shots at a target, 159
  evenly spaced increments of a Brownian
   motion, 159
  see also under Perrin
 inhomogeneity is possible, 163, 166
 see also Umpire-bonus model
 appropriately scaled limit of binomials, 153,
  4.11, **4.12**
 Galton's quincunx, 163, **4.15**
 in scientific sampling and survey research,
  160, 161
 and temperature change, at continental scale,
  106–107, **3.7**
normal equations, in least squares, 140, 230, 233,
 6.10
 as weighted sums of the errors-in-equations, 139
null-hypothesis statistical significance testing
 (NHSST), *see* significance testing
numbers, role in science of, xxvii

obedience, Milgram's experiment on, 266–272,
 285
 "authority of the experimenter," 270
 calibration as consilience, 267–270, **4.50**
 basic dramaturgy, 266–267
 ethical issues, 270–272
 choice of a mensurand, **4.49**, 269–270
observer-expectancy effect, 277
off-diagonal entries
 of a correlation matrix, *see under* factor analysis
 of the inverse of a covariance matrix, *see under*
  precision matrix
Olson, E., 72
On the Mode of Communication of Cholera, *see*
 Snow
organisms, as complex organized systems, 160,
 402
Ornstein–Uhlenbeck process, 316n7
"orphan tsunami," *see* Seattle, earthquakes near
Oxnard, C., xxiv

paleoanthropology, limits of abduction and
 consilience, in, 213, 301

paleomagnetism, *see under* continental drift
Pangæa, Panthalassa, 19
paraboloid, 348–349, **6.2**
 see also hyperbolic paraboloid
parapsychology, as pathological science, 207–208
Partial Least Squares (PLS), 366–372
 and additive conjoint measurement, 379
 by count of separate measurement blocks
 two-block
  as a path analysis, **6.11**
  as an SVD, 371
  iterative algorithm for, 375–376, **6.15**
  interpretation in terms of scores, 371
  example from fetal alcohol psychology, **6.14**
  example from endophrenology, 448–449, **7.24**
 three-block
  algorithm for, 376–377, **6.15**
  multiblock ("domino PLS"), 377
 and passive smoking, according to the U. S. Surgeon General, 254
 for strain energy against shape change, **7.30**
 structured variables in, 372, 402
  see also shape coordinates
 styles of, 366–367
partial warps, 423, 439, **7.7**, **7.17**, **7.18**
passive smoking, *see* environmental tobacco smoke
path analysis, path coefficients, 140, 162, 367–368
 see also under multiple regression, as path analysis
pathological science, 42, 116, 203, 206–212
 Feynman's views on, 212
 Langmuir's and Rousseau's criteria for, 207
pattern languages, 306n5
patterns, *see* complex organized systems
Pauling, L., 214–215, 219
PCA, *see* principal components analysis
PCO, *see* principal coordinates analysis
Pearl, J., xxi, 220, 353
Pearson, K., xxi, xxvii, 15, 34, 53–55, **2.17**, 315, 361, 443, 476, 499
 and civic virtue, 466–471
 and Cromwell's skull, 463–466
 and the inheritance of human height, 168–171, **4.16**, **4.17**, **7.34**, **7.36**
 see also under correlation, formula for; principal components; chisquare; civic virtue
Peirce, C. S., xxviii, 4, 12, 13, 85–86, 91–92, 150, 303, 405, 491
periodic table of the elements, 18, 52–53, **2.14**, **2.15**, **2.16**

permutation testing, 451–452, **7.26**
Perrin, J., study of molecular magnitudes, 4, 11, 14, 37–38, 40–41, 116, 175–186, 284, 285, 490
 abductive in form, 179
 atmosphere, rarefaction of the, 177
 consilience
  as explicit goal of, 176
  as achieved, 183–184
  the kinetic hypothesis, 176
 measurements, design of, 177–178, 185–186
 origins of, in Brownian motion, 176–177
 verified as regards both slope and uncertainty, 181–182
 Einstein's equipartition law for, 182
 random walk, as one component of the study, 179–182, **4.19**
 one-sentence summary of, 3
 variations of experimental conditions, 178, 181
 around a line, themselves lawful, 182
 weight of a particle, as measureable intermediary, 177
photoelectric effect (physics), *see under* Millikan
photography, averaged, *see under* Galton
physics, as a domain specialized for abduction, 96
 role of abduction and consilience in, 143–144
physiognomy, 34, **5.3**
π (pi), unexpected appearances of, 156, 474–476
Planck's constant (h), precision of Millikan's estimate of, 148
plasticity (of organismal form), 44–45
plate tectonics, *see* continental drift
Platt, J., xxiii, 13, 101–104, 482, 491
 on mathematics, 103
 on the social sciences, 103–104
plausible rival hypotheses, 263–266
PLS, *see* Partial Least Squares
Pólya, G., 87–88, 95, 159, 173n8, 324
Popper, K., 477
power laws, for biological growth, 237
pragmatism (American school of philosophy), as context of abduction, 100
precision matrix, 318–321, 369
principal components analysis
 characterizations of
  as uncorrelated extrema of variance, 358
  low-rank least-squares fits to data, *see* singular value decomposition
 for dimension reduction prior to covariance distance analysis, 398–399
 common, 362
 examples
  hominid evolution, 362–366, **6.7**, **6.8**, **6.9**
  rodent skull growth, **7.5**, **7.16**

interpretation as patterns, 361
  scores as weighted averages of structured variable lists, 359–360
  for the patterns of small variance, 361
  for multidimensional random walks, 50–51, **2.13**, 289, 473–474
  and principal cooordinates analysis, 384
  power method for, 376
  statistical aspects
    variation of the principal directions of, Anderson's formula, 359
    Anderson's statistic for meaningfulness of, 322–324, **5.12**, 328, 363n4
    contraindicated when showing clusters known a-priori, 364
  see also singular value decomposition
principal coordinates analysis, 381–384, **6.17**
  and principal components analysis, 384
  examples, using covariance distance, 398–399
  relative warps as principal coordinates of Procrustes shape distance, 415
  see also singular value decomposition; rules of morphometrics, second
prior distribution, 192–193
  in Perrin's study, 178, 184
  see also under inverse probability
principal warps, *see* partial warps
Procrustes (mythical Greek), 410n1
Procrustes average shape
  as coefficient matrix in the projection notation, 419
  as nuisance parameter, 416–417
Procrustes distance
  algorithms for, 411–412, **7.2**
  has no biological content, 404
  as common metric for shape variables, 361, 402, 414, 419
Procrustes form distance, 419–420
Procrustes form space, **6.7**, 420, 424
  see also rules of morphometrics, third
Procrustes registration, 338, 344, 412–414, **7.4**
  restoring the Centroid Size component, 419–420
  see also shape coordinates
Procrustes shape distance, 410–412, **7.2**
Procrustes shape space, 289
  as projection down from digitizing space, 417–419
  subspaces of, 418, **7.17**
proof
  "decisive," 178, 179, 286
  "incontrovertible," 122, 126
  see also mathematics, notes on
Prossinger, H., xxiv

proteomics, *see* Ramachandran plot
psychology, *see* evolutionary psychology, social psychology
psychophysics, *see* Hagen's hypothesis

quadric surfaces, *see* ellipsoid; hyperbolic paraboloid; discrimination, quadratic
quasi-experimental designs, 265
Queteletian fallacy, 162
quincunx, *see* Galton

Raff, A., *see under* Zebra Diagram
Ramachandran plot (proteomics), 293–296, **5.2**, **5.3**, 491, 492
randomness, 14, 312–331
  as a null, vs. clinical inutility, 204
random walk, 45–51, 316–318
  distribution of an excursion statistic, 317–318, **5.10**
  intuitive inaccessibility of, 48–50
  and the Guttman scale model, 473
  in Perrin's study of Avogadro's number, *see under* Perrin
  principal components of, **7.37**
rank, of a matrix, 350
  of Procrustes shape coordinates, 414n5
Rao, C., 158
regression
  and the AIC, **5.19**
  and climate change, *see* Berkeley Earth
  elliptical geometry of, 168, 171–173, **4.18**
  typical Victorian example of, **4.16**, **4.17**
  intercept of the line on the ellipse, 173
  Freedman's scale of practicality of, 233–234
  intercept, *see under* intercept
  on the quincunx, 165–167, **4.14**
  regression line, as prediction or explanation, 167
  not necessarily causal, 238
  limits of the linear argument, 262
  the case of multiple regression, 231
  as one of a pair of lines, 169, **4.17**, 223
  see also under correlation
  slope, *see under* slope
  see also laws, physical, ofter are linear
"regression to the mean," 173, 223
relabeled reflection, 454
relative eigenanalysis, **6.22**
  see also covariance distance
relative warps, 362, 416
  see morphometrics, principal components in
replication, *see* Collins
rete mirabile, as metaphor for complex systems analysis, 495

retroduction, 92n9
  see abduction
Reyment, R., xxi, xxiv, 405n1
ridge curve, 407
  examples, on the human mandible, 434–436, **Pl.7.14**
Rohlf, F. J., xxiv, 402
Rosenthal, R., 277–278
Rothman, K., xxi
Rousseau, D., 42, 207
Ruhla, C., 186–187
rules of morphometrics, 15
  first, 407
    examples, **6.8, 6.25, 6.26, 7.1, 7.14, 7.19, 7.22, 8.1**
  second, 415
    examples, **5.9, 5.20, 5.21, 6.7, 7.5, 7.13, 7.19, 7.22, 7.30, 8.3**
  third, 420
    examples, **6.7, 7.6, 7.13**
  fourth, 426
    examples, **6.9, 7.9, 7.13, 7.15, 7.16, 7.17, 7.18, 7.19, 7.23, 7.24, 7.28, 8.4**
Russell, B., xxvii

sample size, and significance testing, *see under* Jeffreys
Saturday Review, xxii
scaling
  of shape data, *see* Centroid Size
  of variables, 132
  see also under correlations
schizophrenia, morphometric examples on, 328, **5.16, 5.17,** 443–445, **7.19**
sciences, unity of the, 30
scientific fact, definition of, *see under* Fleck
scientific paper, format of, xx
Scorpion, a sunken submarine, 3, 5–11, 204
  finding it
    as an abduction, 97–98
    as the closure of a numerical inference, 10
  visualization and, 18
seafloor spreading, 3, 18–27, **2.6**, 211
  as an abduction, 98
  the Juan de Fuca ridge map and, 22, **2.4**
  magnetic reversals and, 22, 26, **2.3, 2.4, 2.5**
Seattle, earthquakes in or near, xxii, 98–99
Seidler, H., xxiii
selection
  of biometric data, 43
  eigenanalysis as, 309
  emphasis upon, in numerical inference for complex systems, 290
  rules for, in typical complex systems studies, 300
semilandmarks, **Pl.7.10, Pl.7.14,** 426–437, 456–459
  and bilateral symmetry, 457–458, **7.29**
  examples of
    human brain images, **5.20, Pl.7.10**
    human skull growth, **7.13**
  formula for, 427
  interpretation, 431–432
  as an additional parameter for the relabeling group, 427
  single-point prototype for, 428–429, **7.11**
  informal statistic for, 340
  on surfaces, 432
  working from a template, 433–437, **Pl.7.14**
  see also thin-plate spline
Semmelweis, I., principal discovery of, as an abduction, 86–87
Shannon, C., 331
shape coordinates
  Procrustes, 410–415
    see also rules of morphometrics, second
  two-point, 439–443, 491n2
    see also under variables, structured
sickle-cell anemia, evolution of, 212–213
significance testing
  the Bayesian replacement, 193–198
  and health bureaucracies, 201
  domains of unsuitability
    for studies of coin-flipping, 195
    see under Jeffreys paradox
    for studies of complex systems, 304
    replacement by extended distributions, 492
    see under singular value decomposition
  as an impediment to the development of morphometrics, 405
  as responsible for parapsychology, 208
  in a linear multivariate context, 311
  null-hypothesis statistical (NHSST)
    as deplorable, xxii, 83, 116, 198–201, 211
    with the exception of the "strong form," 199, 284
single-factor model, 396; *see also* factor, general
singular value decomposition (SVD), 14, 289, 349–354, 375–376, 482
  as decomposition of the sum of squared matrix entries, 352
  formula for, 351
    associated identities, 352
    interpretation of second and higher singular vectors, 361
    least-squares property, 353

central role in the pattern analysis of organized
systems, 350, 483
regarding abduction, 483–484
regarding consilience, 484–485
rank-1 example, 351, **6.3**
iterative algorithm for, 375–376, **6.15**
rank-2 example (viremia in monkeys), 354–357
salience of the findings, vis-á-vis a PCA, 357
regression coefficients, singular vector elements
as, 353
see also under its special cases: Partial Least
Squares, principal components analysis,
principal coordinates analysis
SIV (simian immunodeficiency virus), example of
an SVD, 354–357
skull, human, in midsagittal section, **Pl.7.10**
Slice, D., xxiv
slide rule, **4.28**, 219, **6.23**, 396
sliding landmarks, *see* semilandmarks
slope (in regression analysis)
abduction on, 260–262
in biology, 100
consilience on, 148, 178, **4.21**
formula for
in simple regression, 140
in multiple regression, *see* multiple
regression, as path model
as weighted average of casewise slopes, 141
small multiples, principle of (visualization), 81,
3.5, 89
Snow, J., on the mode of "communication"
(dissemination) of cholera, 14, 115,
117–126, 211, 249, 284, 285, 490
abduction and, 118, 123, 126
Broad Street pump, role of, in this inference
interpretation of the 1854 epidemic near, 118
map of, **4.1**
natural experiment, logic of the, 122, 200
counterfactual summary of, 126
shoe leather, as characterization of his
methodology, 97, 122
one-sentence summary for, 3
water companies of London, role of, in
inference, 118–126
death rates by neighborhood in 1853, **4.2**
death rates by neighborhood by water
company in 1854, **4.4**, **4.5**
map of, **Pl.4.3**
vis-á-vis the Marshall-Warren ulcer studies, 204
social class
as confound, in studies of fetal alcohol damage,
15, 461–463
as a consideration in Pearson's studies of
inheritance, 169n7, **7.34**, 468–470

social psychology, abduction and consilience in,
272–274, 276–278
social sciences
abduction and consilience in, 15–16
critiques of numerical inferences in, various
examples, 272–280
Kuhn's comments on, 41
types of measurement, 278
Wilson's comments on, 36
sociobiology, 36, 274–276
sociology, limits of abduction and consilience in,
278–280
space, Euclidean models for, 28
spacetime
Galilean models of, 28
role in visualizations, 57–58
sparrows, effect of a snowstorm upon, *see under*
Bumpus
spectrum, index of one type of variable list, 360
splenium intercept point, **8.1**, **8.2**, **8.3**
spline, thin-plate, *see* thin-plate spline
standard deviation, 132
standard error of the mean, 134
standard scores (in correlational analysis), 221
statistics
inutility of standard courses in, xxii
a 20th-century curriculum for, 302–303
a 21st-century curriculum for, 494, 496–500
role in physics, 54–57
Stirling's approximation for $n!$, 152, 174, 196; *see
also* factorial
Stoppard, T., *Rosencrantz and Guildenstern are
Dead,* 93
strain, statistics of, 459–461, **7.30**
strain energy, relation to bending energy, 425, **7.30**
Streissguth, A., xxiii, 296, **5.4**, **5.7**, 372–375
strong inference, 13, 100–103, 211, 280–284
in complex organized systems, 334, 338
definition of, 5
Platt's characterization of, 101
studentized residuals (in multiple regression), 233
subitizing (direct apprehension of small counts),
178
submarine, *see Scorpion*
"Suddenly I became aware" (comment of J.
Watson), 217, 491
sum of squares, 131, 132; *see also* least squares
as an energy term, in interpreting an SVD, 361
surprise, 117, 144, 210, 217, 247, 296, 296n2,
300, 303–304, 312, 394, 481, 484, 496
as a component in the definition of abduction,
xxviii, 86
in a grid, 326, **5.14**
in an image, 330

surprise (*cont.*)
  in an investigation of random walk, 318
  Alvarez's iridium anomaly, 281, **4.53**
  arguing against surprise where it is unwanted, 259
  see also plausible rival hypotheses
SVD, *see* singular value decomposition
swept area (in gravitational physics), 64
Switzerland, 2003 summer temperatures in, 90
symmetry, 26, 158, 190, 226, 361
  bilateral, 454–456, **7.27, 7.28**
  reductionism vs. holism as an example of, 35
  of semilandmarks, 456–458
  substituting the reflection plane for a landmark, 459
symmetry curve, 407, 445, **7.21**

Taleb, N., xxi
template (algorithmic schematic of a semilandmark scheme), 433–436, **Pl.7.14**
tensor, as a report of a uniform shape change, 443
teratology, *see* epidemiology; fetal alcohol
Tetrahymena (protist), 237–239, **4.36**
thin-plate spline, 403, 421–426
  lack of any biological basis for, 425
  bending energy of, 423
  construction of, **7.8**, 424
  kernels, in 2D and 3D, 425
  prototypes, **7.7**
  visual processing of, 425
  see also rules of morphometrics, fourth
Thompson, Sir W., xxvii, 403
time series, patterns of, after an intervention, **4.46**
  as index of a variable type, 361
  see also viremia
total effect, *see under* multiple regression, as path model
trigonometric functions (sine, cosine), 28, 61–62, 473
tsunami, orphan, *see* Seattle, earthquakes near
Tufte, E., 81, 253
Tukey, J., 201
  formula of, for approximating regression coefficients, 226, 231
twins, identical, *see under* Weiss

ulcers, 3, 201–205, 211, 491
  the main abduction in, 204
  two types of graphics for, **4.25**
  one-sentence summary for, 3
ultrasound, neonatal intracranial, 485–488
"Umpire-bonus model," 162
  see also under Galton, F., quincunx machine of

unicorns, and abduction, 210
uniform term, in Procrustes shape space, 418, **7.17, 7.18**, 457
  see also partial warps

validity and invalidity, *see* plausible rival hypotheses
variables, structured, varieties of, 360, 372
variance, 132; *see also* least squares
  "explained" and "unexplained," 222–223
variances of independent processes add, 133
varilocal superpositions, 340
Varmus, H., xxviii, 35–36
Vilmann, H., rodent skull data of, 398–399
  see also Bookstein, examples, rodent skull growth
viremia, SIV-related (example of the SVD), 354–357
Visible Female (NIH avatar), hidden consilience in, **Pl.2.8**, 493
visualization, and numerical inference, *see* inference, numerical, visualization and

Waals, van der, J., estimates of molecular magnitudes by, 175–176
Wanderer über dem Nebelmeer, Der (painting by C. D. Friedrich), front and back covers, 43, 45, 70, 296, 312–313, 484
  as metaphor for multiscale representation of complex processes, 313
warps, *see* principal warps, partial warps, relative warps
Warren, R., 116
  see also under ulcers
water companies of London, *see under* Snow
Watson, J., 4, 14, 99, 116
  see also under double helix
Weaver, W., 302
Wegener, A., 19–22, 42, 491
  see also Atlantic Ocean
weighted average, *see* average, weighted
Weisberg, S., regression textbook of, 169, 220–221, 233, 240
Weiss, P., xxvii, 18, 45, 70–71
Whewell, W., 17
Wigner, E., xxvii, 6, 12, 28, 29, 33, 51, 69, 84, 92, 149, 184–185, 293, 492
  Millikan's study as illustration of, 149–150
Wilkinson skull, *see* Cromwell
Williams, R., 72
Wilson, E. B., xxi
Wilson, E. O., xxvii, 12–13, 15, 17, 29–37, 405
Wilson, J. T., 22, **2.4, 2.5**
wind direction, as a mappable variable, 127, **4.6**

Winn, J., xxiv
Wold, H. and S., 367
Worsley, K., Markov models for images of, 328–330
Wright, S., 59, 72, 160, 229, 231, 367
   leghorn chicken data, correlation matrix of, 319–320

Yucatán, *see* Chicxulub

Zebra Diagram, 22, **2.3**
Ziliak, S., 198–201
Ziman, J., xxi
z-scores (standardized variables), unsuitability for consilient reasoning, 361